D1250890

Introduction to System Sensitivity

Introduction to System Sensitivity Theory

PAUL M. FRANK

Gesamthochschule Duisburg
Fachgebiet Mess- und Regelungstechnik
Duisburg, Germany

ACADEMIC PRESS New York San Francisco London 1978
A Subsidiary of Harcourt Brace Jovanovich, Publishers

ACADEMIC PRESS, INC.
111 Fifth Avenue, New York, New York 10003

United Kingdom Edition published by
ACADEMIC PRESS, INC. (LONDON) LTD.
24/28 Oval Road, London NW1

Library of Congress Cataloging in Publication Data

Frank, Paul M
Introduction to system sensitivity theory.

Includes bibliographical references.
1. System analysis. 2. Control theory.
I. Title.
QA402.F683 003 76-55969
ISBN 0-12-265650-4

PRINTED IN THE UNITED STATES OF AMERICA

To Stefan and Brigitte

Contents

Chapter 8 Sensitivity Analysis of Optimal Systems

Chapter 9 Applications

Preface

This work is an outgrowth of the lecture notes for a two hour per week lecture series for sixth- and eighth-semester students of electrical engineering at the Universität Karlsruhe, conducted by me from 1972 to 1976. Its intention is to serve as an introduction to the basic concepts and methods of sensitivity theory which, I believe, should be part of any engineering education. Major emphasis has been placed on the methods of sensitivity analysis, which can also be considered as the fundamentals of sensitivity synthesis.

The positive response to the German edition "Empfindlichkeitsanalyse dynamischer Systeme" and the fact that I gave this course in English in 1975 at the University of Washington, Seattle, and in 1976 as a summer course at the Centro de Investigación del Instituto Politécnico Nacional in México City encouraged me to publish this text in English. For this edition, the original German version was revised, several parts completed and substantially extended, and a series of new problems and examples included.

Sensitivity considerations of dynamical systems are most important in the everyday life of engineers. The mathematical models of the systems with which they are concerned are idealized, inexactly identified, or the systems themselves are subject to unpredictable changes with time due to environmental, material property, or operational influences so that there is always a discrepancy between the physical reality and the mathematical model. This is of particular importance to modern control engineers whose assignment is to design highly sophisticated systems with prescribed or optimal behavior on the basis of such mathematical models. The results are then useless in practice if they prove to be very sensitive to parameter changes. Sensitivity analysis provides the engineer with methods for investigating or minimizing the effects of such parameter deviations. Furthermore, there are many other problems in control engineering in which sensitivity considerations are useful if not mandatory, e.g., in applications of gradient methods, model tracking in adaptive and self-learning systems, determination of optimal input signals for parameter identification, and system simulations.

An obvious indication of the great need for sensitivity theory is the overwhelming number of papers on sensitivity problems published in the past decade in technical journals or presented at international symposia and conferences. However, there are not many books on this topic, and there is a particular lack at the introductory level. The few books in existence are out of date or devoted and restricted to special subjects of application or they make use of advanced mathematical theories, which make them less suitable for use by engineers.

This situation has induced me to compile the most important concepts and basic methods of sensitivity analysis in a form that engineers can easily understand. I have, with intention, confined myself to the fundamental principles of sensitivity theory and have tried to dispense with all details which may be found in special literature and are not needed for conceptual clarity.

The first chapter gives a general introduction to the objectives, a historical review, and the basic definition of sensitivity theory. In Chapter 2, the basic concepts of sensitivity theory are defined, the parameters classified, and the mathematical tools required for later chapters given.

The third chapter contains the definitions and interpretations of the sensitivity functions and measures of the time and frequency domains and the performance index. The following subjects are discussed in detail: output sensitivity, trajectory sensitivity, sensitivity of time response overshoot, sensitivity with respect to structural changes, eigenvalue sensitivity, Bode sensitivity, Horowitz sensitivity, comparison sensitivity, root sensitivity, sensitivity measure of Biswas–Kuh, sensitivity of frequency response overshoot, performance-index sensitivity, and sensitivity of discrete systems.

Chapter 4 presents the most important methods for the calculation and measurement of output sensitivity functions for linear and nonlinear, continuous and sampled-data systems with constant and time-varying parameter variations which may or may not change the order of the system. The Laplace transform method (Chang) and, based on the sensitivity equation and the sensitivity model, the method of network simulation (Bikhovski), the method of variable components (Kokotović–Rutman), and the sensitivity points method (Kokotović) are described in detail.

The fifth chapter deals with the methods of calculation and measurement of the trajectory sensitivity functions for the same types of systems and parameter changes as considered in Chapter 4. The theorems of Wilkie and Perkins, which define the minimum order sensitivity model necessary for a simultaneous determination of all trajectory sensitivity functions, are described, proven, and demonstrated by simple examples. For sampled-data systems and more general discrete systems such as pulse frequency modulated systems, the trajectory sensitivity equation is derived for changes of all kinds of parameters including the sampling period. In Chapter 6 the method of calculation and measurement of the sensitivity functions of the frequency domain are given.

Methods for comparing the sensitivities of open- and closed-loop systems are discussed in Chapter 7. The first section deals with the sensitivity comparison in the frequency domain for single-input, single-output systems as well as for multivariable and state-feedback systems. Satisfactory conditions are then given for a sensitivity comparison in the time domain, using a norm of the trajectory sensitivity functions. The final section deals with the performance index and trajectory sensitivity comparison of optimized feedback systems. The Pagurek–Witsenhausen theorem is stated and proven.

Chapter 8 is devoted to sensitivity analysis in optimal systems. First, elementary methods for the determination of the performance-index sensitivity, which is the natural sensitivity measure in optimal systems, are presented. Next, it is shown how the performance-index sensitivity can be calculated by means of a Hamilton–Jacobi equation. This method is then applied to the followup and linear regulator problem. In the subsequent section, the relative sensitivity measure, called optimality loss, first proposed by Rohrer and Sobral, is discussed. This sensitivity measure is well suited for the design of optimal low sensitivity control systems.

In Chapter 9, four examples of application are considered: (1) automatic optimization and parameter identification using a sensitivity model; (2) optimization of the input signal for the identification of the parameters of a system; (3) design of an optimal insensitive feedback control system by including the trajectory sensitivity function in the performance index; and (4) a method for the design of minimally sensitive sampled-data systems according to the dead-beat criterion.

For a more rigorous discussion of some subtle and intriguing questions, a number of references are given.

This book is intended to stimulate lectures on sensitivity theory. It will serve the student as a textbook by introducing the basic concepts and methods of sensitivity analysis. Many examples and problems and their solutions are given as exercises. On the other hand, it may provide the advanced engineer with the tools for understanding special literature in this field, and thereby make available current sensitivity methods for the solution of his problems.

Much of the manuscript preparation was done during my stay as a guest professor at the University of Washington in Seattle, which was sponsored by the Stiftung Volkswagenwerk and the Fulbright Commission. I would like to express my thanks to both institutions. I am especially grateful to Dipl.-Ing. Kai-Shing Yeung for his help in translating the text and to Mrs. Liselotte Huber for drawing the figures. I would also like to thank Mrs. Rita Bellm who did her utmost in typing the text.

PAUL M. FRANK

DUISBURG

Chapter 1

Introduction

1.1 THE SIGNIFICANCE OF SENSITIVITY THEORY

The sensitivity of a dynamic system to variations of its parameters is one of the basic aspects in the treatment of dynamic systems. The question of parameter sensitivity particularly arises in the fields of engineering, where mathematical models are used for the purposes of analysis and synthesis. In order to be able to give a unique formulation of the mathematical problem, the mathematical model is usually assumed to be known exactly. This assumption is, strictly speaking, unrealistic since there is always a certain discrepancy between the actual system and its mathematical model. This is due to the following reasons:

(1) A real system cannot be identified exactly because of the restricted accuracy of the measuring devices.

(2) A theoretical concept cannot be implemented exactly because of manufacturing tolerances.

(3) The behavior of any real system changes with time in an often unpredictable way caused by environmental, material property, or operational influences.

(4) Mathematical models are often simplified or idealized intentionally in order to simplify the mathematical problem or to make it solvable at all.

For these reasons, the results of mathematical syntheses need not necessarily be practicable. They may even be very poor, e.g., if there are considerable parameter deviations between the real system and the mathematical model and the solution is very sensitive to the parameters. Therefore, it should be part of the solution to a practical problem to know the parameter sensitivity prior to its implementation or to reduce the sensitivity systematically if this turns out to be necessary.

This is of particular importance if optimization procedures are involved, since it is in the nature of optimization to extremize a certain performance index for the special set of parameters. Furthermore, there are many other problems where sensitivity considerations are either useful or mandatory. Some examples are the application of gradient methods, adaptive and self-learning systems, the design of insensitive and suboptimal control systems, the determination of allowed tolerances in the design of networks, the calculation of optimal input signals for parameter identification, analog and digital simulation of dynamic systems, and so forth.

It is the objective of this book to give an introduction to the basic concepts and methods developed to solve problems of this nature.

1.2 HISTORICAL REVIEW

Sensitivity considerations have long been of concern in connection with dynamic systems. The study of the influence of the coefficients of a differential equation on its solution started with the origins of differential equations. However, for a long period of time, those considerations were merely of mathematical interest [104].

This situation has changed basically with the development of the highly sophisticated methods of modern control theory and their applications by engineers. Historically, sensitivity considerations have provided a fundamental motivation for the use of feedback and are largely responsible for its development into what is called modern control theory, implying the principles of optimization and adaptation.

Therefore, it is quite natural that the basic concepts in this area were already given in the fundamental literature on feedback control systems thirty years ago. Bode [1] was the first to establish the significance of sensitivity in the design of feedback control systems. He has introduced a proper sensitivity definition on the basis of the frequency domain.

In its subsequent development it seemed that automatic control theory should include the study of sensitivity as an essential component. However, with few exceptions, the sensitivity problem was not even discussed in the academic texts on automatic control in the following decade. It was mainly the problem of accuracy in network-analyzers and analog computers that gave new impulses to the theory of sensitivity during the fifties [68, 69]. Many basic methods were also worked out in connection with the design of electric networks [2, 24, 25]. Toward the end of this period the ideas of Bode were rediscovered in control engineering with the appearance of adaptive systems, more precisely, as a reaction to their appearance. Horowitz [11] has developed the methods of frequency domain to a high extent and has applied them with great success to the design of low sensitivity conventional feedback control systems (see also Horowitz and Shaked [45]).

Beginning in the period 1958–1960, the number of publications devoted to sensitivity considerations in the time domain rose considerably due to the development of state space methods in control engineering and the availability of the digital computer. This also gave rise to a new interest in the general sensitivity problem in automatic control systems with an overwhelming number of papers [56, 74] and even some book publications [5, 18, 19]. In particular, the essential contributions of Kokotović and co-workers [53–57], Perkins and Cruz [30–32, 77–80], and Kreindler [58–60] should be mentioned in this connection.

In 1963 Dorato [36] called attention to the problem of parameter sensitivity of the performance index of optimal control systems. In the sequel, many papers were published clarifying certain unexpected problems emerging from this definition of sensitivity [75, 86, 98, 109, 110]. Excellent reviews of the significant publications of that period are given in Kokotović and Rutman [56] and Nuguyen Thuong Ngo [74].

1.3 SENSITIVITY ANALYSIS AND SYNTHESIS

The essential ideas of sensitivity hitherto published can be traced back to a few principles and basic concepts of a general theory, called sensitivity theory. This theory can be seen as parallel to the signal theory already well developed for dynamic systems. Thus, sensitivity theory can be interpreted as a section of a general system theory, taking into account parameter variations as inputs instead of signals. It is the major objective of this book to take the initiative in setting up and introducing such a general sensitivity theory, which can also be applied to fields other than technical ones such as economics or social sciences.

As in the case of signal theory, it is useful to subdivide sensitivity theory into two categories: sensitivity analysis and synthesis. *Sensitivity analysis* provides the basic methods to study the sensitivity of a system to parameter variations. On the other hand, *sensitivity synthesis* is defined as the design of dynamic systems, especially feedback systems, with due regard to sensitivity specifications, say, to obtain minimal or (in some cases) maximal sensitivity to parameter variations. This book is mainly concerned with the methods of sensitivity analysis, which, however, is the basis of sensitivity synthesis.

In order to outline the sensitivity analysis in more detail, recall that, in general, the dynamics of a system can be represented by a single block (Fig. 1.3-1), which will, for short, be called the system. From a mathematical point of view, what we call a system is the explicitly or implicitly given relationship between the input signal $\mathbf{u}(t)$ and the output signal $\mathbf{y}(t)$. In general, $\mathbf{u}(t)$ and $\mathbf{y}(t)$ can be vectors. The character of this relationship is commonly called the

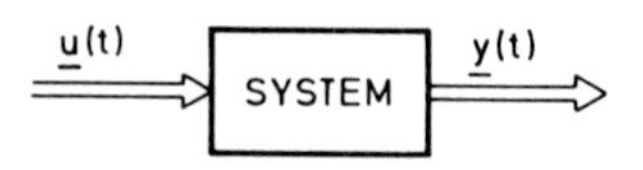

FIG. 1.3-1. General representation of a dynamic system.

structure of the system. For example, the structure of the system may be characterized by

the *order* of a differential or difference equation,
linearity or nonlinearity,
the order of the numerator and denominator of a rational transfer function, and
the rationality or irrationality of the transfer function.

The quantitative properties of the system are characterized by the system *parameters*. Typical parameters are

initial conditions,
time-invariant or time-variant coefficients,
natural frequencies, pulse frequencies,
sampling periods, sampling instants,
pulse width or magnitude, and
dead times (or time delays).

Dynamic processes in a system, say, the change of the state or of the output variable with time, can be caused (see Fig. 1.3-2) by

(1) the influence of input signals,
(2) the change of parameters.

While studying the influence of input signals, the dynamics of the system are usually considered only as a function of the input signals, assuming that the relationship is qualitatively and quantitatively unchanged. This is the subject matter of conventional system theory.

While studying the influence of parameters, the dynamics of the system are considered as a function of changes in the parameters (or of the structure of the system, because the change of system parameters can also change the system structure). The dependence of the system dynamics on the parameters is called sensitivity. Strictly, parameter sensitivity can be defined as follows:

Definition 1.3-1 *Parameter sensitivity* is the effect of parameter changes on the dynamics of a system, say, the time response, the state, the transfer function, or any other quantity characterizing the system dynamics.

With regard to the mathematical treatment of the sensitivity problem, it is useful to distinguish between two types of parameter deviations:

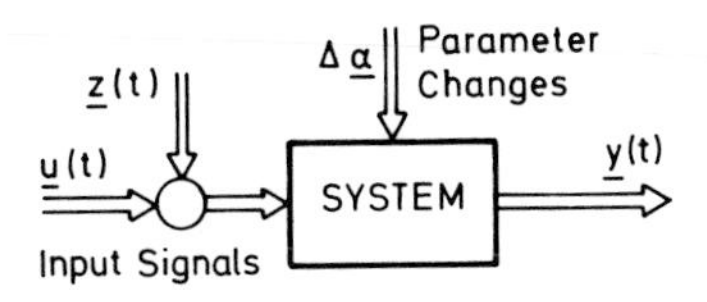

FIG. 1.3-2. Quantities affecting the dynamics of a system.

(1) errors and tolerances of the underlying mathematical model (those parameter changes that are time-invariant), and slowly varying (quasi-constant) parameters; and
(2) *changes* in the parameters with time.

Parameter changes of the first category can be caused by

tolerances of manufacturing (when realizing a system),
measurement errors (when identifying a system),
approximations (when setting up the mathematical model), and
seasoning of elements (erosion, abrasion, wear, etc.).

Parameter changes of the second category can be caused by

seasoning of elements (erosion, abrasion, wear, etc.),
changes in environmental conditions (temperature, humidity, gravitation, etc.), and
changes in operating conditions (load changes, change of inertia by fuel consumption, influence of nonlinearities, etc.).

Either of the two categories of parameter changes requires its own methods of treatment.

Parameter changes of both categories appear in any engineering system. Thus, sensitivity analysis can be regarded, along with signal analysis, as a necessary tool in the treatment of engineering systems.

Chapter 2

Basic Concepts and Tools of Sensitivity Theory

2.1 INTRODUCTION

The basis of all sensitivity considerations in the case of time-invariant parameter variations is the so-called *sensitivity function*. (In the case of time-variant parameter variations, the corresponding expression is the first variation, as will be shown in Chapter 5.) If the sensitivity function is known, it will be easy to calculate the change in the system behavior from given parameter deviations and, conversely, to calculate allowable parameter deviations from a given or preassigned system behavior. The latter problem is often referred to as the inverse sensitivity problem.

In this chapter the basic idea of sensitivity theory is outlined, the various ways of defining the sensitivity function are shown, and mathematical rules governing the sensitivity functions and their relations are given for use in later chapters. Furthermore, the different kinds of parameter variations will be classified according to a proposal of Miller and Murray [68].

2.2 THE BASIC IDEA OF SENSITIVITY THEORY

The mathematical problem to be solved in sensitivity theory is the calculation of the change in the system behavior due to the parameter variations. Let the parameters of the system be represented by a vector $\boldsymbol{\alpha} = [\alpha_1\ \alpha_2 \cdots \alpha_r]^T$. As outlined in the previous chapter, the mathematical model of a system relates the parameter vector $\boldsymbol{\alpha}$ to a quantity characterizing its dynamic behavior in some way. In general, this quantity can be a vector and hence shall be denoted by a bold-faced symbol, say $\boldsymbol{\zeta}$. For example, $\boldsymbol{\zeta}$ can represent the state $\boldsymbol{x}$ of the system.†

Let us explain the basic idea of sensitivity theory by means of this example.

†In the figures, vectors will be denoted by underlining.

It is assumed that the mathematical model of the (possibly nonlinear) system is given by the general vector differential equation

$$\dot{\boldsymbol{x}} = \boldsymbol{f}(\boldsymbol{x}, \boldsymbol{\alpha}, t, \boldsymbol{u}), \qquad \boldsymbol{x}(t_0) = \boldsymbol{x}^0, \tag{2.2–1}$$

where $\boldsymbol{x}$ represents the state vector with the initial state $\boldsymbol{x}(t_0) = \boldsymbol{x}^0$, and $\boldsymbol{u}$ represents the input vector. Among other things, this equation relates the state vector $\boldsymbol{x}$ to the parameter vector $\boldsymbol{\alpha}$. In terms of set theory, this relation can also be interpreted as a mapping $\boldsymbol{\alpha} \to \boldsymbol{\zeta}$.

Generally, in mathematics, a unique relationship between the parameter vector and the state vector is assumed. However, this is not possible in engineering practice due to the reasons outlined at the beginning. Here the parameter vector of the mathematical model means a *nominal* parameter vector that will be denoted by $\boldsymbol{\alpha}_0$ in the sequel, whereas the parameter vector of the actual system is $\boldsymbol{\alpha} = \boldsymbol{\alpha}_0 + \Delta\boldsymbol{\alpha}$, in the sequel called the *actual* parameter vector.

In order to study the influence of the parameter deviations $\Delta\boldsymbol{\alpha}$ on the behavior of the system, let us define

R_α as the subspace of the parameter variations $\Delta\boldsymbol{\alpha}$ around $\boldsymbol{\alpha}_0$, and
R_x as the corresponding subspace of the state vector.

By this definition the mapping $\boldsymbol{\alpha} \to \boldsymbol{x}$ can be replaced by the mapping $R_\alpha \to R_x$ as shown in Fig. 2.2-1.

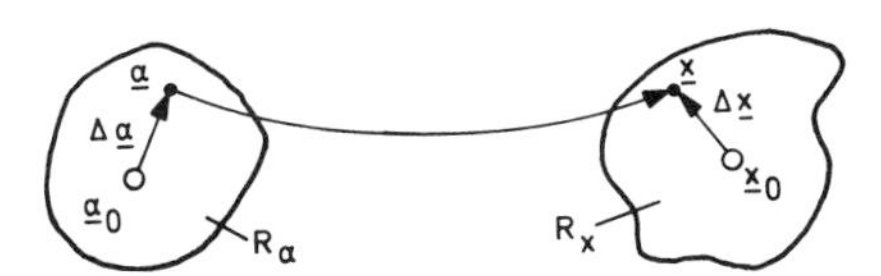

FIG. 2.2-1. Mapping of the parameter space into the state space.

R_x is uniquely determined by Eq. (2.2–1) if R_α is known. However, for a number of reasons, it is not reasonable to characterize the sensitivity in terms of Eq. (2.2–1): first, since the direct solution of Eq. (2.2–1) for all elements of R_α requires an infinite number of solutions and depends on the definition of R_α, and, second, since the result for small parameter variations $\|\Delta\boldsymbol{\alpha}\| \ll \|\boldsymbol{\alpha}_0\|$ would be very inaccurate if approximations are applied for the evaluation of this equation. For example, this would be true in the case of numerical or analog computation.

Therefore, it is a common practice in sensitivity theory to define a so-called *sensitivity function* $\boldsymbol{S}$ which, under certain continuity conditions, relates the elements of the set of the parameter deviations $\Delta\boldsymbol{\alpha}$ to the elements of the set of the parameter-induced errors of the system function $\Delta\boldsymbol{x}$ by the linear equation

$$\Delta\boldsymbol{x} \approx \boldsymbol{S}(\boldsymbol{\alpha}_0)\, \Delta\boldsymbol{\alpha}. \tag{2.2–2}$$

This relation is a linear approximation of Eq. (2.2–1) and is valid only for small parameter variations, i.e., $\|\Delta\boldsymbol{\alpha}\| \ll \|\boldsymbol{\alpha}_0\|$. Since $\boldsymbol{S}$ depends on the nominal parameter vector, it can be calculated from nominal parameter values. How this is done will be shown later. In the particular case, $\boldsymbol{S}$ is a matrix function known as the trajectory sensitivity matrix $\boldsymbol{\lambda}$ (see Chapter 5).

With such an approach, the sensitivity analysis is reduced in practice to the determination (calculation or measurement) of the sensitivity function $\boldsymbol{S}(\boldsymbol{\alpha}_0)$ or equivalent sensitivity measures as well as the subspace R_α. It is evident that, because of the use of a linearized equation of the type (2.2–2), the sensitivity theory is restricted to small parameter variations only. Then the first variation is $\delta\boldsymbol{\zeta} \approx \Delta\boldsymbol{\zeta}$. If the parameter variations are infinitesimally small, then Eq. (2.2–2) holds exactly and $\delta\boldsymbol{\zeta} = \Delta\boldsymbol{\zeta}$.

This approach forms the basis of the treatment of the sensitivity theory throughout this book.

Besides the sensitivity approach as defined above, similar problems of parameter variation have been tackled in the Soviet Union by the so-called *invariance theory*. In this approach the relationship between $\Delta\boldsymbol{\alpha}$ and $\Delta\boldsymbol{x}$ is defined as

$$\Delta\boldsymbol{x} = \boldsymbol{S}(\boldsymbol{\alpha}_0)\,\Delta\boldsymbol{\alpha} + \boldsymbol{R}(\Delta\boldsymbol{\alpha}), \tag{2.2–3}$$

where the expression $\boldsymbol{R}(\Delta\boldsymbol{\alpha})$ represents the higher order terms of $\Delta\boldsymbol{\alpha}$. This relation can be interpreted as a complete series expansion of the original relationship (2.2–1), and, hence, $\Delta\boldsymbol{\alpha}$ may take on any orbitrary value. In these terms, parameter invariance means $\Delta\boldsymbol{x} = \boldsymbol{0}$ for arbitrary $\Delta\boldsymbol{\alpha}$ and all t.

In practice, the evaluation of Eq. (2.2–3) can be extremely cumbersome, since the mapping is no longer linear. In order to simplify the evaluation at the expense of obtaining accurate results, the conception of *zero sensitivity* was introduced. Here, only the first variation of Eq. (2.2–3), i.e., of

$$\Delta\boldsymbol{x} = \delta\boldsymbol{x} + \delta\boldsymbol{x}^2 + \cdots = \boldsymbol{S}\,\Delta\boldsymbol{\alpha} + \boldsymbol{R}(\Delta\boldsymbol{\alpha}), \tag{2.2–4}$$

is taken into account without imposing a corresponding restriction on $\Delta\boldsymbol{\alpha}$. Hence, the mapping is defined, instead of Eq. (2.2–3), as

$$\delta\boldsymbol{x} = \boldsymbol{S}\,\Delta\boldsymbol{\alpha} \tag{2.2–5}$$

for arbitrary $\Delta\boldsymbol{\alpha}$. In these terms a system is said to have zero sensitivity if $\boldsymbol{S} \equiv \boldsymbol{0}$ for arbitrary $\Delta\boldsymbol{\alpha}$ and $t > 0$. Note that the only difference between the latter and the sensitivity approach as treated throughout this book is that, in the latter, arbitrary $\Delta\boldsymbol{\alpha}$ are allowed, whereas in the sensitivity approach, $\Delta\alpha$ is assumed to be (infinitesimally) small.

In contrast to the sensitivity function $\boldsymbol{S}(\boldsymbol{\alpha}_0)$, which *approximately* relates the variation $\Delta\boldsymbol{x}$ to parameter variations $\Delta\boldsymbol{\alpha}$, the so-called *pseudosensitivity function* is defined so as to *exactly* relate $\Delta\boldsymbol{x}$ to $\Delta\boldsymbol{\alpha}$, even if $\Delta\boldsymbol{x}$ and $\Delta\boldsymbol{\alpha}$ are finite. That is,

$$\Delta\boldsymbol{x} \equiv \boldsymbol{S}^*(\boldsymbol{\alpha})\,\Delta\boldsymbol{\alpha} \tag{2.2–6}$$

for arbitrary $\Delta\boldsymbol{\alpha}$ and $\Delta\mathbf{x}$. The pseudosensitivity function approaches the sensitivity function as $\Delta\boldsymbol{\alpha}$ approaches zero, that is, $\boldsymbol{S}(\boldsymbol{\alpha}_0) = \lim_{\Delta\boldsymbol{\alpha}\to 0} S^*(\boldsymbol{\alpha})$.

Pseudosensitivity functions are extremely useful in the design of efficient noniterative parameter identification, optimization, and adaptation devices.

2.3 BASIC DEFINITIONS

There are several ways to define quantities for the characterization of the parameter sensitivity of a system. These definitions will be summarized in this section.

Let the behavior of the dynamic system be characterized by a quantity $\boldsymbol{\zeta} = \boldsymbol{\zeta}(\boldsymbol{\alpha})$, called a system function, which, among other dependences, is a function of the parameter vector $\boldsymbol{\alpha} = [\boldsymbol{\alpha}_1\ \boldsymbol{\alpha}_2 \cdots \boldsymbol{\alpha}_r]^{\mathrm{T}}$. For example, $\boldsymbol{\zeta}$ can represent any time domain or frequency domain property or a performance index. Let the nominal parameter vector be denoted by $\boldsymbol{\alpha}_0 = [\boldsymbol{\alpha}_{10}\ \boldsymbol{\alpha}_{20} \cdots \boldsymbol{\alpha}_{r0}]^{\mathrm{T}}$ and the nominal system function by $\boldsymbol{\zeta}_0 \triangleq \boldsymbol{\zeta}(\boldsymbol{\alpha}_0)$. Then, under certain continuity conditions, the following general definitions hold.

Definition 2.3-1 *(Absolute) sensitivity function.*

$$\boldsymbol{S}_j \triangleq \left.\frac{\partial \boldsymbol{\zeta}(\boldsymbol{\alpha})}{\partial \alpha_j}\right|_{\boldsymbol{\alpha}0} = \boldsymbol{S}_j(\boldsymbol{\alpha}_0) \qquad j = 1, 2, \ldots, r. \tag{2.3–1}$$

The subscript $\boldsymbol{\alpha}_0$ shall indicate that the partial derivative expressed by ∂ is taken at nominal parameter values. Besides depending on $\boldsymbol{\alpha}_0$, $\boldsymbol{S}_j$ can also depend on other variables, such as the time t or the frequency ω. Therefore, $\boldsymbol{S}_j$ shall be termed a function rather than a coefficient as is done in some former publications [18]. If $\boldsymbol{S}_j$ is a function only of $\boldsymbol{\alpha}_0$, we shall term it sensitivity for short (e.g., "performance-index sensitivity").

Definition 2.3-2 *Parameter-induced error of the system function.*

$$\Delta\zeta \triangleq \sum_{j=1}^{r} \boldsymbol{S}_j\, \Delta\alpha_j. \tag{2.3–2}$$

Definition 2.3-3 *Maximum error of the system function.*

$$|\Delta\boldsymbol{\zeta}| \triangleq \sum_{j=1}^{r} |\boldsymbol{S}_j|\,|\Delta\alpha_j|. \tag{2.3–3}$$

The vertical bars in combination with a vector shall indicate that the absolute values of the elements of the corresponding vector are to be taken; for example, $|\boldsymbol{S}_j| = [|S_{1j}|\,|S_{2j}| \cdots |S_{nj}|]^{\mathrm{T}}$.

Definition 2.3-4 *Relative (logarithmic) sensitivity function.*

$$\bar{\boldsymbol{S}}_j \triangleq \left.\frac{\partial \ln \boldsymbol{\zeta}}{\partial \ln \alpha_j}\right|_{\boldsymbol{\alpha}0} = \bar{\boldsymbol{S}}_j(\boldsymbol{\alpha}_0), \qquad j = 1, 2, \ldots, r. \tag{2.3–4}$$

Note that $\ln \boldsymbol{\zeta}$ means the vector of the logarithms of the elements of $\boldsymbol{\zeta}$. Hence,

$\partial \ln \boldsymbol{\zeta} = [\partial\zeta_1/\zeta_1 \; \partial\zeta_2/\zeta_2 \; . \; . \; . \; \partial\zeta_n/\zeta_n]^{\mathrm{T}}$. The ith element of $\bar{\boldsymbol{S}}_j$ can be expressed by

$$\bar{S}_{ij} = \left.\frac{\partial\zeta_i/\zeta_i}{\partial\alpha_j/\alpha_j}\right|_{\boldsymbol{\alpha}0} = S_{ij}\frac{\alpha_{j0}}{\zeta_{i0}}, \qquad \begin{array}{l} i = 1, 2, \ldots, n, \\ j = 1, 2, \ldots, r, \end{array} \tag{2.3–5}$$

where S_{ij} is the ith element of the absolute sensitivity function $\boldsymbol{S}_j$.

Definition 2.3-5 *Relative error of the system function.* The ith element of the relative error of the system function is defined as

$$\frac{\Delta\zeta_i}{\zeta_{i0}} \triangleq \sum_{j=0}^{r} \bar{S}_{ij}\frac{\Delta\alpha_j}{\alpha_{j0}}, \qquad i = 1, 2, \ldots, n. \tag{2.3–6}$$

Definition 2.3-6 *Maximum relative error of the system function.* The ith element is defined as

$$\left|\frac{\Delta\zeta_i}{\zeta_{i0}}\right| \triangleq \sum_{j=1}^{r} |\bar{S}_{ij}|\left|\frac{\Delta\alpha_j}{\alpha_{j0}}\right|, \qquad i = 1, 2, \ldots, n. \tag{2.3–7}$$

Besides these definitions, the so-called semirelative sensitivity functions are in use. There are two ways to define a semirelative sensitivity function.

Definition 2.3-7 *Semirelative (semilogarithmic) sensitivity function.*

(a) $$\underset{\sim}{S}_j \triangleq \left.\frac{\partial \ln \boldsymbol{\zeta}}{\partial\alpha_j}\right|_{\boldsymbol{\alpha}0}, \qquad j = 1, 2, \ldots, r. \tag{2.3–8}$$

For the ith components of $\underset{\sim}{\boldsymbol{S}}_j$,

$$\underset{\sim}{S}_{ij} = \left.\frac{\partial\zeta_i/\zeta_i}{\partial\alpha_j}\right|_{\boldsymbol{\alpha}0} = \frac{1}{\zeta_{i0}}S_{ij}, \qquad \begin{array}{l} i = 1, 2, \ldots, n, \\ j = 1, 2, \ldots, r. \end{array}$$

(b) $$\tilde{\boldsymbol{S}}_j \triangleq \left.\frac{\partial\zeta}{\partial \ln \alpha_j}\right|_{\boldsymbol{\alpha}0} = \alpha_{j0}\,\boldsymbol{S}_j(\boldsymbol{\alpha}_0), \qquad j = 1, 2, \ldots, r. \tag{2.3–9}$$

Note that all sensitivity functions are defined for nominal parameter values. If not indicated otherwise, the cross bar on top is employed to characterize a relative sensitivity function and a tilde will characterize a semirelative one.

2.4 GENERAL RULES

By the definition of the sensitivity functions, it is evident that the basic calculus applying to sensitivity functions is the calculus of partial differentiation. In this section some basic rules of differentiation needed for the calculation and manipulation of the sensitivity functions throughout this book are reviewed.

Let us write the absolute sensitivity function of a system function $\zeta = \zeta(\alpha)$ with respect to α in the following form:

$$S_\alpha^\zeta \triangleq \left.\frac{\partial\zeta(\alpha)}{\partial\alpha}\right|_{\alpha 0}. \tag{2.4–1}$$

Now, if $\zeta_1 = \zeta_1(\alpha)$ and $\zeta_2 = \zeta_2(\alpha)$, and the derivatives exist, the following relations hold:

$$S_\alpha^{\zeta_1\zeta_2} = \zeta_{10}\, S_\alpha^{\zeta_2} + \zeta_{20}\, S_\alpha^{\zeta_1} \qquad \text{(product rule)}, \tag{2.4–2}$$

$$S_\alpha^{\zeta_1/\zeta_2} = \frac{1}{\zeta_{20}}\, S_\alpha^{\zeta_1} - \frac{\zeta_{10}}{\zeta_{20}^2}\, S_\alpha^{\zeta_2} \qquad \text{(quotient rule)}. \tag{2.4–3}$$

If $\zeta = f(\beta)$ and $\beta = g(\alpha)$, then we have

$$S_\alpha^\zeta = S_\beta^\zeta\, S_\alpha^\beta \qquad \text{(chain rule)}. \tag{2.4–4}$$

For the relative sensitivity functions denoted by

$$\bar{S}_\alpha^\zeta \triangleq \left.\frac{\partial \ln \zeta(\alpha)}{\partial \ln \alpha}\right|_{\alpha_0} = \left.\frac{\alpha_0}{\zeta_0}\frac{\partial \zeta}{\partial \alpha}\right|_{\alpha_0},$$

the following relations hold:

$$\bar{S}_\alpha^{\zeta_1\zeta_2} = \bar{S}_\alpha^{\zeta_1} + \bar{S}_\alpha^{\zeta_2} \qquad \text{(product rule)}, \tag{2.4–5}$$

$$\bar{S}_\alpha^{\zeta_1/\zeta_2} = \bar{S}_\alpha^{\zeta_1} - \bar{S}_\alpha^{\zeta_2} \qquad \text{(quotient rule)}; \tag{2.4–6}$$

and with the presumptions of Eq. (2.4–4), we have

$$\bar{S}_\alpha^\zeta = \bar{S}_\beta^\zeta\, \bar{S}_\alpha^\beta \qquad \text{(chain rule)}. \tag{2.4–7}$$

From these relations which can easily be proven by using the differential notation, it is obvious that it is an advantage to operate with the relative sensitivity functions rather than with the absolute ones. This is especially true for ζ representing a frequency function, such as the transfer function, because the Bode diagram technique can then be applied with all its benefits.

Now some *chain rule applications* will be presented. To treat the problem quite generally, let us consider a continuous function of the variables $x_1, x_2, \ldots, x_r$,

$$f = f(x_1, x_2, \ldots, x_r). \tag{2.4–8}$$

The total differential of f is defined by the relationship

$$df \triangleq \frac{\partial f}{\partial x_1}\, dx_1 + \frac{\partial f}{\partial x_2}\, dx_2 + \cdots + \frac{\partial f}{\partial x_r}\, dx_r. \tag{2.4–9}$$

In the following, several types of dependences among the variables $x_1, x_2, \ldots, x_r$ are considered. In each of the cases to be studied, the continuity of all derivatives is assumed.

(1) Suppose that $x_1, x_2, \ldots, x_r$ are functions of a single variable, say a, i.e., $x_1(a), x_2(a), \ldots, x_r(a)$. Then f may be considered as a function of a,

$$f = g(a). \tag{2.4–10}$$

Note that g represents the same physical quantity as f. Therefore, in engineering literature, no different symbol is commonly used to express the different dependence. This is done by different arguments only. For simplicity of notation, we will follow this convention in most of this book. However, for the sake of clarity, it seems to be necessary to employ the more exact mathematical notation in this section.

Now, with the aid of the total differential, the total derivative of $g(a)$ with respect to a is obtained as

$$\frac{dg}{da} = \frac{\partial f}{\partial x_1}\frac{dx_1}{da} + \frac{\partial f}{\partial x_2}\frac{dx_2}{da} + \cdots + \frac{\partial f}{\partial x_r}\frac{dx_r}{da}. \tag{2.4–11}$$

The variables x_i are also known as intermediate variables.

(2) Suppose that $x_1, x_2, \ldots, x_r$ are functions of another set of independent variables, say, $a_1, a_2, \ldots, a_s$,

$$\begin{aligned} x_1 &= x_1(a_1, a_2, \ldots, a_s), \\ &\vdots \\ x_r &= x_r(a_1, a_2, \ldots, a_s). \end{aligned} \tag{2.4–12}$$

Then f may be considered as a function of the a_j's,

$$f = g(a_1, a_2, \ldots, a_s). \tag{2.4–13}$$

In this case, dg/da_1 is no longer defined. Instead of the total derivative, only the partial derivatives can be taken:

$$\begin{aligned} \frac{\partial g}{\partial a_1} &= \frac{\partial f}{\partial x_1}\frac{\partial x_1}{\partial a_1} + \frac{\partial f}{\partial x_2}\frac{\partial x_2}{\partial a_1} + \cdots + \frac{\partial f}{\partial x_r}\frac{\partial x_r}{\partial a_1}, \\ &\vdots \\ \frac{\partial g}{\partial a_s} &= \frac{\partial f}{\partial x_1}\frac{\partial x_1}{\partial a_s} + \frac{\partial f}{\partial x_2}\frac{\partial x_2}{\partial a_s} + \cdots + \frac{\partial f}{\partial x_r}\frac{\partial x_r}{\partial a_s}. \end{aligned} \tag{2.4–14}$$

It is understood that in the above equations all independent variables are considered constant except the one with respect to which the derivative is taken.

(3) Now the variables $x_1, x_2, \ldots, x_{r-1}$ may be functions of x_r,

$$\begin{aligned} x_1 &= x_1(x_r), \\ &\vdots \\ x_{r-1} &= x_{r-1}(x_r). \end{aligned} \tag{2.4–15}$$

Recall that f is a function of the intermediate variables $x_1, x_2, \ldots, x_{r-1}$ and the independent variable x_r. Considering f as a function of x_r only,

$$f = g(x_r), \tag{2.4–16}$$

and taking the total derivative of $g(x_r)$ with respect to x_r, we obtain

$$\frac{dg}{dx_r} = \frac{\partial f}{\partial x_1}\frac{dx_1}{dx_r} + \frac{\partial f}{\partial x_2}\frac{dx_2}{dx_r} + \cdots + \frac{\partial f}{\partial x_{r-1}}\frac{dx_{r-1}}{dx_r} + \frac{\partial f}{\partial x_r}. \tag{2.4–17}$$

(4) Suppose that $x_1, x_2, \ldots, x_k$ are functions of $x_{k+1}, \ldots, x_r$, where $x_{k+1}, \ldots, x_r$ are independent variables,

$$\begin{aligned} x_1 &= x_1(x_{k+1}, \ldots, x_r), \\ &\vdots \\ x_k &= x_k(x_{k+1}, \ldots, x_r). \end{aligned} \tag{2.4–18}$$

Then, considering f as a function of $x_{k+1}, \ldots, x_r$, we define

$$f = g(x_{k+1}, \ldots, x_r). \tag{2.4–19}$$

As in case (2), the total derivative is not defined. Taking the partial derivatives yields the expressions

$$\begin{aligned} \frac{\partial g}{\partial x_{k+1}} &= \frac{\partial f}{\partial x_1}\frac{\partial x_1}{\partial x_{k+1}} + \ldots + \frac{\partial f}{\partial x_k}\frac{\partial x_k}{\partial x_{k+1}} + \frac{\partial f}{\partial x_{k+1}} \\ &\vdots \\ \frac{\partial g}{\partial x_r} &= \frac{\partial f}{\partial x_1}\frac{\partial x_1}{\partial x_r} + \ldots + \frac{\partial f}{\partial x_k}\frac{\partial x_k}{\partial x_r} + \frac{\partial f}{\partial x_r} \end{aligned} \tag{2.4–20}$$

(5) Finally, suppose that

$$\begin{aligned} x_1 &= x_1(x_{k+1}, \ldots, x_r), \\ &\vdots \\ x_{k-1} &= x_{k-1}(x_{k+1}, \ldots, x_r), \\ x_k &= x_k(x_r). \end{aligned} \tag{2.4–21}$$

Again we define the dependence on $x_{k+1}, \ldots, x_r$ by a new symbol,

$$f = g(x_{k+1}, \ldots, x_r). \tag{2.4–22}$$

Taking the partial derivatives, we obtain the same expressions as in case (4) except for the last equation which now reads

$$\frac{\partial g}{\partial x_r} = \frac{\partial f}{\partial x_1}\frac{\partial x_1}{\partial x_r} + \ldots + \frac{\partial f}{\partial x_k}\frac{d x_k}{d x_r} + \frac{\partial f}{\partial x_r}. \tag{2.4–23}$$

The term $\partial x_k/\partial x_r$ of Eq.(2.4–20) has been replaced by the term dx_k/dx_r, since $\partial x_k/\partial x_r$ would not make any sense in the above case.

2.4.1 Some Definitions of Derivatives in the Vector Space

It is important to establish certain rules of differentiation in the case where the variables are vectors or vector functions.

Definition 2.4–1 *The gradient vector.* If ζ is a scalar function of an $r \times 1$ vector $\boldsymbol{\alpha} = [\alpha_1\ \alpha_2\ \ldots\ \alpha_r]^{\mathrm{T}}$, i. e.,

$$\zeta = \zeta(\alpha_1, \alpha_2, \ldots, \alpha_r) = \zeta(\boldsymbol{\alpha}), \tag{2.4–24}$$

then we define the partial derivative of ζ with respect to $\boldsymbol{\alpha}$ as the $1 \times r$ row vector

$$\frac{\partial \zeta}{\partial \boldsymbol{\alpha}} \triangleq \begin{bmatrix} \frac{\partial \zeta}{\partial \alpha_1} & \frac{\partial \zeta}{\partial \alpha_2} & \cdots & \frac{\partial \zeta}{\partial \alpha_r} \end{bmatrix} = \Delta \zeta = \boldsymbol{\zeta}_{\boldsymbol{\alpha}}. \tag{2.4–25}$$

This vector is known as the gradient vector. In some books this vector is defined as a column vector. Notice that we will use it as a row vector throughout the book.

Definition 2.4–2 *The Jacobian (matrix).* If $\boldsymbol{\zeta}$ is an $n \times 1$ vector function of the $r \times 1$ parameter vector $\boldsymbol{\alpha} = [\alpha_1\ \alpha_2 \ldots \alpha_r]^{\mathrm{T}}$, i.e.,

$$\boldsymbol{\zeta} = \boldsymbol{\zeta}(\alpha_1, \alpha_2, \ldots, \alpha_r) = \boldsymbol{\zeta}(\boldsymbol{\alpha}), \tag{2.4–26}$$

then we define the partial derivative of $\boldsymbol{\zeta}$ with respect to $\boldsymbol{\alpha}$ as the $n \times r$ matrix

$$\frac{\partial \boldsymbol{\zeta}}{\partial \boldsymbol{\alpha}} \triangleq \begin{bmatrix} \frac{\partial \zeta_1}{\partial \alpha_1} & \frac{\partial \zeta_1}{\partial \alpha_2} & \cdots & \frac{\partial \zeta_1}{\partial \alpha_r} \\ \vdots & \vdots & & \vdots \\ \frac{\partial \zeta_n}{\partial \alpha_1} & \frac{\partial \zeta_n}{\partial \alpha_2} & \cdots & \frac{\partial \zeta_n}{\partial \alpha_r} \end{bmatrix}. \tag{2.4–27}$$

This matrix is called the Jacobian (matrix).

Note that by these definitions the rows of the Jacobian matrix are the gradient vectors of the elements of $\boldsymbol{\zeta}$. This is an advantage of defining the gradient vector as a row vector, since then the Jacobian of a scalar ζ reduces to the gradient vector automatically without the need of transposition. Hence, no transpositions are required in carrying out the partial derivatives as is necessary in those publications which define the gradient vector as a column vector.

2.4.2 Calculation of the Sensitivity Function with Respect to One Set of Parameters from the Sensitivity Function with Respect to Another One

In theory, the coefficients of the mathematical model are usually regarded as the parameters of the system. In practice, however, the actually independent parameters are certain physical quantities, such as temperature, pressure, resistance etc., and the mathematical parameters are functions of these physical parameters. Therefore, the question arises how the sensitivity functions due to one set of parameters, say the physical parameters, can be calculated from the sensitivity functions in terms of another set of parameters, say the mathematical parameters. In this section the corresponding formula will be given.

Let the set of mathematical parameters be denoted by α_j ($j = 1, 2, \ldots, r$). Suppose that the sensitivity functions with respect to these parameters,

$$S_{\alpha_j}^{\zeta} = \left.\frac{\partial \zeta}{\partial \alpha_j}\right|_{\alpha_0}, \qquad j = 1, 2, \ldots, r, \tag{2.4-28}$$

are known. On the other hand, let the physical parameters be denoted by a_i ($i = 1,2,\ldots, s$), and assume that the α_j are continuous functions of the a_i

$$\alpha_j = \alpha_j(a_1, a_2, \ldots, a_s). \tag{2.4-29}$$

The question is: How can the sensitivity functions

$$S_{a_i}^{\zeta} = \left.\frac{\partial \zeta}{\partial a_i}\right|_{\boldsymbol{a}_0}, \qquad i = 1, 2, \ldots, s, \tag{2.4-30}$$

be determined in terms of the $S_{\alpha_j}^{\zeta}$?

By the application of the chain rule [see Eq. (2.4–14)], we obtain

$$\frac{\partial \zeta}{\partial a_i} = \frac{\partial \zeta}{\partial \alpha_1}\frac{\partial \alpha_1}{\partial a_i} + \frac{\partial \zeta}{\partial \alpha_2}\frac{\partial \alpha_2}{\partial a_i} + \cdots + \frac{\partial \zeta}{\partial \alpha_r}\frac{\partial \alpha_r}{\partial a_i}, \qquad i = 1, 2, \ldots, s. \tag{2.4-31}$$

Now setting $\boldsymbol{\alpha} = \boldsymbol{\alpha}_0$ and $\boldsymbol{a} = \boldsymbol{a}_0$ yields

$$S_{a_i}^{\zeta} = \sum_{j=1}^{r} S_{\alpha_j}^{\zeta} \left.\frac{\partial \alpha_j}{\partial a_i}\right|_{\boldsymbol{\alpha}_0}, \qquad i = 1, 2, \ldots, s. \tag{2.4-32}$$

Thus, if the dependence of the α_j on the a_i as well as the sensitivity functions $S_{\alpha_j}^{\zeta}$, are known, the sensitivity functions with respect to the a_i can be calculated from Eq. (2.4–32).

In addition, by substituting the above expression for $S_{a_i}^{\zeta}$ in

$$\Delta\zeta = \sum_{i=1}^{s} S_{a_i}^{\zeta}\, \Delta a_i, \tag{2.4-33}$$

the parameter-induced error can be given as

$$\Delta\zeta = \sum_{i=1}^{s}\sum_{j=1}^{r} S_{\alpha_j}^{\zeta} \left.\frac{\partial \alpha_j}{\partial a_i}\right|_{\boldsymbol{\alpha}_0} \cdot \Delta a_i \tag{2.4-34}$$

This shows that the restriction to mathematical parameters, which is a common practice throughout this book, does not mean any loss of generality.

Example 2.4–1 Suppose that the differential equation of a linear system is given by $\alpha_2 \dot{y}(t) + \alpha_1 y(t) = u(t)$, where $u(t)$ represents the input signal and $y(t)$ the output signal; α_1 and α_2 are the mathematical parameters. Let us assume that the sensitivity functions of $y(t)$ with respect to α_1 and α_2, denoted by S_{α_1} and S_{α_2}, have been determined by means of one of the methods described in later chapters. Hence, the parameter-induced error of the output signal can be calculated from

$$\Delta y = S_{\alpha_1}\,\Delta\alpha_1 + S_{\alpha_2}\,\Delta\alpha_2. \tag{2.4-35}$$

Furthermore, we will assume that α_1 and α_2 are not independent of each other, so that $\Delta\alpha_1$ and $\Delta\alpha_2$ cannot be chosen independently. Suppose that

$$\alpha_1 = a_1, \tag{2.4-36}$$

$$\alpha_2 = a_1 a_2, \tag{2.4-37}$$

where a_1 and a_2 are independent physical parameters.

Thus, to calculate the parameter-induced error due to changes in a_1 and a_2, we obtain from Eq. (2.4–34)

$$\Delta y = S_{\alpha_1}\left(\frac{\partial\alpha_1}{\partial a_1}\Delta a_1 + \frac{\partial\alpha_1}{\partial a_2}\Delta a_1\right) + S_{\alpha_2}\left(\frac{\partial\alpha_2}{\partial a_1}\Delta a_1 + \frac{\partial\alpha_2}{\partial a_2}\Delta a_2\right), \tag{2.4-38}$$

which has to be evaluated for nominal parameter values a_{10} and a_{20}. In the special case,

$$\left.\frac{\partial\alpha_1}{\partial a_1}\right|_{\boldsymbol{\alpha}_0} = 1, \qquad \left.\frac{\partial\alpha_1}{\partial a_2}\right|_{\boldsymbol{\alpha}_0} = 0, \qquad \left.\frac{\partial\alpha_2}{\partial a_1}\right|_{\boldsymbol{\alpha}_0} = a_{20}, \qquad \left.\frac{\partial\alpha_2}{\partial a_2}\right|_{\boldsymbol{\alpha}_0} = a_{10}. \tag{2.4-39}$$

Hence, the parameter-induced error to be determined is given by

$$\Delta y = (S_{\alpha_1} + a_{20}S_{\alpha_2})\,\Delta a_1 + S_{\alpha_2}a_{10}\,\Delta a_2. \tag{2.4-40}$$

Note that $S_{a_1} = S_{\alpha_1} + a_{20}S_{\alpha_2}$ and $S_{a_2} = S_{\alpha_2}a_{10}$.

2.5 MILLER–MURRAY CLASSIFICATION OF PARAMETER VARIATIONS

In order to attack the sensitivity problem from a quite general point of view, it is useful to classify the parameter variations with due regard to the different methods required for their treatment. There are only a few categories that have to be distinguished. For continuously acting systems, Miller and Murray [68,69] have proposed such a classification in 1953, which has proved useful up to the present, and which will be followed in this book. According to this, the following three categories of parameter variations may be distinguished in continuous systems.

Definition 2.5-1 *α-Errors* (*α-variations*). α-errors are parameter variations around a nominal value $\boldsymbol{\alpha}_0$ that do not affect the order of the mathematical model. A necessary condition for errors to be α-errors is that $\boldsymbol{\alpha}_0 \neq 0$.

Sources of α-errors may be identification inaccuracies, manufacturing tolerances, uncertainty of the knowledge, or changes, of environmental and operating conditions.

Because of the validity of the existence theorem of the differential equations which says that the output of a system is continuous in $\boldsymbol{\alpha}$ if the corresponding differential equation is continuous in $\boldsymbol{\alpha}$, the system function can be considered an analytic function of $\boldsymbol{\alpha}_0$. Parameters of this kind will be referred to as α-parameters.

Definition 2.5-2 *β-Errors (β-variations).* β-errors are variations of the initial conditions from their nominal values β_0.

Sources of β-errors may be measurement inaccuracies, inexact adjustments, or the presence of noise in simulating the system, for example, on an analog computer. Since the output of a continuous system is always continuous in the initial conditions, the system function in case of β-errors can generally be considered an analytic function of $\boldsymbol{\beta}_0$. Parameters of this type will be called β-parameters.

Difinition 2.5-3 *λ-Errors (λ-variations, singular perturbations).* λ-errors are parameter variations from a nominal value $\boldsymbol{\lambda}_0 = \mathbf{0}$ that affect the order of the mathematical model.

Sources of λ-errors may be idealizations in the development of the mathematical model (e.g., neglecting parasitic capacitances or inductances) or erronious detection of the order of the system. Since the dependence of the output of a system on the parameters is not defined for nominal parameter values, the sensitivity function is not an analytic function in case of λ-errors. The term "singular perturbation," which is also used for λ-errors, indicates the singularity of the system function at nominal values $\boldsymbol{\lambda} = \mathbf{0}$. Parameters of this type will be termed λ-parameters.

To understand the motive of λ-errors, it should be noticed that λ-errors play an important role in the investigation of the sensitivity of a mathematical model with respect to neglected parameters. It is a common habit to neglect small parameters when setting up a mathematical model of a system. There is a good reason for this simplification: It reduces the order of the mathematical model, thereby simplifying the mathematical treatment of the problem. Many procedures of modern control theory, such as optimization, require such a simplification because they are extremely cumbersome or not even applicable in case of high order mathematical models.

However, there is a certain risk in this procedure. As is well known, the neglect of parameters can give rise to considerable errors and even to instability when the mathematical results are applied to the actual system. How critical the situation will be depends on how sensitive the system behavior is with respect to the neglected parameter. If the sensitivity of the system is high, the application of sensitivity theory enables us to take into account the neglected parameters without repeating the design for a higher order model or, in other words, to reintroduce the neglected parameters at a lower cost than by repeating the design.

In the case of discontinuous systems, such as sampled-data systems or systems with discontinuous right-hand sides of their differential equations, the sampling instants and sampling periods have to be considered as parameters as well. They can be treated in a similar way as α-parameters.

Example 2.5-1 Let the nominal differential equation of a system be given by

$$a_2\ddot{y} + a_1\dot{y} + a_0 y = u(t)$$

with the initial conditions $\dot{y}(0) = y_1{}^0$, $y(0) = y_0{}^0$. Suppose that the parameters of the actual system are such that the corresponding differential equation reads

$$\Delta a_3 \dddot{y} + (a_2 + \Delta a_2)\ddot{y} + (a_1 + \Delta a_1)\dot{y} + (a_0 + \Delta a_0)y = u(t)$$

with the initial conditions $\ddot{y}(0) = \Delta y_2{}^0$, $\dot{y}(0) = y_1{}^0 + \Delta y_1{}^0$, and $y(0) = y_0{}^0 + \Delta y_0{}^0$. Classify the parameter variations.

Solution According to the above definitions,

$\Delta a_0, \ldots, \Delta a_2$	are α-variations,
Δy_0^0, $\Delta y_1{}^0$	are β-variations,
Δa_3	is a λ-variation.

PROBLEMS

2.1 Show the validity of Eqs. (2.4–2)–(2.4–7).

2.2 Develop the equations corresponding to Eqs. (2.4–2)–(2.4–4) for the semirelative sensitivity functions

$$\tilde{S}_\alpha{}^\zeta \triangleq (\partial\zeta/\partial \ln \alpha)_{\alpha 0} \qquad \text{and} \qquad \underset{\sim}{S}_\alpha{}^\zeta \triangleq (\partial \ln \zeta/\partial\alpha)_{\alpha 0}.$$

2.3 Consider the function $f(x,y,z) = x^2 + xz + 2y^2$.

(a) Let x,y,z be functions of t and determine df/dt.

(b) Consider y and z as functions of x and determine df/dx.

(c) Consider z as a function of x and y and determine $\partial f/\partial x$.

2.4 The differential equation of an RC low-pass is of the form $\alpha_2\dot{y} + \alpha_1 y = u$, where u represents the input and y the output voltage. The coefficients are $\alpha_1 = K$ and $\alpha_2 = KRC$, where K is the amplification factor, R denotes the resistance, and C the capacitance. Suppose that the sensitivity functions of y with respect to α_1 and α_2, $S_{\alpha 1}^y$ and $S_{\alpha 2}^y$ are known.

(a) Determine the sensitivity functions $S_K{}^y$ and $S_R{}^y$ as well as Δy assuming that $\Delta K \ll K_0$, $\Delta R \ll R_0$, and C is unperturbed.

(b) Determine $S_K{}^y$, $S_R{}^y$, and $S_C{}^y$ as well as Δy for given variations $\Delta K \ll K_0$, $\Delta R \ll R_0$, and $\Delta C \ll C_0$.

2.5 Suppose that a pendulum clock of pendulum length l_0 keeps correct time at temperature v_0. The period of oscillation, generally given by $T = 2\pi\sqrt{l_0/g}$, where g is the gravitation constant, is 1 second. At temperature v_1 the length will be $l = l_0\,(1 + \alpha\,\Delta v)$, where $\alpha = 10^{-5}/°\mathrm{C}$.

(a) Determine the (absolute) sensitivity function of T with respect to Δv, denoted by $S_v{}^T$.

(b) How many seconds per day does the clock lose if the temperature rises 10° C?

2.6 Let the state equation of a system be given by $\dot{\boldsymbol{x}} = \boldsymbol{f}[\boldsymbol{x}(t, \boldsymbol{\alpha}), \boldsymbol{u}(t), t, \boldsymbol{\alpha}]$, where $\boldsymbol{x}$ is the state vector, $\boldsymbol{u}$ the input vector, t the time, and $\boldsymbol{\alpha}$ the parameter vector. Take the partial derivative of the above equation with respect to $\boldsymbol{\alpha}$.

2.7 Given the ordinary differential equation of an nth order system in the general form

$$f[y^{(n)}(t, \boldsymbol{\alpha}), \ldots, \dot{y}(t, \boldsymbol{\alpha}), y(t, \boldsymbol{\alpha}), u(t), \boldsymbol{\alpha}] = 0.$$

Find the partial derivative of this equation with respect to the parameter vector $\boldsymbol{\alpha}$.

2.8 In a tangent galvanometer, the relationship between the intensity of the current I and the angle of deflection of the magnetic needle α is given by $I = c \tan \alpha$, where $0 \leq \alpha \leq \pi/2$ and c is a characteristic constant of the apparatus.

(a) Determine the relationship between the percentage changes of α and I, and give the value of α_0 for which $\Delta\alpha/\alpha_0$ as a function of $\Delta I/I_0$ is maximal.

(b) Suppose that there is a constant deflection error $\Delta\alpha$ superimposed upon the angle α_0. Determine the value of α_0 for which the accuracy of indication reaches its maximum value.

2.9 Consider a direct-reading capacitive micromanometer. The relationship between the voltage u and the pressure p to be measured is given by $u = a\sqrt{p}$, where $0 \leq u \leq 10$ V and a is a characteristic constant.

(a) Determine the percentage change of the voltage u due to a percentage change of p.

(b) Determine the semirelative sensitivity function $S_u{}^p$ and give the value of u_0 for which the accuracy of indication is maximal (i.e., for which the percentage change of p due to a constant perturbation Δu is minimal).

Chapter 3

Definitions of Sensitivity Functions and Measures

3.1 INTRODUCTION

Dynamic systems can be characterized in several ways: in the time domain, in the frequency domain, or in terms of a performance index. There is evidently an adequate number of ways to define the sensitivity function of a dynamic system. The definition that is actually used depends on the form of the mathematical model as well as on the purpose of consideration. For example, if the system is represented by a transfer function, the sensitivity will be defined on the basis of the parameter-induced change of the transfer function; whereas in case of a state space representation, the natural basis of the sensitivity definition will be the parameter-induced change of the trajectory.

Thus, the sensitivity functions can be classified into the following three categories:

(1) sensitivity functions in the time domain,
(2) sensitivity functions in the frequency or z-domain, and
(3) performance-index sensitivity.

In this chapter the most important representatives of each category are introduced and discussed. Besides these sensitivity functions there are so-called sensitivity measures that are defined on the entirety of the sensitivity functions and, therefore, allow for a global characterization of the sensitivity by a single number. Some of them will also be given in this chapter.

The oldest definition of a sensitivity function was given by Bode [1]. This definition is based on the transfer function and was restricted to infinitesimal parameter deviations. In the sequel, Horowitz [11] gave a different interpretation of *Bode's sensitivity function* and also used it with great success for the design of control systems in the frequency domain [45, 46]. Perkins and Cruz

[77] extended Bode's sensitivity function in different directions, also establishing its significance for time domain considerations.

In connection with simulations on network analyzers and analog computers, the *output sensitivity functions* were introduced in the fifties mainly by Bikhovski [24] and Miller and Murray [68,69]. In the early sixties this definition was extended to the state space, resulting in the so-called *trajectory sensitivity function* [14, 18, 30]. The discussion of the merit of the time domain sensitivity functions has not yet come to an end [45]. However, there is no doubt that they play an important role in the comparison of open- and closed-loop systems as well as in the design of optimal controls.

In 1963 Dorato [36] introduced the so-called *performance-index sensitivity*. The practical importance of this sensitivity definition is still questionable, since it has been found to be useless for the comparison of open- and closed-loop optimal control systems, for which purpose it was mainly defined.

Besides the sensitivity functions mentioned above there are various special sensitivity definitions, such as the sensitivity of the *overshoot* in the time or frequency domain, the *eigenvalue* (*pole or zero*) *sensitivity*, and so on. Definitions such as these may be very helpful in the characterization of the sensitivity of a system in a certain aspect such as its relative stability. However, they are not nearly so important as the sensitivity functions, and, therefore, less attention will be paid to them in this book.

3.2 SENSITIVITY FUNCTIONS IN THE TIME DOMAIN

3.2.1 The Output Sensitivity Function of Continuous Systems

Consider the input–output behavior of a continuous, possibly nonlinear, single-variable system described by an ordinary differential equation of the type

$$f[y^{(n)}, y^{(n-1)}, \ldots, y, t, \alpha_0] = 0 \tag{3.2–1}$$

with the initial conditions $y^{(i)}(t_0) \triangleq y_i^{\,0}\ (i = 0, 1, \ldots, n-1)$. y denotes the output signal, t the time, and α a single *time-invariant* or slowly varying α-parameter that has the *nominal* value α_0.

In general, f is a function of the input u as well. However, if u is an external input which does not depend on α_0, the dependence of f on u is not relevant for further considerations and hence will be dropped for ease of notation. Note that this is not allowed if u represents a feedback signal, thus depending on α_0. This case, excluded here, will be considered later.

Let us suppose that the above *nominal* differential equation has the unique solution

$$y_0 = y(t, \alpha_0), \tag{3.2–2}$$

which we shall call the *nominal solution*. For the same reasons as above, u is dropped in the argument of y.

Let us now assume that the parameter changes from α_0 to $\alpha = \alpha_0 + \Delta\alpha$, where $\Delta\alpha$ is time-invariant or slowly varying with time. α is called the *actual parameter* value. The corresponding *actual differential equation* can then be written as

$$f[y^{(n)}, y^{(n-1)}, \ldots, y, t, \alpha] = 0. \tag{3.2–3}$$

Note that by this change of α_0 into α the initial conditions remain unchanged, namely $y^{(i)}(t_0) = y_i^0$. The corresponding solution is

$$y = y(t, \alpha), \tag{3.2–4}$$

which we shall call the *actual* (or *perturbed*) *solution*.

It is assumed that $y(t, \alpha)$ is of the same type as $y(t, \alpha_0)$, and $y(t, \alpha)$ deviates infinitesimally from $y(t, \alpha_0)$ if α deviates infinitesimally from α_0 or, in other words, that $y(t, \alpha)$ is a continuous function of α. The conditions for fulfilling this requirement are given in the mathematical literature (see, e.g., the condition of Lipschitz as given in [16]). For our purpose, it is sufficient to know that y is continuous in α if f is continuous in y which is true for all continuous systems and $\alpha_0 \neq 0$.

With the above assumptions the actual solution $y(t, \alpha_0 + \Delta\alpha)$ can be expanded into a Taylor series around α_0, yielding

$$y(t, \alpha) = y(t, \alpha_0) + \left.\frac{\partial y}{\partial \alpha}\right|_{\alpha 0} \Delta\alpha + \left.\frac{1}{2}\frac{\partial^2 y}{\partial \alpha^2}\right|_{\alpha 0} \Delta\alpha^2 + \cdots \tag{3.2–5}$$

or

$$\Delta y = \left.\frac{\partial y}{\partial \alpha}\right|_{\alpha 0} \Delta\alpha + \frac{\Delta\alpha^2}{2}\frac{\partial^2 y(\varphi)}{\partial \varphi^2}, \tag{3.2–6}$$

where $\varphi = \alpha_0 + \theta\,\Delta\alpha$ $(0 < \theta < 1)$ is an intermediate value which does not need to be more precisely known. If $\Delta\alpha \ll \alpha_0$, the Taylor series can be truncated at the linear term. This gives

$$y(t,\alpha) = y(t,\alpha_0) + \left.\frac{\partial y}{\partial \alpha}\right|_{\alpha 0} \Delta\alpha. \tag{3.2–7}$$

For finite values of $\Delta\alpha$, this expression can be considered a first-order approximation of $y(t, \alpha)$. For infinitesimal parameter changes, however, Eq.(3.2–7) is exact and can be used for the definition of the actual solution. In practice, one has to evaluate from time to time the range of $\Delta\alpha$ for which Eq.(3.2–7) can be used. Now we define the following:

Definition 3.2-1 *Output sensitivity function.* The limit

$$\sigma(t, \alpha_0) \triangleq \left.\frac{\partial y(t, \alpha)}{\partial \alpha}\right|_{\alpha 0} = \lim_{\Delta\alpha \to 0} \frac{y(t, \alpha_0 + \Delta\alpha) - y(t, \alpha_0)}{\Delta\alpha} \tag{3.2–8}$$

is defined as the (absolute) output sensitivity function of the system.

The output sensitivity function is a time function whose shape depends on the shape of the input signal u. Besides this, σ is a function of the nominal parameter value α_0. If we assume that α_0 is a constant and the parameter changes around α_0 are small, σ can be regarded as a function of a single variable t that can be represented in a single plane. However, if the parameter α_0 changes over a wide range, σ has to be regarded as a function of two variables, t and α_0, and the three-dimensional space or a group of curves are required for the representation of σ. In this case, σ is referred to as the *global sensitivity function* [18]. If σ depends on r parameters $\alpha_{10}, \ldots, \alpha_{r0}$, an $(r + 2)$ — dimensional space is needed for the representation of the global sensitivity function [16]. Since we restrict ourselves to small parameter changes, we shall not have to cope with this problem.

Introducing Definition 3.2–1 into Eq.(3.2–7), the actual output can be written as

$$y(t, \alpha) \triangleq \underbrace{y(t, \alpha_0)}_{\text{nominal output}} + \underbrace{\sigma(t, \alpha_0)\, \Delta\alpha}_{\text{parameter-induced output error}}. \tag{3.2–9}$$

The parameter-induced output error is in these terms

$$\Delta y(t, \alpha) \triangleq \sigma(t, \alpha_0)\, \Delta\alpha, \tag{3.2–10}$$

which coincides with Eq.(2.2–2) or Eq.(2.3–2) for $r = 1$.

Example 3.2-1 Consider a linear third-order system described by the differential equation

$$\dddot{y} + 3a\ddot{y} + 3a^2\dot{y} + a^3 y = u$$

with all initial conditions zero. Let u be the impulse function $\delta(t)$. The solution of this equation by means of the Laplace transform yields the actual output function

$$y(t, a) = \tfrac{1}{2} t^2 e^{-at}.$$

Let the nominal parameter be a_0. Then the output sensitivity function becomes

$$\sigma(t, a_0) = \left.\frac{\partial y(t, a)}{\partial a}\right|_{a0} = -\tfrac{1}{2} t^3 e^{-a_0 t}.$$

Substituting for $\sigma(t,a_0)$ in Eq.(3.2–10) yields the parameter-induced output error for small variations Δa around a_0,

$$\Delta y(t, a) = -\tfrac{1}{2}\, \Delta a\, t^3 e^{-a_0 t}$$

and, with the aid of Eq. (3.2–9), the actual output function

$$y(t, a) = \tfrac{1}{2} t^2 e^{-a_0 t}(1 - \Delta a\, t).$$

The *multiparameter case* can be treated in an analogous manner. Let the system behavior depend on r parameters represented by the vector $\boldsymbol{\alpha} = [\alpha_1 \; \alpha_2 \cdots \alpha_r]^T$. Then the *nominal* differential equation can be written as

$$f[y^{(n)}, y^{(n-1)}, \ldots, y, t, \boldsymbol{\alpha}_0] = 0, \tag{3.2–11}$$

and the *actual* differential equation is given by

$$f[y^{(n)}, y^{(n-1)}, \ldots, y, t, \boldsymbol{\alpha}_0 + \Delta\boldsymbol{\alpha}] = 0. \tag{3.2–12}$$

The actual system output is then

$$y(t, \boldsymbol{\alpha}) = y(t, \boldsymbol{\alpha}_0) + \sum_{j=1}^{r} \sigma_j(t, \boldsymbol{\alpha}_0)\, \Delta\alpha_j, \tag{3.2–13}$$

where

$$\sigma_j(t, \boldsymbol{\alpha}_0) \triangleq \left.\frac{\partial y(t, \boldsymbol{\sigma})}{\partial \alpha_j}\right|_{\boldsymbol{\alpha}0}, \qquad j = 1, 2, \ldots, r. \tag{3.2–14}$$

Now defining the *output sensitivity vector*

$$\boldsymbol{\sigma}_1 \triangleq \left.\frac{\partial y}{\partial \boldsymbol{\alpha}}\right|_{\boldsymbol{\alpha}_0} = \left[\frac{\partial y}{\partial \alpha_1} \;\; \frac{\partial y}{\partial \alpha_2} \;\; \cdots \;\; \frac{\partial y}{\partial \alpha_r}\right]_{\boldsymbol{\alpha}0}, \tag{3.2–15}$$

which is identical with the gradient vector (Definition 2.3–8), the parameter-induced output error can be written as

$$\Delta y(t, \boldsymbol{\alpha}) = \sigma_1(t, \boldsymbol{\alpha}_0)\, \Delta\boldsymbol{\alpha}. \tag{3.2–16}$$

The extension of this procedure to systems with *multiple outputs*, where q is the number of outputs, yields

$$y_k(t, \boldsymbol{\alpha}) = y_k(t, \boldsymbol{\alpha}_0) + \boldsymbol{\sigma}_k(t, \boldsymbol{\alpha}_0)\, \Delta\boldsymbol{\alpha}, \qquad k = 1, 2, \ldots, q. \tag{3.2–17}$$

The row vector

$$\boldsymbol{\sigma}_k(t, \boldsymbol{\alpha}_0) \triangleq \left.\frac{\partial y_k}{\partial \boldsymbol{\alpha}}\right|_{\boldsymbol{\alpha}0} = \left[\frac{\partial y_k}{\partial \alpha_1} \;\; \frac{\partial y_k}{\partial \alpha_2} \;\; \cdots \;\; \frac{\partial y_k}{\partial \alpha_r}\right]_{\boldsymbol{\alpha}0} \tag{3.2–18}$$

is the output sensitivity vector of the kth component y_k of the output vector $\mathbf{y} = [y_1 \, y_2 \cdots y_q]^T$ with respect to $\boldsymbol{\alpha}$.

Definition 3.2-2 *Output sensitivity matrix (α-parameters).* If the $q \times 1$ output vector $\mathbf{y}$ of a system depends on an $r \times 1$ parameter vector $\boldsymbol{\alpha}$, the Jacobian

$$\boldsymbol{\sigma} \triangleq \left.\frac{\partial \mathbf{y}}{\partial \boldsymbol{\alpha}}\right|_{\boldsymbol{\alpha}0} = \left[\frac{\partial y_i}{\partial \alpha_k}\right]_{\boldsymbol{\alpha}0}, \qquad \begin{array}{l} i = 1, 2, \ldots, q, \\ k = 1, 2, \ldots, r, \end{array} \tag{3.2–19}$$

is defined as the output sensitivity matrix with respect to α-parameter variations.

With this definition, the parameter-induced output error can be written as

$$\Delta \mathbf{y}(t, \boldsymbol{\alpha}) = \boldsymbol{\sigma}(t, \boldsymbol{\alpha}_0)\, \Delta\boldsymbol{\alpha}. \tag{3.2–20}$$

Notice that the rows of the output sensitivity matrix $\boldsymbol{\sigma}$ are the output sensitivity vectors $\boldsymbol{\sigma}_k$ of the components y_k defined by Eq.(3.2–18).

So far we have considered time-invariant α-parameters. It is quite clear that similar definitions hold in the case of *β-parameters.*

Definition 3.2-3 *Output sensitivity vector* (*β-parameters*). Consider an nth order single input single output system with the output y. Suppose that $\boldsymbol{\beta} = [\beta_0, \beta_1, \ldots, \beta_{n-1}]^T$ is the vector of initial conditions, where $\beta_\nu = y^{(\nu)}(0)$, and $\boldsymbol{\beta}_0$ denotes its nominal value. Then the partial derivative

$$\boldsymbol{\sigma}_\beta(t, \boldsymbol{\beta}_0) \triangleq \left.\frac{\partial y(t, \boldsymbol{\beta})}{\partial \boldsymbol{\beta}}\right|_{\beta_0} \tag{3.2–21}$$

is defined as the output sensitivity vector with respect to variations of the initial conditions. The corresponding relationship for Δy is the same as Eq. (3.2–20). The multivariable case is better treated in the state space.

Let us finally consider the output sensitivity function in the case of *time-variant* parameter variations. We shall again assume that the parameters slightly change around a nominal value α_0 but now according to the relation

$$\alpha(t) = \alpha_0 + \varepsilon g(t), \tag{3.2–22}$$

where ε is a constant (with respect to time), $g(t)$ is assumed to be a uniformly bounded, integrable function. The term "uniform boundedness' is a generalization of the term "boundedness", known from time-invariant systems, for time-variant systems. A boundedness is called uniform if the area in which the initial value $g(t_0)$ must lie in order to satisfy $|g(t) - g_e| < \infty$ (g_e steady state value) is independent of the initial time t_0.

In this case, the output sensitivity function is defined as follows:

Definition 3.2-4 *Output sensitivity function (in the time-variant case):*

$$\sigma(t, \alpha_0, g) \triangleq \lim_{\varepsilon \to 0} \frac{y[(\alpha_0 + \varepsilon g), t] - y(\alpha_0, t)}{\varepsilon} \triangleq \sigma_\varepsilon . \tag{3.2–23}$$

Notice that in the above definition, the function $g(t)$ appears additionally in the argument of σ. Hence, in the time-variant case, the character of the parameter perturbation $g(t)$ must be known in addition to α_0.

The output sensitivity vector and matrix are defined in an analogous manner. It is evident that all these definitions also hold for β-variations.

3.2.2 Interpretation of the Output Sensitivity Function

In order to illustrate the interpretation of the output sensitivity function, let us consider a system with the differential equation

$$\dddot{y} + a_2 \ddot{y} + a_1 \dot{y} + a_0 y = u \tag{3.2–24}$$

with the initial conditions $y(0) = \cdots = \ddot{y}(0) = 0$. Let the input signal u be the unit step function denoted by $1(t)$. The nominal values of the parameters are assumed to be $a_{00} = 20$, $a_{10} = 15$, $a_{20} = 5$. The shapes of the output y and of the three output sensitivity functions $(\partial y/\partial a_0)_{\boldsymbol{\alpha}0}$, $(\partial y/\partial a_1)_{\boldsymbol{\alpha}0}$, $(\partial y/\partial a_2)_{\boldsymbol{\alpha}0}$ are plotted in Fig. 3.2–1a as functions of time on different scales.

By means of these plots, the following observations can be made:

(1) The plot of $(\partial y/\partial a_0)_{\boldsymbol{\alpha}0}$ indicates that a parameter change Δa_0 affects primarily the steady state of $y(t)$, having little effect on the rise time or overshoot.

(2) Since $(\partial y/\partial a_1)_{\boldsymbol{\alpha}0}$ is largest at the time where the overshoots of y occur, it indicates that a change Δa_1 most strongly affects the overshoots of y.

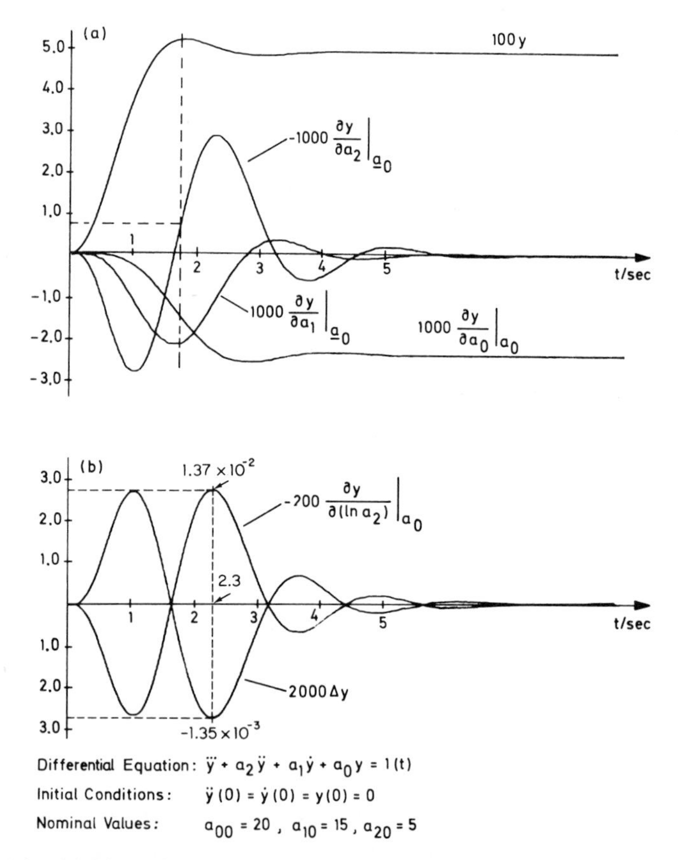

Fig. 3.2-1. (a) Plots of the output y and the output sensitivity functions of the third-order system of example 3.2–2. (b) Comparison of the parameter-induced output error obtained by the sensitivity approach with the exact result.

(3) The plot of $(\partial y/\partial a_2)_{\boldsymbol{a}0}$ indicates that a change of a_2 most strongly affects the slopes of y, having no effect on the overshoots or the steady state.

From these plots the parameter-induced change of the output function Δy can readily be calculated for small parameter changes according to the formula

$$\Delta y \approx \left.\frac{\partial y}{\partial a_0}\right|_{\boldsymbol{a}0} \Delta a_0 + \left.\frac{\partial y}{\partial a_1}\right|_{\boldsymbol{a}0} \Delta a_1 + \left.\frac{\partial y}{\partial a_2}\right|_{\boldsymbol{a}0} \Delta a_2. \tag{3.2–25}$$

As has been pointed out in Section 2.3, it is often more convenient to use percentage changes, rather than absolute changes. For example, this is true if the sensitivity is to be compared with respect to the various parameters. Multiplying the numerators and denominators of the terms on the right-hand side of Eq. (3.2–25) by the corresponding nominal parameter values, we obtain

$$\Delta y\,(t, \boldsymbol{a}) = \sum_{j=0}^{2} a_{j0} \left.\frac{\partial y}{\partial a_j}\right|_{\boldsymbol{a}_0} \frac{\Delta \boldsymbol{a}_j}{a_{j0}}. \tag{3.2–26}$$

This can also be written as

$$\Delta y(t, \boldsymbol{a}) = \sum_{j=0}^{2} \left.\frac{\partial y}{\partial \ln a_j}\right|_{\boldsymbol{a}_0} \frac{\Delta \boldsymbol{a}_j}{a_{j0}}, \tag{3.2–27}$$

or, in abbreviated form,

$$\Delta y(t, \boldsymbol{a}) = \sum_{j=0}^{2} \tilde{\sigma}_{a_j}(t, \boldsymbol{a}_0) \frac{\Delta a_j}{a_{j0}}, \tag{3.2–28}$$

where $\tilde{\sigma}_{a_j}$ is the semirelative output sensitivity function with respect to the parameter a_j. In an analogous manner, the relative change of y can be obtained,

$$\frac{\Delta y(t, \boldsymbol{a})}{y(t, \boldsymbol{a}_0)} = \sum_{j=0}^{2} \bar{\sigma}_{a_j}(t, \boldsymbol{a}_0) \frac{\Delta a_j}{a_{j0}}, \tag{3.2–29}$$

where $\bar{\sigma}_{a_j}$ is the relative sensitivity function defined in Section 2.3 (Definition 2.3–4). The merit of this formulation is that only dimensionless expressions occur on both sides of Eq. (3.2–29).

In order to illustrate the utility of the first-order approximation, suppose that it is desired to estimate the change in y due to a 10% increase in a_2, leaving a_{00} and a_{10} unchanged. Fig. 3.2–1b shows the negative shape of the semirelative sensitivity function with respect to a_2 and also the exact change of y due to the 10% increase of a_2, i.e., $\Delta a_2/a_{20} = 0.1$. Comparing both curves, we see that, for example, at $t = 2.3$, the value of the sensitivity function is $\sigma = -1.37 \times 10^{-2}$. Hence, multiplying with $\Delta a_2/a_{20} = 0.1$, we obtain $\Delta y = -1.37 \times 10^{-3}$. The exact value of Δy at $t = 2.3$, taken from the plot of Δy, is -1.35×10^{-3}. Thus we see that the result obtained by the use of the sensitivity function is a very accurate estimate of the actual value of Δy.

Experiences like this have verified that in many cases the first-order approximations are satisfactory for parameter changes up to 30% [72].

Example 3.2-2 Determine the output sensitivity function $\sigma(t, \omega_0) = (\partial y/\partial\omega)_{\omega 0}$ and the parameter-induced output error Δy of the system with the differential equation

$$\ddot{y} + \omega^2 y = \omega u,$$

where the input u is a unit impulse $\delta(t)$ and the initial conditions $y(0)$ and $\dot{y}(0)$ are both equal to zero.

Using the Laplace transform, we obtain the actual output function

$$y(t, \omega) = \sin \omega t.$$

The partial differentiation of y with respect to ω yields for $\omega = \omega_0$

$$\sigma(t, \omega_0) = t \cos \omega_0 t,$$

and with this result we obtain

$$\Delta t(t, \omega) = \sigma(t, \omega_0)\, \Delta\omega = t\, \Delta\omega \cos \omega_0 t.$$

Apparently, the envelopes of $\sigma(t, \omega_0)$ and Δy diverge with time. This implies that, for a constant change $\Delta\omega$, the amplitude of the parameter-induced output error is increasing with time. On the one hand, this property of the sensitivity function generates severe problems when it is used in adaptive systems; on the other hand, this makes it an excellent tool to detect steady oscillations, i.e., instability of a system.

3.2.3 The Trajectory Sensitivity Function of Continuous Systems

A continuous, possibly nonlinear system of nth order can, in general, be described in the state space by a vector differential equation of the form

$$\dot{\boldsymbol{x}} = \boldsymbol{f}(\boldsymbol{x}, t, \boldsymbol{u}, \boldsymbol{\alpha}_0), \qquad \boldsymbol{x}(t_0) = \boldsymbol{x}^0. \tag{3.2–30}$$

Here $\boldsymbol{x}$ is an $n \times 1$ state vector, $\boldsymbol{f}$ an $n \times 1$ vector function, $\boldsymbol{u}$ an input vector, $\boldsymbol{\alpha}_0$ a nominal $r \times 1$ parameter vector (assumed to be an α-parameter), and $\boldsymbol{x}^0$ is the $n \times 1$ initial condition vector or initial state. Equation (3.2–30) is called the *nominal* state equation.

Assuming that the parameter vector deviates from the nominal value $\boldsymbol{\alpha}_0$ by $\Delta\boldsymbol{\alpha}$, we have

$$\dot{\boldsymbol{x}} = \boldsymbol{f}(\boldsymbol{x}, t, \boldsymbol{u}, \boldsymbol{\alpha}), \qquad \boldsymbol{x}(t_0) = \boldsymbol{x}^0, \tag{3.2–31}$$

with the initial conditions $\boldsymbol{x}^0$ unchanged. This equation is called the *actual* state equation.

Now it is assumed that Eq. (3.2–31) has a unique solution $\boldsymbol{x} = \boldsymbol{x}(t, \boldsymbol{\alpha})$ for all admissible initial conditions and parameter values. $\boldsymbol{x}$ is of course a function of $\boldsymbol{u}$, $\boldsymbol{x}^0$ and t_0 as well. However, this dependence is not needed for the following considerations and will, therefore, be dropped again for ease of notation. Furthermore, the solution $\mathbf{x}$ is assumed to be a bounded continuous function in t and $\boldsymbol{\alpha}$. It is known from mathematical literature that this property is guaranteed if $\boldsymbol{f}$ is a bounded continuous function satisfying the Lipschitz condition [16]. It is sufficient for us to know that this condition is fulfilled for all continuous systems with nominal parameter values not equal to zero (α-parameters).

If the parameter takes on its nominal value $\boldsymbol{\alpha}_0$, the *nominal* solution $\boldsymbol{x}_0 = \boldsymbol{x}(t, \boldsymbol{\alpha}_0)$ is obtained. If, on the other hand, the *actual* solution is given by $\boldsymbol{x} \triangleq \boldsymbol{x}(t, \boldsymbol{\alpha})$, then the parameter-induced change of the state vector is defined as

$$\Delta \boldsymbol{x}\,(t, \boldsymbol{\alpha}) \triangleq \boldsymbol{x}(t, \boldsymbol{\alpha}) - \boldsymbol{x}(t, \boldsymbol{\alpha}_0) \tag{3.2–32}$$

A first-order approximation of $\Delta \boldsymbol{x}$ can be written by use of a Taylor expansion in the form

$$\Delta \boldsymbol{x}(t, \boldsymbol{\alpha}) = \sum_{j=1}^{r} \left.\frac{\partial \boldsymbol{x}}{\partial \alpha_j}\right|_{\boldsymbol{\alpha}0} \Delta\alpha_j . \tag{3.2–33}$$

This equation can be viewed as a definition of the parameter-induced trajectory deviation. Now the following definitions are used:

Definition 3.2-5 *Trajectory sensitivity vector.* Let the state $\boldsymbol{x}$ of a continuous system be a continuous function of a time-invariant parameter vector $\boldsymbol{\alpha} = [\alpha_1\ \alpha_2\ \cdots\ \alpha_r]^{\mathrm{T}}$. Then the partial derivative

$$\boldsymbol{\lambda}_j(t, \boldsymbol{\alpha}_0) \triangleq \left.\frac{\partial \boldsymbol{x}(t, \boldsymbol{\alpha})}{\partial \alpha_j}\right|_{\boldsymbol{\alpha}0}, \qquad j = 1, 2, \ldots, r, \tag{3.2–34}$$

is called the trajectory sensitivity vector with respect to the jth parameter.

Note that the trajectory sensitivity vector is of the same dimension as the state vector, namely, n. Its components are the *trajectory sensitivity functions*

$$\lambda_{ij}(t, \boldsymbol{\alpha}_0) \triangleq \left.\frac{\partial x_i(t, \boldsymbol{\alpha})}{\partial \alpha_j}\right|_{\boldsymbol{\alpha}0}, \qquad i = 1, 2, \ldots, n, \qquad j = 1, 2, \ldots, r, \tag{3.2–35}$$

i.e., the partial derivatives of the ith state variable to the jth parameter. Hence,

$$\boldsymbol{\lambda}_j = [\lambda_{1j}\ \cdots\ \lambda_{nj}\ ]^{\mathrm{T}} = \left[\frac{\partial x_1}{\partial \alpha_j}\ \cdots\ \frac{\partial x_n}{\partial \alpha_j}\right]^{\mathrm{T}}_{\boldsymbol{\alpha}0}. \tag{3.2–36}$$

Definition 3.2–6 *Trajectory sensitivity matrix.* The entirety of all $n \times r$ trajectory sensitivity functions form the trajectory sensitivity matrix

$$\boldsymbol{\lambda} = [\boldsymbol{\lambda}_1\ \cdots\ \boldsymbol{\lambda}_r) \triangleq \left.\frac{\partial \boldsymbol{x}}{\partial \boldsymbol{\alpha}}\right|_{\boldsymbol{\alpha}0} \tag{3.2–37}$$

$$= \begin{bmatrix} \dfrac{\partial x_1}{\partial \alpha_1} & \cdots & \dfrac{\partial x_1}{\partial \alpha_r} \\ \vdots & & \vdots \\ \dfrac{\partial x_n}{\partial \alpha_1} & \cdots & \dfrac{\partial x_n}{\partial \alpha_r} \end{bmatrix}_{\boldsymbol{\alpha}0}. \tag{3.2–38}$$

The columns of $\boldsymbol{\lambda}$ are the trajectory sensitivity vectors $\boldsymbol{\lambda}_j$. Here $\boldsymbol{\lambda}$ is the Jacobian matrix of the state vector $\boldsymbol{x}$ with respect to the parameter vector $\boldsymbol{\alpha}$, taken at nominal parameter values.

With these definitions the parameter-induced change of the trajectory can be rewritten, for small $\Delta\alpha_j$'s, as

$$\Delta \boldsymbol{x}(t, \boldsymbol{\alpha}) = \boldsymbol{\lambda}(t, \boldsymbol{\alpha}_0)\, \Delta\boldsymbol{\alpha} = \sum_{j=1}^{r} \boldsymbol{\lambda}_j\, \Delta\alpha_j. \tag{3.2–39}$$

This result is again in agreement with the general relations which we have already used in Sections 3.2.2 and 3.2.3. A graphical representation of the parameter-induced trajectory error is given in Fig. 3.2–2 for a second-order system.

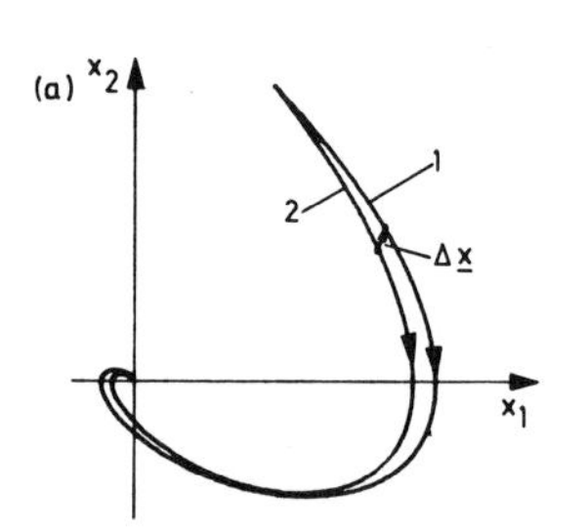

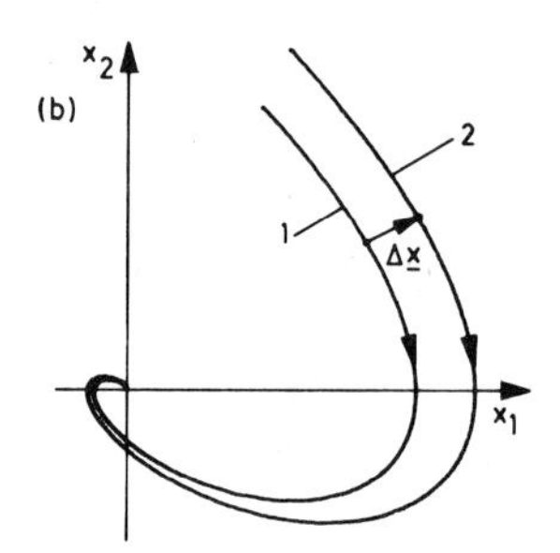

Curve 1: Nominal trajectory; parameters: $a_{10} = 0.08$, $a_{20} = 0.05$

Curve 2: Actual trajectory, parameters: $a_1 = 0.096$, $a_2 = 0.06$

Curve 1: Nominal trajectory; parameters: $y_0(0) = 0.2$, $\dot{y}_0(0) = 0.4$

Curve 2: Actual trajectory, parameters: $\dot{y}(0) = 0.48$, $y(0) = 0.24$

FIG. 3.2-2. Graphical representation of the parameter-induced trajectory error for the second-order system $\ddot{y} + a_1\dot{y} + a_0 y = Ku$. (a) Initial conditions fixed ($y(0) = 0.2$, $\dot{y}(0) = 0.4$, a_1, a_2 parameters). (b) a_1, a_2 fixed ($a_1 = 0.08$, $a_2 = 0.05$, initial conditions parameters).

It should be noted that the corresponding relative and semirelative trajectory sensitivity functions (and vectors and matrices) are defined in the same way as generally shown in Section 3.2.3.

Example 3.2-3 A dc motor in a positioning control system can be described by the ordinary differential equation

$$\ddot{y} + a_1\dot{y} = ku,$$

or by the state equations

$$\begin{bmatrix} \dot{x}_1 \\ \dot{x}_2 \end{bmatrix} = \begin{bmatrix} 0 & 1 \\ 0 & -a_1 \end{bmatrix} \begin{bmatrix} x_1 \\ x_2 \end{bmatrix} + \begin{bmatrix} 0 \\ k \end{bmatrix} u,$$

$$y = x_1.$$

Suppose that the nominal value of the parameter a_1 is $a_{10} = 1$. Further, let u be $u = 0$, and let the initial conditions be $y(0) = 0$, $\dot{y}(0) = 1$ or, in terms of the state equation, $x_1{}^0 = 0$, $x_2{}^0 = 1$. Determine the trajectory sensitivity vector

$$\boldsymbol{\lambda}_1 = \left[\frac{\partial x_1}{\partial a_1}\ \frac{\partial x_2}{\partial a_1}\right]^{\mathrm{T}}_{a_{10}}$$

and the trajectory deviation $\Delta\boldsymbol{x}$ due to a 10% increase of a_1.

Applying the Laplace transform to the original differential equation, we obtain with $Y = \mathscr{L}\{y\}$

$$s^2Y - sy(0) - \dot{y}(0) + a_1 sY(s) - a_1 y(0) = 0.$$

The substitution of $y(0) = 0$ and $\dot{y}(0) = 1$ yields

$$Y(s^2 + a_1 s) = 1.$$

The corresponding time functions are

$$x_1(t, a_1) = y(t, a_1) = \frac{1}{a_1}(1 - e^{-a_1 t}),$$

$$x_2(t, a_1) = \dot{x}_1(t, a_1) = e^{-a_1 t}.$$

Taking partial derivatives, we obtain the trajectory sensitivity vector

$$\boldsymbol{\lambda}_1(t, a_{10}) = \begin{bmatrix} \lambda_{11}(t, a_{10}) \\ \lambda_{21}(t, a_{10}) \end{bmatrix} = \begin{bmatrix} \dfrac{1}{a^2{}_{10}}((1 + a_{10}t)e^{-a_{10}t} - 1) \\ -te^{-a_{10}t} \end{bmatrix}.$$

The parameter-induced trajectory deviation is now found by the aid of Eq. (3.2–39):

$$\Delta\boldsymbol{x}(t, a_1) = \boldsymbol{\lambda}_1\,\Delta a_1 = \begin{bmatrix} \dfrac{\Delta a_1}{a_{10}^2}((1 + a_{10}t)e^{-a_{10}t} - 1) \\ -\Delta a_1\, te^{-a_{10}t} \end{bmatrix},$$

where $a_{10} = 1$, $\Delta a_1 = 0.1$.

In order to give a graphical interpretation, we first eliminate time as the independent variable in the nominal state variables by writing

$$\frac{dx_{20}}{dt} = -a_{10}\frac{dx_{10}}{dt},$$

$$\frac{dx_{20}}{dx_{10}} = -a_{10},$$

$$a_{10}\,dx_{10} + dx_{20} = 0,$$

$$a_{10}\int_0^{x_{10}} d\tilde{x}_1 + \int_1^{x_{20}} d\tilde{x}_2 = a_{10}x_{10} + x_{20} - 1 = 0,$$

or

$$a_{10}x_{10} + x_{20} = 1,$$

which is the equation of a straight line, as shown in Fig. 3.2–3 by a solid line. In the same way the actual trajectory is determined, which is shown in Fig. 3.2–3 by a broken line.

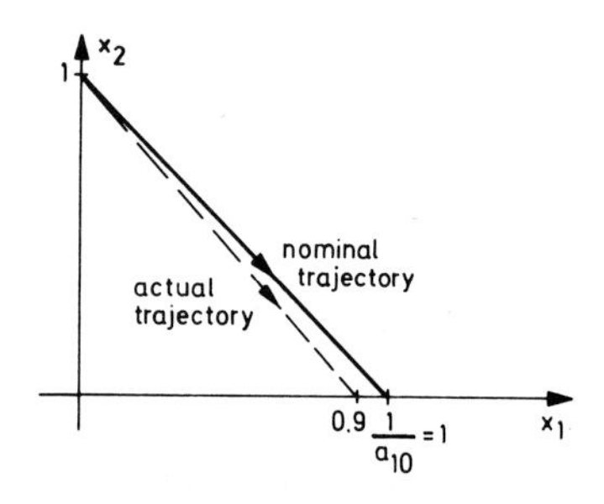

FIG. 3.2-3. Plot of the trajectory of Example 3.2–3.

3.2.4 Sensitivity Functions with Respect to Structural Changes

So far we have dealt with α- and β-variations. They do not affect the structure of the system. By the definition of λ-variations, however, they also change the structure of the system. This will be illustrated by the following example.

Consider the system with the actual differential equation

$$\lambda\ddot{y} + T\dot{y} + y = 1(t), \qquad \dot{y}(0) = y(0) = 0, \tag{3.2–40}$$

where λ and T are the actual parameters. Let us assume that λ is small enough to be neglected. Setting the nominal value of λ equal to $\lambda_0 = 0$, the nominal differential equation degenerates to the first-order differential equation $T_0\dot{y} + y = 1(t)$, $y(0) = 0$. The corresponding step response is an exponential function starting at $t = 0$ with a finite slope, Fig. 3.2–4. An infinitesimal perturbation of λ_0 qualitatively alters the structure and the behavior of the system to

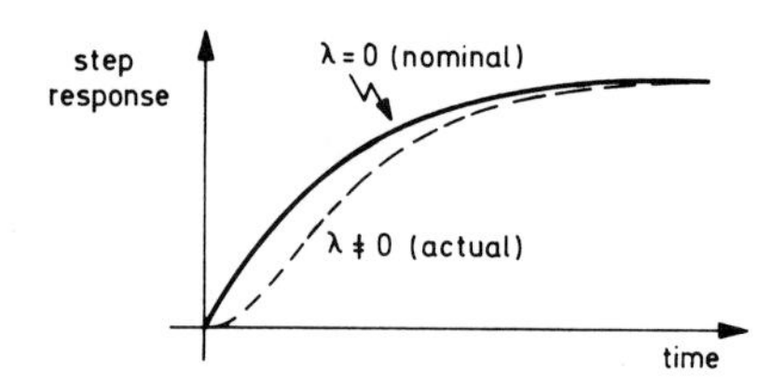

FIG. 3.2-4. Nominal step response ($\lambda = 0$) and actual step response ($\lambda \neq 0$) of a first-order system with a λ-variation.

that of a second-order system, Eq. (3.2–40), whose step response is a sum of two exponential functions, starting at $t = 0$ with a horizontal slope, as shown in Fig. 3.2–4. Therefore, the system is said to have an unstable structure for $\lambda = \lambda_0 = 0$. Conversely, a structurally stable system is one whose structure and behavior remain qualitatively the same if its parameters change slightly.

Thus, in connection with λ-variations the question of *structural stability* arises. This term, defined by A. Andronov in 1937, basically differs from what is known as Liapunov stability. Without giving a precise mathematical definition, the difference may be outlined in the following way [19]:

Liapunov stability means that the system will return to the steady state or at least will not remain much displaced from it after a perturbation of the state.

Structural stability means that the behavior of the system remains qualitatively the same after an infinitesimal parameter perturbation.

Now the question is how the sensitivity function of the system to structural changes can be defined. Do we need the nonreduced actual model, or is it possible to define the sensitivity function on the basis of the reduced nominal model in which the parameter under consideration does not even appear?

In order to answer this question, let us consider a continuous system with a single λ-variation. To avoid confusion with the trajectory sensitivity matrx λ, the λ-parameter shall be denoted by α, keeping in mind that $\alpha_0 = 0$. The fact that the order of the actual system decreases by r if α is set equal to zero can be expressed by the following set of vector differential equations [12]:

$$\dot{\boldsymbol{x}} = \boldsymbol{f}(\boldsymbol{x}, \boldsymbol{z}, t, \boldsymbol{u}, \alpha), \qquad \boldsymbol{x}(t_0) = \boldsymbol{x}^0, \tag{3.2–41}$$

$$\alpha\dot{\boldsymbol{z}} = \boldsymbol{f}_1(\boldsymbol{x}, \boldsymbol{z}, t, \boldsymbol{u}, \alpha), \qquad \boldsymbol{z}(t_0) = \boldsymbol{z}^0, \tag{3.2–42}$$

$$\boldsymbol{y} = \boldsymbol{g}(\boldsymbol{x}, \boldsymbol{z}, t, \boldsymbol{u}, \alpha). \tag{3.2–43}$$

$\boldsymbol{x} = [x_1\ x_2\ \cdots\ x_n]^{\mathrm{T}}$ is an $n \times 1$ state vector and $\boldsymbol{z} = [x_{n+1}\ \cdots\ x_{n+r}]^{\mathrm{T}}$ the vector of the r state variables by which the state increases when α deviates from its nominal value $\alpha_0 = 0$. $\boldsymbol{y}$ is the $q \times 1$ output vector, and $\boldsymbol{f}, \boldsymbol{f}_1, \boldsymbol{g}$ are n- or r- or q-dimensional vector functions, respectively. Equations (3.2–41)–(3.2–43) are the *actual state and output equations* of the system.

If α is set equal to zero in the above equations, the state diminishes from

$$\begin{bmatrix} \boldsymbol{x} \\ \boldsymbol{z} \end{bmatrix} = [x_1\ \cdots\ x_{n+r}]^{\mathrm{T}} \qquad \text{to} \quad \boldsymbol{x}_0 = [x_{10}\ \cdots\ x_{n0}]^{\mathrm{T}}.$$

To show this, we set α equal to zero. Denoting the corresponding symbols by the subscript 0, the state equations become

$$\dot{\boldsymbol{x}}_0 = \boldsymbol{f}(\boldsymbol{x}_0, \boldsymbol{z}_0, t, \boldsymbol{u}, 0), \tag{3.2–44}$$

$$\boldsymbol{0} = \boldsymbol{f}_1(\boldsymbol{x}_0, \boldsymbol{z}_0, t, \boldsymbol{u}, 0). \tag{3.2–45}$$

Equation (3.2–45) is an algebraic or transcendental equation that is assumed to have at least one real solution

$$\boldsymbol{z}_0 = \boldsymbol{\varphi}(\boldsymbol{x}_0, \boldsymbol{u}, t). \tag{3.2–46}$$

Note that the initial value $\boldsymbol{z}_0^{\ 0} = \boldsymbol{\varphi}(\boldsymbol{x}_0, \boldsymbol{u}, t_0)$ may differ from the original $\mathbf{z}^0$. If this solution is substituted into Eq. (3.2–44), the *nominal state equation* reads

$$\dot{\boldsymbol{x}}_0 = \boldsymbol{f}[\boldsymbol{x}_0, \boldsymbol{\varphi}(\boldsymbol{x}_0, \boldsymbol{u}, t), t, 0] \triangleq \boldsymbol{f}_0(\boldsymbol{x}_0, \boldsymbol{u}, t), \qquad \boldsymbol{x}_0^{\ 0} = \boldsymbol{x}^0. \tag{3.2–47}$$

As a result of this procedure, the $r \times 1$ state vector $\mathbf{z}$ completely disappears in the nominal state equation. This means that for $\alpha = 0$ the order is reduced by r, or conversely, that for $\alpha \neq 0$ the order increases from n to $n + r$. Hence, Eqs. (3.2–41)–(3.2–43) are indeed a proper means to represent the effect of λ-parameters.

Example 3.2–4 *Armature controlled dc motor.* As a simple example, we consider the mathematical model of an armature controlled dc motor whose equivalent circuit diagram is shown in Fig. 3.2–5. The mathematical model of this motor is given by the equations

$$\theta\dot{\omega} = ki_a, \tag{3.2–48}$$

$$L_a\dot{i}_a = v_a - r_a i_a - k\omega, \tag{3.2–49}$$

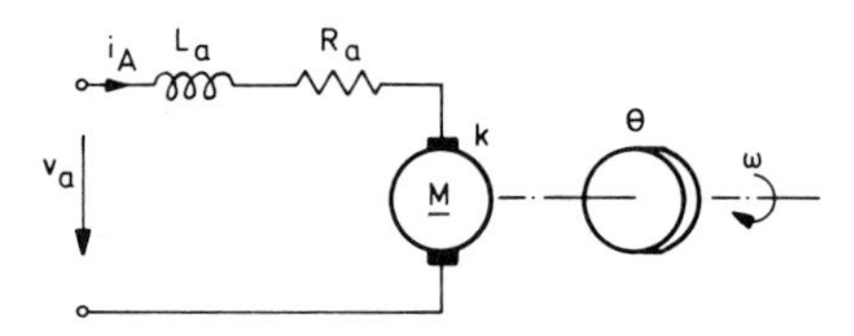

FIG. 3.2-5. Network diagram of an armature controlled dc motor.

where v_a represents the armature voltage, i_a the current, ω the angular speed, R_a the armature resistance, L_a the armature inductance, θ the moment of inertia, and k the motor constant. With the abbreviation $k/\theta \triangleq a$ and the symbols used previously, $\omega \triangleq x$, $i_a \triangleq z$, $v_a = u$, we obtain

$$\dot{x} = az, \tag{3.2–50}$$

$$L_a\dot{z} = -kx - R_a z + u. \tag{3.2–51}$$

These equations are the *actual state equations* of the motor. Usually, L_a is small enough to be neglected in the mathematical model. Therefore, L_a has the meaning of a λ-parameter. Setting $L_a = 0$, we obtain

$$\dot{x}_0 = az_0, \tag{3.2–52}$$

$$0 = -kx_0 - R_a z_0 + u. \tag{3.2–53}$$

Solving Eq. (3.2–53) for z_0,

$$z_0 = \frac{1}{R_a}(u - kx_0), \tag{3.2–54}$$

and substituting this expression into Eq. (3.2–52), we obtain the *degenerated state equation* of the motor,

$$\dot{x}_0 = \frac{a}{R_a}(u - kx_0). \tag{3.2–55}$$

Thus, by setting $L_a = 0$, the order of the state equation reduces from 2 to 1 and z completely disappears from the nominal state equation, which is in agreement with the general result obtained in Eq. (3.2–47).

In the case of a system with λ-parameters, there are in principle two different definitions of the trajectory sensitivity vector:

(1) For the nonreduced mathematical model ($\alpha \neq 0$),

$$\left.\frac{\partial \boldsymbol{x}}{\partial \alpha}\right|_{\alpha 0} \triangleq \boldsymbol{\lambda}^*, \qquad \left.\frac{\partial \boldsymbol{z}}{\partial \alpha}\right|_{\alpha 0} \triangleq \boldsymbol{\eta}^*. \tag{3.2–56}$$

(2) For the degenerated mathematical model ($\alpha = 0$),

$$\left.\frac{\partial \boldsymbol{x}_0}{\partial \alpha}\right|_{\alpha 0} \triangleq \boldsymbol{\lambda}, \qquad \left.\frac{\partial \boldsymbol{z}_0}{\partial \alpha}\right|_{\alpha 0} = \boldsymbol{\eta}. \tag{3.2–57}$$

The problem is to determine under which conditions the sensitivity functions $\boldsymbol{\lambda}$ and $\boldsymbol{\eta}$ of the nominal (degenerated) model can be used for the specification of the sensitivity to structural changes. Evidently, this is possible if the limits

$$\lim_{\alpha \to 0} \frac{\partial \boldsymbol{x}}{\partial \alpha} = \left.\frac{\partial \boldsymbol{x}_0}{\partial \alpha}\right|_{\alpha 0} = \boldsymbol{\lambda}, \tag{3.2–58}$$

$$\lim_{\alpha \to 0} \frac{\partial \boldsymbol{z}}{\partial \alpha} = \left.\frac{\partial \boldsymbol{z}_0}{\partial \alpha}\right|_{\alpha 0} = \boldsymbol{\eta} \tag{3.2–59}$$

exist. It is not easy to show under which conditions these limits exist, since the right-hand side of the equation

$$\dot{\boldsymbol{z}} = \frac{1}{\alpha}\boldsymbol{f}_1(\boldsymbol{x}, \boldsymbol{z}, t, \boldsymbol{u}, \alpha), \tag{3.2–60}$$

which is equivalent to Eq. (3.2–42), tends to infinity as α tends to zero. Therefore, instead of the classical existence and continuity theorems, the singular perturbation theory must be applied. In 1948 and later, the Russian mathematicians Tikhonov and Vasileva gave the general conditions under which, for $\alpha = 0$,

(1) the solutions of the nonreduced mathematical model $\boldsymbol{x}$, $\boldsymbol{z}$ converge toward the solutions $\boldsymbol{x}_0$, $\boldsymbol{z}_0$ of the degenerated model; and

(2) the limits of Eqs. (3.2–58) and (3.2–59) exist.

The result will be summarized by the following theorem.

Theorem 3.2–1. *(Tikhonov 1948).* Consider the auxiliary equation

$$\frac{d\boldsymbol{z}(\tau)}{d\tau} = \boldsymbol{f}_1[\boldsymbol{x}, \boldsymbol{z}(\tau), t, \boldsymbol{u}, 0], \tag{3.2–61}$$

where $\boldsymbol{x}$ and t are considered as fixed parameters, $\boldsymbol{f}_1$ being defined by the state equation (3.2–42). Then the solutions $\boldsymbol{x}(t, \alpha)$, $\boldsymbol{z}(t, \alpha)$ of the nonreduced system equations (3.2–41) and (3.2–42) tend to the solutions $\boldsymbol{x}_0(t)$, $\boldsymbol{z}_0(t)$ of the degenerated model equation (3.2–47) as α tends to zero if the auxiliary equation (3.2–61) has an asymptotically stable solution in the Liapunov sense. In other words, let $\boldsymbol{z} = \boldsymbol{\varphi}(\boldsymbol{x}, t, \boldsymbol{u})$ be an asymptotically stable equilibrium point of Eq. (3.2–61) in the region R of the space of variables $\boldsymbol{x}$, $\boldsymbol{z}$, t. Then the above limits exist if the initial state $\boldsymbol{x}^0$, $\boldsymbol{z}^0$ of the nonreduced system and the initial instant t_0 lie in the region R and if the solution $\boldsymbol{x}_0$, $\boldsymbol{z}_0$ of the degenerated equations (3.2–46), (3.2–47) with the initial condition $\boldsymbol{x}_0{}^0 = \boldsymbol{x}^0$ lies in the region R for all $t \geq t_0$. The limit $\lim_{\alpha \to 0} \boldsymbol{z} = \boldsymbol{z}_0$ exists only for $t > t_0$, since $\boldsymbol{z}^0$, in general, does not satisfy Eq. (3.2–47), a fact that will be treated later in more detail.

A *sufficient condition* for the asymptotic stability of Eq. (3.2–61) is that $\partial \boldsymbol{f}_1/\partial \boldsymbol{z}$ have all its eigenvalues with negative real parts in the region R.

Example 3.2–5 *Nonlinear second-order plant.* Consider a nonlinear plant with a block diagram as shown in Fig. 3.2–6. The nonlinearity consists of a cubic characteristic $v = z^3$, k is a gain factor assumed to be very large. Setting $k = 1/\alpha$, the actual state equations of this second-order system read

$$\dot{x} = z, \tag{3.2–62}$$

$$\alpha \dot{z} = -x - \alpha z - z^3 + u. \tag{3.2–63}$$

If k tends to infinity, α tends to zero, leading to the following nominal (degenerate) state equations:

$$\dot{x}_0 = z_0, \tag{3.2–64}$$

$$0 = -x_0 - z_0{}^3 + u, \tag{3.2–65}$$

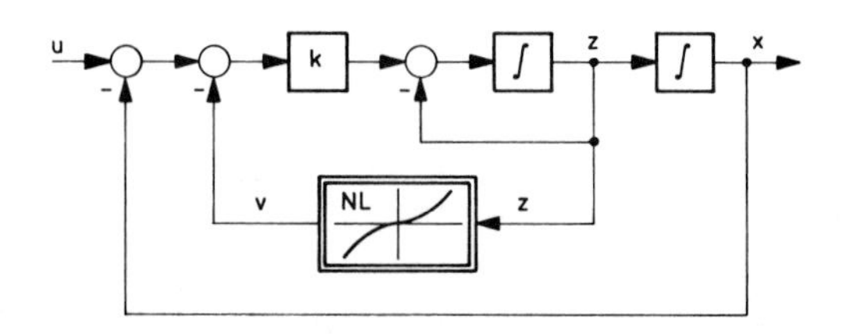

FIG. 3.2–6. Block diagram of the actual nonlinear second-order plant of Example 3.2–5.

or, in abbreviated form,

$$\dot{x}_0 = \sqrt[3]{-x_0 + u}. \tag{3.2–66}$$

The auxiliary equation (3.2–61) becomes

$$\frac{dz(\tau)}{d\tau} = -x - z^3(\tau) + u \equiv f_1. \tag{3.2–67}$$

The partial derivative $\partial f_1/\partial z = -3z^2$ turns out to be negative definite for all x, z, t. Hence, the auxiliary equation is asymptotically stable, and thereby it is proven that, for $k = \infty$,

(1) x and z tend to x_0 and z_0, respectively, and

(2) the limits exist and the sensitivity functions of the degenerated model can be used to characterize the sensitivity with respect to a change from $k = \infty$ to finite values.

3.2.5 Sensitivity Functions of Discontinuous Systems

In this section, we shall discuss the trajectory sensitivity functions of systems that can be described either by vector difference equations or by vector differential equations with discontinuous right-hand sides. Mathematical models of this kind can represent *sampled-data systems* in which continuous operations are sampled at certain instants only, as well as general *systems with switching functions,* such as time optimal or pulse-frequency modulated systems. (There are a great number of publications dealing with the sensitivity of discontinuous systems [22, 33, 51, 72, 79, 89, 90, 100, 101, 102].)

Sampled-data systems in which the continuous state vector $\boldsymbol{x}(t, \boldsymbol{\alpha})$ is sampled at certain instants t_k, $k = 0, 1, \ldots$ can be described by state equations of the form

$$\boldsymbol{x}(t_{k+1}, \boldsymbol{\alpha}) = \boldsymbol{f}[\boldsymbol{x}(t_k, \boldsymbol{\alpha}), t_k, \boldsymbol{u}(t_k), \boldsymbol{\alpha}], \qquad \boldsymbol{x}(0) = \boldsymbol{x}^0, \tag{3.2–68}$$

where $\boldsymbol{x}$ means an $n \times 1$ state vector, $\boldsymbol{f}$ an n-dimensional vector function, $\boldsymbol{u}$ an input vector generated by a sample and hold circuit, and $\boldsymbol{\alpha}$ an $r \times 1$ α-parameter vector.

If we assume that the sampling instants t_k are independent of $\boldsymbol{\alpha}$, then it is evident that the trajectory sensitivity definitions with respect to the α-, β-, and λ-parameters are quite the same as for continuous systems (see Section 3.2.3). No continuity problem arises from the discrete nature of the system. However, the sampling instants t_k have to be considered as additional parameters.

To define the trajectory sensitivity function with respect to inaccuracies of the sampling instants, let us assume that $\boldsymbol{x}$ and $\boldsymbol{f}$ are continuous functions in t_k. Moreover, no α-parameters are considered to be present, i.e., $\boldsymbol{x}$ is assumed to be only a function of t_k. The trajectory sensitivity with respect to t_k is then defined as follows:

Definition 3.2–7 *Local trajectory sensitivity vector.* If t_{k0} denotes the nominal kth sampling instant, the $n \times 1$ vector

$$\boldsymbol{\lambda}(t_{k0}) \triangleq \left.\frac{d\boldsymbol{x}(t_k)}{dt_k}\right|_{t_{k0}} = \lim_{\Delta t \to 0} \frac{\boldsymbol{x}(t_{k0} + \Delta t) - \boldsymbol{x}(t_{k0})}{\Delta t} = \left.\frac{d\boldsymbol{x}(t)}{dt}\right|_{tk0} \tag{3.2–69}$$

represents the local trajectory sensitivity vector with respect to inaccuracies of the sampling instant t_k.

This definition shows that the local trajectory sensitivity vector is identical with the ordinary derivative of the state vector with respect to time, taken at t_{k0}. Since such a vector has to be taken for all sampling instants t_{k0} ($k = 0, 1, \ldots$), it is clear that a total of $k + 1$ local sensitivity vectors is required to characterize the sensitivity up to the kth sampling instant. All these vectors may be combined to form a matrix, the *local trajectory sensitivity matrix* $[\boldsymbol{\lambda}(t_{10}) \ldots \boldsymbol{\lambda}(t_{k0})] \triangleq \boldsymbol{\lambda}_k$.

Very often, in practice, the nominal sampling period, i.e., $T_0 = t_{k+1,0} - t_{k0}$, is fixed. In this case the actual state equation can be written as

$$\boldsymbol{x}[(k + 1)T] = \boldsymbol{f}[\boldsymbol{x}(kT), \boldsymbol{u}(kT), kT], \qquad \boldsymbol{x}(0) = \boldsymbol{x}^0. \tag{3.2–70}$$

The trajectory sensitivity function with respect to a change in the sampling period T_0 is then defined as follows:

Definition 3.2–8 *Global (or sampling-period) trajectory sensitivity vector.* If T_0 is the nominal value of the sampling period T and if the state $\boldsymbol{x}$ is a continuous function of T, then the total derivative

$$\boldsymbol{\lambda}(kT_0) \triangleq \left.\frac{d\boldsymbol{x}\,(kT)}{dT}\right|_{T_0} = \lim_{\Delta T \to 0} \frac{\boldsymbol{x}[k(T_0 + \Delta T)] - \boldsymbol{x}(kT_0)}{\Delta T} \tag{3.2–71}$$

is defined as the sampling-period trajectory sensitivity vector.

The above requirement of the continuity of $\boldsymbol{x}$ in T is fulfilled whenever the input signal $\boldsymbol{u}$ is generated by the sampling of a continuous signal.

The term "global sensitivity vector" may indicate that by this definition the effect of the changes of all sampling instants $t_0, \ldots, t_k$ up to the present are taken into account. If T deviates from its nominal value T_0 by ΔT, the sampling instants $t_0, \ldots, t_k$ deviate from their nominal values $t_{k0} = kT_0$ by the amount $k\,\Delta T$ increasing with time. Thus, the effect of all previous changes up to the present are added up, resulting in a divergence of $\boldsymbol{\lambda}(kT_0)$ with time. Therefore, the employment of $\boldsymbol{\lambda}(kT_0)$ is only useful for small numbers of k, and it is only utilized for the extrapolation of the state of the system. In adaptive systems, however, the local sensitivity vector is employed instead because the latter only accounts for the present deviation of t_k, therefore, not diverging with time.

If $\boldsymbol{\lambda}(kT_0)$ is known, the state deviation caused by inaccuracies ΔT can again be calculated approximately by the linear relation $\Delta \boldsymbol{x}\,(kT) = \boldsymbol{\lambda}(kT_0)\,\Delta T$.

Sometimes it may be interesting to know the trajectory sensitivity vector with respect to ΔT *between sampling instants.* Between sampling instants, the actual sampled-data system with a zero-order hold is described by the vector difference-differential equation

$$\dot{\boldsymbol{x}}(kT + \tau) = \boldsymbol{f}[\boldsymbol{x}(kT + \tau), \boldsymbol{u}(kT), kT + \tau], \qquad \boldsymbol{x}(kT) = \mathbf{x}_k{}^0, \quad (3.2\text{–}72)$$

where $k = 0, 1, \ldots, 0 \leq \tau < T$, and the dot now indicates differentiation with respect to τ. The boundary condition at $\tau = 0$ is obtained from Eq. (3.2–70). Now the sensitivity with respect to ΔT is defined as follows:

Definition 3.2–9 *Continuous sampling-period sensitivity vector.* Let $\boldsymbol{x}(t,T)$ be the solution of Eq. (3.2–72) and T_0 the nominal value of the sampling period. Then the partial derivative

$$\boldsymbol{\lambda}(t,T_0) \triangleq \left.\frac{\partial \boldsymbol{x}(t, T)}{\partial T}\right|_{T_0} \quad (3.2\text{–}73)$$

defines the sampling-period sensitivity vector between the sampling instants.

Finally, we will devote our attention to the sensitivity definitions in the case of *systems with switching functions*. Generalizing Eq.(3.2–72), we can describe a wide class of such systems by different vector differential equations defined on different time intervals, i.e.,

$$\dot{\boldsymbol{x}}(t, \boldsymbol{\alpha}) = \boldsymbol{f}_k[\boldsymbol{x}(t, \boldsymbol{\alpha}), \boldsymbol{u}, t, \boldsymbol{\alpha}], \qquad t_{k-1} < t < t_k, \quad (3.2\text{–}74)$$

where $k = 1, 2, \ldots$ and the $\boldsymbol{f}_k$'s are (possibly nonlinear) vector functions. Furthermore, the initial conditions may be given by $\boldsymbol{x}^0 \triangleq \boldsymbol{x}(t_0, \boldsymbol{\alpha})$, $t_0 = t_0(\boldsymbol{\alpha})$ being the initial time and $\boldsymbol{\alpha}$ an α-parameter vector.

The switching instants t_k, defining the boundaries of the intervals of validity of the corresponding state equations, may be determined either implicitly by a relation of the form

$$\theta_k(\boldsymbol{x}_k{}^-, t_k, \boldsymbol{\alpha}) = 0, \qquad k = 0, 1, \ldots, \quad (3.2\text{–}75)$$

where θ_k is a scalar, possibly a nonlinear function, and $\boldsymbol{x}_k{}^- = \boldsymbol{x}(t_k{}^-, \boldsymbol{\alpha})$ is the state at the left-hand side of t_k, or explicitly by a relation of the form

$$t_k = t_{k-1} + T(\boldsymbol{x}_{k-1}, \boldsymbol{\alpha}), \qquad k = 0, 1, \ldots, \quad (3.2\text{–}76)$$

where T is a scalar, possibly nonlinear function.

In addition, when passing from one interval to the next at t_k, the state vector may be changed according to the relationship

$$\boldsymbol{x}_k{}^+ = \boldsymbol{\Phi}_k(\boldsymbol{x}_k{}^-, t_k, \boldsymbol{\alpha}), \qquad k = 0, 1, \ldots, \quad (3.2\text{–}77)$$

where the $\boldsymbol{\Phi}_k$'s are possibly nonlinear vector functions and $\boldsymbol{x}_k{}^-$ and $\boldsymbol{x}_k{}^+$ are the states at the left- and right-hand side of t_k, respectively. Note that both $\boldsymbol{x}_k = \boldsymbol{x}(t_k, \boldsymbol{\alpha})$ and $t_k = t_k(\boldsymbol{\alpha})$ depend on $\boldsymbol{\alpha}$.

With the assigned intial conditions $\boldsymbol{x}^0$, Eq. (3.2–74)–(3.2–77) determine, in general, a solution $\boldsymbol{x} = \boldsymbol{x}(t, \boldsymbol{\alpha})$, which is a discontinuous function of t and $\boldsymbol{\alpha}$ at the switching instants t_k. The equations can be solved successively.

The *trajectory sensitivity matrix* $\boldsymbol{\lambda}(t, \boldsymbol{\alpha}_0)$ and the *trajectory sensitivity vectors* $\boldsymbol{\lambda}_j(t, \boldsymbol{\alpha}_0)$ are defined in the same way as in the case of continuous systems (Section 3.2.3), but only in the *interior* of the intervals. However, we have to bear in mind that they are discontinuous at the boundaries of these intervals t_k. Therefore, by virtue of classical mathematics, they can be determined only in the open intervals $t_k < t < t_{k+1}$, whereas at the switching instants t_k jump conditions have to be evaluated as was done by Rozenvasser [89] and Rozenvasser and Yusupov [90]. Alternatively, an approach using distributions as suggested by Tsypkin and Rutman [101] can be employed.

It should be noted that discontinuities of $\boldsymbol{\lambda} = (\partial \boldsymbol{x}/\partial \alpha_j)_{\boldsymbol{\alpha}0}$ at t_k may occur even if $\boldsymbol{x}(t, \boldsymbol{\alpha})$ has a continuous transition at t_k. This will be illustrated intuitively by Fig. 3.2–7; a more rigorous mathematical derivation will be given in Chapter 5.

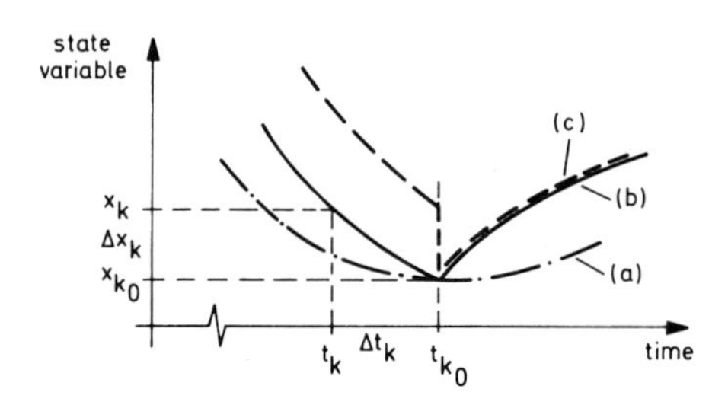

FIG. 3.2–7. Illustration of the discontinuity of $\boldsymbol{\lambda}$ at t_{k0}. (a) **u** and **x** continuous, (b) **u** discontinuous but **x** continuous, (c) **u** and **x** discontinuous.

To illustrate the above statement, let the nominal value of t_k be denoted by t_{k0}. Furthermore, it is assumed that the state vector $\boldsymbol{x}$ is *not* changed discontinuously at t_k according to Eq.(3.2–77). Let us first assume that the input $\boldsymbol{u}(t)$ starts continuously (i.e., without a step) at t_{k0}. It is well known from state space theory that all components of the state vector $\boldsymbol{x}$ are continuous with time (outlined in Fig. 3.2–7 by the dashed and dotted line). Because of the dependence of t_k on $\boldsymbol{\alpha}$, a change of $\boldsymbol{\alpha}$ causes a change $\Delta t_k = t_k - t_{k0}$. In the present case, the derivative of $\boldsymbol{x}$ at t_{k0} is continuous and, therefore

$$\boldsymbol{\lambda} = \left.\frac{\partial \boldsymbol{x}}{\partial \boldsymbol{\alpha}}\right|_{\boldsymbol{\alpha}0} = \left.\frac{\partial \boldsymbol{x}}{\partial t_k}\right|_{t_{k0}} \left.\frac{\partial t_k}{\partial \boldsymbol{\alpha}}\right|_{\boldsymbol{\alpha}0} + \cdots \tag{3.2–78}$$

is continuous. However, if the input begins at t_{k0} with a discontinuity, as for example in case of a rectangular pulse, then there is at least one state variable x_k having a corner at t_{k0} (curve in solid lines). Now the derivative of $\boldsymbol{x}$ at t_{k0} is discontinuous and, therefore, if t_{k0} changes slightly with $\boldsymbol{\alpha}$, at least one component of $\boldsymbol{\lambda}$ (Eq.3.2–78) is discontinuous. This is, of course, especially true in the case for which $\boldsymbol{x}$ is changed discontinuously at t_{k0} (dashed lines).

Examples of such systems are pulse-frequency or pulse-width modulated feedback systems or optimal control loops, in which switching operations are performed as soon as the state has reached a preassigned switching curve or switching hyperplane.

The above sensitivity functions are not only functions of the nominal parameter vector $\boldsymbol{\alpha}_0$ but also functions of time. Sometimes, however, it is desirable to characterize the sensitivity of a system simply by a number, rather than by a time function. Sensitivity definitions of this kind shall be termed *sensitivity measures*.

There are various ways to exclude time from the sensitivity definition, depending on what sort of study of the dynamic system we are making. Three sensitivity measures of the time domain are presented in the sequel.

3.2.6 Overshoot Sensitivity (Peak-Response Sensitivity)

In many technical applications, a dynamic system has to be designed with respect to the overshoot as a design criterion. It is clear that in this case the overshoot sensitivity is the natural sensitivity measure, giving information about the change of relative stability of the system as well.

The overshoot sensitivity can be deduced from the output sensitivity function $\sigma(t, \boldsymbol{\alpha}_0) = (\partial y(t, \boldsymbol{\alpha})/\partial \alpha_j)_{\boldsymbol{\alpha}0}$ in the following way. Suppose that the input signal is a step function. Then the output $y(t,\boldsymbol{\alpha})$ means the step response, denoted by $h(t,\boldsymbol{\alpha})$, that may have a shape such as shown in Fig. 3.2–8.

Definition 3.2–10 *Overshoot (peak-response) sensitivity.* Let t_m be the time of the overshoot, i.e., the time where $h(t, \boldsymbol{\alpha}_0)$ takes on its maximal value $h_m = \max_t h(t, \boldsymbol{\alpha}_0)$. Then

$$\sigma(t_m, \boldsymbol{\alpha}_0) \triangleq \left.\frac{\partial h(t, \boldsymbol{\alpha})}{\partial \alpha_j}\right|_{t_m, \boldsymbol{\alpha}0} \tag{3.2–79}$$

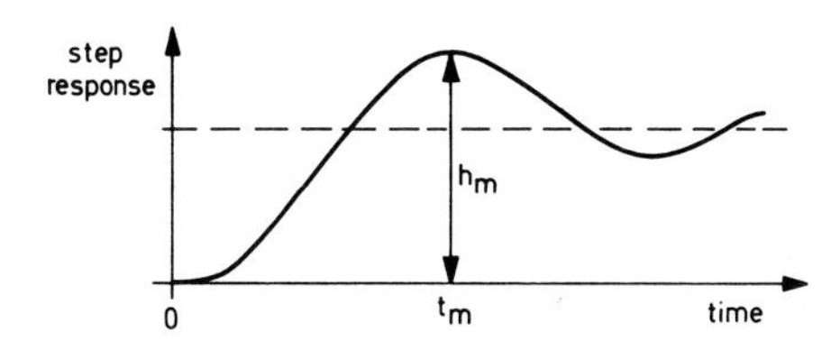

FIG. 3.2–8. Shape of the step response used for the difinition of σ_m.

represents the overshoot sensitivity with respect to the parameter α_j, also called peak-response sensitivity [4].

Note that σ_m is no longer a function of time. It is simply determined from the output sensitivity function due to a step input by setting $t = t_m$. There are, of course, cases where the overshoot sensitivity is not defined. This measure is primarily used to characterize the sensitivity of feedback control systems designed with respect to overshoot.

Example 3.2–6 For the system treated in Section 3.2.2 with the differential equation

$$\dddot{y} + a_2\ddot{y} + a_1\dot{y} + a_0 y = 1(t), \qquad y(0) = \cdots = \ddot{y}(0) = 0,$$

determine the overshoot sensitivity with respect to the coefficients a_0, a_1, a_2, assuming that their nominal values are $a_{00} = 20$, $a_{10} = 15$, $a_{20} = 5$.

The definition of the overshoot sensitivity reveals that it can be determined by taking the value of the output sensitivity function at the time t_m of the overshoot. Thus we can find overshoot sensitivities by the aid of the plots in Fig. 3.2–1a, taking the values at $t_m = 1.78$ sec. By this procedure, we obtain the desired values of the sensitivity functions $\sigma_0(t_m, \boldsymbol{a}_0) = -1.6 \times 10^{-3}$, $\sigma_1(t_m, \boldsymbol{a}_0) = -2.4 \times 10^{-3}$, $\sigma_2(t_m, \boldsymbol{a}_0) = -0.75 \times 10^{-3}$.

To compare the sensitivities with respect to the various parameters, let us form the values of the semirelative sensitivity functions by multiplying the above results with the corresponding nominal parameter values according to Definition 2.3–9. This gives $\tilde{\sigma}_0(t_m, \boldsymbol{a}_0) = -32 \times 10^{-3}$; $\tilde{\sigma}_1(t_m, \boldsymbol{a}_0) = -36 \times 10^{-3}$; $\tilde{\sigma}_2(t_m, \boldsymbol{a}_0) = -3.75 = 10^{-3}$. The comparison of these numbers shows that the overshoot is most sensitive to changes of a_1; however, we see that there is also a considerable influence of a_0.

3.2.7 L_2-Norm

A more general way to define a sensitivity measure of the time domain is to take the L_2-norm of the sensitivity vectors or system errors. This measure allows a global sensitivity characterization of the system in terms of the sensitivity or the errors of the output or state.

Let us assume that the sensitivity functions (or vectors) or the parameter-induced errors of the output or state variables of a system are known. They are denoted by $\boldsymbol{e}(t)$ for short.

Definition 3.2–11 *L_2-norm.* The definite integral of the quadratic form,

$$I_M \triangleq \int_{t_0}^{t_1} \boldsymbol{e}^T(t)\boldsymbol{Z}\,\boldsymbol{e}(t)\,dt, \tag{3.2–80}$$

where $\boldsymbol{Z}$ is a symmetrical positive definite weighting matrix and $t_0 \geq 0$, $t_1 > t_0$, two positive instants, is called the L_2-norm.

To characterize the sensitivity of a system by the L_2-norm, the following quantities have to be substituted for $\boldsymbol{e}(t)$:

(1) in the case of time-invariant parameter variations: the (absolute, relative, or semirelative) output or trajectory sensitivity vectors $\boldsymbol{\sigma}_j, \bar{\boldsymbol{\sigma}}_j, \tilde{\boldsymbol{\sigma}}_j$ or $\boldsymbol{\lambda}_j, \bar{\boldsymbol{\lambda}}_j, \tilde{\boldsymbol{\lambda}}_j$ with respect to one parameter α_j,

(2) in the case of time-variant and/or large parameter variations: the vector of parameter-induced output or state erros $\Delta \boldsymbol{y}$ or $\Delta \boldsymbol{x}$, respectively.

This sensitivity measure has proved to be *very useful* for the comparison of the sensitivity of open- and closed-loop systems even in the case of optimal systems. The reason for the application in optimal systems is that the performance index sensitivity, which is the natural sensitivity measure in this case, is of no use.

A special case of the L_2-norm is the so-called *system error integral*

$$I_M \triangleq \int_0^{\infty} \Delta y^2(t, \alpha)\, dt, \tag{3.2–81}$$

suggested by Mazer in 1960 for single variable systems [66]. (See also [93].) Here $\Delta y(t, \alpha)$ is the parameter-induced output error of the system. With the aid of Parseval's theorem, this integral may be written as

$$I_M = \frac{1}{2\pi} \int_{-\infty}^{\infty} |\Delta Y(j\omega, \alpha)|^2\, d\omega. \tag{3.2–82}$$

If $\Delta\alpha \ll \alpha_0$, the Laplace transform of the system error ΔY can be expressed by

$$\begin{aligned} \Delta Y(j\omega, \alpha) &= Y(j\omega, \alpha) - Y(j\omega, \alpha_0) \\ &= [G(j\omega, \alpha) - G(j\omega, \alpha_0)]U(s) = \Delta G(j\omega, \alpha)\, U(s), \end{aligned} \tag{3.2–83}$$

where G represents the transfer function of the system and $U(s)$ the Laplace transform of the input. Further, we have

$$\Delta G = \left.\frac{\partial G}{\partial \alpha}\right|_{\alpha 0} \Delta\alpha = \left.\frac{\partial \ln G}{\partial \ln \alpha}\right|_{\alpha 0} \frac{G_0}{\alpha_0} \Delta\alpha = \bar{S}_\alpha\, G_0 \frac{\Delta\alpha}{\alpha_0}, \tag{3.2–84}$$

where $G_0 \triangleq G(j\omega, \alpha_0)$, and $\bar{S}_\alpha^G$ is the Bode sensitivity function (treated in more detail in the next section). Substituting Eq.(3.2–84) into (3.2–83) and (3.2–83) into (3.2–82), we obtain

$$I_M = I_R \left(\frac{\Delta\alpha}{\alpha_0}\right)^2, \tag{3.2–85}$$

where

$$I_R = \frac{1}{2\pi} \int_{-\infty}^{\infty} |\bar{S}_\alpha^G\, G(j\omega, \alpha_0)\, U(j\omega)|^2\, d\omega. \tag{3.2–86}$$

I_R can be interpreted as a weighting factor for $\Delta\alpha/\alpha_0$. Thus the sensitivity with respect to the relative change of various parameters may be compared by means of I_R. For example, it turns out that in minimum phase control systems the parameter yielding the largest value of I_R is the gain factor K_S of the plant. This means that, in the sense of I_M, the control loop is most sensitive to changes of K_S [93].

A similar connection between time and frequency domain can be discovered in the general case of the L_2-norm, as will be shown in Chapter 6.

Example 3.2–7 Given a system described by the state equations

$$\dot{x}_1 = x_2, \qquad x_1(o) = 0,$$

$$\dot{x}_2 = -(a_1 + a_2)x_2 - a_1 a_2 x_1 + u, \qquad x_2(o) = 0.$$

Assuming that u is the impulse function $\delta(t)$ and that the nominal values of a_1, a_2 are given by $a_{10} = 2$, $a_{20} = 1$, determine the L_2-norm

$$I_M = \int_0^\infty \boldsymbol{\lambda}_1^{\mathrm{T}}(t, \boldsymbol{a}_0)\boldsymbol{Z}\,\boldsymbol{\lambda}_1(t, \boldsymbol{a}_0)\,dt,$$

where $\boldsymbol{\lambda}_1$ is the trajectory sensitivity vector with respect to a_1, and $\boldsymbol{Z} = \boldsymbol{I}$ is the unity matrix.

The solution of the above state equations yields

$$x_1(t, \boldsymbol{a}) = \frac{1}{a_2 - a_1}(e^{-a_1 t} - e^{-a_2 t}),$$

$$x_2(t, \boldsymbol{a}) = \frac{1}{a_1 - a_2}(a_1 e^{-a_1 t} - a_2 e^{-a_2 t}).$$

Taking the partial derivatives with respect to a_1 and then setting a_1,a_2 equal to $a_{10} = 2$, $a_{20} = 1$, respectively, we obtain the trajectory sensitivity vector

$$\boldsymbol{\lambda}_1(t, \boldsymbol{a}_0) = \begin{bmatrix} \lambda_{11}(t, \boldsymbol{a}_0) \\ \lambda_{21}(t, \boldsymbol{a}_0) \end{bmatrix} = \begin{bmatrix} (1 + t)e^{-2t} - e^{-t} \\ -(1 + 2t)\,e^{-2t} + e^{-t} \end{bmatrix}.$$

I_M, as specified above, can be written as

$$I_M = \int_0^\infty [\lambda_{11}^2(t, \boldsymbol{a}_0) + \lambda_{21}^2(t, \boldsymbol{a}_0)]\,dt.$$

The evaluation of this integral finally yields $I_M = 0.21$.

3.2.8 Eigenvalue Sensitivity and Eigenvector Sensitivity

It is well known from state space theory that the free motions of a dynamic system governed by the vector state equation

$$\dot{\boldsymbol{x}}(t) = \boldsymbol{A}\boldsymbol{x}(t), \qquad \boldsymbol{x}(0) = \boldsymbol{x}^0, \tag{3.2–87}$$

where $\boldsymbol{A}$ has distinct eigenvalues λ_i $(i = 1, 2, \ldots, n)$, are given by an expression of the form

$$\boldsymbol{x}(t) = \sum_{1=i}^{n} C\boldsymbol{v}_i e^{\lambda_i t}, \tag{3.2–88}$$

where $\boldsymbol{x}$ is the $n \times 1$ state vector of the system, $\boldsymbol{A}$ the $n \times n$ system matrix, and the $\boldsymbol{v}_i$ $(i = 1, 2, \ldots, n)$ represent the linearly independent eigenvectors of $\boldsymbol{A}$. Substituting the ith term of Eq.(3.2–88) into Eq.(3.2–87) gives

$$\boldsymbol{A}\boldsymbol{v}_i = \lambda_i \boldsymbol{v}_i. \tag{3.2–89}$$

C is a scalar that can be written as $C = \boldsymbol{w}_i^{\mathrm{T}}\boldsymbol{x}^0$, where the $\boldsymbol{w}_i$ $(i = 1, 2, \ldots, n)$ are the eigenvectors of A^{T} satisfying the equation

$$\boldsymbol{A}^{\mathrm{T}}\boldsymbol{w}_i = \lambda_i \boldsymbol{w}_i. \tag{3.2–90}$$

Thus the free motion as well as the stability of the system are uniquely defined by the eigenvalues λ_i of the matrix $\boldsymbol{A}$.

Applying the Laplace transform to Eq.(3.2–87) and solving it for $\mathscr{L}\{\boldsymbol{x}(t)\} = \boldsymbol{X}(\lambda)$ gives

$$\boldsymbol{X}(\lambda) = (\lambda\boldsymbol{I} - \boldsymbol{A})^{-1}\boldsymbol{x}(0) = \frac{\operatorname{Adj}(\lambda\boldsymbol{I} - \boldsymbol{A})}{\det(\lambda\boldsymbol{I} - \boldsymbol{A})}\,\boldsymbol{x}^0. \tag{3.2–91}$$

The poles of $\boldsymbol{X}(\lambda)$ are identical with the eigenvalues λ_i of $\boldsymbol{A}$. Thus, the λ_i values can be determined by solving the characteristic equation

$$\det(\lambda\boldsymbol{I} - \boldsymbol{A}) = 0. \tag{3.2–92}$$

The sensitivity of the free motion of a system as well as the sensitivity of its relative stability with respect to any parameters of $\boldsymbol{A}$ can now be characterized by the sensitivity of the eigenvalues λ_i with respect to the parameters. We define the following:

Definition 3.2–12 *Eigenvalue sensitivity.* Let the $n \times n$ matrix $\boldsymbol{A} = \boldsymbol{A}(\alpha)$ be a function of any $r \times 1$ parameter vector $\boldsymbol{\alpha}$ and let $\lambda_i = \lambda_i(\boldsymbol{\alpha})$ be the ith simple eigenvalue of $\boldsymbol{A}$; then the partial derivative

$$S_{\alpha_j}^{\lambda_i} = \left.\frac{\partial \lambda_i}{\partial \alpha_j}\right|_{\boldsymbol{\alpha}0} \tag{3.2–93}$$

is called the (absolute) eigenvalue sensitivity. The relative and semirelative definitions $\bar{S}_{\alpha_j}^{\lambda_i} = (\partial \ln \lambda_i/\partial \ln \alpha_j)_{\boldsymbol{\alpha}0}$ and $\tilde{S}_{\alpha_j}^{\lambda_i} = (\partial\lambda_j/\partial \ln \alpha_j)_{\boldsymbol{\alpha}0}$ are used as well. In the case of complex eigenvalues $\lambda_i = \sigma_i \pm j\omega_i$, the normalization of λ_i in $\bar{S}_{\alpha_j}^{\lambda_i}$ has to be done componentwise, that is, $\bar{S}_{\alpha_j}^{\lambda_i} = (\partial \ln \sigma_i/\partial \ln \alpha_j)_{\boldsymbol{\alpha}0} \pm j(\partial \ln \omega_i/\partial \ln \alpha_j)_{\boldsymbol{\alpha}0}$.

Sometimes it is desirable to know the sensitivity of the *eigenvectors* with respect to changes in $\boldsymbol{A}$. This is defined as follows.

Definition 3.2–13 *Eigenvector sensitivity.* Let $\boldsymbol{v}_i$ $(i = 1, 2, \ldots, n)$ be the eigenvectors of the system matrix $\boldsymbol{A} = \boldsymbol{A}(\boldsymbol{\alpha})$. Then the partial derivative $(\partial \boldsymbol{v}_i / \partial \alpha_j)_{\boldsymbol{\alpha}0}$ is called the eigenvector sensitivity of the matrix $\boldsymbol{A}$ with respect to α_j.

An efficient concept to characterize the sensitivity of an eigenvalue λ_i of $\boldsymbol{A}$ due to the effects of a simultaneous variation of all elements a_{jk} of $\boldsymbol{A}$ in the worst case is the summed semirelative eigenvalue sensitivity defined as follows:

Definition 3.2–14 *Summed semirelative eigenvalue sensitivity.* Consider a square matrix $\boldsymbol{A}$ having only real-valued elements a_{jk}. Let $\Delta \ln a_{jk} = \Delta a_{jk}/a_{jk0}$ be the relative deviation of the element a_{jk} from the nominal value a_{jk0}, and let $\Delta\lambda_i\,(\Delta \ln a_{jk})$ denote the corresponding absolute deviation of the ith eigenvalue of $\boldsymbol{A}$, then the limit

$$\tilde{S}_A^{\lambda_i} \triangleq \lim_{\Delta\to 0} \frac{1}{\Delta} \operatorname*{supc}_{|\Delta \ln a_{jk}| \leq \Delta} \sum_{j,k}^{n} \Delta\lambda_i(\Delta \ln a_{jk}) \tag{3.2–94}$$

is defined as the *summed semirelative* eigenvalue sensitivity of $\boldsymbol{A}$ at the eigenvalue λ_i $(i = 1, 2, \ldots, n)$.

In the above definition, supc $\Delta\lambda_i$ is the complex supremum of $\Delta\lambda_i$ referring to the possibly complex values $\Delta\lambda_i^*$ of $\Delta\lambda_i$ for which $|\Delta\lambda_i^*| = \sup \Delta\lambda_i$. It is worth noting that, unlike earlier definitions [97], this definition provides a summation of all *semirelative* (instead of absolute) eigenvalue sensitivity functions of λ_i, thus taking into account percentage tolerances of the parameters.

Example 3.2–8 Find the semirelative eigenvalue sensitivities, the eigenvector sensitivities, and the summed semirelative eigenvalue sensitivities of the system governed by the state equation

$$\dot{\boldsymbol{x}} = \begin{bmatrix} 0 & 1 \\ -\alpha_1 & -\alpha_2 \end{bmatrix} \boldsymbol{x},$$

where α_1 and α_2 are considered to be the parameters having the nominal values $\alpha_{10} = 3$, $\alpha_{20} = 4$. (The sensitivities with respect to the zero and unity element are, by definition, zero.)

Solution The characteristic equation becomes

$$\det[\lambda \boldsymbol{I} - \boldsymbol{A}] = \det \begin{bmatrix} \lambda & -1 \\ \alpha_1 & \lambda + \alpha_2 \end{bmatrix} = \lambda^2 + \alpha_2 \lambda + \alpha_1 = 0.$$

Therefore,

$$\lambda_1 = -\frac{\alpha_2}{2} + \left[\frac{\alpha_2{}^2}{4} - \alpha_1\right]^{1/2}; \qquad \lambda_2 = -\frac{\alpha_2}{2} - \left[\frac{\alpha_2{}^2}{4} - \alpha_1\right]^{1/2},$$

$$\lambda_{10} = -1 \qquad , \qquad \lambda_{20} = -3.$$

Using Eq.(3.2–93) yields the required eigenvalue sensitivities

$$\tilde{S}^{\lambda_1}_{\alpha 1} = -\frac{\alpha_{10}}{2}\left[\frac{\alpha_{20}^2}{4} - \alpha_{10}\right]^{-1/2} = -\frac{3}{2},$$

$$\tilde{S}^{\lambda_1}_{\alpha 2} = -\frac{\alpha_{20}}{2} + \frac{\alpha_{20}^2}{4}\left[\frac{\alpha_{20}^2}{4} - \alpha_{10}\right]^{-1/2} = 2,$$

$$\tilde{S}^{\lambda_2}_{\alpha 1} = -S^{\lambda_1}_{\alpha 1} = \frac{3}{2},$$

$$\tilde{S}^{\lambda_2}_{\alpha 2} = -\frac{\alpha_{20}}{2} - \frac{\alpha_{20}^2}{4}\left(\frac{\alpha_{20}^2}{4} - \alpha_{10}\right)^{-1/2} = -6.$$

Substituting λ_1 or λ_2 into Eq.(3.2–89), we obtain

$$[\lambda_i \boldsymbol{I} - \boldsymbol{A}]\boldsymbol{v}_i = \begin{bmatrix} \lambda_i & -1 \\ \alpha_1 & \lambda_i + \alpha_2 \end{bmatrix}\begin{bmatrix} v_{i1} \\ v_{i2} \end{bmatrix} = \boldsymbol{0}, \qquad i = 1,2.$$

This yields the eigenvectors

$$\boldsymbol{v}_1 = \begin{bmatrix} 1 \\ \lambda_1 \end{bmatrix}, \qquad \boldsymbol{v}_2 = \begin{bmatrix} 1 \\ \lambda_2 \end{bmatrix}$$

or any vector proportional to these vectors. Partial differentiation gives the eigenvector sensitivities

$$\left.\frac{\partial \boldsymbol{v}_1}{\partial \alpha_1}\right|_{\boldsymbol{\alpha}0} = \begin{bmatrix} 0 \\ S^{\lambda_1}_{\alpha 1} \end{bmatrix} = \begin{bmatrix} 0 \\ -\frac{1}{2} \end{bmatrix}, \qquad \left.\frac{\partial \boldsymbol{v}_1}{\partial \alpha_2}\right|_{\boldsymbol{\alpha}0} = \begin{bmatrix} 0 \\ S^{\lambda_1}_{\alpha 2} \end{bmatrix} = \begin{bmatrix} 0 \\ \frac{1}{2} \end{bmatrix}$$

$$\left.\frac{\partial \boldsymbol{v}_2}{\partial \alpha_1}\right|_{\boldsymbol{\alpha}0} = \begin{bmatrix} 0 \\ S^{\lambda_2}_{\alpha 1} \end{bmatrix} = \begin{bmatrix} 0 \\ \frac{1}{2} \end{bmatrix}, \qquad \left.\frac{\partial \boldsymbol{v}_1}{\partial \alpha_2}\right|_{\boldsymbol{\alpha}0} = \begin{bmatrix} 0 \\ S^{\lambda_2}_{\alpha 2} \end{bmatrix} = \begin{bmatrix} 0 \\ -\frac{3}{2} \end{bmatrix}.$$

The summed semirelative eigenvalue sensitivities simply become in this case

$$\tilde{S}^{\lambda_1}_{A} = |\tilde{S}^{\lambda_1}_{\alpha 1}| + |\tilde{S}^{\lambda_1}_{\alpha 2}| = \tfrac{3}{2} + 2 = 3.5,$$

$$\tilde{S}^{\lambda_2}_{A} = |\tilde{S}^{\lambda_2}_{\alpha 1}| + |\tilde{S}^{\lambda_2}_{\alpha 2}| = \tfrac{3}{2} + 6 = 7.5.$$

Many expressions relating the eigenvalue sensitivity and eigenvector sensitivity of the matrix $\boldsymbol{A}$ to changes of the coefficients of $\boldsymbol{A}$ have been developed by numerous authors since the early work of Jacobi in 1846 [48]. Thus the eigenvalue sensitivity problem has been treated as a problem of numerical analysis, as a problem of perturbation theory, and as a problem of linear system theory [29, 70, 87, 103]. There are even attempts to apply this approach to the design of feedback systems [70], although this is a tedious algebriac process with much less transparency compared to the frequency domain approach [45]. Formulas for the calculation of eigenvalue sensitivities are given in Section 5.8.

3.3 SENSITIVITY FUNCTIONS IN THE FREQUENCY DOMAIN

One of the shortcomings of the sensitivity definitions in the time domain is that they depend on the actual form of the input signals. Therefore, if they are to be used for the characterization or the comparison of sensitivity with respect to several parameters, one will have to be restricted to applying standardized input signals such as step or impulse functions.

In the face of the study and the comparison of the parameter sensitivity of open- and closed-loop systems, as well as of the synthesis of parameter-insensitive systems, it is desirable to have a sensitivity definition that is independent of the form of the input signal. In other words, it should be defined in terms of the *structure* of the system rather than in terms of signals. In fact, this is true for the sensitivity definitions in the frequency domain that are based on the transfer function or transfer matrix of the system. The most important definitions are

(1) the sensitivity function of Bode (classical sensitivity function),

(2) the sensitivity function of Horowitz (finally equivalent to Bode's sensitivity function), and

(3) the sensitivity operator of Perkins and Cruz, comparison sensitivity function (extension of Bode's sensitivity function).

Since definitions (1) and (2) are based on the transfer functions and matrices, they are restricted to linear systems. However, definition (3), being formulated in terms of operators, may readily be extended to general, possibly nonlinear and time-varying, systems, as, in fact, was done by Perkins and Cruz [5].

Besides these sensitivity functions the most important sensitivity measures of the frequency domain are presented in this section.

3.3.1 Sensitivity Function of Bode

Consider a linear time-invariant single-variable system whose transfer function G is a continuous function not only of the complex frequency $s = \delta + j\omega$ but also of the $r \times 1$ parameter vector $\boldsymbol{\alpha} = [\alpha_1 \ldots \alpha_r]^{\mathrm{T}}$, see Fig. 3.3–1.

In these terms, the relative (logarithmic) partial derivative of a parameter α_j with respect to the transfer function G was originally defined by Bode as the

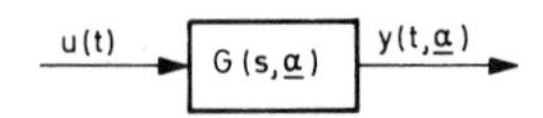

FIG. 3.3–1. Block diagram of a system whose transfer function depends on the parameter vector $\boldsymbol{\alpha}$.

sensitivity function [1]. Today the Bode sensitivity function is defined as the reciprocal ratio.

Definition 3.3–1 *Bode's sensitivity function.* Let $G = G(s, \boldsymbol{\alpha})$ and $G_0 = G(s, \boldsymbol{\alpha}_0)$ be the actual or nominal transfer functions, respectively, and $\boldsymbol{\alpha}_0$ the nominal parameter vector. Then the logarithmic partial derivative

$$S_{\alpha_j}^G(s) \triangleq \left.\frac{\partial \ln G}{\partial \ln \alpha_j}\right|_{\boldsymbol{\alpha}0} = \left.\frac{\partial G/G}{\partial \alpha_j/\alpha_j}\right|_{\boldsymbol{\alpha}0} = \left.\frac{\partial G}{\partial \alpha_j}\right|_{\boldsymbol{\alpha}0} \frac{\alpha_{j0}}{G_0} \tag{3.3–1}$$

is called the sensitivity function of Bode or the classical sensitivity function. Following the general usage, we will drop the cross bar on top of $\bar{S}_{\alpha_j}^G$ that should be made according to our earlier agreement.

In order to discover the relationship between the Laplace transform of the output sensitivity function $\Sigma(s, \boldsymbol{\alpha}_0) \triangleq \mathscr{L}\{\sigma(t, \boldsymbol{\alpha}_0)\}$ and Bode's sensitivity function, let us consider the relationship for $Y(s, \boldsymbol{\alpha})$, which is, due to Fig. 3.3–1,

$$Y(s, \boldsymbol{\alpha}) = G(s, \boldsymbol{\alpha}\ )U(s). \tag{3.3–2}$$

Taking the partial derivative with respect to one of the parameters α_j and assuming that the order of the performance of the Laplace transformation and the partial derivative may be reversed, we obtain

$$\Sigma_j(s, \boldsymbol{\alpha}_0) = \left.\frac{\partial G(s, \boldsymbol{\alpha})U(s)}{\partial \alpha_j}\right|_{\boldsymbol{\alpha}0} = S_{\alpha_j}^G(s)\, \frac{G(s, \boldsymbol{\alpha}_0)}{\alpha_{j0}}\, U(s). \tag{3.3–3}$$

Bode's sensitivity definition can formally be extended to the characterization of the sensitivity of the transfer function G with respect to the transfer function of a subsystem G_1, called the variable component. Suppose that $G = G(s, G_1)$, where $G_1 = G_1(s, \boldsymbol{\alpha})$ is a function of $\boldsymbol{\alpha}$, $G_{10} = G_1(s, \boldsymbol{\alpha}_0)$ being its nominal value. Then

$$S_{G_1}^G(s) \triangleq \left.\frac{\partial \ln G}{\partial \ln G_1}\right|_{G10} = \left.\frac{\partial G/G}{\partial G_1/G_1}\right|_{G10} = \left.\frac{G_{10}}{G_0}\frac{\partial G}{\partial G_1}\right|_{G10}. \tag{3.3–4}$$

Using this definition, the Laplace transform of the output sensitivity function σ_j becomes, by the aid of the chain rule,

$$\Sigma_j(s, \boldsymbol{\alpha}_0) = \frac{G(s, G_{10})}{\alpha_{j0}}\, S_{G_1}^G(s) S_{\alpha_j}^{G_1}(s) U(s). \tag{3.3–5}$$

Furthermore, the actual output can be written as

$$Y(s, \boldsymbol{\alpha}) = Y(s, \boldsymbol{\alpha}_0) + \sum_{j=1}^{r} S_{G_1}^G(s) S_{\alpha_j}^{G_1}(s) Y(s, \boldsymbol{\alpha}_0) \frac{\Delta\alpha_j}{\alpha_{j0}}, \tag{3.3–6}$$

and for the relative parameter-induced output error we obtain

$$\frac{\Delta Y(s, \boldsymbol{\alpha})}{Y(s, \boldsymbol{\alpha}_0)} = S_{G_1}^G(s) \sum_{j=1}^{r} S_{\alpha_j}^{G_1}(s) \frac{\Delta\alpha_j}{\alpha_{j0}}. \tag{3.3–7}$$

This equation relates the relative output error to the relative changes of the parameters by means of the Bode sensitivity functions.

Notice that $S_{\alpha_j}^{G_1}(s)$ specifies the given component with respect to the parameter and, in general, cannot be modified anymore. On the other hand, $S_{G_1}^{G}(s)$ specifies the configuration of the system with respect to the variable component G_1. This implies that $S_{G_1}^{G}(s)$ can be modified over a wide range by structural manipulations and, therefore, it is this quantity that underlies the design of parameter-insensitive feedback systems in the frequency domain. In this case, G has the meaning of the transfer function of the control loop and G_1 that of the parameter-dependent transfer function of the plant.

Example 3.3–1 Consider the classical control loop shown in Fig. 3.3–2, where the transfer function of the plant is given by $G_1(s,\boldsymbol{\alpha}) = 1/(1 + \alpha_1 s)$. Denote the overall transfer function $Y(s, \boldsymbol{\alpha})/U(s)$ by $G(s,\alpha_1)$ and determine $S_{\alpha_1}^{G_1}$, $S_{G_1}^{G}$, and the relative output variation $\Delta Y/Y_0$ for a relative change $\Delta\alpha_1/\alpha_{10}$. How does $S_{G_1}^{G}$ change if the control loop is opened at A?

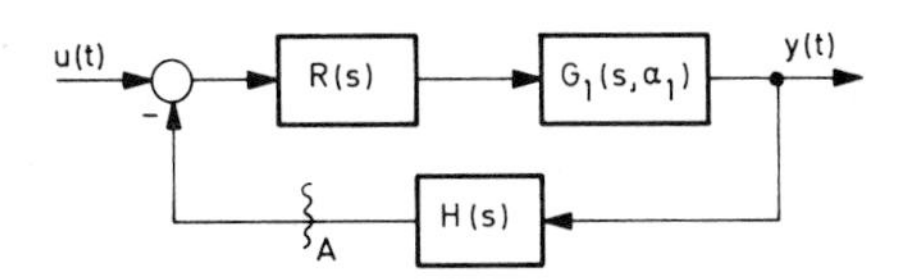

FIG. 3.3–2. Classical control loop of Example 3.3–1.

For $S_{\alpha_1}^{G_1}(s)$, we obtain

$$S_{\alpha_1}^{G_1}(s) = \left.\frac{\partial(1 + \alpha_1 s)^{-1}}{\partial\alpha_1}\right|_{\alpha_{10}} (1 + \alpha_{10}s)\,\alpha_{10} = -\frac{\alpha_{10}s}{1 + \alpha_{10}s}$$

The transfer function of the closed loop is

$$G = \frac{RG_1}{1 + HRG_1}.$$

Taking the partial derivative, we obtain

$$S_{G_1}^{G} = \left.\frac{\partial G}{\partial G_1}\right|_{G_{10}} \frac{G_{10}}{G_0} = \frac{1}{1 + HRG_{10}}.$$

Therefore,

$$\frac{\Delta Y(s, \alpha_1)}{Y(s, \alpha_{10})} = S_{G_1}^{G}(s)\,S_{\alpha_1}^{G_1}(s)\,\frac{\Delta\alpha_1}{\alpha_{10}} = -\frac{\Delta\alpha_1\, s}{1 + \alpha_{10}s + H(s)R(s)}.$$

If the control loop is opened at A, $G = RG_1$ and $S_{G_1}^{G} = 1$.

If $S_{G_1}^{G}$ is a real number, it can be interpreted as the ratio of the relative change of the overall transfer factor G to the relative parameter-induced change of the transfer factor G_1 of the variable component.

Example 3.3-2 Consider the two systems illustrated in Fig. 3.3–3. For nominal values $K_{10} = K_{20} = 100$, both circuits have the same transfer function, that is,

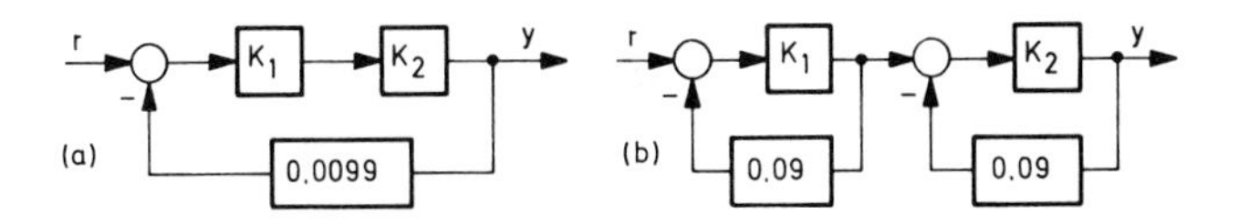

FIG. 3.3-3. The two control loops of Example 3.3–2.

$$\text{(a)}\ G_{10} = \frac{K_{10}K_{20}}{1 \times 0.0099K_{10}K_{20}} = 100,$$

$$\text{(b)}\ G_{20} = \frac{K_{10}}{1 + 0.09K_{10}} \frac{K_{20}}{1 + 0.09K_{20}} = 100.$$

Compare the sensitivity of both configurations with respect to K_2 for nominal values $K_{10} = K_{20} = 100$.

For the system (a), we obtain

$$S_{K2}^{G1} = \frac{\partial}{\partial K_2}\left(\frac{K_{10}K_2}{1 + 0.0099K_{10}K_2}\right)\Bigg|_{K20} \frac{1 + 0.0099K_{10}K_{20}}{K_{10}K_{20}} K_{20}$$

$$= \frac{1}{1 + 0.0099K_{10}K_{20}} = \frac{1}{100} = 0.01.$$

For system (b),

$$S_{K2}^{G2} = \frac{\partial}{\partial K_2}\left(\frac{K_2}{1 + 0.09K_2}\right)\Bigg|_{K20} \frac{(1 + 0.09K_{20})}{K_{20}} K_{20}$$

$$= \frac{1}{1 + 0.09K_{20}} = \frac{1}{10} = 0.1.$$

This reveals that a 10% variation in K_2 will approximately produce a 0.1% variation in G_{10} and a 1% variation in G_{20}. Thus system (b) is 10 times more sensitive to variations of K_{20} than is system (a).

In general, S_α^G is a function of the complex frequency $s = \delta + j\omega$. If $s = j\omega$ is a pure imaginary number, S_α^G can be split up into

$$S_\alpha^G(j\omega) = |S_\alpha^G(j\omega)|(\cos\psi + j\sin\psi), \tag{3.3–8}$$

where

$$\psi = \tan^{-1}\frac{\mathrm{Im}(S_\alpha^G(j\omega))}{\mathrm{Re}(S_\alpha^G(j\omega))}. \tag{3.3–9}$$

On the other hand, we have

$$S_\alpha^G(j\omega) = \mathrm{Re}(S_\alpha^G(j\omega) + j\,\mathrm{Im}(S_\alpha^G(j\omega)). \tag{3.3–10}$$

Writing

$$G(j\omega) = |G(j\omega)|e^{j\varphi(\omega)} \tag{3.3–11}$$

and applying the product rule [Eq.(2.4–5)], we obtain

$$S_\alpha^{\ G} = S_\alpha^{|G|} + S_\alpha^{e^{j\varphi}}. \tag{3.3–12}$$

The second term on the right-hand side of Eq.(3.3–12) can be written, by use of the chain rule [Eq.(2.4–7)],

$$S_\alpha^{e^{j\varphi}} = S_\varphi^{e^{j\varphi}} S_\alpha^{\varphi} = j\varphi S_\alpha^{\ \varphi}, \tag{3.3–13}$$

giving

$$S_\alpha^{\ G}(j\omega) = S_\alpha^{|G|}(\omega) + j\varphi S_\alpha^{\ \varphi}(\omega). \tag{3.3–14}$$

Comparing Eq.(3.3–8) with (3.3–14), we see that

$$\mathrm{Re}(S_\alpha^{\ G}(j\omega)) = S^G(\omega), \tag{3.3–15}$$

$$\mathrm{Im}(S_\alpha^{\ G}(j\omega)) = \varphi(\omega) S_\varphi^{\ \alpha}(\omega). \tag{3.3–16}$$

This result shows that the Bode sensitivity function of G can be split up into a real part, which is simply the Bode sensitivity function of the magnitude of G, and an imaginary part, which is a product of the phase and the Bode sensitivity function of the phase of G.

Example 3.3-3 Given a pure dead-time system with the transfer function

$$G(s,T) = Ke^{-Ts},$$

where K is a constant and T is the dead-time. Determine the Bode sensitivity functions $S_T^{\ G}(j\omega)$, $S_T^{|G|}(j\omega)$, $S_T^{\ \varphi}(j\omega)$.

The Bode sensitivity function of G with respect to T is

$$S_T^{\ G}(j\omega) = \left.\frac{\partial(Ke^{-j\omega T})}{\partial T}\right|_{T_0} \frac{T_0}{Ke^{-j\omega T_0}} = -j\omega T_0.$$

The Bode sensitivity function of magnitude $|G| = K$ with respect to T is

$$S_T^{\ |G|} = \left.\frac{\partial |G|}{\partial T}\right|_{T0} \frac{T_0}{K} = 0.$$

The Bode sensitivity function of the phase $\varphi = -\omega T$ with respect to T is

$$S_T^{\ \varphi} = \left.\frac{\partial \varphi}{\partial T}\right|_{T_0} \frac{T_0}{\varphi_0} = -\omega \frac{T_0}{-\omega T_0} = 1.$$

This verifies Eq.(3.3–14) since the application of Eq.(3.3–14) gives

$$S_T^{\ G} = S_T^{\ |G|} + j\varphi S_T^{\ \varphi} = -j\omega T.$$

Very often in practice the transfer function of the system under consideration is a fraction of the form

$$G(s, \alpha) = \frac{N(s, \alpha)}{D(s, \alpha)}, \tag{3.3–17}$$

where $N(s,\alpha)$ and $D(s,\alpha)$ are real continuous functions of s and α. In this case, the Bode sensitivity function can be written as

$$S_\alpha{}^G = \left.\frac{\partial \ln (N/D)}{\partial \ln \alpha}\right|_{\alpha 0} = \left.\frac{\partial \ln N}{\partial \ln \alpha}\right|_{\alpha 0} - \left.\frac{\partial \ln D}{\partial \ln \alpha}\right|_{\alpha_0} = S_\alpha{}^N - S_\alpha{}^D. \quad (3.3\text{–}18)$$

This formula facilitates the evaluation of the Bode sensitivity function in practice. In particular, Eq.(3.3–17) includes the class of rational fractions, where $N(s, \alpha)$ and $D(s, \alpha)$ are polynomials in s.

Example 3.3-4 Given a system with the block diagram shown in Fig.(3.3–4). Determine the Bode sensitivity function of the overall transfer function $G(s,\alpha) = Y(s,\alpha)/U(s)$ with respect to $P(s,\alpha)$ of the plant.

The overall transfer function is

$$G = \frac{HP + RLP}{1 + LP}.$$

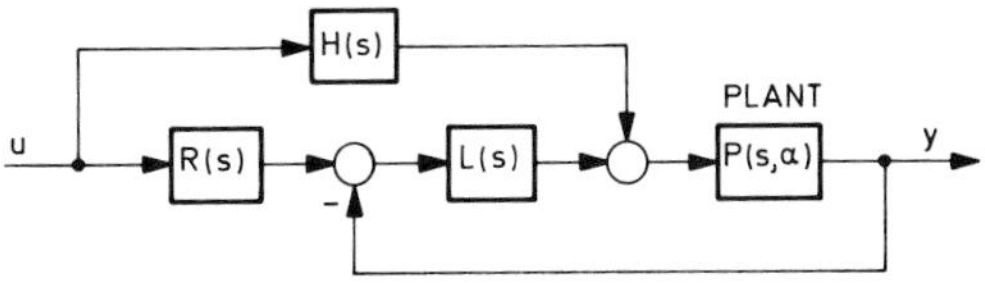

FIG. 3.3-4. Block diagram for Example 3.3–4.

Rewriting G as $G = N/D$, where $N = HP + RLP$ and $D = 1 + LP$, and applying formula (3.3–18), we obtain

$$S_P{}^N = \left.\frac{\partial(HP + RLP)}{\partial P}\right|_{P0} \frac{P_0}{HP_0 + RLP_0} = 1,$$

$$S_P{}^D = \left.\frac{\partial(1 + LP)}{\partial P}\right|_{P0} \frac{P_0}{1 + LP_0} = \frac{LP_0}{1 + LP_0}$$

Therefore,

$$S_P{}^G = S_P{}^N - S_P{}^D = 1 - \frac{LP_0}{1 + LP_0} = \frac{1}{1 + LP_0}.$$

The surprising fact that the sensitivity does not depend on $R(s)$ and $H(s)$ is very important for the design of insensitive systems. Thus the sensitivity can be assigned by a proper choice of $L(s)$, whereas by $R(s)$ and/or $H(s)$, a certain signal behavior can be realized independently.

A special but very important class of system transfer functions is of the form

$$G(s, \alpha) = \frac{G_1(s) + \alpha G_2(s)}{G_3(s) + \alpha G_4(s)} \quad (3.3\text{–}19)$$

where the parameter α occurs explicitly as a factor and G_1, G_2, G_3, G_4 are solely functions of s. Many transfer functions can be brought into this form for the parameter of interest.

Making use of the formula (3.3–18), the Bode sensitivity function of the above experssion with respect to α can be given as

$$S_\alpha^G = \left.\frac{\alpha(G_2G_3 - G_1G_4)}{(G_3 + \alpha G_4)(G_1 + \alpha G_2)}\right|_{\alpha 0}. \tag{3.3–20}$$

This formula is most useful in evaluating the Bode sensitivity functions in practice.

Example 3.3-5 The transfer function of the feedback system shown in Fig. 3.3–5 is

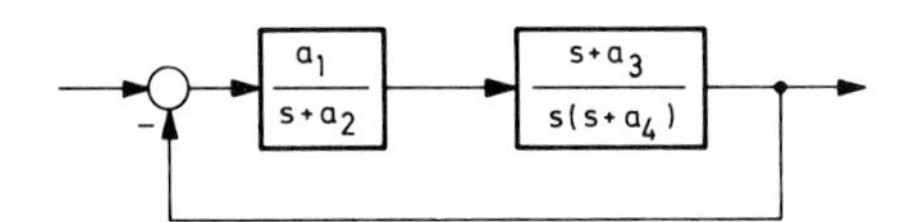

FIG. 3.3-5. Feedback system of Example 3.3–5.

$$G = \frac{a_1 s + a_1 a_3}{s^3 + (a_2 + a_4)s^2 + (a_1 + a_2 a_4)\, s + a_1 a_3}.$$

Dertermine the Bode sensitivity functions with respect to a_1 and a_4.

If a_1 is the parameter of interest, the terms in G are arranged as follows

$$G = \frac{a_1(s + a_3)}{[s^3 + (a_2 + a_4)s^2 + a_2 a_4 s] + a_1(s + a_3)}.$$

Comparing G with Eq.(3.3–19), we see that $G_1 = 0$, $G_2 = s + a_3$, $G_3 = s^3 + (a_2 + a_4)s^2 + a_2 a_4 s$, $G_4 = s + a_3$. Therefore, it follows immediately from Eq.(3.3–20) that

$$S_{a_1}^G = \frac{a_{10}(s + a_{30})[s^3 + (a_{20} + a_{40})s^2 + a_{20}a_{40}s)]}{a_{10}(s^3 + (a_{20} + a_{40})s^2 + a_{20}a_{40}s + a_{10}(s + a_{30})](s + a_{30})}$$

$$= \left[1 + \frac{a_{10}(s + a_{30})}{s^3 + (a_{20} + a_{40})s^2 + a_{20}a_{40}s}\right]^{-1}.$$

If a_4 is the parameter of interest, G can be rewritten as

$$G = \frac{a_1(s + a_3)}{[s^3 + a_2 s^2 + a_1 s + a_1 a_3] + a_4(s^2 + a_2 s)}.$$

Again comparing this expression with Eq.(3.3–29), we observe that $G_1 =$

$a_1(s + a_3)$, $G_2 = 0$, $G_3 = s^3 + a_2 s^2 + a_1 s + a_1 a_3$, $G_4 = s(s + a_2)$. Hence, it follows from Eq.(3.3–20) that

$$S_{a_4}^G = \frac{-a_{40}a_{10}s(s + a_{30})(s + a_{20})}{[s^3 + a_{20}s^2 + a_{10}s + a_{10}a_{30} + a_{40}s(s + a_{20})]a_{10}(s + a_{30})}$$

$$= -\left[1 + \frac{s^3 + a_{20}s^2 + a_{10}s + a_{10}a_{30}}{a_{40}s(s + a_{20})}\right]^{-1}.$$

In general, the Bode sensitivity function is a complex function. It is difficult to understand what sensitivity means in terms of a complex function. Therefore, to characterize the sensitivity, a certain attribute of the Bode sensitivity function such as the real or imaginary part [Eq.(3.3–15) or Eq.(3.3–16)] or a certain norm of $S_\alpha{}^G$ can be used. Most frequently the magnitude of $S_\alpha{}^G$,

$$|S_\alpha{}^G(j\omega)| = [S_\alpha{}^G(j\omega)S_\alpha{}^G(-j\omega)]^{1/2}, \qquad (3.3\text{–}21)$$

is utilized which is considered over the whole frequency range $0 \leq \omega \leq \infty$. It gives more insight into the sensitivity of the system than the real part of $S_\alpha{}^G$ which is simply the sensitivity function of the magnitude of G. The underlying idea is that, for that part of the frequency range for which the magnitude of $S_\alpha{}^G$ is large, the influence of α on the transfer function is large and vice versa.

This definition also lacks completeness since the phase of $S_\alpha{}^G$ is neglected. However, it allows for a comparison of the sensitivity of different systems with a very interesting interpretation in the time domain. This interpretation will be given in Chapter 6.

The frequency characteristics $|S_\alpha{}^G(j\omega)|$ of a number of dynamic standard elements are listed in Table 3.3–1 (from Schmidt [93]). From these plots, the following conclusions may be drawn.

Variations of the gain factor affect the transfer function with constant intensity over the whole frequency range $0 \leq \omega \leq \infty$.

The effects of dead-time variations decrease with decreasing frequencies from $\omega = 1/T_{t0}$ on and increase with frequency for $\omega > T_{t0}^{-1}$. They do not affect the steady state ($\omega = 0$).

Variations of time constants act like dead-times for small frequencies ($\omega \leq T_{i0}^{-1}$), and therefore are less critical than variations of the gain factor K. For $\omega > T_{i0}^{-1}$, they act like gain factors.

Thus, variations of gain factors and dead-times have a stronger influence on the transfer function than time constants, and in many situations it is sufficient to consider the latter ones only.

It is emphasized that $|S_\alpha{}^G(j\omega)| \neq S_\alpha^{|G|}(\omega)$, that is, the magnitude of the sensitivity function is not identical with the sensitivity of the magnitude of G). This can easily be shown by the aid of Eq.(3.3–15) which gives

$$S_\alpha^{|G|}(\omega) = \mathrm{Re}(S_\alpha{}^G(j\omega)) = |S_\alpha{}^G(j\omega)|\cos\psi \neq |S_\alpha{}^G(j\omega)|, \qquad (3.3\text{–}22)$$

thus proving the above statement.

TABLE 3.3-1

Characteristic Shape of the Magnitude $|S_\alpha{}^G(j\omega)|$ of the Bode Sensitivity Function $S_\alpha{}^G(j\omega)$ of Some Standard System Components

Parameter α		Transfer function $G = G(s)$	Sensitivity function	$\ln|S_\alpha^G(j\omega)| = \ln S$
Gain	K	$G = K$ $G = K(1 + T_i s)^{\pm 1}$	$S_K{}^G = 1$ $S_K{}^G = 1$	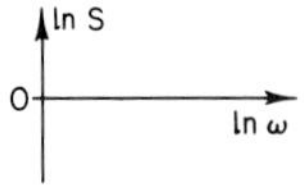
Real zero, Real pole	δ_i	$G = K(s - \delta_i)^{\pm 1}$	$S_{\delta i}^G = \mp \dfrac{\delta_{io}}{s - \delta_{io}}$	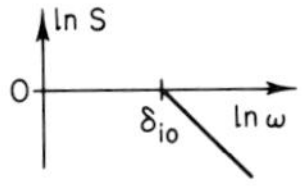
Damping ratio of a complex pair of poles	D	$G = \dfrac{K\omega_K{}^2}{s^2 + 2D\omega_K s + \omega_K{}^2}$	$S_D{}^G = -\dfrac{2D_0\omega_K s}{s^2 + 2D_0\omega_K s + \omega_K{}^2}$	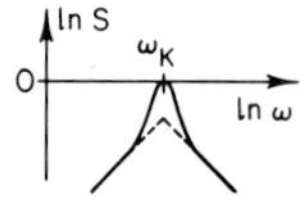
Time constant (lead or lack)	T_i	$G = K(1 + T_i s)^{\pm 1}$	$S_{Ti}^G = \pm \dfrac{T_{io}s}{1 + T_{io}s}$	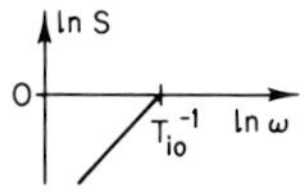
Dead-time	T_t	$G = Ke^{-T_t s}$	$S_{Tt}^G = -T_{to}s$	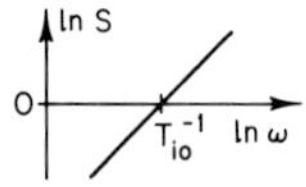

3.3.2 Sensitivity Function of Horowitz

The sensitivity function of Bode applies, strictly speaking, only to infinitesimal parameter variations. For moderate and large parameter variations, Horowitz has proposed the following definition.

Definition 3.3-2 *Sensitivity function of Horowitz.* Let G_0 represent the nominal transfer function of the system and α_0 the nominal value of the parameter under consideration, and let G, α represent the corresponding actual values, that is,

$$G = G_0 + \Delta G, \qquad \alpha = \alpha_0 + \Delta\alpha.$$

Then the Horowitz sensitivity function is defined as

$$H_\alpha{}^G \triangleq \frac{\Delta G/G}{\Delta\alpha/\alpha} = \frac{(G - G_0)/G}{(\alpha - \alpha_0)/\alpha}. \tag{3.3–23}$$

If G contains a variable component G_1 that is a function of the parameter α, the sensitivity with respect to G_1 is defined in an analogous manner:

$$H_{G_1}^G \triangleq \frac{\Delta G/G}{\Delta G_1/G_1} = \frac{(G - G_0)/G}{(G_1 - G_{10})/G_1}. \tag{3.3–24}$$

Notice that contrary to Bode's definition the variations are related to the *actual* values.

In terms of the above definitions, the ratio G_0/G of the transfer function due to a relative change of the parameter α or of the variable component G_1 is given by

$$\frac{G_0}{G} = 1 - \frac{\Delta\alpha}{\alpha} H_\alpha{}^G \quad \text{or} \quad \frac{G_0}{G} = 1 - \frac{\Delta G_1}{G_1} H_{G_1}^G, \qquad (3.3\text{–}25)$$

respectively. Horowitz has used this expression with great success in designing parameter-insensitive feedback systems [11].

Example 3.3-6 Determine the Horowitz sensitivity function $H_{G_1}^G$ and G_0/G for the classical control loop shown in Fig. 3.3–2.

Substituting the closed-loop transfer functions

$$G = \frac{G_1 R}{1 + G_1 RH} \quad \text{and} \quad G_0 = \frac{G_{10} R}{1 + G_{10} RH}$$

in the definition of $H_{G_1}^G$ [Eq.(3.3–24)], we obtain

$$H_{G_1}^G = \frac{G - G_0}{G_1 - G_{10}} \frac{G_1}{G} = \frac{G_1 - (G_0 G_1/G)}{G_1 - G_{10}} = \frac{1}{1 + G_{10} RH}.$$

Using this result, G_0/G becomes

$$\frac{G_0}{G} = 1 - \frac{G_1 - G_{10}}{G_1} \frac{1}{1 + G_{10} RH} = \frac{(G_{10}/G_1) + G_{10} RH}{1 + G_{10} RH}.$$

As we see, the expression obtained above for $H_{G_1}^G$ is identical with the corresponding expression of the Bode sensitivity function in Example 3.3–1. Thus, the Bode and Horowitz sensitivity functions of the classical control loop have the same values, although they are defined differently.

The formula found for G_0/G can be utilized for the design of freedback control systems that are insensitivie to finite variations of the plant transfer function G_1 caused by finite variations of the plant parameters [11]. The idea is to choose the nominal open-loop transfer function $L_0 = G_{10} RH$ such that, for given values of G_{10}/G_1 at certain frequencies the corresponding ratios of the closed loop G_0/G take on preassigned values.

3.3.3 Sensitivity Operator of Perkins and Cruz (Comparison Sensitivity)

The sensitivity definition of Bode applies only to linear time-invariant single-variable systems. For the comparison of open- and closed-loop systems, Perkins and Cruz generalized these definitions such that, on the one hand, arbitrary parameter changes, and, on the other hand, multivariable, time-variant, and even nonlinear systems are comprised [5].

Fundamental to the comparison of two system configurations is the fact that they are equivalent in a certain sense. In the present case, it is reasonable to require nominal equivalence defined as follows.

Definition 3.3–3 *Nominal equivalence.* Two system configurations are said to be nominally equivalent if, for nominal parameter values, the outputs due to the same inputs are equal.

The above definition of nominal equivalence implies that the two systems being compared have the same nominal trajectory. In some cases, the requirement of full trajectory equivalence is not necessary and too strong. Therefore, Kreisselmeier and Grübel [61] have developed a suitably relaxed equivalence requirement resulting in a generalized comparison sensitivity concept. The basic idea is that for a meaningful comparison of the sensitivity of two admissible control systems merely some relevant nominal response characteristics have to be the same or lie within given bounds. A suitable equivalence performance criterion is $J_E[\boldsymbol{x}(t,\boldsymbol{\alpha}_0)]$ so that Definition 3.3–3 can be modified as follows:

Definition 3.3-3(a) *Equivalence performance.* Two control systems with nominal trajectories $\boldsymbol{x}_R(t, \boldsymbol{\alpha}_0)$ and $\boldsymbol{x}_S(t, \boldsymbol{\alpha}_0)$ are called equivalent if the relation

$$J_E[\boldsymbol{x}_S(t, \boldsymbol{\alpha}_0)] = J_E[\boldsymbol{x}_R(t, \boldsymbol{\alpha}_0)]$$

holds where $\boldsymbol{\alpha}_0$ denotes the nominal parameter vector. In particular, J_E may be defined as the L_2-norm of the nominal trajectory, that is,

$$J_E = \int_{t_0}^{t_1} \boldsymbol{x}^{\mathrm{T}}(t, \boldsymbol{\alpha}_0)\boldsymbol{Z}\, \mathbf{x}(t, \boldsymbol{\alpha}_0)\, dt,$$

or as the integral of the squared control error of a control system.

The relaxed equivalence performance defined above may primarily be employed as an equivalence constraint in procedures for the sensitivity reduction of control systems. A successful application is described by Kreisselmeier and Grübel [61]. In the following derivation, full trajectory equivalence as defined in Definition 3.3–3 will be needed.

The sensitivity function of Perkins and Cruz will first be derived for *linear time-invariant single-variable* systems. For this purpose consider the open-loop system and the closed-loop system shown in Fig. 3.3–6. In both cases, $P(s, \boldsymbol{\alpha})$ represents the same plant which is affected by the parameter vector $\boldsymbol{\alpha}$, and $F(s)$, $R(s)$, $H(s)$ are parameter-independent transfer functions. The subscripts S and R shall specify the signals of the open- and closed-loop configurations, respectively.

The objective is to compare the sensitivity of both configurations due to certain (not necessarily small) parameter changes $\boldsymbol{\alpha}$ around $\boldsymbol{\alpha}_0$ or, which will be the same, due to the parameter-induced change of $P \triangleq P(s, \boldsymbol{\alpha})$ around $P_0 \triangleq P(s, \boldsymbol{\alpha}_0)$. It is therefore assumed that both configurations are *nominally*

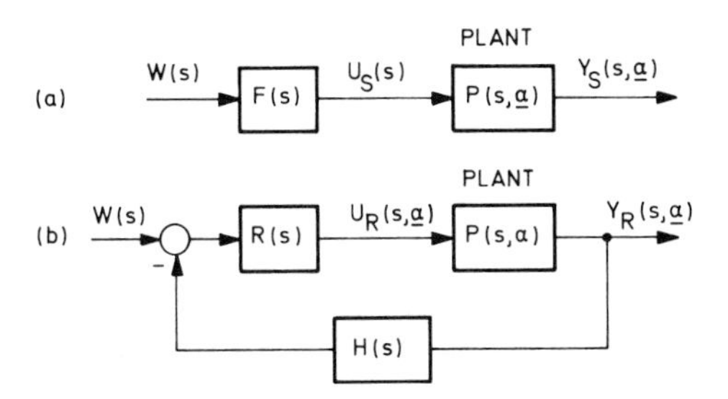

FIG. 3.3-6. Open-loop (a) and closed-loop configuration (b) for the derivation of comparison sensitivity.

equivalent. This implies that the output Y as well as the input U of the plant have to be the same for both systems for $\boldsymbol{\alpha} = \boldsymbol{\alpha}_0$ and the same system input W. That is, in terms of the Laplace transforms,

$$Y_R(s, \boldsymbol{\alpha}_0) = Y_S(s, \boldsymbol{\alpha}_0), \tag{3.3–26}$$

$$U_R(s, \boldsymbol{\alpha}_0) = U_S(s). \tag{3.3–27}$$

Equations (3.3–26) and (3.3–27) require that $F(s)$ of the open-loop configuration must be $F = R/(1 + P_0RH)$. (The arguments are dropped for ease of notation). With this choice of F, the nominal output signal becomes in the open-loop case

$$Y_S(s, \boldsymbol{\alpha}_0) \triangleq Y_{S0} = \frac{P_0R}{1 + P_0RH} W, \tag{3.3–28}$$

and in the closed-loop case

$$Y_R(s, \boldsymbol{\alpha}_0) \triangleq Y_{R0} = \frac{P_0R}{1 + P_0RH} W, \tag{3.3–29}$$

thus verifying the property of nominal equivalence.

Now letting $\boldsymbol{\alpha}_0$ change to $\boldsymbol{\alpha} = \boldsymbol{\alpha}_0 + \Delta\boldsymbol{\alpha}$, we obtain in the open-loop case

$$Y_S(s, \boldsymbol{\alpha}) \triangleq Y_S = \frac{PR}{1 + P_0RH} W \tag{3.3–30}$$

and in the closed-loop case

$$Y_R(s, \boldsymbol{\alpha}) \triangleq Y_R = \frac{PR}{1 + PRH} W. \tag{3.3–31}$$

Using these expressions, the corresponding parameter-induced output errors can be calculated. This yields in the open-loop case

$$\Delta Y_S \triangleq Y_S - Y_{S0} = \frac{PR}{1 + P_0RH} W - \frac{P_0R}{1 + P_0RH} W = \frac{(P - P_0)R}{1 + R_0RH} W \tag{3.3–32}$$

and in the closed-loop case

$$\Delta Y_R \triangleq Y_R - Y_{R0} = \frac{PR}{1 + PRH} W - \frac{P_0 R}{1 + P_0 PH} W$$
$$= \frac{(P - P_0)R}{(1 + PRH)(1 + P_0 RH)} W. \qquad (3.3\text{–}33)$$

Comparing the parameter-induced errors of the open- and closed-loop configurations given by Eqs.(3.3–32) and (3.3–33), we obtain the simple relation

$$\Delta Y_R(s, \boldsymbol{\alpha}) = \frac{1}{1 + P(s, \boldsymbol{\alpha})\, R(s)H(s)} \Delta Y_S(s, \boldsymbol{\alpha}), \qquad (3.3\text{–}34)$$

no matter how large the parameter variations are.

Generalizing this result, we define the following:

Definition 3.3-4 *Comparison sensitivity function.* The complex function

$$S_p(s, \boldsymbol{\alpha}) \triangleq \frac{\Delta Y_R(s, \boldsymbol{\alpha})}{\Delta Y_S(s, \boldsymbol{\alpha})}, \qquad (3.3\text{–}35)$$

relating the parameter-induced output error of a system configuration (R) to another nominally equivalent configuration (S), is called the comparison sensitivity function or sensitivity function of Perkins and Cruz.

From this definition, it follows that if (R) is a closed-loop configuration and (S) an open-loop configuration, the comparison sensitivity function becomes

$$S_p(s, \boldsymbol{\alpha}) = \frac{1}{1 + P(s, \boldsymbol{\alpha})\, R(s)H(s)}. \qquad (3.3\text{–}36)$$

Notice that the comparison sensitivity function S_p in Eq.(3.3–36) involves the *actual* plant transfer function $P(s, \boldsymbol{\alpha})$. Moreover, since no restrictions have been imposed upon the size of the parameter variations, Eq.(3.3–36) is valid for any $\boldsymbol{\alpha}$ and any $P(s,\boldsymbol{\alpha})$, no matter how large the deviation respectively from $\boldsymbol{\alpha}_0$ and $P(s, \boldsymbol{\alpha}_0)$ will be.

To demonstrate the difference between comparison and Bode sensitivity, let us assume infinitesimal parameter variations, i.e., $\Delta\boldsymbol{\alpha} = d\boldsymbol{\alpha} \to 0$. Then $P(s, \boldsymbol{\alpha})$ can be replaced by $P(s, \boldsymbol{\alpha}_0)$ so that the comparison sensitivity function becomes

$$S_p(s, \boldsymbol{\alpha}_0) = \frac{1}{1 + P(s, \boldsymbol{\alpha}_0)R(s)H(s)}. \qquad (3.3\text{–}37)$$

Comparing this expression with the Bode sensitivity function $S_P{}^G(s)$ of the same closed-loop configuration (see Example 3.3–1), we see that both sensitivity functions are identical. Thus, the Bode sensitivity function may be interpreted so as to relate the parameter-induced output error of the closed loop to that of the nominally equivalent open loop in the case of infinitesimal parameter deviations. In this case, ΔY_R and ΔY_S in Eq.(3.3–34) can be written as ∂Y_R and ∂Y_S, respectively. Dividing Eq.(3.3–34) by $\boldsymbol{\alpha}$ and setting $\boldsymbol{\alpha} = \boldsymbol{\alpha}_0$, we obtain

$$\left.\frac{\partial Y_R(s, \boldsymbol{\alpha})}{\partial \boldsymbol{\alpha}}\right|_{\boldsymbol{\alpha}0} = \frac{1}{1 + P(s, \boldsymbol{\alpha}_0)R(s)H(s)} \left.\frac{\partial Y_S(s, \boldsymbol{\alpha})}{\partial \boldsymbol{\alpha}}\right|_{\boldsymbol{\alpha}0}. \tag{3.3–38}$$

The same relationship is obtained by dividing Eq.(3.3–34) by ∂P instead of $\partial \boldsymbol{\alpha}$. This reveals that the Bode sensitivity function may also be interpreted so as to relate the Laplace transforms of the output sensitivity functions of the open- and closed-loop configurations, and this also reveals that this is true with respect to any parameter $\boldsymbol{\alpha}_j$ and $\boldsymbol{\alpha}$, as well as with respect to $P(s, \boldsymbol{\alpha})$.

Conversely, the comparison sensitivity function may be understood as a generalization of Bode's sensitivity function for the case of large parameter variations.

In order to develop a relation between the comparison sensitivity and Horowitz sensitivity functions, let us rewrite the definition of the comparison sensitivity function in the form

$$S_p(s, \boldsymbol{\alpha}) \triangleq \frac{\Delta Y_R(s, \boldsymbol{\alpha})}{\Delta Y_S(s, \boldsymbol{\alpha})} = \frac{\Delta G_R(s, \boldsymbol{\alpha})}{\Delta G_S(s, \boldsymbol{\alpha})}, \tag{3.3–39}$$

where G_R and G_S denote the overall transfer functions of the configurations (R) and (S), respectively, due to $Y_R = G_R U$ and $Y_S = G_S U$ and ΔG_R, ΔG_S their variations. Multiplying numerator and denominator by $G_R G_S P \Delta P$, where P denotes the actual transfer function of the plant and ΔP its parameter-induced change, we have

$$S_p = \frac{\Delta G_R}{G_R} \frac{P}{\Delta P} \frac{\Delta P}{\Delta G_S} \frac{G_S}{P} \frac{G_R}{G_S}. \tag{3.3–40}$$

Recall now that the first fraction is the Horowitz sensitivity function of the closed loop $(H_P{}^G)_R$, and the second one is the Horowitz sensitivity function of the open loop $(H_P{}^G)_S$. Thus, the desired relationship can be written as

$$S_p = \frac{(H_p{}^G)_R}{(H_P{}^G)_S} \frac{G_R}{G_S}. \tag{3.3–41}$$

Equation(3.3–41) reveals that the comparison sensitivity function is identical with the ratio of the Horowitz sensitivity functions of the closed loop and open loop multiplied by G_R/G_S.

Example 3.3-7 From the results of Examples 3.3–1 and 3.3–6, and Eq.(3.3–36), we see that for a control loop of the form shown in Fig. 3.3–6b, we obtain

the Bode sensitivity function

$$S_p{}^G = \frac{1}{1 + P_0 RH},$$

the Horowitz sensitivity function

$$H_p^G = \frac{1}{1 + P_0RH},$$

the comparison sensitivity function

$$S_p = \frac{1}{1 + PRH}.$$

Since S_p^G and H_p^G of the open loop, Fig. 3.3–6a, are equal to one, whatever F is like, we see that the relations $S_p^G = H_p^G$, $(S_p^G)_R/(S_p^G)_S = (H_p^G)_R/(H_p^G)_S$, and Eq.(3.3–41) are satisfied.

Note that all these relations not only apply to the comparison of open- and closed-loop configurations but also to any pair of configurations to be compared.

Example 3.3-8 Suppose the two systems shown in Fig. 3.3–7 are nominally equivalent, that is $F = (1 + P_0^{-1}G)$. Compare the sensitivity by means of the sensitivity functions of Bode, Horowitz, and Perkins and Cruz.

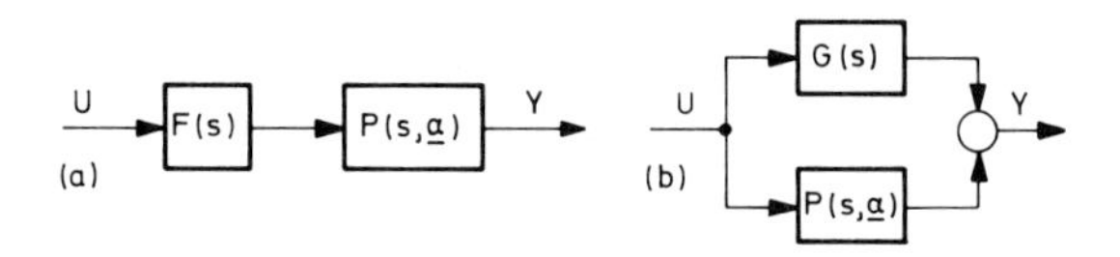

FIG. 3.3-7. The two systems whose sensitivities are compared in Example 3.3–8.

The nominal and actual transfer functions of configuration (a) are $G_{a0} = P_0 + G$ and $G_a = (1 + P_0^{-1}G)P$, respectively. The corresponding transfer functions of configuration (b) are $G_{b0} = P_0 + G$ and $G_b = P + G$. The Bode sensitivity functions of configurations (a) and (b) become $(S_p^G)_a = 1$, $(S_p^G)_b = P_0/(P_0 + G)$, whence $(S_p^G)_b/(S_p^G)_a = P_0/(P_0 + G)$. For the Horowitz sensitivity functions, we obtain $(H_p^G)_a = 1, (H_p^G)_b = P/(P + G)$, and therefore $(H_p^G)_b/(H_p^G)_a = P(P + G)$. The comparison sensitivity function becomes $S_p = P_0/(P_0 + G)$. This again verifies Eq. (3.3–41); it shows, however, that in this case for configuration (*b*), $S_p = S_p^G$.

The comparison sensitivity function may readily be extended to multivariable systems [30]. In this case, the transfer functions $R(s)$, $H(s)$, $P(s, \boldsymbol{\alpha})$ have to be replaced by the corresponding transfer matrices denoted by $\boldsymbol{R}(s)$, $\boldsymbol{H}(s)$, $\boldsymbol{P}(s, \boldsymbol{\alpha})$, the derivation itself being entirely analogous to the single-variable case. By this procedure we obtain, if it exists, the matrix counterpart of Eq. (3.3–36),

$$\boldsymbol{S}_p(s, \boldsymbol{\alpha}) \triangleq [\boldsymbol{I} + \boldsymbol{P}(s, \boldsymbol{\alpha})\boldsymbol{R}(s)\boldsymbol{H}(s)]^{-1} \tag{3.3–42}$$

which is called the *comparison sensitivity matrix*. As in the single-variable case, $\boldsymbol{S}_p(s, \boldsymbol{\alpha})$ relates the parameter-induced output errors of the open- and closed-loop configuration by

$$\Delta \boldsymbol{Y}_R(s, \boldsymbol{\alpha}) = \boldsymbol{S}_p(s, \boldsymbol{\alpha})\, \Delta \boldsymbol{Y}_S(s, \boldsymbol{\alpha}). \tag{3.3–43}$$

This equation may be considered as an implicit definition of $\boldsymbol{S}_p(s, \boldsymbol{\alpha})$.

For small parameter variations, the actual plant transfer matrix $\boldsymbol{P}(s, \boldsymbol{\alpha})$ in Eq. (3.3–42) can be replaced by the nominal transfer matrix $\boldsymbol{P}(s, \boldsymbol{\alpha}_0)$. Thus the corresponding comparison sensitivity matrix $\boldsymbol{S}_p(s, \boldsymbol{\alpha}_0)$ can be interpreted as a *Bode sensitivity matrix* for the multiparameter case.

To compare the sensitivity of open- and closed-loop multivariable systems (as a function of frequency) by a real number rather than by the complex quantity $\boldsymbol{S}_p$, the expression

$$\boldsymbol{S}_p^{\mathrm{T}}(-j\omega)\boldsymbol{Z}\, \boldsymbol{S}_p(j\omega) \tag{3.3–44}$$

is used, where $\boldsymbol{Z}$ represents a fixed positive definite weighting matrix. This expression can be interpreted as a generalization of the magnitude $|S_\alpha^G(j\omega)|$ [Eq. (3.3–21)] in the single-variable case. It will be shown in Chapter 7 what role this expression plays in the comparison of open- and closed-loop systems.

Example 3.3–9 Referring to Example 3.3–4, determine the comparison sensitivity matrix $\boldsymbol{S}_p$ of the configuration shown in Fig. 3.3–4, assuming that the several blocks have the following transfer matrices:

$$\boldsymbol{P}(s, \boldsymbol{\alpha}_1) = \begin{bmatrix} P_{11}(s, \boldsymbol{\alpha}) & P_{12}(s, \boldsymbol{\alpha}) \\ P_{21}(s, \boldsymbol{\alpha}) & P_{22}(s, \boldsymbol{\alpha}) \end{bmatrix}, \qquad \boldsymbol{L}(s) = \begin{bmatrix} L_{11}(s) & L_{12}(s) \\ L_{21}(s) & L_{22}(s) \end{bmatrix},$$

$$\boldsymbol{R}(s) = \begin{bmatrix} R_{11}(s) & R_{12}(s) \\ R_{21}(s) & R_{22}(s) \end{bmatrix}, \qquad \boldsymbol{H}(s) = \begin{bmatrix} H_{11}(s) & H_{12}(s) \\ H_{21}(s) & H_{22}(s) \end{bmatrix}.$$

Denoting the Laplace transforms of the input and output vectors by $\boldsymbol{U}$ and $\boldsymbol{Y}$, respectively, and the nominal quantities by the subscript zero, we obtain for the nominal parameters

$$\boldsymbol{Y}_0 = (\boldsymbol{I} + \boldsymbol{P}_0\boldsymbol{L})^{-1}\boldsymbol{P}_0(\boldsymbol{H} + \boldsymbol{L}\,\boldsymbol{R})\,\boldsymbol{U},$$

where $\boldsymbol{I}$ represents the 2×2 unity matrix. By the definition of nominal equivalence, this equation must also hold for the nominally equivalent open-loop configuration.

Assuming actual parameters, we have for the given closed-loop configuration,

$$\boldsymbol{Y}_R = (\boldsymbol{I} + \boldsymbol{P}\,\boldsymbol{L})^{-1}\boldsymbol{P}(\boldsymbol{H} + \boldsymbol{L}\,\boldsymbol{R})\boldsymbol{U},$$

and for the open-loop configuration

$$\boldsymbol{Y}_L = (\boldsymbol{I} + \boldsymbol{P}\,\boldsymbol{L}_0)^{-1}\boldsymbol{P}(\boldsymbol{H} + \boldsymbol{L}\,\boldsymbol{R})\boldsymbol{U}.$$

Taking the differences $\Delta\boldsymbol{Y}_R = \boldsymbol{Y}_R - \boldsymbol{Y}_0$ and $\Delta\boldsymbol{Y}_S = \mathrm{Y}_S - \boldsymbol{Y}_0$, we find

$$\Delta\boldsymbol{Y}_R = \boldsymbol{S}_p\, \Delta\boldsymbol{Y}_S$$

where

$$\boldsymbol{S}_p = (\boldsymbol{I} + \boldsymbol{P}\,\boldsymbol{L})^{-1} = \frac{\operatorname{adj}(\boldsymbol{I} + \boldsymbol{P}\,\boldsymbol{L})}{\det(\boldsymbol{I} + \boldsymbol{P}\,\boldsymbol{L})}$$

and

$$\operatorname{Adj}(\boldsymbol{I} + \boldsymbol{P}\,\boldsymbol{L}) = \begin{bmatrix} 1 + P_{21}L_{12} + P_{22}L_{22} & -(P_{11}L_{12} + P_{12}L_{22}) \\ -(P_{21}L_{11} + P_{22}L_{21}) & 1 + P_{11}L_{11} + P_{12}L_{21} \end{bmatrix},$$

$$\det(\boldsymbol{I} + \boldsymbol{P}\,\boldsymbol{L}) = (1 + P_{11}L_{11} + P_{12}L_{21})(1 + P_{21}L_{12} + P_{22}L_{22}) - (P_{11}L_{12} + P_{12}L_{22})(P_{21}L_{11} + P_{22}L_{21}).$$

In general, all elements P_{ij} and L_{ij} are functions of s.

Note that all the above derivations based on transfer functions or matrices are also valid for z-transfer functions without any restriction; they are, therefore, applicable to linear sampled-data systems as well.

Example 3.3–10 Find the Bode sensitivity function of the z-transfer function of the closed loop with respect to the z-transfer function of the plant for the digital feedback control system shown in Fig. 3.3–8. Give the comparison sensitivity function that relates the sensitivity of the z-transform of the output with respect to the gain K of the plant to the corresponding sensitivity of the nominally equivalent open-loop configuration. Assume that the algorithm of the PI-controller that is programmed on the digital computer is given by

$$u_k = K_c\left[v_k + \frac{T}{T_N}\sum_{\nu=0}^{k} v_\nu\right], \qquad k = 0, 1, \ldots,$$

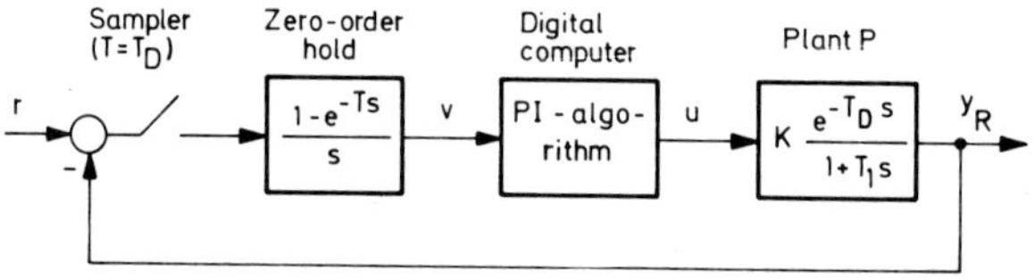

FIG. 3.3-8. Digital feedback control system of example 3.3–10.

where K_c, T_N are the characteristic constants of the controller and T is the sampling period. It is further assumed that $T = T_D$ and that T_N is chosen as

$$T_N = \frac{Te^{-T/T_1}}{1 - e^{-T/T_1}}.$$

If we apply the z-transform to the above relation, we obtain the z-transfer function of the controller:

$$G_z(z) = \frac{K_c}{a}\frac{z - a}{z - 1}, \qquad a = e^{-T/T_1}.$$

The z-transfer function of the open loop ("loop transmission") is now determined by including the transfer functions of Fig. 3.3–8 and again applying the z-transform. This produces

$$F_z(z) = (1 - z^{-1})\frac{K_c}{a}\frac{z - a}{z - 1}\frac{K}{T_1}z^{-1}\mathscr{Z}\left\{\frac{1}{s(s + 1/T_1)}\right\} = \frac{A}{z(z - 1)},$$

where $A = K\,K_c(e^{-T/T_1} - 1)$. Using this expression, the z-transfer function of the closed loop becomes

$$G_z(z) = \frac{F_z(z)}{1 + F_z(z)} = \frac{A}{z^2 - z + A}.$$

According to the result of Example 3.3–4, the desired Bode sensitivity of $G_z(z)$ with respect to $F_z(z)$ is now

$$S_{Fz}^{Gz}(z) = \frac{1}{1 + F_{z0}(z)} = \frac{z^2 - z}{z^2 - z + A_0},$$

where $A_0 = K_0K_c(e^{-T/T_1} - 1)$, and K_0 is the nominal value of K. By virtue of Eq. (3.3–38), the above expression can be interpreted as the desired comparison sensitivity $S_{pz}(z)$ that relates the sensitivity of the z-transform of the output, $(\partial Y_R(z)/\partial K)_{K_0}$ of the considered feedback system, to $(\partial Y_S(z)/\partial K)_{K_0}$ of its nominally equivalent open-loop configuration, that is,

$$S_{pz}(z) = \frac{(\partial Y_R(z)/\partial K)_{K_0}}{(\partial Y_S(z)/\partial K)_{K_0}} = \frac{z^2 - z}{z^2 - z + A_0}.$$

For finite parameter changes of K, A_0 in the comparison sensitivity function has to be replaced by A. Note that we obtain the same results when T_1 is the parameter of interest.

It is possible to extend the idea of comparison sensitivity to even more complex systems, such as *nonlinear, time-variant, general discrete,* or *distributed parameter* systems [5]. In these situations, it is useful to define the comparison sensitivity as an operator.

Definition 3.3-5 *Comparison sensitivity operator.* The operator $\boldsymbol{S}_p$ relating the first-order parameter-induced output error $\Delta\boldsymbol{y}_R$ of a certain configuration (R) to the corresponding error $\Delta\boldsymbol{y}_S$ of a nominally equivalent configuration (S) by

$$\Delta\boldsymbol{y}_R(t, \boldsymbol{\alpha}) \triangleq \boldsymbol{S}_p\,\Delta\boldsymbol{y}_S(t, \boldsymbol{\alpha}) \tag{3.3–45}$$

is called a comparison sensitivity operator or a sensitivity operator of Perkins and Cruz.

The comparison sensitivity operator $\boldsymbol{S}_p$ may be time-variant (in the case of time-variant systems), but if $\Delta\boldsymbol{y}_R$ and $\Delta\boldsymbol{y}_S$ are taken as the first-order approximations, $\boldsymbol{S}_p$ is *always linear* even in the case of nonlinear systems.

The most important case in practice is that of (R) being a closed loop and (S) being the nominally equivalent open loop, as shown in Fig. 3.3–6 with $\boldsymbol{R}$, $\boldsymbol{H}$, and $\boldsymbol{P}$ now representing the operators of the components. In this case $\boldsymbol{S}_p$ simply becomes

$$\boldsymbol{S}_p = (\boldsymbol{I} + \boldsymbol{P}_1\boldsymbol{R}_1\boldsymbol{H}_1)^{-1}, \tag{3.3–46}$$

where $\boldsymbol{P}_1$, $\boldsymbol{R}_1$, and $\mathbf{H}_1$ represent the first-order approximations of the operators $\boldsymbol{P}$, $\boldsymbol{R}$, $\boldsymbol{H}$ with respect to their inputs. For further details see Cruz [5].

3.3.4 Root Sensitivity

A problem of fundamental importance in system theory, especially in the analysis and design of feedback systems, is that of stability. When the system is considered from the frequency domain point of view, the stability can most vividly be characterized by the *location of the poles* of the corresponding transfer functions or transfer matrices in the complex s-plane. Many techniques of analysis and design, such as Nyquist and root locus, are based on the root locations. When the system is described in the *state space,* the poles are defined by the *eigenvalues* of the system matrix.

In order to study the influence of parameter changes on the stability of the system, it is therefore useful to determine the sensitivity of the pole locations or the eigenvalues with respect to the parameter changes. A very popular method based on this idea is, for example, the root locus method which studies the change of the root locations due to a finite change of a single parameter, namely the gain of the loop transmission in a closed loop. Unfortunately this approach cannot be extended to more than one parameter change without loosing its simplicity [16].

In many situations, it is sufficient to know the sensitivity of the root locations to various infinitesimal rather than large parameter variations. The *root sensitivity* characterizes both the direction and, to the first order, the amount of shifting of the root locations due to simultaneous small changes of all parameters of interest.

When this problem is treated in the frequency domain, the parameters of interest are the coefficients of the transfer function (or matrix); when treated in the state space, the parameters of interest are the coefficients of the matrices of the vector differential equations. The latter approach of *eigenvalue sensitivity* is primarily useful for the study of time domain properties such as how to find a state representation that is least sensitive to inaccurate implementation of the coefficients of the matrices. It is, however, less suitable for the design of insensitive linear feedback systems, since it is more tedious and less transparent than the frequency domain approach [45]. Therefore, before using one or the other approach, one should carefully ventilate the question of usefulness for the special purpose.

In this section, root sensitivity is defined and discussed from the transfer function point of view.

Consider a system described by the transfer function

$$G(s, \boldsymbol{\alpha}) = \frac{\sum_{\mu=0}^{m} b_\mu(\boldsymbol{\alpha})\, s^\mu}{\sum_{\nu=0}^{n} a_\nu(\boldsymbol{\alpha})\, s^\nu}, \qquad m \leq n, \tag{3.3–47}$$

where the coefficients are continuous functions of a parameter vector $\boldsymbol{\alpha} = [\alpha_1\ \alpha_2\ .\ .\ .\ \alpha_r]^{\mathrm{T}}$. It is well known that this transfer function can be brought into the form

$$G(s, \boldsymbol{\alpha}) = K(\boldsymbol{\alpha}) \frac{\prod_{\mu=1}^{m} (s - s_\mu{}^0(\boldsymbol{\alpha}))}{\prod_{\nu=1}^{n} (s - s_\nu(\boldsymbol{\alpha}))}, \qquad m \leq n, \tag{3.3–48}$$

by determing the roots of the equation $\sum_{\mu=0}^{m} b_\mu(\boldsymbol{\alpha}) s^\mu = 0$ and of the characteristic equation

$$\sum_{\nu=0}^{n} a_\nu(\boldsymbol{\alpha}) s^\nu = 0. \tag{3.3–49}$$

The roots $s_\mu{}^0(\boldsymbol{\alpha})$ are the zeros and $s_\nu(\boldsymbol{\alpha})$ the poles of the system transfer function, and $K(\boldsymbol{\alpha})$ is the gain factor. They all depend continuously on $\boldsymbol{\alpha}$.

Let us first consider the case in which parameter changes do not affect the order of the characteristic equation, thus referring to *α-variations.* This implies that the nominal values of the coefficients a_ν are unlike zero.

The sensitivities of the root locations have to be defined in different manners depending on whether the roots are real or complex conjugate. Consider first *real roots,* abbreviated by s_i, $i = 1, 2, . . . , n + m$. We define the following:

Definition 3.3-6 *Real-root sensitivity.* Let s_i represent the real roots s_i of $G(s, \boldsymbol{\alpha})$ in Eq. (3.3–48). Then the semirelative partial derivative

$$\tilde{S}_{\alpha_j}^{s_i} \triangleq \left.\frac{\partial s_i}{\partial \ln \alpha_j}\right|_{\boldsymbol{\alpha}0} = \left.\frac{\partial s_i}{\partial \alpha_j / \alpha_j}\right|_{\boldsymbol{\alpha}0}, \qquad \begin{array}{l} i = 1, 2, . . . , m + n, \\ j = 1, 2, . . . , r, \end{array} \tag{3.3–50}$$

is called the (semirelative) real-root sensitivity of the system with respect to α_j. Analogously, the relative real-root sensitivity can be defined as $\bar{S}_{\alpha_j}^{s_i} = (\partial \ln s_i / \partial \ln \alpha_j)_{\boldsymbol{\alpha}0}$. Note that $\tilde{S}_{\alpha_j}^{s_i}$ and $\bar{S}_{\alpha_j}^{s_i}$ are real numbers.

Using the above definition, the effect of a sufficiently small relative parameter variation $\Delta\alpha_j/\alpha_{j0}$ on the root location s_i can be determined approximately by either of the following relations

$$\Delta s_i = \sum_{j=1}^{r} \tilde{S}_{\alpha_j}^{s_i} \frac{\Delta\alpha_j}{\alpha_{j0}}, \qquad \frac{\Delta s_i}{s_{i0}} = \sum_{j=1}^{r} \bar{S}_{\alpha_j}^{s_i} \frac{\Delta\alpha_j}{\alpha_{j0}}. \tag{3.3–51}$$

Since the sensitivity functions $\tilde{S}_{\alpha_j}^{s_i}$ and $\bar{S}_{\alpha_j}^{s_i}$ are real numbers, the parameter-induced errors given by Eq. (3.3–51) are also real numbers and thus indicate the absolute or relative amount of the shifting of the root locations along the real axis.

Example 3.3-11 For the network shown in Fig. 3.3–9a, determine the pole sensitivities of the complex input impedance $Z(s, R)$ with respect to R, assuming that $L = C = 1$ and the nominal value of R is $R_0 = 2.5$ (normalized quantities). What happens when $R_0 = 2$ or less?

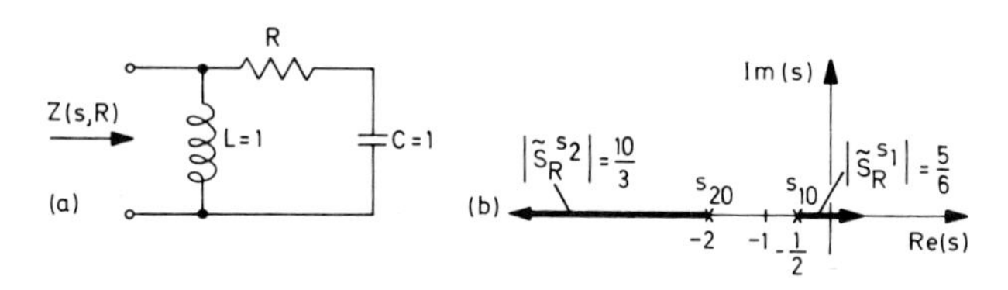

FIG. 3.3-9. Network diagram (a) and corresponding pole sensitivities (b) of Example 3.3–11.

The complex input impedance is

$$Z(s, R) = \frac{LCRs^2 + Ls}{LCs^2 + CRs + 1} = \frac{Rs^2 + s}{s^2 + Rs + 1}.$$

The poles of $Z(s, R)$ are

$$s_1 = -\frac{R}{2} + \left(\frac{R^2}{4} - 1\right)^{1/2}, \qquad s_2 = -\frac{R}{2} - \left(\frac{R^2}{4} - 1\right)^{1/2}$$

with the nominal values $s_{10} = -1/2$, $s_{20} = -2$. According to Definition 3.3–6, the pole sensitivities become

$$\tilde{S}_R^{s_1} = \left.\frac{\partial s_1}{\partial R}\right|_{R_0} R_0 = -\frac{R_0}{2} + \frac{R^2{}_0}{4}\left(\frac{R^2{}_0}{4} - 1\right)^{-1/2} = \frac{5}{6},$$

$$\tilde{S}_R^{s_2} = \left.\frac{\partial s_2}{\partial R}\right|_{R_0} R_0 = -\frac{R_0}{2} - \frac{R_0{}^2}{4}\left(\frac{R_0{}^2}{4} - 1\right)^{-1/2} = -\frac{10}{3}.$$

This result indicates that the amount of pole shifting due to an incremental change in R is four times larger for the pole s_2 than it is for the pole s_1. In other words, the pole location of s_2 is four times more sensitive to a change in R than that of s_1. The different signs show that the poles tend to opposite directions. This is illustrated in Fig. 3.3–9b.

For $R_0 = 2$ a double real pole occurs at $s_{1,2} = -1$. The real root sensitivities then both go to infinity. This result can be interpreted such that for an incremental increase of R from $R_0 = 2$ the double real pole totally disappears. Instead, two single poles appear that move along the real axis for $R > 2$ and in a perpendicular direction to the real axis for $R < 2$. In other words, the root locus has a breakaway point at $s_{10} = s_{20} = -1$, which is the reason for the above difficulty.

This shows that the real root sensitivity does not apply to *double real roots.* The root sensitivity of double real roots should rather be determined as the sensitivity of the pair of complex conjugate roots with the imaginary part set equal to zero.

In the case of *complex conjugate* roots, a *complex* root sensitivity has to be determined characterizing both the amount and the direction (angle in the complex s-plane) of the shifting of the root locations due to an incremental parameter change. In order to establish the corresponding definitions, let us assume that $s_i = \sigma_i \pm j\omega_i$ represents the ith pair of roots either of the numerator or denominator of $G(s, \boldsymbol{\alpha})$ in Eq. (3.3–48). Then the following definitions hold:

Definition 3.3-7 *Sensitivity of the real part $\sigma_i(\boldsymbol{\alpha})$ to α_j is*

$$\tilde{S}_{\alpha_j}^{\sigma_i} \triangleq \left.\frac{\partial \sigma_i(\boldsymbol{\alpha})}{\partial \ln \alpha_j}\right|_{\boldsymbol{\alpha}0}. \qquad (3.3\text{–}52)$$

Definition 3.3-8 *Sensitivity of the frequency $\omega_i(\boldsymbol{\alpha})$ to α_j is*

$$\tilde{S}_{\alpha_j}^{\omega_i} = \left.\frac{\partial \omega_i(\boldsymbol{\alpha})}{\partial \ln \alpha_j}\right|_{\boldsymbol{\alpha}0}. \qquad (3.3\text{–}53)$$

The corresponding *complex root sensitivity is then given by* $\tilde{S}_{\alpha_j}^{s_i} = \tilde{S}_{\alpha_j}^{\sigma_i} + j\tilde{S}_{\alpha_j}^{\omega_i}$. The magnitude $|\tilde{S}_{\alpha_j}^{s_i}| = [(\tilde{S}_{\alpha_j}^{\sigma_i})^2 + (\tilde{S}_{\alpha_j}^{\omega_i})^2]^{1/2}$ characterizes the vehemence of the parameter-induced shifting of the root location at nominal parameter values, and the phase angle $\rho = \tan^{-1}(\tilde{S}_{\alpha_j}^{\omega_i}/\tilde{S}_{\alpha_j}^{\sigma_i})$ gives its direction. This direction is identical with the direction of the root locus at nominal parameter values.

The ith pair of roots is often written in the form

$$s_{i,i+1} = -\omega_{in}\xi_i \pm j\omega_{in}(1 - \xi_i^2)^{1/2} = \sigma_i \pm j\omega_i, \qquad (3.3\text{–}54)$$

where ω_{in} is the (undamped) natural frequency and ξ_i is the damping ratio. Then it is convenient to use the following definitions:

Definition 3.3–9 *Sensitivity of the damping ratio ξ_i to α_j is*

$$\bar{S}_{\alpha_j}^{\xi_i} \triangleq \left.\frac{\partial \ln \xi_i(\boldsymbol{\alpha})}{\partial \ln \alpha_j}\right|_{\boldsymbol{\alpha}0}. \qquad (3.3\text{–}55)$$

Definition 3.3-10 *Sensitivity of the natural frequency ω_{in} to α_j is*

$$\bar{S}_{\alpha_j}^{\omega_{in}} \triangleq \left.\frac{\partial \ln \omega_{in}(\boldsymbol{\alpha})}{\partial \ln \alpha_j}\right|_{\boldsymbol{\alpha}0}. \qquad (3.3\text{–}56)$$

Making use of the notation $\delta_i = -\omega_{in}\xi_i$, $\omega_i = \omega_{in}(1 - \xi_i^2)^{1/2}$, it can be shown (see Exercise) that the following relations hold:

$$\bar{S}_{\alpha j}^{\omega_{in}} = \xi_{i0}^2 \bar{S}_{\alpha j}^{\delta_i} - (1 - \xi_{i0}^2)\bar{S}_{\alpha j}^{\omega_i}, \tag{3.3–57}$$

$$\bar{S}_{\alpha_j}^{\xi_i} = (1 - \xi_{i0}^2)(\bar{S}_{\alpha j}^{\delta_i} - \bar{S}_{\alpha j}^{\omega_i}), \tag{3.3–58}$$

where ξ_{i0} is the nominal value of the damping ratio, and $\bar{S}_{\alpha j}^{\delta_i}$ and $\bar{S}_{\alpha j}^{\omega_i}$ are the relative sensitivities normalized on δ_{i0} and ω_{i0}. respectively.

Example 3.3-12 Given a system with the transfer function

$$G(s) = K\frac{s + b}{s^2 + a_1 s + a_0},$$

where the nominal values of the parameters are $b_0 = 1$, $a_{00} = 5$, $a_{10} = 2$: Find the semirelative real root sensitivity of the zero s_1 and the corresponding complex root sensitivities of the poles s_2 and s_3 with respect to the parameter a_1.

The zero is $s_1 = -b$ and its nominal value is $s_{10} = -1$. The desired real root sensitivity is then $\tilde{S}_{a1}^{s1} = (\partial s_1/\partial a_1)_{\alpha 0} a_{10} = 0$. Evaluating the characteristic equation $s^2 + a_1 s + a_0 = 0$ yields the poles $s_2 = \sigma_2 + j\omega_2$, $s_3 = \sigma_3 + j\omega_3$, where

$$\sigma_2 = \sigma_3 = -a_1/2, \qquad \omega_2 = -\omega_3 = [a_0 - (a_1{}^2/4)]^{1/2}.$$

The corresponding nominal values are $\sigma_{20} = \sigma_{30} = -1$ and $\omega_{20} = 2$, $\omega_{30} = -2$. Evaluating Eqs. (3.3–52) and (3.3–53) with these expressions yields the required complex root sensitivities of s_2 and s_3 with respect to a_1:

$$\tilde{S}_{a1}^{s2} = \left.\frac{\partial \sigma_2}{\partial a_1}\right|_{\boldsymbol{a}0} a_{10} + j\left.\frac{\partial \omega_2}{\partial a_1}\right|_{\boldsymbol{a}0} a_{10} = -1 - 1/2j,$$

$$\tilde{S}_{a1}^{s3} = \left.\frac{\partial \sigma_3}{\partial a_1}\right|_{\boldsymbol{a}0} a_{10} + j\left.\frac{\partial \omega_3}{\partial a_1}\right|_{\boldsymbol{a}0} a_{10} = -1 + 1/2j.$$

Thus the magnitudes of the root sensitivities become $|\tilde{S}_{a1}^{s1}| = 0$ and $|\tilde{S}_{a1}^{s2}| = |\tilde{S}_{a1}^{s3}| = 1.11$. The corresponding angles are $\rho_1 = \angle\tilde{S}_{a1}^{s1} = 0°$, $\rho_2 = \angle\tilde{S}_{a1}^{s2} = 206°$, $\rho_3 = \angle\tilde{S}_{a1}^{s3} = 154°$. The vectors of the complex root sensitivities are shown in Fig. 3.3–10 along with the root locus with respect to a_1. The result shows that the zero does not change with a_1 at all; the vectors of *complex* root sensitivities give an idea of the vehemence of the change of the pole locations due to a change in a_1 and indicate the corresponding direction which is in accordance with the direction of the root locus at a_{10}. Note that the latter does not hold if the root sensitivity is defined as a relative sensitivity measure as was done, e.g, in [16]. This is why the above root sensitivity is defined as a semirelative sensitivity measure.

The total of all root sensitivities of a system characterizes the overall sensitivity property of the system from the root locus point of view. This leads to the following definition:

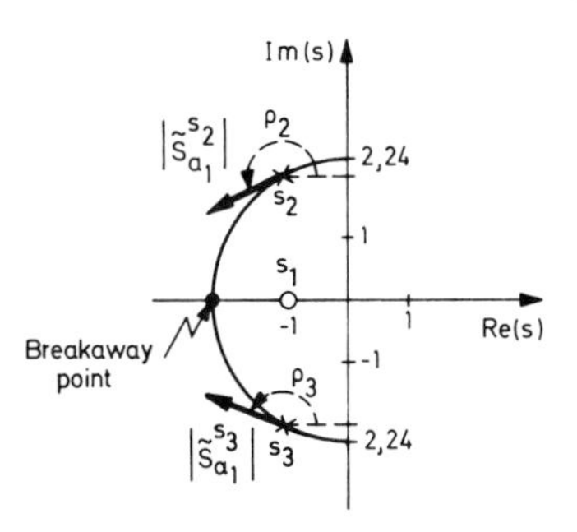

FIG. 3.3-10. Root locus and root sensitivities of the zero s_1 and the two poles s_2 and s_3 of Example 3.3–12.

Definition 3.3-11 *Root sensitivity matrix.* Suppose that a system transfer function $G(s, \boldsymbol{\alpha})$ in the form of Eq. (3.3–48) has $m + n$ simple roots s_i $(i = 1, 2, \ldots, m + n)$ that are continuous functions of r parameters α_j $(j = 1, 2, \ldots, r)$. Then the $(m + n) \times r$ matrix

$$\tilde{\boldsymbol{S}}_r = \begin{bmatrix} \tilde{S}_{\alpha_1}^{s_1} & \cdots & \tilde{S}_{\alpha_r}^{s_1} \\ \tilde{S}_{\alpha_1}^{s_{m+n}} & \cdots & \tilde{S}_{\alpha_r}^{s_{m+n}} \end{bmatrix} \tag{3.3–59}$$

is called the (semirelative) root sensitivity matrix.

Let us now consider *λ-variations.* In this case a variation of the parameter increases the order of the characteristic equation. The corresponding poles go to infinity, and the pole sensitivities become infinite.

To show this, consider first the simple case that the poles are real and negative and that the order of the nominal characteristic equation

$$Q_0(s) = a_n s^n + a_{n-1} s^{n-1} + \ldots + a_0 = 0 \tag{3.3–60}$$

increases by one as λ deviates from $\lambda_0 = 0$ by an infinitesimal amount $\Delta\lambda$. Thus, actually,

$$Q(s, \Delta\lambda) = \Delta\lambda\, s^{n+1} + a_n s^n + \ldots + a_0 = 0, \tag{3.3–61}$$

where $\lim_{\Delta\lambda \to 0} a_\nu = a_{\nu 0}$, $\nu = 0, 1, \ldots, n$. Dividing by $\Delta\lambda s^n$ gives

$$s = -\frac{1}{\Delta\lambda}\left(a_n + a_{n-1}\frac{1}{s} + \cdots + a_0\frac{1}{s^n}\right) \tag{3.3–62}$$

Now letting $\Delta\lambda$ approach zero and denoting the corresponding pole by s_{n+1}, we obtain

$$\begin{aligned} s_{n+1} &= \lim_{\Delta\lambda \to 0}\left[-\frac{1}{\Delta\lambda}\left(a_n + \frac{a_{n-1}}{s_{n+1}} + \cdots + \frac{a_0}{s_{n+1}^n}\right)\right] \\ &= \lim_{\Delta\lambda \to 0}\left(-\frac{a_n}{\Delta\lambda}\right) = -\infty. \end{aligned} \tag{3.3–63}$$

Thus for sufficiently small $\Delta\lambda$ we can write

$$s_{n+1} \approx -a_n/\Delta\lambda \triangleq s_{n+1}^*. \tag{3.3–64}$$

More generally, if the parameter-induced increase of the order of the characteristic equation is greater than one, say m, then the actual characteristic polynomial $Q(s, \Delta\lambda)$ may be split up into two portions:

$$Q(s, \Delta\lambda) = \underbrace{a_0 + a_1 s + \cdots + a_n s^n}_{Q_0(s)} + \underbrace{a_{n+1}(\Delta\lambda)s^{n+1} + \cdots + a_{n+m}(\Delta\lambda)s^{n+m}}_{Q_1(s,\Delta\lambda)}, \tag{3.3–65}$$

where the $a_i(\Delta\lambda)$ $(i = n+1, \ldots, n+m)$ vanish for $\Delta\lambda = 0$. Thus

$$Q(s, \Delta\lambda) = Q_0(s) + Q_1(s, \Delta\lambda) = 0, \tag{3.3–66}$$

where $Q_0(s)$ is the nominal characteristic polynomial and $Q_1(s, \Delta\lambda)$ is the parameter-induced part of $Q(s, \Delta\lambda)$ containing the m additional terms. Now we make the following assumptions:

(1) $Q_1(s, \Delta\lambda) \to 0$ as $\Delta\lambda \to 0$.
(2) The coefficients of $Q_1(s, \Delta\lambda)$ are analytic in the vicinity of $\lambda_0 = 0$.

Then, as $\Delta\lambda$ approaches zero,
(i) n roots of $Q(s, \Delta\lambda)$ approach the roots of $Q_0(s)$, and
(ii) the magnitudes of the remaining m roots approach infinity.

Proof It follows from the first assumption that $Q_1(s, \Delta\lambda)$ approaches zero as $\Delta\lambda$ approaches zero. This proves statement (i). To prove (ii), consider the roots of $Q(s_j, \Delta\lambda)$. If there should be any *finite* root s_j other than the roots of $Q_0(s)$, then

$$Q(s_j, \Delta\lambda) = Q_0(s_j) + Q_1(s_j, \Delta\lambda) = 0. \tag{3.3–67}$$

Since for $\Delta\lambda \to 0$ all the coefficients of $Q_1(s_j, \Delta\lambda)$ approach zero and s_j is finite, $Q_1(s_j, \Delta\lambda)$ must approach zero. Hence $Q_0(s_j) = 0$, which contradicts the assumption that s_j is not a root of $Q_0(s)$. Since the m extra roots must go somewhere except to finite values, they must go to infinity, which proves the second part of the statement.

In addition, as $\Delta\lambda$ approaches zero, the magnitudes of the m extra roots, as well as their real and imaginary parts, approach powers or fractional powers of $1/\Delta\lambda$, and the residues of the m roots cannot approach infinity faster than a finite power of $1/\Delta\lambda$. This is due to the fact that by assumption (2) every coefficient of $Q_1(s, \Delta\lambda)$ can be approximated by the first nonvanishing power in its Laurent series expansion.

The same considerations apply to the numerator polynomial of the transfer function $G(s,\Delta\lambda)$.

Now the semirelative root sensitivities are defined as

$$\tilde{S}_{\lambda}^{s_j} \triangleq \lim_{\Delta\lambda \to 0} \frac{\partial s_j^*}{\partial \ln \lambda}, \tag{3.3–68}$$

where the $s_j^{*\prime}$s represent the above mentioned approximations by the powers or fractional powers of the Laurent series expansion. With these approximations, the above limit approaches infinity as $\Delta\lambda$ approaches zero. Therefore, $|\tilde{S}_{\lambda}^{s_j}| = \infty$ for λ-variations. For example, the semirelative root sensitivity of s_{n+1} in Eq. (3.3–64) is

$$\tilde{S}_{\lambda}^{s_{n+1}} = \lim_{\Delta\lambda \to 0} \frac{a_n \Delta\lambda}{\Delta\lambda^2} = \lim_{\Delta\lambda \to 0} \frac{a_n}{\Delta\lambda} = \infty. \tag{3.3–69}$$

Example 3.3-13 Suppose the nominal polynomial $Q_0(s)$ of the characteristic equation is

$$Q_0(s) = 1 + 2Ds + s^2.$$

For $\Delta\lambda \neq 0$, the polynomial $Q_1(s, \Delta\lambda)$ may take any of the following forms:

(1) $\Delta\lambda(s^2 + s^3)$,
(2) $\Delta\lambda(s^3 + s^4)$,
(3) $\Delta\lambda\, s^3 + \Delta\lambda^2 s^4$,
(4) $\Delta\lambda s^5$.

Determine the roots s_j and the root sensitivities for $\Delta\lambda = 0$. The results are listed in Table 3.3–2. This result verifies that the root sensitivities become infinite in the case of λ-variations.

There is an important relationship between Bode's sensitivity function $S_{\alpha}^{G}(j\omega)$ and the root sensitivities of a system whose transfer function $G(s, \boldsymbol{\alpha})$ is in the form of Eq. (3.3–48). In order to discover this relationship, let us

TABLE 3.3-2

Case	Approximation for $Q(s, \Delta\lambda)$	Approximation of roots for $\Delta\lambda \to 0$, $s \to \infty$	Root sensitivity for $\Delta\lambda \to 0$
(1)	$s^2 + \Delta\lambda\, s^3$	$-\frac{1}{\Delta\lambda}$	Infinite
(2)	$s^2 + \Delta\lambda\, s^3 + \Delta\lambda\, s^4$	$-\frac{1}{2} \pm j\sqrt{\frac{1}{\Delta\lambda}}$	Infinite
(3)	$s^2 + \Delta\lambda\, s^3 + \Delta\lambda^2\, s$	$-\frac{1}{2\Delta\lambda} \pm j\frac{\sqrt{3}}{2\Delta\lambda}$	Infinite
(4)	$s^2 + \Delta\lambda\, s^5$	$-\Delta\lambda^{-1/3}$	
		$\frac{1}{2}\Delta\lambda^{-1/3} \pm j\frac{\sqrt{3}}{1}\Delta\lambda^{-1/3}$	Infinite

assume that the roots of G are continuous functions of a parameter vector $\boldsymbol{\alpha}$ (α-variations). Taking the logarithms on both sides of Eq. (3.3–48), we have

$$\ln G(s, \boldsymbol{\alpha}) = \ln K(\boldsymbol{\alpha}) + \sum_{\mu=1}^{m} \ln (s - s_\mu{}^0(\boldsymbol{\alpha})) - \sum_{\nu=1}^{n} \ln (s - s_\nu(\boldsymbol{\alpha})). \quad (3.3\text{–}70)$$

Partial differentiation with respect to $\ln \alpha_j$ yields

$$\frac{\partial \ln G(s, \boldsymbol{\alpha})}{\partial \ln \alpha_j} = \frac{\partial \ln K(\boldsymbol{\alpha})}{\partial \ln \alpha_j} + \sum_{\mu=1}^{m} \frac{1}{s - s_\mu^0} \frac{-\partial s_\mu{}^0}{\partial \ln \alpha_j} - \sum_{\nu=1}^{n} \frac{1}{s - s_\nu} \frac{-\partial s_\nu}{\partial \ln \alpha_j}. \quad (3.3\text{–}71)$$

Now setting $\boldsymbol{\alpha} = \boldsymbol{\alpha}_0$ and making use of Definition 3.3–6, we obtain

$$S_{\alpha_j}^G = S_{\alpha_j}^K - \sum_{\mu=1}^{m} \frac{1}{s - s_\mu^0} \tilde{S}_{\alpha_j}^{s_\mu^0} + \sum_{\nu=1}^{n} \frac{1}{s - s_{\nu 0}} \tilde{S}_{\alpha_j}^{s_\nu}, \quad (3.3\text{–}72)$$

where $s_{\mu 0}^0 = s_\mu{}^0(\boldsymbol{\alpha}_0)$ and $s_{\nu 0} = s_\nu(\boldsymbol{\alpha}_0)$ are the nominal roots of $G(s, \boldsymbol{\alpha}_0)$.

This form of $S_{\alpha_j}^G$ may be interpreted as the partial fraction expansion of the Bode sensitivity function, where the root sensitivities appear as its residues at $s = s_\mu{}^0$ and $s = s_\nu$, respectively. In a shorter notation, Eq. (3.3–72) can be rewritten as

$$S_{\alpha_j}^G(s) = \sum_{i=1}^{m+n} \frac{h_{ij}}{s - s_i} + h_{m+n+1,j}, \qquad j = 1, 2, \ldots, r, \quad (3.3\text{–}73)$$

where the h_{ij}'s ($i = 1, 2, \ldots, m + n$) are the residues of $S_{\alpha j}^G$ at the root locations $s = s_i$, i.e., the corresponding root sensitivities, and $h_{m+n+1,j}$ is a constant term.

Equations (3.3–72) and (3.3–73) reveal that each root sensitivity contributes to the Bode sensitivity function, weighted by a frequency-dependent factor that tends to infinity as s tends to the corresponding root location. In other words, the effect of a root sensitivity $S_{\alpha_j}^{s_i}(j\omega)$ is strongest in the close vicinity around the corresponding root location.

When some of the roots of $G(s, \boldsymbol{\alpha})$ are repeated, the root sensitivities are not uniquely defined, since a multiple root coincides with a breakaway point in the root locus. Nevertheless, Eqs. (3.3–72) and (3.3–73) also hold in this case, but the poles s_i ($i = 1, 2, \ldots, m + n$) of the Bode sensitivity function are now the $m + n$ distinct roots of the transfer function $G(s, \boldsymbol{\alpha})$.

Formulas for the evaluation of the root sensitivities are given in Chapter 6.

Finally, it should be noted that the above results are valid whether the system is described a priori in terms of inputs and outputs or by its state equations

$$\dot{\boldsymbol{x}} = \boldsymbol{A}\boldsymbol{x} + \boldsymbol{B}u, \qquad \boldsymbol{x}(0) = \boldsymbol{x}^0 = \boldsymbol{0}, \quad (3.3\text{–}74)$$

$$y = \boldsymbol{C}\boldsymbol{x}, \quad (3.3\text{–}75)$$

where in general $\boldsymbol{A} = \boldsymbol{A}(\boldsymbol{\alpha})$, $\boldsymbol{B} = \boldsymbol{B}(\boldsymbol{\alpha})$, $\boldsymbol{C} = \boldsymbol{C}(\boldsymbol{\alpha})$. The corresponding transfer function of the system can easily be determined by means of the Laplace transform:

$$G(s, \boldsymbol{\alpha}) = \boldsymbol{C}(s\boldsymbol{I} - \boldsymbol{A})^{-1} \boldsymbol{B}. \tag{3.3–76}$$

This expression may be used alternatively instead of Eq. (3.3–48). However, the situation is now different insofar as the parameters in the latter case are defined in terms of the matrices $\boldsymbol{A}$, $\boldsymbol{B}$, $\boldsymbol{C}$. Therefore, the corresponding pole sensitivities are identical with the eigenvalue sensitivities treated in Section 3.2.8. Formulas for the evaluation of the eigenvalue sensitivities are given in Chapter 5; additional formulas for the determination of the root sensitivities to the parameters of the frequency domain are presented infection 6.3.

3.3.5 Scalar Sensitivity Measure of Biswas and Kuh

In 1971 Biswas and Kuh [27] proposed a multiparameter sensitivity measure that takes into account all root sensitivities of the system and thereby allows the global estimation of the system sensitivity over the whole frequency range. This sensitivity measure will be derived in this section.

Let $\Delta G(s, \boldsymbol{\alpha})$ be the change of the transfer function $G(s, \boldsymbol{\alpha}_0)$ due to small variations $\Delta \boldsymbol{\alpha}_j (j = 1, 2, \ldots, r)$ of its parameters. Then the first-order approximation of the relative parameter-induced change of G is

$$\frac{\Delta G(s, \boldsymbol{\alpha})}{G(s, \boldsymbol{\alpha}_0)} = \sum_{j=1}^{r} S_{\alpha j}^{G}(s) \frac{\Delta \alpha_j}{\alpha_{j0}}. \tag{3.3–77}$$

Denoting

$$\varepsilon_j \triangleq \frac{\Delta \alpha_j}{\alpha_{j0}}$$

and writing $S_{\alpha j}^{G}(s)$ in the form of Eq. (3.3–73), we obtain

$$\frac{\Delta G(s, \boldsymbol{\alpha})}{G(s, \boldsymbol{\alpha}_0)} = \sum_{j=1}^{r} \left(\sum_{i=1}^{m+n} \frac{h_{ij}}{s - s_i} + h_{m+n+1,j} \right) \varepsilon_j$$

$$= \sum_{i=1}^{m+n} \sum_{j=1}^{r} \frac{h_{ij}}{s - s_i} \varepsilon_j + \sum_{j=1}^{r} h_{m+n+1,j}\, \varepsilon_j. \tag{3.3–78}$$

With the abbreviation

$$f_i \triangleq \sum_{j=1}^{r} h_{ij} \varepsilon_j, \qquad i = 1, 2, \ldots, m + n + 1, \tag{3.3–79}$$

Eq. (3.3–78) can be written as

$$\frac{\Delta G(s, \boldsymbol{\alpha})}{G(s, \boldsymbol{\alpha}_0)} = \sum_{i=1}^{m+n} \frac{f_i}{s - s_i} + f_{m+n+1}. \tag{3.3–80}$$

Note that f_i give the displacements of the locations of the simple roots s_i of $G(s, \boldsymbol{\alpha})$. In matrix notation, Eq. (3.3–79) may be rewritten as

$$\boldsymbol{f} = \boldsymbol{H}\,\boldsymbol{\varepsilon}, \tag{3.3–81}$$

where

$$\boldsymbol{f} \triangleq [f_1 \quad f_2 \quad \cdots \quad f_{m+n+1}]^{\mathrm{T}},$$
$$\boldsymbol{\varepsilon} \triangleq [\varepsilon_1 \quad \varepsilon_2 \quad \cdots \quad \varepsilon_r]^{\mathrm{T}}$$

are the root perturbation vector and the parameter perturbation vector, respectively, and the $(m + n + 1 \times r)$ matrix

$$\boldsymbol{H} \triangleq \begin{bmatrix} h_{11} & \cdots & h_{1r} \\ \vdots & & \vdots \\ h_{m+n+1,1} & \cdots & h_{m+n+1,r} \end{bmatrix} \tag{3.3–82}$$

is the extended root sensitivity matrix of Definition 3.3–11 which characterizes the overall sensitivity property of the complete system in terms of the root sensitivities and the gain factor sensitivity with respect to all parameters. Note that, in general, $\boldsymbol{H}$ is a complex matrix, since the h_{ij} may be complex.

A scalar sensitivity measure of the effect of all parameter variations on the system behavior must be a norm of the vector $\boldsymbol{f}$. The simplest one is the Euclidean norm

$$\|\boldsymbol{f}\| \triangleq (|f_1|^2 + |f_2|^2 + \cdots + |f_{m+n+1}|^2)^{1/2}. \tag{3.3–83}$$

In this definition, all root sensitivities are equally weighted. In practice, however, it might be advantageous to assign different weights to the different roots. Therefore, the following sensitivity measure will be used:

Definition 3.3-12 *Scalar sensitivity measure of Biswas–Kuh.* Let $\mathrm{Re}\,\boldsymbol{f}$ and $\mathrm{Im}\,\boldsymbol{f}$ denote the real and imaginary parts of $\boldsymbol{f}$ in Eq. (3.3–81), respectively, and let

$$\boldsymbol{A} = \mathrm{diag}\,(a_1, a_2, \ldots, a_{m+n+1}),$$
$$\boldsymbol{B} = \mathrm{diag}\,(b_1, b_2, \ldots, b_{m+n+1})$$

be two diagonal weighting matrices with all elements real and positive. Then the following norm of $\boldsymbol{f}$,

$$m \triangleq [(\mathrm{Re}\,\boldsymbol{f})^{\mathrm{T}} \boldsymbol{A}(\mathrm{Re}\,\boldsymbol{f}) + (\mathrm{Im}\,\boldsymbol{f})^{\mathrm{T}} \boldsymbol{B}(\mathrm{Im}\,\boldsymbol{f})]^{1/2}, \tag{3.3–84}$$

is called the sensitivity measure of Biswas–Kuh.

The advantage of having two separate weighting matrices $\boldsymbol{A}$ and $\boldsymbol{B}$ for the real and imaginary parts of $\boldsymbol{f}$ is that by a suitable choice of $\boldsymbol{A}$ and $\boldsymbol{B}$, one can not only assign different weights to the displacements of different roots but can also discriminate between different directions of the root displacements in the complex s-plane. However, whether a system proves to be sensitive or not strongly depends on the choice of the weighting matrices. Therefore, the proper choice of $\boldsymbol{A}$ and $\boldsymbol{B}$ is problematical in practice.

3.3.6 Maximum-Modulus Sensitivity

In a manner similar to the way time was excluded from the sensitivity function in the time domain, frequency can be excluded from the sensitivity function in the frequency domain by defining sensitivity in terms of the maximum modulus of the frequency response.

Consider the magnitude of the frequency response versus frequency ω as shown in Fig. 3.3–11. Then we define the following:

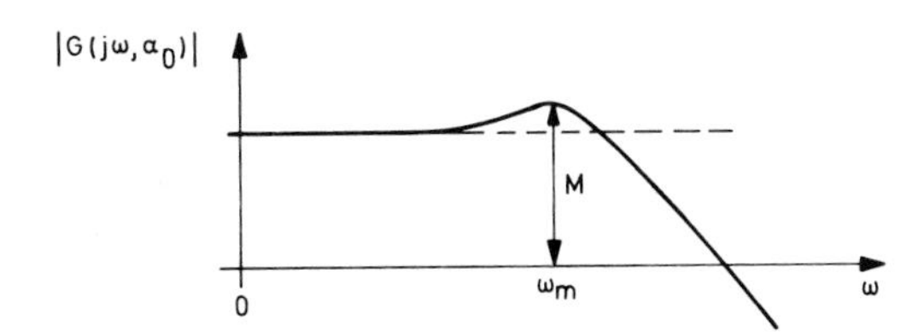

FIG. 3.3-11. Definition of the maximum modulus M.

Definition 3.3-13 *Maximum-modulus sensitivity.* Let $|G(j\omega, \boldsymbol{\alpha}_0)|$ be the nominal magnitude of the frequency response $G(j\omega, \boldsymbol{\alpha})$ and let $M = \max_\omega |G(j\omega, \boldsymbol{\alpha}_0)|$ be its maximum value with respect to ω. Denote the frequency corresponding to M by ω_m. Then the partial derivative

$$S_m \triangleq \left.\frac{\partial M}{\partial \ln \alpha_j}\right|_{\boldsymbol{\alpha}_0} = \left.\frac{\partial |G(j\omega_m, \boldsymbol{\alpha})|}{\partial \ln \boldsymbol{\alpha}_j}\right|_{\boldsymbol{\alpha}_0} \tag{3.3–85}$$

is called the maximum-modulus sensitivity of the system.

This sensitivity definition is less informative than overshoot sensitivity since a system is not completely described by using the magnitude of $G(j\omega, \boldsymbol{\alpha})$ only. Moreover, it is not easy to make conclusions from S_m for the system behavior in the time domain.

Let G and $\bar{G}$ represent $G(j\omega_m, \boldsymbol{\alpha})$ and $G(-j\omega_m, \boldsymbol{\alpha})$, respectively. Then

$$\begin{aligned} S_m &= \left.\frac{\partial M}{\partial \ln \alpha_j}\right|_{\boldsymbol{\alpha}_0} = \left.\frac{1}{2M}\frac{\partial (G\bar{G})}{\partial \ln \alpha_j}\right|_{\boldsymbol{\alpha}_0} = -\left.\frac{M}{2}\frac{\partial \ln (G^{-1}\bar{G}^{-1})}{\partial \ln \alpha_j}\right|_{\boldsymbol{\alpha}_0} \\ &= -\frac{M}{2}\left[\bar{G}\frac{\partial \bar{G}^{-1}}{\partial \ln \alpha_j} + G\frac{\partial G^{-1}}{\partial \ln \alpha_j}\right]_{\boldsymbol{\alpha}_0} \\ &= -M \text{ real part of } \left(\frac{\partial G^{-1}}{\partial \ln \alpha_j}\right)_{\boldsymbol{\alpha}_0}. \end{aligned} \tag{3.3–86}$$

This formula may be helpful in evaluating S_m.

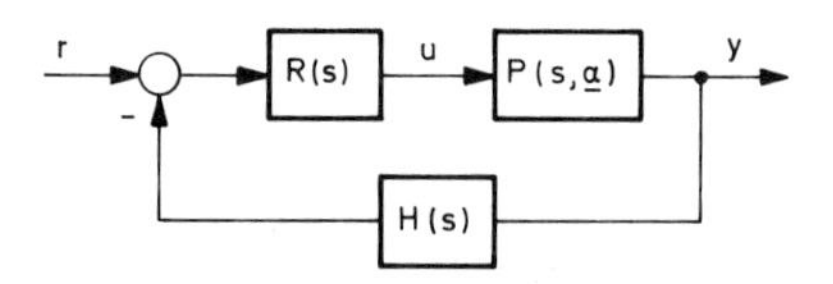

FIG. 3.3-12. Feedback control system of Example 3.3-14.

Example 3.3-14 Given the feedback control system shown in Fig. 3.3–12 with $H(s) = 1$ and the open-loop transfer function

$$L(s,\ \boldsymbol{\alpha}) = R(s)P(s,\ \boldsymbol{\alpha}) = \frac{10(1 + 0.5s)\ (1 + s)}{(1 + 0.1s)\ (1 + 10s)} \frac{k}{(1 + T_1 s)(1 + T_2 s)}.$$

Let $\boldsymbol{\alpha} = [k\ T_1\ T_2]^{\mathrm{T}}$ be the parameter vector, where the nominal parameter values are $k_0 = 10$, $T_{10} = 0.5$, and $T_{20} = 1$. Determine the maximum-modulus sensitivity of the closed loop with respect to a change in T_1.

With the above nominal parameter values, the open-loop transfer function becomes

$$L(s,\ \boldsymbol{\alpha}_0) = \frac{100}{(1 + 0.1s)\ (1 + 10s)},$$

and the corresponding closed-loop transfer function is

$$G(s,\ \boldsymbol{\alpha}_0) = \frac{100}{s^2 + 10.1s + 101}.$$

Therefore, $\omega_0 = \sqrt{101} = 10.05,\ \zeta = 0.5025$; whence

$$\omega_m = \omega_0\sqrt{1 - 2\zeta^2} = 7.08$$

and

$$M = \frac{100}{|(j\omega_m)^2 + 10.1(j\omega_m) + 101|} = 1.14.$$

For the control loop under consideration, Eq. (3.3–86) can be written as

$$S_m = \left.\frac{\partial M}{\partial \ln \alpha_j}\right|_{\boldsymbol{\alpha}_0} = -M \times \text{real part of} \left(G \frac{\partial L^{-1}}{\partial \ln T_1}\right)_{\boldsymbol{\alpha}_0}$$

$$= M \times \text{real part of} \left[\frac{1}{1 + L(j\omega_m, \boldsymbol{\alpha})} \frac{\partial \ln L(j\omega_m,\ \boldsymbol{\alpha})}{\partial \ln \alpha_j}\right]_{\boldsymbol{\alpha}_0}.$$

Evaluating

$$L(j\omega_m,\ \boldsymbol{\alpha}_0) = \frac{100}{(1 + j0.708)\ (1 + j70.8)} = 1.153\underline{/-124.5}$$

and

$$\left.\frac{\partial \ln L(j\omega_m,\ \boldsymbol{\alpha})}{\partial \ln T_1}\right|_{\boldsymbol{\alpha}_0} = \frac{-j\omega_m T_{10}}{1 + T_{10} j\omega_m} = 0.96\underline{/-164.2},$$

one obtains finally

$$S_m = M \times \text{real part of } (0.95 < -94.2) = -0.08.$$

The result indicates that the effect of a relative change in T_1 on the maximum modulus M is $\Delta M = -0.08\ \Delta T_1/T_{10}$.

3.4 PERFORMANCE-INDEX SENSITIVITY

Dynamic systems, especially control systems, are often designed such that a certain performance index takes on either an optimal or preassigned value. In these cases the quality of the system is characterized by the performance index, and it is quite logical that then the sensitivity measure of interest is the so-called performance-index sensitivity which is defined as follows:

Definition 3.4-1 *Performance-index sensitivity.* Let J be the performance index used for the design of the system, and assume that J is, among other dependences, a continuous function of a parameter α. Then the partial derivative

$$J_\alpha \triangleq \left.\frac{\partial J}{\partial \alpha}\right|_{\alpha 0} \tag{3.4-1}$$

is called the performance-index sensitivity. If J depends on a parameter vector $\boldsymbol{\alpha} = [\alpha_1\ \alpha_2\ \ldots\ \alpha_r]^{\mathrm{T}}$, then the row vector

$$\boldsymbol{J}_{\boldsymbol{\alpha}} \triangleq \left[\left.\frac{\partial J}{\partial \alpha_1}\right|_{\boldsymbol{\alpha}_0} \cdots \left.\frac{\partial J}{\partial \alpha_r}\right|_{\boldsymbol{\alpha}_0}\right] = [J_{\alpha 1} \cdots J_{\alpha_r}] \tag{3.4-2}$$

is called the performance- index sensitivity vector.

Performance-index sensitivity was introduced in 1963 by Dorato [36]. Subsequently, it was the subject of many publications [5, 75, 98, 109, 110] which exposed a number of problems created by the use of this sensitivity measure. Two problems will be outlined here.

In Definition 3.4-1, J is taken quite generally. For example, J can even represent the overshoot of the step response when the system is designed to produce minimum overshoot. However, the most important performance indices, especially in control engineering, are the integral criteria or cost functionals. Hence performance-index sensitivity is commonly used in connection with integral criteria or cost functionals. The most frequently employed integral criteria are

(1) the integral squared error, that is, the integral of the squared control error $\varepsilon = y - r$ (y output, r reference input) of a feedback control system,

$$J = \int_{t0}^{tf} \varepsilon^2\, dt, \tag{3.4-3}$$

or similar expressions such as

$$J = \int_{t_0}^{t_f} |\varepsilon|\, dt \qquad \text{or} \qquad J = \int_{t_0}^{t_f} \varepsilon^2 t\, dt;$$

(2) the general cost functional of the *Bolza type*,

$$J(\boldsymbol{u}, \boldsymbol{\alpha}) = M(\boldsymbol{x}, t, \boldsymbol{\alpha})_{t_f} + \int_{t_0}^{t_f} L(\boldsymbol{x}, \boldsymbol{u}, t, \boldsymbol{\alpha})\, dt, \tag{3.4–4}$$

where $\boldsymbol{x}$ is the state vector, $\boldsymbol{u}$ the input vector, $\boldsymbol{\alpha}$ a parameter vector of the system to be optimized, and M and L are scalar functions. Mostly M and L are defined as quadrratic forms. In this case, Eq. (3.4–4) reads

$$J(\boldsymbol{u}, \boldsymbol{\alpha}) = \tfrac{1}{2}\, \boldsymbol{x}_f{}^{\mathrm{T}}\, \boldsymbol{S}\, \boldsymbol{x}_f + \tfrac{1}{2} \int_{t_0}^{t_f} (\boldsymbol{x}^{\mathrm{T}} \boldsymbol{Q}\, \boldsymbol{x} + \boldsymbol{u}^{\mathrm{T}} \boldsymbol{R}\, \boldsymbol{u})\, dt, \tag{3.4–5}$$

where $\boldsymbol{S}$, $\boldsymbol{Q}$ are symmetrical positive semidefinite weighting matrices, $\boldsymbol{R}$ is a symmetrical positive definite weighting matrix, and $\boldsymbol{x}_f$ is the final value of $\boldsymbol{x}$. In general, $\boldsymbol{x} = \boldsymbol{x}(t,\boldsymbol{\alpha})$, $\boldsymbol{u} = \boldsymbol{u}(t,\boldsymbol{\alpha})$, $\boldsymbol{S} = \boldsymbol{S}(\boldsymbol{\alpha})$, $\boldsymbol{Q} = \boldsymbol{Q}(t,\boldsymbol{\alpha})$, and $\boldsymbol{R} = \boldsymbol{R}(t,\boldsymbol{\alpha})$ may be functions of $\boldsymbol{\alpha}$. The α-dependence of $\boldsymbol{x}$ and $\boldsymbol{u}$ is of interest in the case of systems with varying parameters, whereas the α-dependence of $\boldsymbol{S}$, $\boldsymbol{Q}$, $\boldsymbol{R}$ may be of concern when favorable values for $\boldsymbol{S}$, $\boldsymbol{Q}$, $\boldsymbol{R}$ are being sought.

Special forms of the general expression (3.4–5) are the cost functional for *final value optimization* (Meyer problem), where $L = 0$, i.e.,

$$J(\boldsymbol{u}, \boldsymbol{\alpha}) = \boldsymbol{x}^{\mathrm{T}}(t_f, \boldsymbol{\alpha}) \boldsymbol{S}(\alpha) \boldsymbol{x}(t_f, \boldsymbol{\alpha}), \tag{3.4–6}$$

the cost functional for *optimum linear regulators* (Lagrange problem), where $M = 0$, i.e.,

$$J(\boldsymbol{u}, \boldsymbol{\alpha}) = \int_{t_0}^{t_f} (\boldsymbol{x}^{\mathrm{T}} \boldsymbol{Q}\, \boldsymbol{x} + \boldsymbol{u}^{\mathrm{T}} \boldsymbol{R}\, \boldsymbol{u})\, dt, \tag{3.4–7}$$

and the cost functional for *minimum energy consumption* (special Lagrange problem), where $M = 0$ and $\boldsymbol{Q} = \boldsymbol{O}$, i.e.,

$$J(\boldsymbol{u}, \boldsymbol{\alpha}) = \int_{t_0}^{t_f} \mathbf{u}^{\mathrm{T}} \mathbf{R}\, \mathbf{u}\, dt. \tag{3.4–8}$$

The cost function for time optimality characterized by $M = 0$ and $L = 1$ cannot be used in Definition 3.4–1 since in this case J_α or $\boldsymbol{J}_\alpha$ do not make any sense. This already reveals the difficulties that can arise in evaluating the performance-index sensitivity. Another shortcoming is that performance-index sensitivity is of no use for the purpose of comparison of optimal open- and closed-loop control systems, even if these configurations are nominally equivalent. This result, first discovered by Pagurek [75] and later generalized by Witsenhausen [109], and therefore known as the Pagurek–Witsenhausen paradox, will be treated in more detail in Chapter 7.

PROBLEMS

3.1 Show the validity of Eq. (3.3–20).

3.2 Show that the system with the transfer function $G(s) = 1/(1 + s^2/\omega^2)$ and the unit step function $u(t) = 1(t)$ as a driving function has an output sensitivity function $\sigma = (\partial y/\partial\omega)_{\omega 0}$ that diverges with time.

3.3 Given a system with the differential equation $\ddot{y} + 2\delta\omega\dot{y} + \omega^2 y = \omega^2 1(t)$, $\dot{y}(0) = y(0) = 0$. Determine the values of the damping ratio δ for which the output sensitivity function $\sigma = (\partial y/\partial\omega)_{\omega 0}$ diverges. Discuss the result in terms of stability.

3.4 Given the control loop shown in Fig. 3.P-1, where T_0 represents the nominal time constant of the plant: Determine the relative change of the output $\Delta y/y_0$ due to a relative change of the time constant $\Delta T/T_0$.

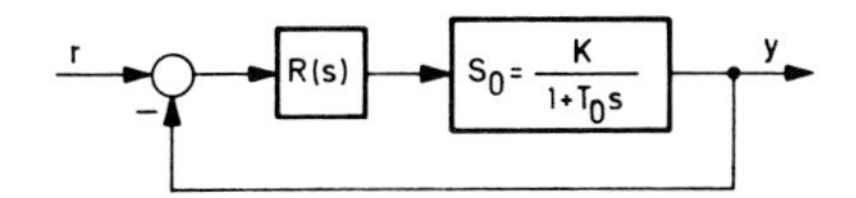

FIG. 3.P-1. Control loop of Problem 3.4.

3.5 Referring to Example 3.3–5, determine the Bode sensitivity functions with respect to a_2 and a_3.

3.6 Determine the Bode sensitivity function $S_C{}^G$ of the transfer function $G(s, C) = U_2(s,C)/U_1(s)$ of the network shown in Fig. 3.P-2 with respect to the capacitance C.

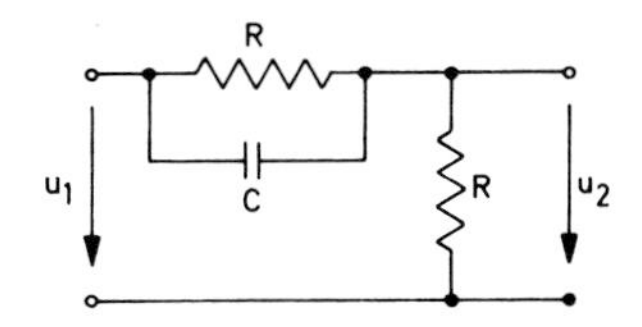

FIG. 3.P-2. Network of Problem 3.6.

3.7 For the feedback configurations (a), (b), and (c) in Fig. 3.P-3, where G_1, G_2, G_3, G_4 are transfer functions and K is a constant,
(i) determine the Bode sensitivity functions of $G(s) = Y(s)/U(s)$ with respect to G_2,
(ii) determine the comparison sensitivity function of $G(s) = Y(s)/U(s)$ with respect to G_2,
(iii) give the conditions for zero sensitivity.

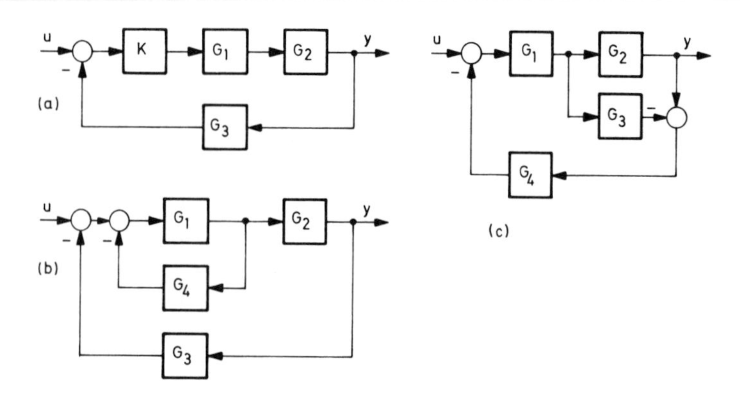

FIG. 3.P-3. Feedback configurations of Problem 3.7.

3.8 Given configuration (a) in Fig. 3.P-4. Suppose that the gain factor K_1 varies between $K/2 \leq K_1 \leq 2K$. Determine h and K_2 of configuration (b) such that the overall transfer function $G(s) = Y(s)/U(s)$ varies by 20% as K_1 varies from $K/2$ to $2K$ and both configurations are equivalent for $K_1 = K$.

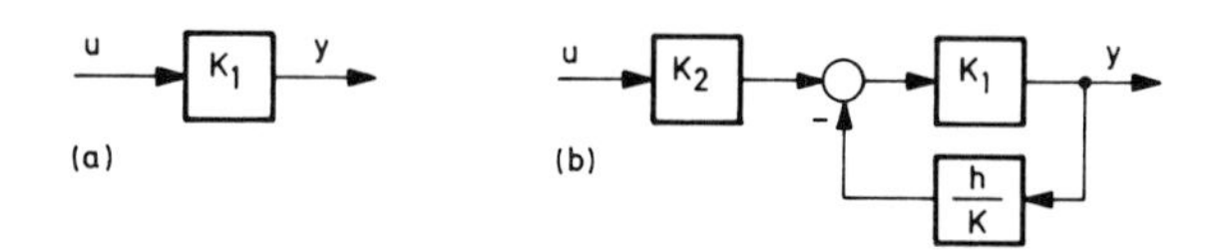

FIG. 3.P-4. Two configurations of Problem 3.8 that are equivalent for $K_1 = K$.

3.9 A double-terminated two-port network characterized by its h-parameters couples a voltage source of internal impedance Z_S to a load Z_L (Fig. 3.P-5). Determine the Bode sensitivity function of the overall gain $K = E_L/E_S$ with respect to the h-parameters h_{11}, h_{12}, h_{21}, h_{22}.

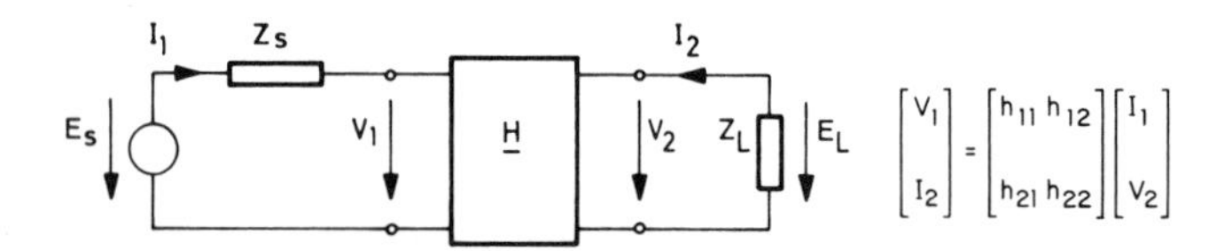

FIG. 3.P-5. Network diagram and mathematical model of the network of Problem 3.9.

3.10 Given the low-pass filter shown in Fig. 3.P-6. It is assumed that the tolerance of C is close compared to R. Determine and sketch the output sensitivity function with respect to R for the following two cases:

(a) $u_1 = \delta(t)$ (impulse function),

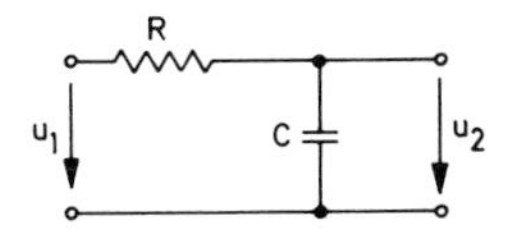

FIG. 3.P-6. Low pass of Problem 3.10.

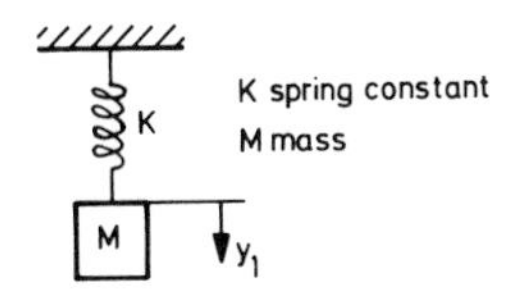

FIG. 3.P-7. Mass–spring system of Problem 3.11.

(b) $u_1 = 1(t)$ (unit step function).
Discuss the results.

3.11 Given the mass–spring system of Fig. 3.P-7, where the friction and mass of the spring are neglected. Let the output vector $\mathbf{y} = [y_1\ y_2]^{\mathrm{T}}$ be defined by the displacement y_1 and the speed $y_2 = \dot{y}_1$. Determine the output sensitivity matrix with respect to the α-parameters K and M, provided that the initial conditions are $y_1(0) = 0$, $y_2(0) = 5$.

3.12 In the bridge circuit shown in Fig. 3.P-8, R_1 is a resistance strain gauge that changes its resistance according to $R_1 = R_{10} + \Delta R_1$. Consider the bridge voltage as the output and determine

(a) the output sensitivity function $\sigma(R_{10})$,
(b) the parameter-induced output error $\Delta U_{12}(R_1)$,
(c) the semirelative output sensitivity function $\tilde{\sigma}(R_{10})$,
(d) Given a certain R_{10}, determine the value of R for which the output sensitivity function $\sigma(R_{10})$ is maximal.

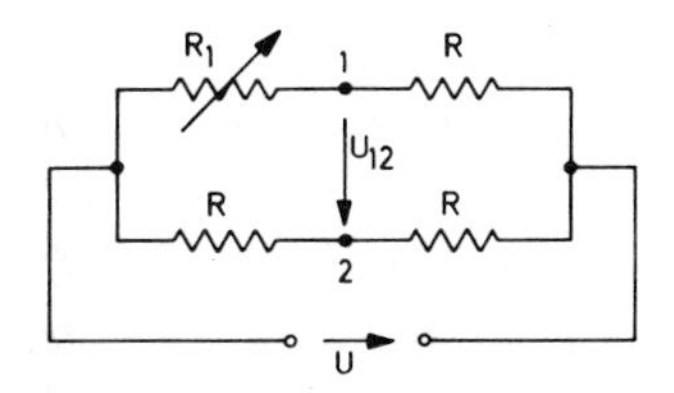

FIG. 3.P-8. Bridge circuit of Problem 3.12.

3.13 For the block diagram shown in Fig. 3.P-9 with the state vector defined by $\mathbf{x} = [x_1\ x_2]^{\mathrm{T}}$ and the parameter vector defined as $\boldsymbol{\alpha} = [K\ T]^{\mathrm{T}}$, determine the trajectory sensitivity matrix $\boldsymbol{\lambda} = (\partial\mathbf{x}/\partial\boldsymbol{\alpha})_{\boldsymbol{\alpha}0}$, assuming that $d < 1$ and all initial conditions are equal to zero.

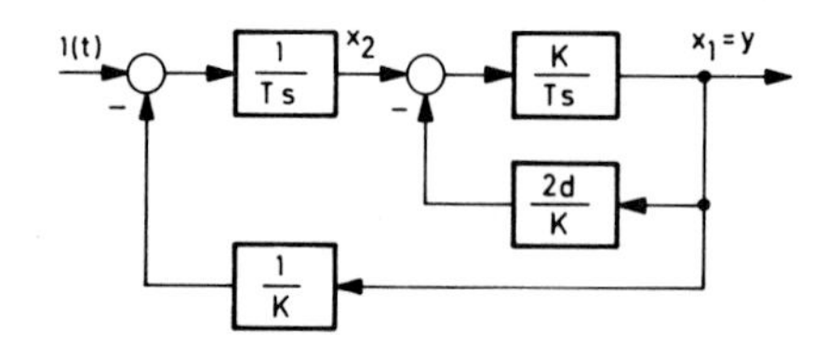

FIG. 3.P-9. Block diagram of Problem 3.13.

3.14 Consider the normalized analog computer diagram of an oscillator shown in Fig. 3.P-10.

(a) Determine the output function $y(t,\boldsymbol{\alpha})$, where $\boldsymbol{\alpha} = [\alpha_1\ \alpha_2\ \alpha_3]^{\mathrm{T}}$.

(b) Calculate the output sensitivity vector $\boldsymbol{\sigma}_1(t,\boldsymbol{\alpha}_0)$ with respect to all parameters.

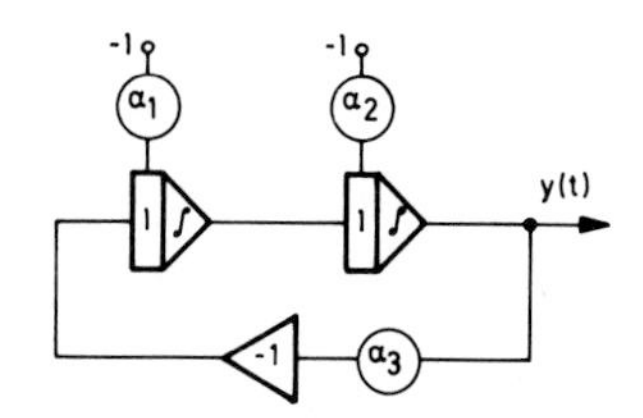

FIG. 3.P-10. Analog computer diagram of Problem 3.14.

3.15 Suppose the nominal differential equation of a system is $a_2\ddot{y} + a_1\dot{y} + a_0 y = u(t)$, where $u(t)$ is a step function $1(t)$, and the actual differential equation is $\alpha y^{(4)} + a_2\ddot{y} + a_1\dot{y} + a_0 y = u(t)$).

(a) Determine the two additional roots of the characteristic equation due to the increase of the order for $s \to \infty$ and $\alpha \to 0$.

(b) Calculate the root sensitivity.

3.16 Referring to Example 3.3–11, determine the scalar sensitivity measure m

of Biswas and Kuh of the complex input impedance $Z(s,R)$ with respect to R. Let $\boldsymbol{A}$ and $\boldsymbol{B}$ be the unity matrix $\boldsymbol{I}$.

3.17 Referring to Example 3.3–12, determine the scalar sensitivity measure of Biswas and Kuh with respect to the parameter a_1. Choose $\boldsymbol{A}$ and $\boldsymbol{B}$ to be the unity matrix $\boldsymbol{I}$.

Chapter 4

Methods for Calculation or Measurement of Output Sensitivity Functions

4.1 INTRODUCTION

This chapter presents the methods of calculation and measurement of output sensitivity functions by means of analytical, numerical, and simulation techniques.

A quite natural approach to the calculation of the output sensitivity function seems to consist of the following procedure. Solve the input–output differential equation of the system twice, namely, first for nominal parameters α_0 and then for slightly changed (actual) parameters $\alpha = \alpha_0 + \Delta\alpha$ with $\Delta\alpha \ll \alpha_0$. Take the difference of both solutions, that is, the parameter-induced output error $\Delta y(t,\alpha) = y(t,\alpha) - y(t,\alpha_0)$, and from this determine the output sensitivity function according to

$$\sigma(t,\alpha_0) = \frac{\Delta y(t,\alpha)}{\Delta\alpha}. \tag{4.1–1}$$

It has been mentioned earlier that there are several objections against this direct procedure. The crucial point is the problem of accuracy. Commonly, the equations can be solved merely by means of numerical integration or analog computation associated with calculation errors. Suppose that the error of $y(t,\boldsymbol{\alpha}_0)$ is ε_1 and the error of $y(t,\boldsymbol{\alpha})$ is ε_2. Then the parameter-indcued output error $\Delta\hat{y}$ is given by

$$\Delta\hat{y}(t,\boldsymbol{\alpha}) = [y(t,\boldsymbol{\alpha}) + \varepsilon_1] - [y(t,\boldsymbol{\alpha}_0) + \varepsilon_2] = \Delta y + \varepsilon_1 - \varepsilon_2. \tag{4.1–2}$$

The error $\varepsilon = \varepsilon_1 - \varepsilon_2$ should, of course, be small compared to the true value Δy to keep the error of $\sigma(t,\alpha_0)$ small. However, this requirement is hard to achieve since by the nature of this procedure $\Delta\alpha$ and hence Δy ought to be small too, so that, in general, Δy is of the same order as ε. Therefore, the resulting sensitivity function.

$$\hat{\sigma}(t, \alpha_0) = \frac{\Delta y(t, \alpha)}{\Delta\alpha} + \frac{\varepsilon}{\Delta\alpha} = \sigma(t, \alpha_0) + \frac{\varepsilon}{\Delta\alpha} \tag{4.1–3}$$

will be subject to large errors since $\Delta\alpha$ is small.

Nevertheless, this method has certain competence in digital computer applications. Here the calculation error is a systematic (reproducible) one which changes little as α is changed slightly, so that the difference $\varepsilon = \varepsilon_1 - \varepsilon_2$ remains very small. In this case, the direct method can in fact be employed without hesitation.

However, the direct method does not apply in the case of analog computers or other analog simulators, where the error ε is of the same order as Δy. In this case one should calculate the sensitivity function σ directly rather than calculating the much smaller error Δy. Another reason for the direct calculation of σ is that in some applications, for example in adaptive or self-optimizing systems, the sensitivity function is needed continuously with time (online) without knowing the input signal in advance.

In both situations it is necessary to calculate or measure the sensitivity function directly, so that the calculation error will not be divided by the small number $\Delta\alpha$ and σ will be available continuously.

There are various methods to determine $\sigma(t, \alpha_0)$ directly. The basic idea is to take the partial derivative with respect to α of the mathmatical model, for example, the differential equation, prior to its solution. This leads to a mathematical model or differential equation in terms of the sensitivity function σ instead of y, which is called the sensitivity model or sensitivity equation. The solution of this equation gives the sensitivity function σ.

Besides analytical methods based on the evaluation of the sensitivity equation which will be treated in this chapter, simulation methods have been developed making use of the structural properties of the sensitivity model. The basic method was first published by Bikhovski and Meissinger in connection with analog computers and network analyzers [24]. Kokotović and Rutman [56], Tomović [18], and Vusković and Circić [106] have extended this idea to more general systems. Their methods known as the variable component method and the method of sensitivity points will also be described in this chapter.

4.2 DIRECT CALCULATION OF $\sigma(t,\alpha_0)$ BY USE OF LAPLACE TRANSFORM

On principle, the sensitivity function can be calculated by solving the differential equation and then taking the partial derivative of the solution $y(t,\sigma)$ with respect to the parameter α. If the system is linear, this procedure can be facilitated considerably by the use of the Laplace transform.

If the Laplace transform is applied to the solution of a differential equation, the solution is, in a certain stage, given as

$$y(t, \alpha) = \mathscr{L}^{-1}\{G(s, \alpha)U(s)\}, \tag{4.2–1}$$

where $G(s,\alpha)$ represents the transfer function of the system and $U(s)$ is the Laplace transform of the input signal. In these terms, the output sensitivity function is

$$\sigma(t, \alpha_0) \triangleq \left.\frac{\partial y(t, \alpha)}{\partial \alpha}\right|_{\alpha 0} = \frac{\partial}{\partial \alpha}[\mathscr{L}^{-1}\{G(s, \alpha)U(s)\}]_{\alpha 0}. \tag{4.2–2}$$

Let us assume that the partial derivative exists and that the order of the limiting process $\alpha \to \alpha_0$ and the Laplace transformation may be reversed. Then Eq. (4.2–2) can be rewritten as

$$\sigma(t, \alpha_0) = \mathscr{L}^{-1}\left\{\left.\frac{\partial G(s, \alpha)}{\partial \alpha}\right|_{\alpha 0} U(s)\right\} \tag{4.2–3}$$

or, by the aid of $G(s, \alpha)U(s) = Y(s, \alpha)$,

$$\sigma(t, a_0) = \mathscr{L}^{-1}\left\{S_\alpha^{\ G}(s)\frac{Y(s, \alpha_0)}{\alpha_0}\right\}, \tag{4.2–4}$$

where $S_\alpha^{\ G}(s)$ is the Bode sensitivity function (Definition 3.3–1). Both equations can be employed to calculate $\sigma(t, \alpha_0)$.

Example 4.2-1 Determine the output sensitivity function with respect to T of an RC network with the differential equation

$$T\dot{y} + y = 1(t), \qquad y(0) = 0,$$

where $1(t)$ is the unit step function and $T = RC$.

Solution The nominal and actual transfer functions are, respectively,

$$G(s, T_0) = \frac{1}{1 + T_0 s}, \qquad G(s, T) = \frac{1}{1 + Ts}.$$

Partial differentiation yields

$$\left.\frac{\partial G(s, T)}{\partial T}\right|_{T0} = \frac{\partial}{\partial T}\left(\frac{1}{1 + Ts}\right)_{T0} = -\frac{s}{(1 + T_0 s)^2}.$$

Hence, due to Eq. (4.2–3),

$$\sigma(t, T_0) = \mathscr{L}^{-1}\left\{-\frac{s}{(1 + T_0 s)^2}\frac{1}{s}\right\} = -\frac{t}{T_0^{\ 2}} e^{-t/T_0}.$$

The question of the reversibility of the order of the limiting process and the Laplace transformation according to

$$\lim_{\alpha \to \alpha_0}\left(\frac{1}{2\pi j}\int_C \frac{\partial G(s, \alpha)}{\partial \alpha} U(s)e^{ts}\, ds\right) = \frac{1}{2\pi j}\int_C \lim_{\alpha \to \alpha_0} \frac{\partial G(s, \alpha)}{\partial \alpha} U(s)e^{ts}\, ds \tag{4.2–5}$$

has been examined by Chang [4]. The results can be summarized as follows:

(1) In the case of α-variations, the reversion is admissible.
(2) For β-variations, this problem does not exist at all.

(3) In the case of λ-variations, the reversion is admissible only if all real parts of the additional roots due to the increase of the order are located in the left half of the s-plane.

Statements (1) and (2) are obvious. To show (3) suppose that the actual differential equation due to a variation of the parameter λ is

$$a_{n+m}y^{(n+m)} + \cdots + a_{n+1}y^{(n+1)} + a_n y^{(n)} + \cdots + a_0 y = u(t). \quad (4.2\text{–}6)$$

The actual transfer function is then

$$G(s, \lambda) = (a_{n+m}s^{n+m} + \cdots + a_{n+1}s^{n+1} + a_n s^n + \cdots + a_0)^{-1}, \quad (4.2\text{–}7)$$

with the actual characteristic equation

$$Q(s, \lambda) = \underbrace{a_0 + \cdots + a_n s^n}_{Q_0(s)} + \underbrace{a_{n+1}s^{n+1} + \cdots + a_{n+m}s^{n+m}}_{Q_1(s,\lambda)} = 0$$

or

$$Q(s, \lambda) = Q_0(s) + Q_1(s, \lambda) = 0. \quad (4.2\text{–}8)$$

Now taking the partial derivative $\partial G/\partial\lambda$ as required in Eq. (4.2–5), we obtain

$$\frac{\partial G(s, \lambda)}{\partial \lambda} = \frac{-\partial Q_1(s, \lambda)/\partial\lambda}{[Q_0(s) + Q_1(s, \lambda)]^2}. \quad (4.2\text{–}9)$$

This reveals that the characteristic equation of $\partial G/\partial\lambda$ is the same as that of $G(s, \lambda)$, Eq.(4.2–8). It was shown in Section 3.3 [see Eqs.(3.3–60)–(3.3–69)] that, as λ approaches zero, n roots of $Q(s, \lambda)$ approach the roots of $Q_0(s)$ and the magnitudes of the additional m roots s_j approach infinity.

Now it depends on the real parts of the roots s_j whether or not the reversion is admissible:

(1) If $\mathrm{Re}(s_j) = -\infty$, the reversion is admissible.

(2) If there is at least one root for which $-\infty < \mathrm{Re}(s_j) < 0$ but no s_j has a real part that is equal to zero or positive, then there exists a transient oscillation of very high frequency which is neglected when the limit is carried out prior to the Laplace transform. However, since those oscillations are not measurable, the reversion is admissible.

(3) If at least one of the roots has a real part equal to zero or positive, sustained oscillations appear in the real system which have been neglected by the reversion. Hence, the reversion is not admissible in this case.

Example 4.2-2 Referring to Example 3.3–13, check in which of the four cases the reversion is admissible.

The reversion is admissible in cases (1), (2), and (3), but it is prohibited in case (4), since in the latter case there exist sustained oscillations.

This elementary method for the determination of $\sigma(t, \alpha_0)$ is practicable at a tolerable expense in simple cases only; for system orders higher than 3, the procedure becomes rather cumbersome, not being applicable at all to nonlinear systems. Furthermore, it gives little insight into the physical background of the problem. Therefore, other methods were developed some time ago. They are based on the sensitivity equation which will be treated now.

4.3 THE OUTPUT SENSITIVITY EQUATION

4.3.1 Sensitivity Equation for Time-Invariant α-Variations

Consider the ordinary differential equation of a general linear or nonlinear, time-invariant or time-variant system

$$f(y^{(n)}, \ldots, \dot{y}, y, t, \alpha) = 0, \tag{4.3–1}$$

with the initial conditions $y^{(i)}(t_0) = \beta_i$, $i = 0, 1, \ldots, n-1$. y denotes the output signal and α is a parameter with a nominal value different from zero. Note that the solution as well as its derivatives are functions of the time t, the parameter α, and the initial conditions $\beta_0, \ldots, \beta_{n-1}$; that is,

$$\begin{aligned} y &= y(t, \alpha, \beta_0, \ldots, \beta_{n-1}), \\ &\vdots \\ y^{(n)} &= y^{(n)}(t, \alpha, \beta_0, \ldots, \beta_{n-1}). \end{aligned} \tag{4.3–2}$$

Let us now differentiate Eq.(4.3–1), including the initial conditions, with respect to $\alpha, \beta_0, \ldots, \beta_{n-1}$. For this purpose define the function

$$g(\alpha, \beta_0, \ldots, \beta_{n-1}) \triangleq f(y^{(n)}, y^{(n-1)}, \ldots, y, t, \alpha). \tag{4.3–3}$$

It is well known that if $f, y, \dot{y}, \ldots, y^{(n)}$ are continuous functions in all arguments, g has a continuous first-order partial derivative that can be determined by application of the chain rule. According to Section 2.4, Case (4) (see also Problem 2.7), the partial differentiation of g with respect to α yields

$$\frac{\partial f}{\partial y^{(n)}} \frac{\partial y^{(n)}}{\partial \alpha} = \frac{\partial f}{\partial y^{(n-1)}} \frac{\partial y^{(n-1)}}{\partial \alpha} + \cdots + \frac{\partial f}{\partial y} \frac{\partial y}{\partial \alpha} + \frac{\partial f}{\partial \alpha} + 0. \tag{4.3–4}$$

This equation is valid for all parameter values. For *nominal* parameter values $\alpha = \alpha_0$, one obtains

$$\begin{aligned} \left.\frac{\partial y}{\partial \alpha}\right|_{\alpha 0} &= \sigma(t, \alpha_0), \\ \left.\frac{\partial \dot{y}}{\partial \alpha}\right|_{\alpha 0} &= \left.\frac{d}{dt} \frac{\partial y}{\partial \alpha}\right|_{\alpha 0} = \dot{\sigma}(t, \alpha_0), \\ &\vdots \\ \left.\frac{\partial y^{(n)}}{\partial \alpha}\right|_{\alpha 0} &= \left.\frac{d^n}{dt^n} \frac{\partial y}{\partial \alpha}\right|_{\alpha 0} = \sigma^{(n)}(t, \alpha_0), \end{aligned} \tag{4.3–5}$$

and therefore

$$\left.\frac{\partial f}{\partial y^{(n)}}\right|_{\alpha 0} \sigma^{(n)} + \left.\frac{\partial f}{\partial y^{(n-1)}}\right|_{\alpha 0} \sigma^{(n-1)} + \cdots + \left.\frac{\partial f}{\partial y}\right|_{\alpha 0} \sigma = -\left.\frac{\partial f}{\partial \alpha}\right|_{\alpha 0}. \quad (4.3\text{–}6)$$

This is a differential equation from which the output sensitivity function σ can be determined. It is therefore called the *sensitivity equation.* The above form of the sensitivity equation especially applies to time-invariant α-variations.

In the case of r parameters characterized by the vector $\boldsymbol{\alpha} = [\alpha_1 \ldots \alpha_r]^{\mathrm{T}}$, the original differential equation reads

$$f(y^{(n)}, y^{(n-1)}, \ldots, y, t, \boldsymbol{\alpha}) = 0, \qquad y^{(i)}(t_0) = \beta_i, \quad i = 0, 1, \ldots, n-1. \quad (4.3\text{–}7)$$

Partial differentiation with respect to one of the parameters α_j yields

$$\frac{\partial f}{\partial y^{(n)}} \frac{\partial y^{(n)}}{\partial \alpha_j} + \frac{\partial f}{\partial y^{(n-1)}} \frac{\partial y^{(n-1)}}{\partial \alpha_j} + \cdots + \frac{\partial f}{\partial y} \frac{\partial y}{\partial \alpha_j} = -\frac{\partial f}{\partial \alpha_j}. \quad (4.3\text{–}8)$$

With $\boldsymbol{\alpha} = \boldsymbol{\alpha}_0$ and the definition of the output sensitivity function σ_j according to Eq.(3.2–14), the output sensitivity equation for the jth parameter becomes

$$\left.\frac{\partial f}{\partial y^{(n)}}\right|_{\alpha 0} \sigma_j^{(n)} + \left.\frac{\partial f}{\partial y^{(n-1)}}\right|_{\alpha 0} \sigma_j^{(n-1)} + \cdots + \left.\frac{\partial f}{\partial y}\right|_{\alpha 0} \sigma_j = -\left.\frac{\partial f}{\partial \alpha_j}\right|_{\alpha 0}. \quad (4.3\text{–}9)$$

The initial conditions of Eq.(4.3–9) are found by partial differentiation of the original initial conditions β_i. Since the β_i's are not functions of $\boldsymbol{\alpha}$, all derivatives vanish, so that

$$\sigma_j(t_0) = \dot{\sigma}_j(t_0) = \cdots = \sigma_j^{(n-1)}(t_0) = 0.$$

In single-variable systems there are in general as many sensitivity equations as parameters, namely r, all of them having the same form. They can, therefore, be solved by the same program on a digital computer or by the same setup on an analog computer.

In the case of multivariable systems with q outputs, the number of output sensitivity equations increases to rq.

This shows that the output sensitivity equations with respect to α-variations are *inhomogeneous ordinary differential equations* of the same order as the original system equation with all initial conditions zero. They are always *linear* whether the original system is linear or not. In general, the following cases can be distinguished:

(1) If f is a *linear function* with *constant coefficients*, the sensitivity equation is also a linear differential equation with constant coefficients but with a different driving function.

Example 4.3-1 Given the actual differential equation

$$a_n y^{(n)} + a_{n-1} y^{(n-1)} + \cdots + a_0 y = u, \tag{4.3–10}$$

where $a_0 = a_0(\alpha), \ldots, a_n = a_n(\alpha)$ and $y = y(t, \alpha), \ldots, y^{(n)} = y^{(n)}(t, \alpha)$ are continuous functions of α, and $u(t)$ is a parameter-independent driving function. The partial derivatives of Eq. (4.3–6) are

$$\left.\frac{\partial f}{\partial y^{(i)}}\right|_{\alpha 0} = a_i(\alpha_0) \triangleq a_{i0}.$$

Hence the output sensitivity equation becomes

$$a_{n0}\sigma^{(n)} + a_{n-1,0}\sigma^{(n-1)} + \cdots + a_{00}\sigma = -\left.\frac{\partial f}{\partial \alpha}\right|_{\alpha 0},$$

with all initial conditions equal to zero.

To solve the sensitivity equation, $(\partial f/\partial \alpha)_{\alpha 0}$ has to be evaluated. This requires knowledge of the dependence $f(\alpha)$, that is, the dependence of the coefficients a_i $(i = 0, 1, \ldots, n)$ on α. In the above case, the driving function becomes

$$\left.\frac{\partial f}{\partial \alpha}\right|_{\alpha 0} = \left.\frac{\partial a_n}{\partial \alpha}\right|_{\alpha 0} y^{(n)}(t, \alpha_0) + \cdots + \left.\frac{\partial a_0}{\partial \alpha}\right|_{\alpha 0} y(t, \alpha_0).$$

As a special case, let us determine the sensitivity equation of a linear system with constant parameters governed by Eq.(4.3–10) where the coefficients a_i $(i = 0, 1, \ldots, n)$ themselves are the parameters of interest and all nominal parameter values a_{j0} are different from zero. Then

$$\left.\frac{\partial f}{\partial y^{(j)}}\right|_{aj0} = a_{j0}, \qquad j = 0, 1, \ldots, n,$$

$$\left.\frac{\partial f}{\partial a_j}\right|_{aj0} = y_0^{(j)}, \qquad j = 0, 1, \ldots, n,$$

where y_0 is the solution of the nominal original differential quation and $y_0^{(j)}$ is its jth derivative with respect to time. Using these expressions, the output sensitivity equation with respect to a_j becomes

$$a_{n0}\sigma_j^{(n)} + a_{n-1,0}\sigma_j^{(n-1)} + \cdots + a_{00}\sigma_j = -y_0^{(j)}. \tag{4.3–11}$$

All initial conditions of Eq.(4.3–11) are equal to zero whether the initial conditions of the original differential equation (4.3–10) are zero or not. The result is illustrated by Fig.4.3–1. Note that there are n such sensitivity equations.

Thus we see that the left side of the original differential equation and the sensitivity equation are the same, a fact that will prove to be very important for the solution of the sensitivity equation.

(2) If f is a *nonlinear function*, the sensitivity equation is a linear differential equation of the same order as the original differential equation but with

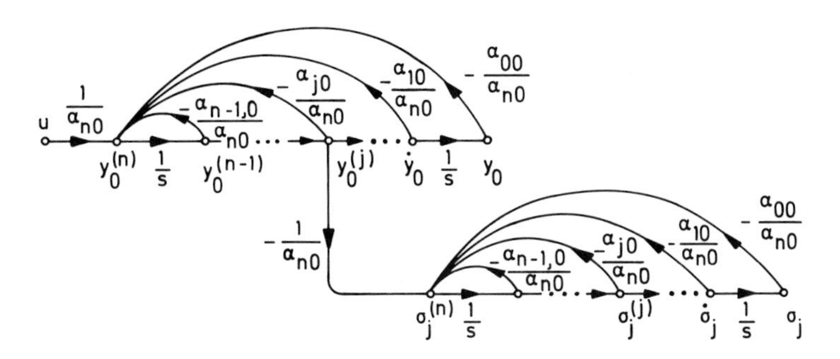

FIG. 4.3-1. Signal flow diagram illustrating the sensitivity equation (4.3–36) of Example 4.3–1.

time-variant coefficients and a different driving function. The dependence on time is due to the fact that $\partial f/\partial y^{(i)}$ contains $y(t, \alpha)$ in some form and is thus dependent on time.

Example 4.3-2 Consider Van der Pol's equation

$$\ddot{y} - \alpha(1 - y^2)\dot{y} + y = 0, \qquad y(0) = y_0^{\,0}, \quad \dot{y}(0) = y_1^{\,0},$$

which provides oscillations of different wave forms depending on the values of the parameter α: If $\alpha = 0$, the differential equation is linear and $y(t)$ is sinusoidal; if $\alpha > 0$, $y(t)$ has the character of relaxation oscillations.

In this case the sensitivity equation takes the form

$$\ddot{\sigma} - \alpha_0(1 - y_0^{\,2})\dot{\sigma} + (2\alpha_0 y_0 \dot{y}_0 + 1)\,\sigma = (1 - y_0^{\,2})\dot{y}_0, \tag{4.3–12}$$

with all initial conditions equal to zero. As we see, the sensitivity equation is linear and of the same order as the original differential equation but has time-variant coefficients found from the nominal differential equation.

(3) If f is a possibly nonlinear function with *time-variant coefficients* (but time-invariant parameters), the sensitivity equation is linear but generally has time-variant coefficients. If f is linear, the coefficients of the sensitivity equation and the original differential equation are the same.

Example 4.3-3 The Mathieu differential equation, which plays a certain role in the theory of parametric amplifiers, is

$$\ddot{y} + (1 - \alpha\cos\omega t)y = 0, \qquad y(0) = y_0^{\,0}, \quad \dot{y}(0) = y_1^{\,0}.$$

Consider α and ω as two α-parameters with nominal values α_0 and ω_0, respectively. Determine the sensitivity equation with respect to ω.

Solution Taking the partial derivative of the above differential equation with respect to ω for nominal parameter values, one obtains, according to Eq.(4.3–6),

$$\ddot{\sigma} + (1 - \alpha_0\cos\omega_0 t)\,\sigma = -y_0 t\alpha_0 \sin\omega_0 t, \qquad \sigma(0) = \dot{\sigma}(0) = 0.$$

$y_0 = y(t, \omega_0)$ is obtained from the nominal original equation. As we see, the sensitivity equation has the same homogeneous part as the original differential equation. The driving function and the initial conditions are different.

4.3.2 Sensitivity Equation for Time-Variant α-Variations

Let us consider α-parameters that change with time around a constant nominal value $\boldsymbol{\alpha}_0$ according to $\boldsymbol{\alpha}(t) = \boldsymbol{\alpha}_0 + \varepsilon \boldsymbol{g}(t)$, where ε is a constant and $\boldsymbol{g}(t)$ a uniformly bounded integrable vector function [see Eq.(3.2–22)]. In this case Eq.(4.3–9) is only an *approximate sensitivity equation* and σ_j is an *approximate sensitivity function.* The reason that this is only an approximation is due to the fact that now $\boldsymbol{\alpha}$ is a function of time, so that the total time derivative becomes

$$\frac{d\sigma_j}{dt} = \frac{\partial \sigma_j}{\partial t} + \sum_{j=1}^{r} \left(\frac{\partial \sigma_j}{\partial \alpha_j} \frac{d\alpha_j}{dt} \right)_{\boldsymbol{\alpha}0} \tag{4.3–13}$$

and so on. Hence Eq.(4.3–9) is a good approximation if the parameters change sufficiently slowly with time. This is frequently the case in model-reference adaptive systems, so that the approximate sensitivity functions are commonly used without serious shortcomings.

However, if the parameter changes in time are fast, the sensitivity equation has to be derived in a different manner. For simplicity let us assume a single parameter $\alpha = \alpha_0 + \varepsilon g(t)$. Then the total differential of f [Eq.(4.3–1)] is

$$\frac{\partial f}{\partial y^{(n)}} dy^{(n)} + \frac{\partial f}{\partial y^{(n-1)}} dy^{(n-1)} + \cdots + \frac{\partial f}{\partial y} dy + \frac{\partial f}{\partial \alpha} d\alpha = 0.$$

Replacing $d\alpha$ by $\varepsilon g(t)$, dividing both sides by ε, and then letting ε go to zero, we obtain

$$\left.\frac{\partial f}{\partial y^{(n)}}\right|_{\alpha 0} \sigma_\varepsilon^{(n)} + \left.\frac{\partial f}{\partial y^{(n-1)}}\right|_{\alpha 0} \sigma_\varepsilon^{(n-1)} + \cdots + \left.\frac{\partial f}{\partial y}\right|_{\alpha 0} \sigma_\varepsilon = -\left.\frac{\partial f}{\partial \alpha}\right|_{\alpha 0} g(t), \tag{4.3–14}$$

where σ_ε is given by Definition 3.2–23. The initial conditions are equal to zero.

This is the output sensitivity equation for a time-variant parameter. As we see, the coefficients $(\partial f/\partial y^{(\nu)})_{\alpha 0}$ are time-variant if f is nonlinear, otherwise they are constant. The driving function consists of $(\partial f/\partial \alpha)_{\alpha 0}$ which can be determined from the nominal original differential equation, and of $g(t)$ which has to be known for solving the sensitivity equation.

Example 4.3-4 Consider the differential equation

$$\ddot{y} + \alpha_1 \dot{y}^2 + \alpha_2 y = u, \qquad y(0) = \beta_0, \quad \dot{y}(0) = \beta_1,$$

where α_1 is a time-variant parameter of the form $\alpha_1 = \alpha_{10} + \varepsilon \sin \omega t$. Then the sensitivity equation with respect to α_1 is

$$\ddot{\sigma}_\varepsilon + 2\alpha_{10}\dot{y}_0\dot{\sigma}_\varepsilon + \alpha_2\sigma_\varepsilon = -\dot{y}_0^{\,2} \sin \omega t, \qquad \sigma(0) = \dot{\sigma}(0) = 0, \tag{4.3–15}$$

where $\sigma_\varepsilon \triangleq \partial y/\varepsilon|_{\varepsilon=0}$. The expressions $\dot{y}_0$ and $\dot{y}_0{}^2$ are obtained by solving the nominal original differential equation.

4.3.3 SENSITIVITY EQUATION FOR β-VARIATIONS

Let us now consider the initial conditions $\beta_0, \ldots, \beta_{n-1}$ as the parameters of the system. In this case the output sensitivity equations are determined in a manner similar to that above, yielding

$$\begin{aligned} \left.\frac{\partial f}{\partial y^{(n)}}\right|_{\boldsymbol{\beta}_0} \sigma_0{}^{(n)} + \left.\frac{\partial f}{\partial y^{(n-1)}}\right|_{\boldsymbol{\beta}_0} \sigma_0{}^{(n-1)} + \cdots + \left.\frac{\partial f}{\partial y}\right|_{\boldsymbol{\beta}_0} \sigma_0 &= 0, \\ &\vdots \\ \left.\frac{\partial f}{\partial y^{(n)}}\right|_{\boldsymbol{\beta}_0} \sigma_{n-1}^{(n)} + \left.\frac{\partial f}{\partial y^{(n-1)}}\right|_{\boldsymbol{\beta}_0} \sigma_{n-1}^{(n-1)} + \cdots + \left.\frac{\partial f}{\partial y}\right|_{\boldsymbol{\beta}_0} \sigma_{n-1} &= 0. \end{aligned} \tag{4.3–16}$$

σ_ν $(\nu = 0, 1, \ldots, n-1)$ denotes the output sensitivity function with respect to the νth initial condition, defined according to Eq.(3.2–21); $\boldsymbol{\beta}_0$ is the nominal initial conditions vector $\boldsymbol{\beta}_0 = [\beta_{00}\ \beta_{10} \cdots \beta_{n-1,0}]^{\mathrm{T}}$.

The corresponding initial conditions are found by differentiation of the original initial conditions $\beta_0, \beta_1, \ldots, \beta_{n-1}$ with respect to the initial condition under consideration. Thus the initial conditions of the first of the above sensitivity equations are $\sigma_0(t_0) = 1$, $\dot{\sigma}_0(t_0) = \cdots = \sigma_0{}^{(n-1)}(t_0) = 0$ and the initial conditions of the last equation are $\sigma_{n-1}(t_0) = \cdots = \sigma_{n-1}^{(n-2)}(t_0) = 0$, $\sigma_{n-1}^{(n-1)}(t_0) = 1$. In general, all initial conditions are zero except the one corresponding to the initial condition under consideration, that is, $\sigma_k{}^{(i)}(t_0)$ is 1 for $i = k$ and 0 for $i \neq k$.

Note that the output sensitivity equations for β-variations are *linear homogeneous ordinary differential equations* of the same order as the original differential equation with nonvanishing initial conditions.

Example 4.3-5 Consider the differential equation

$$\dot{y} + 2ty = 0, \qquad y(0) = y^0, \tag{4.3–17}$$

the solution of which is the Gaussian error function $y(t) = \exp(-t^2)$ provided that the initial condition is taken as $y(0) = 1$. With $y_0(0) = 1$ the nominal parameter value, let us determine the output sensitivity equation with respect to $y(0)$.

Solution Taking the partial derivative of the above equation yields

$$\dot{\sigma} + 2t\sigma = 0,$$

where $\sigma = (\partial y/\partial y(0))_{y_0(0)}$. The initial condition becomes $\sigma(0) = 1$.

This shows that in this case the original system equation and the sensitivity equation, including the initial conditions, are completely the same. Hence, $\sigma(t) = \exp(-t^2)$ or, in other words, the sensitivity function can be determined directly from the nominal original differential equation.

4.3.4 Sensitivity Equation for λ-Variations

In the case of λ-variations the actual differential equation is of higher order than the nominal one. Suppose that by a time-invariant variation of λ the order increases by 1. Then the actual differential equation is of the general form

$$f(y^{(n+1)}, y^{(n)}, \ldots, y, t, \lambda) = 0, \tag{4.3–18}$$

where all terms containing $y^{(n+1)}$ will be multiplied by functions $\Phi_\nu(\lambda)$ that have to satisfy the requirement $\lim_{\lambda \to 0} \Phi_\nu(\lambda) = 0$ for all ν. Taking the derivative of Eq.(4.3–18) with respect to λ, one obtains, by making use of the chain rule,

$$\frac{\partial f}{\partial y^{(n+1)}} \frac{\partial y^{(n+1)}}{\partial \lambda} + \frac{\partial f}{\partial y^{(n)}} \frac{\partial y^{(n)}}{\partial \lambda} + \cdots + \frac{\partial f}{\partial y} \frac{\partial y}{\partial \lambda} + \frac{\partial f}{\partial \lambda} = 0. \tag{4.3–19}$$

For $\lambda = 0$, this equation degenerates to the order n, since the term $\partial f/\partial y^{(n+1)}$ contains multiplicatively the functions $\Phi_\nu(\lambda)$ that vanish for $\lambda = 0$. Hence, with $\lambda = 0$ and $\sigma_\lambda \triangleq (\partial y/\partial \lambda)_{\lambda=0}$, Eq.(4.3–19) assumes the form

$$\left.\frac{\partial f}{\partial y^{(n)}}\right|_{\lambda=0} \sigma_\lambda^{(n)} + \left.\frac{\partial f}{\partial y^{(n-1)}}\right|_{\lambda=0} \sigma_\lambda^{(n-1)} + \cdots + \left.\frac{\partial f}{\partial y}\right|_{\lambda=0} \sigma_\lambda = -\left.\frac{\partial f}{\partial \lambda}\right|_{\lambda=0}. \tag{4.3–20}$$

This is the output sensitivity equation for λ-variations from which σ_λ can be determined. As a most remarkable result we observe that the form of Eq. (4.3–20) is exactly the same as that of Eq.(4.3–6) for α-variations, which is apparently true even if the order of the differential equation increases by more than 1 as λ deviates from 0.

This result points out the important fact that λ-variations can be treated by sensitivity equations of the same kind and same order n as α-variations no matter how much the order of the actual differential equation increases by $\lambda = 0$. There is however a difference as far as the initial conditions are concerned. Before coping with this problem, consider the following example..

Example 4.3-6 Suppose the actual differential equation of a system with a λ-variation is

$$\lambda y^{(n+1)} + (a_n + b\lambda) y^{(n)} + a_{n-1} y^{(n-1)} + \cdots + a_0 y = u, \tag{4.3–21}$$

where u denotes a parameter-independent input signal and $\lambda, b, a_n, \ldots, a_0$ are constants. The corresponding sensitivity equation with respect to λ is

$$a_{n0} \sigma_\lambda^{(n)} + a_{n-1,0} \sigma_\lambda^{(n-1)} + \cdots + a_{00} \sigma_\lambda = -y_0^{(n+1)} - y_0^{(n)}, \tag{4.3–22}$$

where $\sigma_\lambda \triangleq (\partial y/\partial \lambda)_{\lambda=0}$. The subscript 0 indicates nominal values or quantities. The driving function $-y_0^{(n+1)} - y_0^{(n)}$ follows from the nominal original differential equation. Note that the driving function contains derivatives of y_0 of order higher than n(in general, up to the same order as the actual differential equation) which is characteristic for λ-variations.

When determining the sensitivity equation in the above manner, the *initial conditions* in the case of λ-variations are not all equal to zero as in the case of α-variations. This is due to the fact that some of the initial conditions of the original differential equation are neglected when passing over from the actual to the nominal differential equation. For example, if the actual system is of the order $n + 3$ and the nominal system is of the order n, then the initial conditions $y^{(n+2)}(t_0)$, $y^{(n+1)}(t_0)$, and $y^{(n)}(t_0)$ are neglected because of the reduction of the differential equation as λ approaches zero. This negligence can be interpreted as a discontinuity of the corresponding derivatives of y at $t = t_0$, so that their partial derivatives with respect to λ do not exist at $t = t_0$ and $\lambda = \lambda_0$.

Example 4.3-7 Consider a second-order system with the actual differential equation

$$\lambda \ddot{y} + T\dot{y} + y = 1(t) \tag{4.3–23}$$

and the actual initial conditions $y(0, \lambda) = \dot{y}(0, \lambda) = 0$. Now suppose one of the time constants is neglected so that $\lambda = 0$. Hence the nominal differential equation is

$$T\dot{y} + y = 1(t), \tag{4.3–24}$$

with the initial condition $y(0, 0) = 0$.

To show what has happened at $\lambda = 0$ to the second initial condition $\dot{y}$, let us calculate $\dot{y}(0, 0)$ from the degenerate differential equation (4.3–24). The solution of this equation is $y(t, 0) = 1 - \exp(-t/T)$. Taking the derivative with respect to t and setting $t = 0$, we obtain $\dot{y}(0, 0) = -1/T$. On the other hand, for $\lambda \neq 0$ we had $\dot{y}(0, \lambda) = 0$. Thus we see that the initial condition $\dot{y}(0, \lambda)$ jumps at $\lambda = 0$ from $-1/T$ to 0. Consequently, the partial derivative $\sigma = (\partial\dot{y}(t, \lambda)/\partial\lambda)_{\lambda=0}$ does not exist for $t = 0$, that is, σ must have a discontinuity at $t = 0$.

This result is illustrated by Fig. 4.3–2. For $t \neq 0$, the slope of the step response of the first-order system ($\lambda = 0$) can be approximated by the second-order system ($\lambda \neq 0$) with any desired accuracy. At $t = 0$, however, the slope of the second-order system will always remain zero, that is, different from the slope of the first-order system.

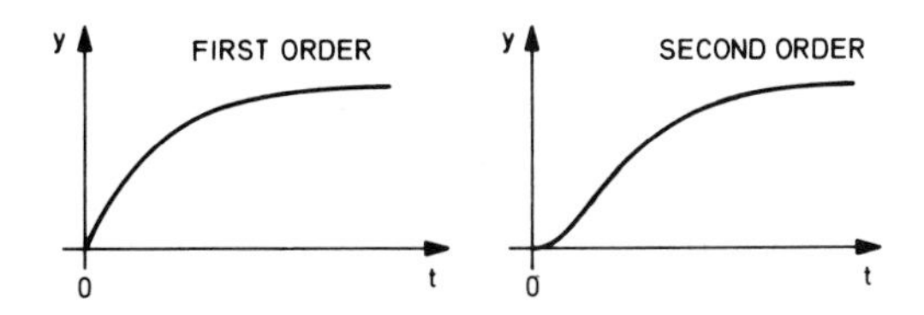

FIG. 4.3-2. Step response of the first and second-order system of Example 4.3–7.

The question now is how to find the initial conditions of the degenerate differential equation. Similar to the application of the Laplace transform to functions that jump at $t = 0$, there are the following two procedures:

(1) Exclude the point $t = 0$ from consideration. This allows us to operate with classical mathematics. However, the initial conditions of the nominal sensitivity equation have to be taken on the right side of the discontinuity, that is, at $t = +0$.

(2) Include the point $t = 0$ in consideration. This requires us to operate with distributions (i.e., step functions and their derivatives) when passing from the actual to the degenerate differential equation. In this case, the "natural" initial conditions, i.e., those of the *actual* system, can be used, which implies that the initial conditions of the sensitivity equation are all zero as in the case of α-variations.

Procedure (1) requires the determination of the initial conditions $\sigma^{(\nu)}(+0, 0)$, $\nu = 0, 1, \ldots, n-1$. A method for finding these initial conditions was given by Chang in 1961 [4]. Chang's method consists of the following steps:

(1) Approximate the system differential equation at the vicinity of $t = 0$ by a linear equation in the y's. Approximate the inputs by functions that are transformable to rational functions in s.

(2) Determine, by Laplace transformation, the expressions

$$Y_i(s, \boldsymbol{\lambda}) \triangleq \mathscr{L}\{y^{(i)}(t, \boldsymbol{\lambda})\} = s^i Y(s, \boldsymbol{\lambda}), \qquad i = 0, 1, \ldots, n-1, \quad (4.3\text{–}25)$$

where $\boldsymbol{\lambda}$ is the λ-parameter vector.

(3) Determine the corresponding sensitivity functions

$$\Sigma_{ik}(s, \mathbf{0}) \triangleq \left.\frac{\partial Y_i(s, \boldsymbol{\lambda})}{\partial \lambda_k}\right|_{\boldsymbol{\lambda}=\mathbf{0}} = \mathscr{L}\{\sigma_{ik}(t, \mathbf{0})\}, \tag{4.3–26}$$

with $k = 1, 2, \ldots, r$ and $i = 0, 1, \ldots, n-1$. These expressions are, by presupposition, rational functions in s. They can be written as

$$\Sigma_{ik}(s, \mathbf{0}) = \frac{N_{ik}(s)}{D_{ik}(s)} + P_{ik}(s), \tag{4.3–27}$$

where $N_{ik}(s)$, $D_{ik}(s)$, and $P_{ik}(s)$ are polynomials in s; $N_{ik}(s)$ is at least one order lower than $D_{ik}(s)$, and $P_{ik}(s)$ may be 0, 1, or any polynomial. The ratio $N_{ik}(s)/D_{ik}(s)$ is the principal part of $\Sigma_{ik}(s)$. Since the inverse Laplace transform of $P_{ik}(s)$ consists of Dirac impulses or their derivatives only, $P_{ik}(s)$ does not contribute to σ_{ik} for $t \neq 0$. Therefore, we have for $t > 0$

$$\sigma_{ik}(t,\mathbf{0}) \triangleq \sigma_k^{(i)}(t,\mathbf{0}) = \mathscr{L}^{-1}\left\{\frac{N_{ik}(s)}{D_{ik}(s)}\right\}. \tag{4.3–28}$$

(4) Now $\sigma_k^{(i)}(t, \mathbf{0})$ is determined for $t = +0$. To find $\sigma_k^{(i)}(+0, \mathbf{0})$, it is not

necessary to carry out the inverse transform. By simply applying the initial value theorem to Eq.(4.3–28), we obtain

$$\sigma_k^{(i)}(+0, \mathbf{0}) = \lim_{s\to\infty}\left[s\frac{N_{ik}(s)}{D_{ik}(s)}\right], \tag{4.3–29}$$

which are the desired initial conditions of the sensitivity equation for λ-variations.

Example 4.3-8 For the second-order system of Example 4.3–6, determine the initial condition $\sigma(+0, 0)$ of the sensitivity equation (4.3–24) using Chang's method and find the solution of the sensitivity equation.

The first step of Chang's method is fulfilled by itself since the system is linear. Step (2) yields, by application of Laplace transformation to Eq.(4.3–23),

$$\lambda s^2 Y + TsY + Y = 1/s,$$

whence

$$Y_0(s, \lambda) = 1/s(\lambda s^2 + Ts + 1).$$

According to step (3) we obtain

$$\Sigma_0(s, 0) = \frac{\partial}{\partial\lambda}\left(\frac{1}{s(\lambda s^2 + Ts + 1)}\right)_{\lambda=0} = -\frac{s^2}{s(1 + Ts)^2} = \frac{N_0(s)}{D_0(s)},$$

where $P_0(s) = 0$. Step (4) finally yields the initial condition

$$\sigma(+0, 0) = \lim_{s\to\infty}\left(-\frac{s^2}{(1 + Ts)^2}\right) = -\frac{1}{T^2}.$$

Now the sensitivity equation reads

$$T\dot{\sigma} + \sigma = -\ddot{y}_0 \qquad \text{with} \quad \sigma(+0, 0) = -1/T^2.$$

$\ddot{y}_0$ is found by solving the nominal original differential equation (4.3–24) and differentiating the solution twice for $t > 0$:

$$\ddot{y}_0 = -\frac{1}{T^2}e^{-t/T}.$$

Thus the sensitivity equation reads

$$T\dot{\sigma} + \sigma = \frac{1}{T^2}e^{-t/T}, \qquad \sigma(+0,0) = -\frac{1}{T^2}.$$

Its solution is

$$\sigma(t, 0) = \left(\frac{t}{T} - 1\right)\frac{1}{T^2}e^{-t/T}.$$

Note that this solution, as well as the differential equation, is valid for $t > 0$ only.

Referring to step (2), the difficulty of determining the initial conditions at $t = +0$ can be avoided by reducing the λ-variation problem to an α-variation problem. Two proposals will be given here; a third one will be discussed in Section 4.5.

(a) In many cases of practical interest, a simple trick can be applied to avoid the initial conditions problem. This will be illustrated by the following example.

Example 4.3-9 Consider again the actual differential equation of Example 4.3–7,

$$\lambda \ddot{y} + T\dot{y} + y = u, \qquad y(0) = y_0{}^0, \quad \dot{y}(0) = y_1{}^0,$$

where λ is a λ-parameter (i.e., $\lambda_0 = 0$) and T is an α-parameter. The sensitivity function $\sigma_\lambda = (\partial y/\partial \lambda)_{\lambda=0}$ is to be found.

Let us take the partial derivatives of the above differential equation with respect to λ and T:

$$\lambda \frac{\partial \ddot{y}}{\partial \lambda} + T\frac{\partial \dot{y}}{\partial \lambda} + \frac{\partial y}{\partial \lambda} = -\ddot{y},$$

$$\lambda \frac{\partial \ddot{y}}{\partial T} + T\frac{\partial \dot{y}}{\partial T} + \frac{\partial y}{\partial T} = -\dot{y}.$$

Now setting $\lambda = 0$, $T = T_0$, and denoting $\sigma_\lambda \triangleq (\partial y/\partial \lambda)_{\lambda=0}$ and $\sigma_T \triangleq (\partial y/\partial T)_{T_0}$, we have

$$T_0\dot{\sigma}_\lambda + \sigma_\lambda = -\ddot{y}_0, \qquad \sigma_\lambda(+0, 0),$$

$$T_0\dot{\sigma}_T + \sigma_T = -\dot{y}_0, \qquad \sigma_T(0, 0) = 0,$$

where y_0 denotes the solution of the nominal original differential equation. Comparing both the sensitivity equations, we see that $\sigma_\lambda \equiv \dot{\sigma}_T$. Consequently, σ_λ can be determined from the sensitivity equation for σ_T with all initial conditions equal to zero.

If, for example, $u = 1(t)$, we have $\dot{y}_0 = (1/T_0)\exp(-t/T_0)$. Substituting this expression into Eq.(4.3–20) and solving it, yields

$$\sigma_T = -\frac{t}{T_0{}^2}\, e^{-t/T_0}.$$

Hence,

$$\sigma_\lambda = \dot{\sigma}_T = \frac{t}{T_0{}^3}\, e^{-t/T_0} - \frac{1}{T_0{}^2}\, e^{-t/T_0}$$

which is in agreement with the result obtained previously.

(b) An alternative way to avoid the initial conditions problem consists in determining the sensitivity function of the *nonreduced* differential equation, i.e., for $\lambda \neq 0$, from $t = 0$ onward rather than from $t = +0$ onward, and then letting λ approach zero. This is now a regular α-parameter problem so that all initial conditions of the sensitivity equation are zero.

The evaluation can be simplified considerably by employment of the Laplace transformation and setting $\lambda = 0$ before the inverse transformation is carried out. By this procedure the discontinuities of the driving function $(\partial f/\partial \lambda)_{\lambda=0}$ at $t = 0$ have to be taken into account, which, is, however, done automatically by application of the Laplace transform. The procedure consists of the following five steps:

(1) Take the Laplace transform of the nonreduced sensitivity equation with zero initial conditions.

(2) Similarly, transform the nonreduced original differential equation and solve it for $\mathscr{L}\{\partial f/\partial \lambda\}$.

(3) Substitute for $\mathscr{L}\{\partial f/\partial \lambda\}$ in the transformed sensitivity equation and solve the latter for $\Sigma_\lambda = \mathscr{L}\{\sigma_\lambda\}$.

(4) Set $\lambda = 0$.

(5) Execute the inverse transform of Σ_λ.

In the case of simple problems, the result is obtained much faster by this procedure than by Chang's method.

Example 4.3-10 The procedure is illustrated by the second-order system of Example 4.3–7, $\lambda \ddot{y} + T\dot{y} + y = 1(t)$, $y(0) = 0$, $\dot{y}(0) = 0$.

The *first* step yields

$$\lambda \ddot{\sigma}_\lambda + T\dot{\sigma}_\lambda + \sigma_\lambda = -\ddot{y}_\lambda, \qquad \sigma(0,0) = 0, \quad \dot{\sigma}(0,0) = 0,$$

where $\sigma_\lambda \triangleq \partial y/\partial \lambda$, and $\ddot{y}_\lambda$ means the second derivative of the solution of the original differential equation, including eventual discontinuities or Dirac pulses at $t = 0$. From this,

$$\lambda s^2 \Sigma_\lambda + Ts\Sigma_\lambda + \Sigma_\lambda = -s^2 Y_\lambda.$$

The *second* step yields

$$\lambda s^2 Y_\lambda + TsY_\lambda + Y_\lambda = 1/s \qquad \text{or} \qquad s^2 Y_\lambda = s/(\lambda s^2 + Ts + 1).$$

Substituting this expression into the transformed sensitivity equation and solving for Σ_λ according to *step 3*, gives

$$\Sigma_\lambda = -s/(\lambda s^2 + Ts + 1)^2.$$

Now set $\lambda = 0$ according to *step* 4:

$$\Sigma_\lambda|_{\lambda=0} = -s/(Ts + 1)^2.$$

The inverse transform according to *step* **5** yields

$$\sigma_\lambda(t, 0) = \left(\frac{t}{T^3} - \frac{1}{T^2}\right)e^{-t/T}$$

which is again in agreement with the previous results.

Notice that since $t = 0$ is included in consideration, eventual discontinuities or Dirac impulses (distributions) of the right-hand side of the sensitivity equation at $t = 0$ have to be taken into account. For example, if y_λ in the above example jumps at $t = 0$, $\ddot{y}_\lambda$ contains step functions δ and δ_1. In the above procedure, they are automatically taken into account by the application of the Laplace transform in step 2. It is this point which makes the essential difference between Chang's method and the one above. In Example 4.3–8, the step and δ functions at $t = 0$ were neglected. Their inclusion in the above method avoids the determination of the initial conditions at $t = +0$.

The results of Section 4.3.3 can be summarized by the following theorem:

Theorem 4.3-1 Regardless of the character of the original system, the output sensitivity equation is always a linear differential equation of the same order as the nominal original system. The initial conditions are 0 for α-parameters, 1 or 0 for β-parameters, and for λ-parameters they have to be determined for $t = +0$ by a separate procedure.

4.3.5 Higher-Order Sensitivity Equations

For certain purposes, such as the study of large parameter variations or automatic optimization including sensitivity functions in the performance index, higher-order sensitivity functions are required in addition to the first-order sensitivity function σ. There are different definitions of second-order sensitivity functions, namely the second derivative $\nu \triangleq (\partial^2 y/\partial\alpha^2)_{\alpha 0}$ with respect to one parameter or mixed-type derivatives of the form $(\partial^2 y/(\partial\alpha_1\, \partial\alpha_2))_{\alpha 0}$ when dealing with more than one parameter. It is a most interesting fact that they can be calculated from higher-order sensitivity equations with the same homogeneous parts as those of the first-order sensitivity equation.

The *second-order* sensitivity equation can be derived by twice differentiating the original differential equation with respect to α [26]. The first differentiation yields

$$\frac{\partial f}{\partial y^{(n)}}\frac{\partial y^{(n)}}{\partial\alpha} + \frac{\partial f}{\partial y^{(n-1)}}\frac{\partial y^{(n-1)}}{\partial\alpha} + \cdots + \frac{\partial f}{\partial y}\frac{\partial y}{\partial\alpha} = -\frac{\partial f}{\partial\alpha}. \tag{4.3–30}$$

When proceeding to the second derivative, one has to take into account that, in general, the expressions $\partial f/\partial y^{(\nu)}$ and $\partial f/\partial\alpha$ may be functions of $y^{(n)}, \ldots, y$, that is,

$$\frac{\partial f}{\partial y^{(\nu)}} \triangleq g_\nu[y^{(n)}(t, \alpha), \ldots, y(t, \alpha), \alpha] \triangleq h_\nu(t, \alpha), \qquad \nu = 0, 1, \ldots, n, \tag{4.3–31}$$

$$\frac{\partial f}{\partial\alpha} \triangleq g_{n+1}[y^{(n)}(t, \alpha), \ldots, y(t, \alpha), \alpha] \triangleq h_{n+1}(t, \alpha). \tag{4.3–32}$$

Now taking the second derivative with respect to α, one obtains

$$\frac{\partial h_n}{\partial \alpha}\frac{\partial y^{(n)}}{\partial \alpha} + \frac{\partial f}{\partial y^{(n)}}\frac{\partial^2 y^{(n)}}{\partial \alpha^2} + \ldots + \frac{\partial h_0}{\partial \alpha}\frac{\partial y}{\partial \alpha} + \frac{\partial f}{\partial y}\frac{\partial^2 y}{\partial \alpha^2} = -\frac{\partial h_{n+1}}{\partial \alpha} \quad (4.3\text{–}33)$$

where

$$\frac{\partial h_\nu}{\partial \alpha} = \frac{\partial g_\nu}{\partial y^{(n)}}\frac{\partial y^{(n)}}{\partial \alpha} + \ldots + \frac{\partial g_\nu}{\partial y}\frac{\partial y}{\partial \alpha} + \frac{\partial g_\nu}{\partial \alpha}, \qquad \nu = 0, 1, \ldots, n + 1. \quad (4.3\text{–}34)$$

Now let α approach α_0. Assuming that all derivatives are continuous, the following substitutions may be introduced:

$$\left.\frac{\partial^2 y}{\partial \alpha^2}\right|_{\alpha 0} \triangleq \nu, \qquad \left.\frac{d}{dt}\frac{\partial^2 y}{\partial \alpha^2}\right|_{\alpha 0} - \dot{\nu}, \qquad \ldots, \qquad \left.\frac{d^n}{dt^n}\frac{\partial^2 y}{\partial \alpha^2}\right|_{\alpha 0} = \nu^{(n)}. \quad (4.3\text{–}35)$$

This yields the desired second-order sensitivity equation

$$\left.\frac{\partial f}{\partial y^{(n)}}\right|_{\alpha 0} \nu^{(n)} + \left.\frac{\partial f}{\partial y^{(n-1)}}\right|_{\alpha 0} \nu^{(n-1)} + \cdots + \left.\frac{\partial f}{\partial y}\right|_{\alpha 0} \nu$$

$$= -\left.\frac{\partial h_0}{\partial \alpha}\right|_{\alpha 0} \sigma - \cdots - \left.\frac{\partial h_n}{\partial \alpha}\right|_{\alpha 0} \sigma^{(n)} - \left.\frac{\partial h_{n+1}}{\partial \alpha}\right|_{\alpha 0} \quad (4.3\text{–}36)$$

with the initial conditions $\nu^{(i)}(t_0) = 0$, $i = 0, 1, \ldots, n - 1$. This procedure can be repeated up to any order, always leading to sensitivity equations of the *same homogeneous part* as the first-order sensitivity equation. The driving function contains the lower-order sensitivity functions or derivatives of the nominal system output y_0. The result is summarized by the following theorem.

Theorem 4.3-2 Output sensitivity equations of any order have the same homogeneous part, that is, they are linear and of the same order as the nominal original system. The excitation consists of lower-order sensitivity functions or derivatives of the nominal output function y_0.

Thus, the higher-order sensitivity function can be calculated recursively from a single type of differential equations which is one of the great advantages of the method of sensitivity equations.

Example 4.3-11 Find the general form of the second-order sensitivity equation of the nth order differential equation

$$a_n y^{(n)} + a_{n-1} y^{(n-1)} + \cdots + a_0 y = u(t)$$

with respect to the coefficients a_j, $j = 0, 1, \ldots, n$.

Differentiating the above equation with respect to a_j and denoting nominal quantities by the subscript 0, we have

$$a_{no}\sigma_j^{(n)} + a_{n-1,0}\sigma_j^{(n-1)} + \cdots + a_{00}\sigma_j = -y_0^{(j)}, \qquad j = 0, 1, \ldots, n.$$

Repeated differentiation with respect to a_j yields

$$a_{no}\nu_j^{(n)} + a_{n-1,0}\nu_j^{(n-1)} + \cdots + a_{00}\nu_j = -\sigma_j^{(j)}, \tag{4.3–37}$$

with all initial conditions equal to zero. Referring to Fig. 4.3–1, a third diagram of the same structure has to be added, the input of which is taken from the point $\sigma_j^{(j)}$ of the second diagram.

4.4 SOLUTION TO THE SENSITIVITY EQUATION

For the determination of the output sensitivity function, the following set of ordinary differential equations is available:

(1) The nominal original system equation

$$f(y^{(n)}, y^{(n-1)}, \ldots, y, t, \boldsymbol{\alpha}_0) = 0, \tag{4.4–1}$$

with the initial conditions $y(0)$, $\dot{y}(0), \ldots, y^{(n-1)}(0)$.

(2) r sensitivity equations of the form

$$\left.\frac{\partial f}{\partial y^{(n)}}\right|_{\boldsymbol{\alpha}0} \sigma_j^{(n)} + \left.\frac{\partial f}{\partial y^{(n-1)}}\right|_{\boldsymbol{\alpha}0} \sigma_j^{(n-1)} + \cdots + \left.\frac{\partial f}{\partial y}\right|_{\boldsymbol{\alpha}0} \sigma_j = -\left.\frac{\partial f}{\partial \alpha}\right|_{\boldsymbol{\alpha}0},$$

$$j = 1, 2, \ldots, r, \tag{4.4–2}$$

where the initial conditions are either equal to zero (α-variations) or partially one and zero (β-variations), or they have to be determined for $t = +0$ (λ-variations). This system of differential equations is often called the *combined system.*

The solution to this system of differential equations can be obtained by either of the following procedures:

(1) analytic and/or numerical methods making use of the digital computer;

(2) simulation methods making use of analog computers or electrical networks.

In this section, both procedures will be discussed briefly.

4.4.1 Analytic Solution

A very important feature of the sensitivity equations is that they are always linear. However, their coefficients may be time-variant in general. Therefore, in general, one has to solve linear time-variant differential equations in addition to the original system equation.

In terms of the differential operator p, the sensitivity equation can be written as

$$H(p, \boldsymbol{\alpha}_0)\,\sigma_j(t, \boldsymbol{\alpha}_0) = u_1(t). \tag{4.4–3}$$

Here $H(p, \boldsymbol{\alpha}_0)$ represents a linear and, in general, time-variant operator containing the initial conditions as well. $u_1(t, \boldsymbol{\alpha}_0) = -(\partial f/\partial \alpha_j)_{\boldsymbol{\alpha}_0}$ is a given time function that is determined from the nominal original differential equation. Writing this function also in operator form, we have

$$u_1(t, \boldsymbol{\alpha}_0) = K(p, \boldsymbol{\alpha}_0)u(t), \tag{4.4-4}$$

where $u(t)$ is the input of the original system and $K(p, \boldsymbol{\alpha}_0)$ denotes an operator that can be nonlinear and/or time-variant.

Thus, the solution to the sensitivity equation is of the form

$$\sigma_j(t, \boldsymbol{\alpha}_0) = H^*(p, \boldsymbol{\alpha}_0)u_1(t, \boldsymbol{\alpha}_0) = H^*(p, \boldsymbol{\alpha}_0)K(p, \boldsymbol{\alpha}_0)u(t), \tag{4.4-5}$$

where $H^*(p, \boldsymbol{\alpha}_0)$ denotes the inverse operator of $H(p, \boldsymbol{\alpha}_0)$, satisfying

$$H^*(p, \boldsymbol{\alpha}_0)\, H(p, \boldsymbol{\alpha}_0) = 1.$$

In the particular case of a linear time-invariant system, $H^* = H^{-1}$ and the operators can be interpreted as transfer functions that may be treated by means of Laplace transformation. Thus in terms of Laplace transformation, the solution is

$$\sigma(t,\boldsymbol{\alpha}_0) = \mathscr{L}^{-1}\left\{\frac{U_1(s, \boldsymbol{\alpha}_0)}{H(s, \boldsymbol{\alpha}_0)}\right\} = \mathscr{L}^{-1}\left\{\frac{K(s, \boldsymbol{\alpha}_0)}{H(s, \boldsymbol{\alpha}_0)}\, U(s)\right\}, \tag{4.4-6}$$

where $U_1(s, \boldsymbol{\alpha}_0)$ and $U(s)$ denote the Laplace transforms of $U_1(t, \boldsymbol{\alpha}_0)$ and $u(t)$, respectively. $U_1(s, \boldsymbol{\alpha}_0)/H(s, \boldsymbol{\alpha}_0)$ is the transformed output sensitivity function $\Sigma(s, \boldsymbol{\alpha}_0)$, and $K(s, \boldsymbol{\alpha}_0)/H(s, \boldsymbol{\alpha}_0) = (\partial G(s, \boldsymbol{\alpha})/\partial \boldsymbol{\alpha}_j)_{\boldsymbol{\alpha}_0}$, where $G(s, \boldsymbol{\alpha})$ is the transfer function of the system.

Example 4.4-1 Consider a system with the transfer function

$$G(s,\tau_0) = \frac{Y(s, \tau_0)}{U(s)} = [(1 + Ts)^2(1 + \tau s)]^{-1}, \tag{4.4-7}$$

where the nominal values of the time constants are $T_0 = 10$ sec, $\tau_0 = 9$ sec. Determine the output sensitivity function with respect to τ for a step input $u(t) = 1(t)$ and the initial conditions equal to zero.

Solution The transformed nominal differential equation is

$$(1 + T_0 s)^2(1 + \tau_0 s)Y(s, \tau_0) = 1/s. \tag{4.4-8}$$

By partial differentiation with respect to τ, the transformed sensitivity equation is obtained as

$$(1 + T_0 s)^2(1 + \tau_0 s)\Sigma(s, \tau_0) = -(1 + T_0 s)^2 s Y(s, \tau_0),$$

with all initial conditions equal to zero, and $\Sigma \triangleq \partial Y/\partial \tau| = \mathscr{L}\{\sigma\}$. Solving the above algebraic equation for $\Sigma(s, \tau_0)$ yields

$$\Sigma(s, \tau_0) = -\frac{s}{1 + \tau_0 s}\, Y(s, \tau_0). \tag{4.4-9}$$

$Y(s, \tau_0)$ can be found from the nominal original differential equation (4.4–8):

$$Y(s, \tau_0) = [s(1 + T_0 s)^2(1 + \tau_0 s)]^{-1}. \tag{4.4–10}$$

Thus for the particular case, the expressions H and K of Eq. (4.4–6) take the form

$$H(p, \boldsymbol{\alpha}_0) = (1 + T_0 s)^2(1 + \tau_0 s),$$

$$K(p, \boldsymbol{\alpha}_0) = \frac{(1 + T_0 s)^2 s}{(1 + T_0 s)^2(1 + \tau_0 s)}.$$

By evaluating Eq. (4.4–6), using the above substitutions, or by substitution of Eq. (4.4–10) into Eq. (4.4–9), we obtain

$$\Sigma(s, \tau_0) = -[(1 + T_0 s)^2(1 + \tau_0 s)^2]^{-1}. \tag{4.4–11}$$

Now carrying out the inverse Laplace transformation of Σ, the desired result is obtained as

$$\sigma(t, \tau_0) = \left[\frac{2T_0\tau_0}{(T_0 - \tau_0)^3} - \frac{t}{(T_0 - \tau_0)^2}\right] e^{-t/T_0} + \left[\frac{2T_0\tau_0}{(\tau_0 - T_0)^3} - \frac{t}{(\tau_0 - T_0)^2}\right] e^{-t/\tau_0} \tag{4.4–12}$$

or, after substituting the normed numerical data,

$$\sigma(t, \tau_0) = (180 - t)\, e^{-t/10} - (180 + t)\, e^{-t/9}.$$

Let us now consider *time-variant parameters.* In this case, the sensitivity equation is of the form of Eq. (4.3–14). Denoting

$$\left.\frac{\partial f}{\partial \alpha}\right|_{\alpha_0} g(t) \triangleq g_1(t) \tag{4.4–13}$$

and defining

$$\Delta\alpha_1 \triangleq \left.\frac{\partial f}{\partial \alpha}\right|_{\alpha_0} \Delta\alpha = \varepsilon g_1(t) \tag{4.4–14}$$

as the new parameter deviation, Eq. (4.3–14) can be rewritten as

$$\mathrm{H}(\mathrm{p}, \alpha_0)\, \sigma_\varepsilon = g_1(t). \tag{4.4–15}$$

Hence,

$$\sigma_\varepsilon = H^*(p, \alpha_0)\, g_1(t), \tag{4.4–16}$$

where H^* (p, α_0) again denotes the inverse operator or H (p, α_0). The parameter-induced change of the output is

$$\Delta y(t, \alpha) = \sigma_\varepsilon \varepsilon = H^*(p, \alpha_0) g_1(t)\, \varepsilon \tag{4.4–17}$$

or, using Eq. (4.4–14),

$$\Delta y(t, \alpha) = H^*(p, \alpha_0)\, \Delta\alpha_1. \tag{4.4–18}$$

This result shows that H^* can be interpreted as a sensitivity operator producing the parameter-induced error Δy as it is applied to the parameter variation $\Delta\alpha_1$. This operator is defined by the left-hand side of the sensitivity equation.

4.4.2 Structural Method; Sensitivity Model

The graphical illustration of the sensitivity equation (see, for example, Fig. 4.3–1) and the formulas obtained by its analytic solution [see Eq. (4.4–6)] allow for an interesting structural interpretation. This gave rise to early attempts to use structural methods for the determination of the sensitivity function associated with analog or digital simulation. In the next section, the general background of the structural method is outlined. Special procedures are presented in Sections 4.5, 4.6, and 4.7.

The background of all strucutral methods is the concept of the sensitivity model that is defined as follows.

Definition 4.4-1 *Sensitivity model.* The system described by the sensitivity equation is referred to as the sensitivity model of the original system.

For each output, a system has as many sensitivity models as parameters of interest. Both the nominal original system and the corresponding sensitivity models form the *combined system.*

Due to the results of Section 4.3 concerning the particular form of the sensitivitiy model, the following theorems hold.

Theorem 4.4-1 The sensitivity model is always linear. In the general case of nonlinear and/or time-variant original systems, its parameters will be time-variant.

Theorem 4.4-2 If $y(t,\alpha)$ is the output of a system, the corresponding sensitivity functions $\sigma_j(t, \boldsymbol{\alpha}_0) = (\partial y/\partial \alpha_j)_{\boldsymbol{\alpha}_0}$ are the outputs of the corrssponding sensitivity models.

Consequently, in order to measure the output sensitivity functions simultaneously, the nominal original system and the sensitivity models have to be simulated (for example, on an analog computer or by analog networks). Then the sensitivity functions can be measured at the outputs of the sensitivity models (e.g., as voltages). The general structural diagram for the case of time-invariant *α-variations* is depicted in Fig. 4.4–1. Due to the results of Section 4.3, this structure is also valid for λ-variations; however, different initial conditions have to be applied. In the case of *β-variations*, only the sensitivity models are required with all inputs equal to zero and, in each case, one initial condition equal to one and all others zero. The double frame used for the original system is to indicate that, in general, the system may be nonlinear, whereas the sensitivity models are always linear.

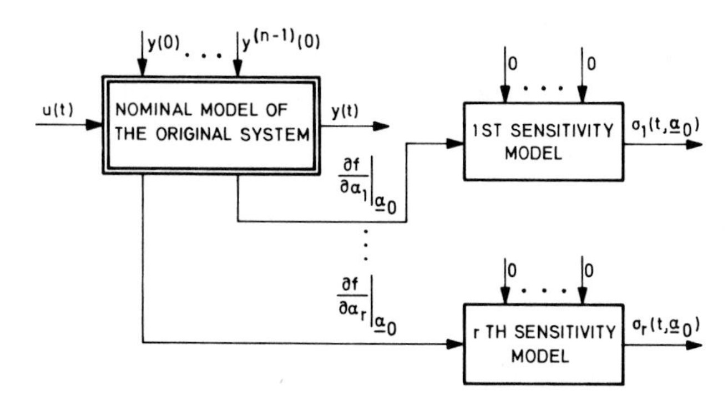

FIG. 4.4-1. General structural diagram of the combined system for the simultaneous measurement of all output sensitivity functions in the case of α-variations.

If the sensitivity functions are to be measured successively, a single sensitivity model is sufficient. This has to be excited successively by different driving functions or initial conditions, respectively.

As an example, consider Fig. 4.4–2 which shows the signal flow diagram for the simultaneous measurement of the output sensitivity functions $\sigma_\lambda = (\partial y/\partial \lambda)_{\lambda=0}$ and $\sigma_T = (\partial y/\partial T)_{T_0}$ of Example 4.3–9. This diagram follows immediately from the structural interpretation of the combined system, making use of the special method of treating λ-problems as shown in Section 4.3.4(a), due to which $\sigma_\lambda = \dot{\sigma}_T$.

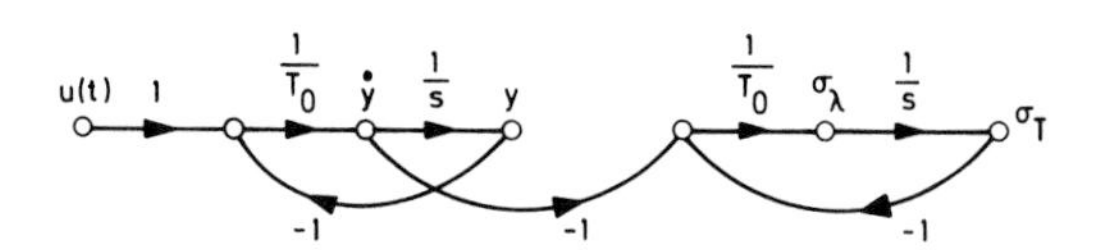

FIG. 4.4-2. Signal flow diagram for the simultaneous measurement of the output sensitivity functions $\sigma_T = (\partial y/\partial T)_{T_0}$ and $\sigma_\lambda = (\partial y/\partial \lambda)_{\lambda=0}$ of the second-order system with the differential equation $\lambda \ddot{y} + T\dot{y} + y = u$, $y(0) = \dot{y}(0) = 0$ (Example 4.3–9).

The structural methods, also called indirect methods, are preferable to the analytic method in the following cases:

(1) if the systems are complex, e.g., nonlinear and/or of high order. Here the analytic solution might sometimes be complicated or even impracticable, whereas the structural method is, in principle, always applicable,

(2) if a continuous determination of the sensitivity function is required, as for example in adaptive or self-optimizing systems;

(3) if certain parts of the system are subject to a detailed study or a more accurate measurement.

The structures found in the way described are not necessarily of minimal structure. In many cases they can be simplified considerably. Therefore, in addition to the above approach, special methods for the measurement of sensitivity functions have been developed completely based on structural considerations. They originate from the early works of Bikhovski [24,25] and Kokotović and co-workers [53–56]. These methods are described in the following sections.

4.5 THE VARIABLE COMPONENT METHOD

The structure for measuring the output sensitivity function of a given system can be found by simple structural considerations. This approach is important in those cases where the mathematical model of the system under consideration is either unknown or only partially known. Perhaps the most important approach to that problem is the method of variable component described by Kokotović and Rutman in the mid sixties [56]. The basic idea comes from Bikhovski [24]. Whereas Bikhovski confined himself to electrical networks, Kokotović and Rutman grounded the method on the more general block diagram, however, still confining themselves to linear systems. The method was later extended to nonlinear systems by Vusković and Circić [106].

This approach is especially important for sensitivity analysis of large systems. A typical example is the control loop, where, in common, the plant is given physically but the analytic description, at least of parts of it as well as of its inputs, may be complicated or unknown.

4.5.1 Derivation for Linear Systems

Consider first a linear *time-invariant* system depending an an *α-parameter*. Assume that the system contains, among other subsystems, only one subsystem with the transfer function $G(s, \alpha)$ that depends on the parameter α. The subsystem will be referred to as the variable component. Suppose that there is no external feedback so that the input u is indepndent of α. Furthermore, let the initial conditions be independent of α. This situation is symbolized by Fig. 4.5–1 for a nominal parameter value α_0.

If the parameter deviates from the nominal value α_0 by $\Delta\alpha$, G changes by ΔG. Because of possible inner feedback, the change ΔG can affect all inner signals (including the input signal u_i of the variable component), thus giving

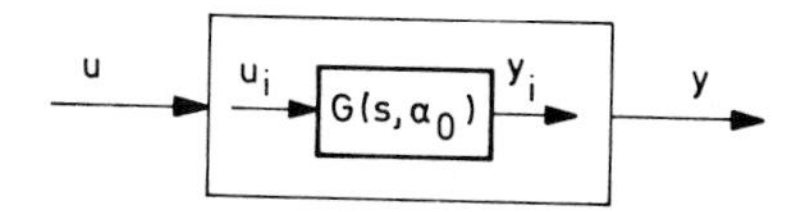

FIG. 4.5-1. Block diagram of the nominal original system.

rise to corresponding deviations Δu_i or Δy_i, respectively. This situation is depicted in Fig. 4.5–2.

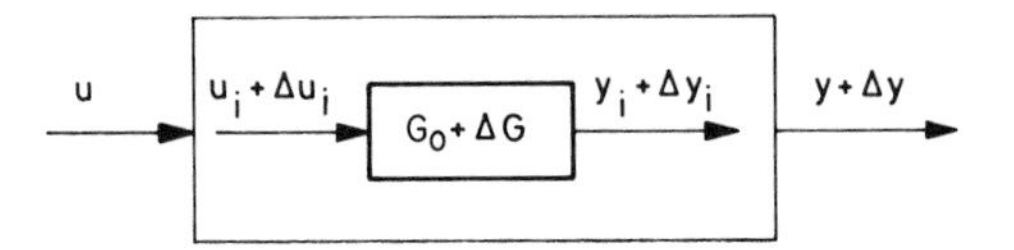

FIG. 4.5-2. Block diagram of the actual original system.

From Fig. 4.5–2, the actual output of the variable component in the frequency domain can be formulated as follows:

$$Y_i + \Delta Y_i = (U_i + \Delta U_i)(G + \Delta G) = (U_i + \Delta U_i)G + (U_i + \Delta U_i)\,\Delta G. \tag{4.5–1}$$

If ΔG is small, the expression $\Delta U_i\,\Delta G$ is second-order small and may be neglected in Eq. (4.5–1). Hence

$$Y_i + \Delta Y_i = (U_i + \Delta U_i)G_0 + U_i\,\Delta G. \tag{4.5–2}$$

The term $\delta Y_i = U_i \Delta G$ can be interpreted as an additional input signal that is produced from U_i by ΔG. Thus, the block diagram of Fig. 4.5–2 can be replaced by the block diagram of Fig. 4.5–3.

Because of $Y_i = GU_i$, and by virtue of the superposition principle, one can consider the signals separately and can cancel out the nominal values y, y_i, u_i, and u in Fig. 4.5–3. This indicates that Δy is generated by u_i or, in other words, that u_i is the new input signal of the system. Since u_i can be taken from the nominal original system, the complete structure for measuring Δy is as shown in Fig. 4.5–4.

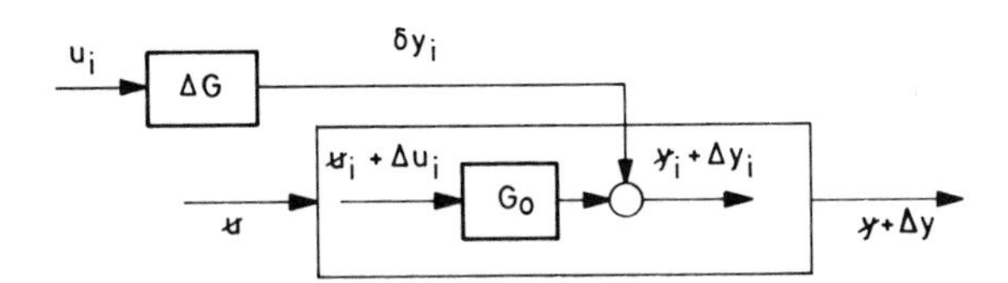

FIG. 4.5-3. Equivalent block diagram of Fig. 4.5–2.

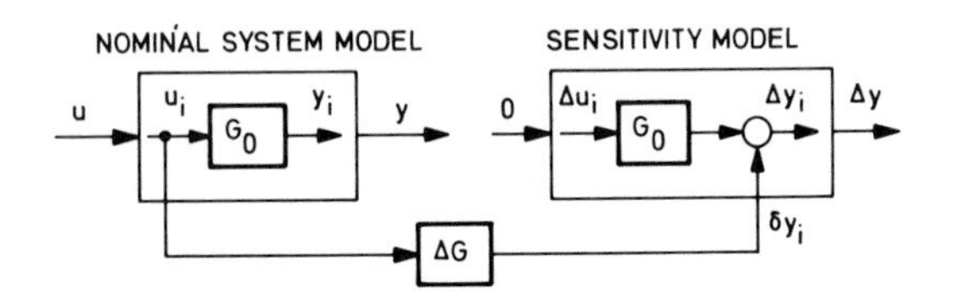

FIG. 4.5-4. Block diagram (combined system) for measuring Δy.

If we now want to measure the sensitivity function $(\partial y/\partial \alpha)_{\alpha 0}$ rather than Δy, we have to divide by $\Delta \alpha$ all parameter-induced deviations and then let $\Delta \alpha$ approach zero. This leads to the final block diagram for the measurement of $\sigma(t,\alpha_0) = (\partial y/\partial \alpha)_{\alpha 0}$, which is shown in Fig. 4.5–5.

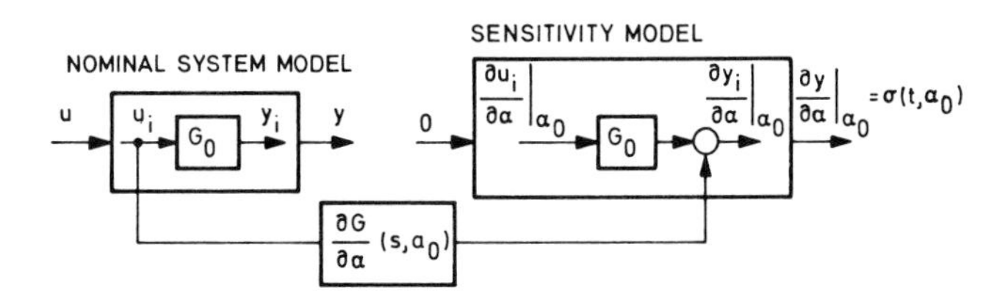

FIG. 4.5-5. Final block diagram (combined system) for measuring the output sensitivity function $\sigma(t, \alpha_0)$.

The resulting block diagram consists of two nominal models of the system (original nominal system model and sensitivity model.) From the input of the variable component in the system model, a connection with the transfer function $(\partial G/\partial \alpha)_{\alpha 0}$ has to be drawn to the output of the variable component in the sensitivity model. If now the input u is applied to the original system model, the output sensitivity function is obtained at the output of the sensitivity model. Furthermore, all sensitivity functions of the outputs of the subsystems can be measured at the corresponding points in the sensitivity model.

So far we have presumed that there is only a single subsystem that depends on the parameter. If several, say q, subsystems depend on the same parameter, the block diagram for the measurement of the output sensitivity function of the system is as shown in Fig. 4.5–6.

Note that a connection has to be established from the *input* of each variable component to the corresponding *output* in the sensitivity model. The input signal and the initial conditions of the sensitivity model are zero. If there are r parameters, one needs a single nominal system model but r sensitivity models with adequate transfer functions in the connection lines.

Example 4.5-1 Consider the system with the block diagram shown in Fig. 4.5–7. Find the block diagram for measuring the output sensitivity function $\sigma = (\partial y/\partial T)_{T_0}$.

Solution According to the general scheme of Fig. 4.5–6, a second nominal system model (sensitivity model) is required with the input signal and the initial conditions equal to zero. The connection link between both models has the transfer function

$$\left.\frac{\partial G}{\partial T}\right|_{T_0} = \frac{\partial}{\partial T}\left(\frac{1}{1 + Ts}\right)_{T_0} = -\frac{s}{(1 + T_0 s)^2}. \tag{4.5–3}$$

The resulting block diagram is shown in Fig. 4.5–8. Note that this diagram can

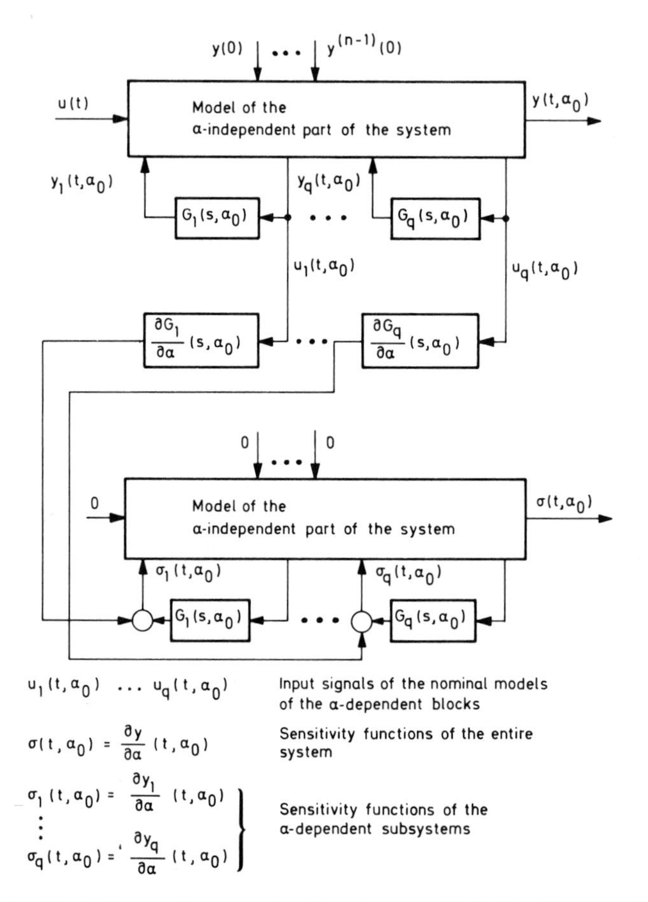

FIG. 4.5-6. Block diagram for measuring the output sensitivity function according to the variable component method in the case of linear systems.

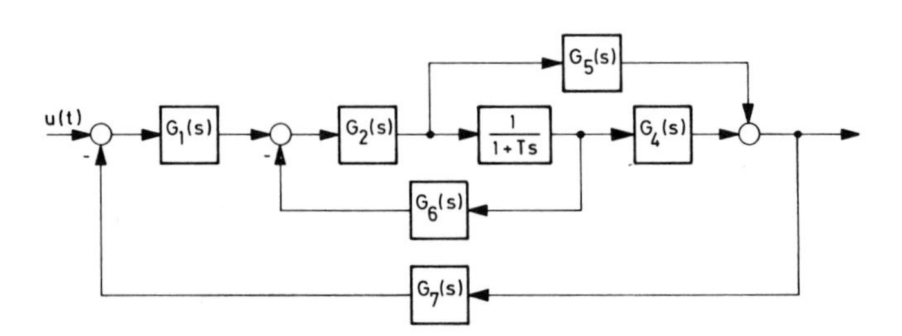

FIG. 4.5-7. Block diagram for the system under consideration in Example 4.5–1.

still be simplified by moving the takeoff point ahead of the block $1/(1 + T_0 s)$ in the original system model.

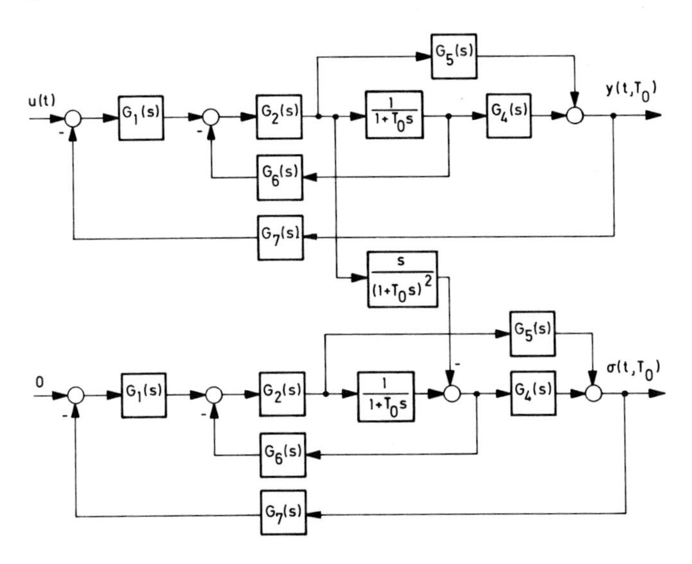

FIG. 4.5-8. Measuring diagram for the output sensitivity function σ of Example 4.5–1.

In order to outline the characteristics of the above procedure, let us derive an analytic expression for the sensitivity function on the basis of the structural diagram obtained. According to the structure of Fig. 4.5–4, the parameter-induced output error becomes, in the frequency domain,

$$\Delta Y(s, \alpha) = U(s)\frac{U_i(s, \alpha)}{U(s)}\Delta G(s, \alpha)\frac{\Delta Y(s, \alpha)}{\delta Y_i(s, \alpha)}. \tag{4.5–4}$$

Using the abbreviations

$$G_2(s, \alpha) \triangleq \frac{\Delta Y(s, \alpha)}{\delta Y_i(s,\alpha)}, \qquad G_1(s, \alpha) \triangleq \frac{U_i(s, \alpha)}{U(s)}, \tag{4.5–5}$$

Eq. (4.5–4) can be rewritten as

$$\Delta Y(s, \alpha) = G_1(s, \alpha)G_2(s, \alpha)\, \Delta G(s, \alpha)\, U(s). \tag{4.5–6}$$

Dividing this equation by $\Delta\alpha$ and then letting $\Delta\alpha$ approach zero, we obtain

$$\Sigma(s, \alpha_0) = \mathscr{L}\{\sigma(t, \alpha_0)\} = G_1(s, \alpha_0)G_2(s, \alpha_0)\left.\frac{\partial G}{\partial \alpha}\right|_{\alpha 0} U(s). \tag{4.5–7}$$

On the other hand, if we derive the transformed sensitivity function $\Sigma(s, \alpha_0)$

directly from the overall transfer function of the system, which shall be denoted by $F(s, \alpha) \triangleq Y(s, \alpha)/U(s)$, then we obtain

$$\Sigma(s, \alpha_0) = \left.\frac{\partial Y(s, \alpha)}{\partial \alpha}\right|_{\alpha 0} = U(s) \left.\frac{\partial F(s, \alpha)}{\partial \alpha}\right|_{\alpha 0} = \left(\frac{\partial F(s, \alpha)}{\partial G(s, \alpha)} \frac{\partial G(s, \alpha)}{\partial \alpha}\right)_{\alpha 0} U(s), \quad (4.5\text{–}8)$$

where $G(s, \alpha)$ denotes the transfer function of the variable component. Comparison of Eqs. (4.5–8) and (4.5–7) yields

$$G_1(s, \alpha_0) G_2(s, \alpha_0) = \left.\frac{\partial F(s, \alpha)}{\partial G(s, \alpha)}\right|_{\alpha 0}. \quad (4.5\text{–}9)$$

The above reveals that the series connection of G_1 and G_2 is equivalent to the change of the overall transfer function of the system ΔF caused by the change of the variable component ΔG. Thus, in order to set up the measuring structure for σ, the dependence of the overall transfer function of the system on the variable component need not be known analytically. It is sufficient to know the structure in terms of a block diagram which may be used idrectly for the simulation. It is this fact which constitutes the advantage of the method of the variable component over the analytical approach in the case of large scale systems.

Another benefit of the variable component method is that its application to *λ-parameter* problems avoids the initial conditions problem and thus simplifies the solution considerably. Again, the underlying idea is to start with $\lambda \neq 0$, thereby treating the problem as an α-parameter problem with all initial conditions of the sensitivity model equal to zero. After the derivative $\partial G/\partial \lambda$ is carried out, λ is set equal to zero, thus yielding automatically the required initial conditions. The procedure shall be demonstrated by the following example.

Example 4.5-2 Consider again the second-order system of Example 4.3–7 with the differential equation

$$\lambda \ddot{y} + T\dot{y} + y = u, \qquad y(0) = \dot{y}(0) = 0,$$

where the nominal value of λ is $\lambda_0 = 0$. For $u(t) = 1(t)$, the sensitivity function with respect to λ is to be found.

Solution As mentioned above, the idea is to start with the actual (nonreduced) system structure which can be treated as an α-parameter problem, and then to set $\lambda = 0$. After rewriting the above actual differential equation in the form

$$y = (1/\lambda)\,(u\text{-}y\text{-}T\dot{y}),$$

the structural diagram can be readily drawn. It is shown in Fig. 4.5–9. Note that $1/\lambda$ is the variable component.

According to the rule derived above, to measure the output sensitivity func-

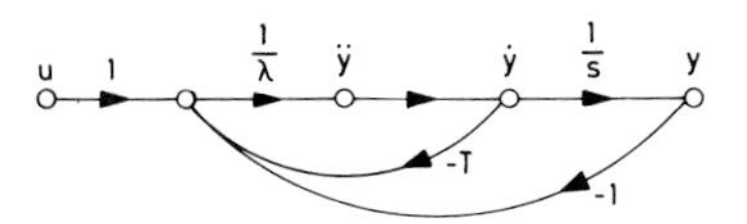

FIG. 4.5-9. Signal flow graph of the actual second-order system of Example 4.5–2.

tion $\sigma(t,0) = (\partial y/\partial\lambda)_{\lambda=0}$, the same diagram has to be drawn once again. A connection has to be established from the input of the variable component $1/\lambda$ in the original model to the output of $1/\lambda$ in the sensitivity model. The transfer function of this connection is

$$\frac{\partial G}{\partial\lambda} = \frac{\partial}{\partial\lambda}\left(\frac{1}{\lambda}\right) = -\frac{1}{\lambda^2}.$$

The corresponding complete structural diagram is shown in Fig. 4.5–10 (solid lines).

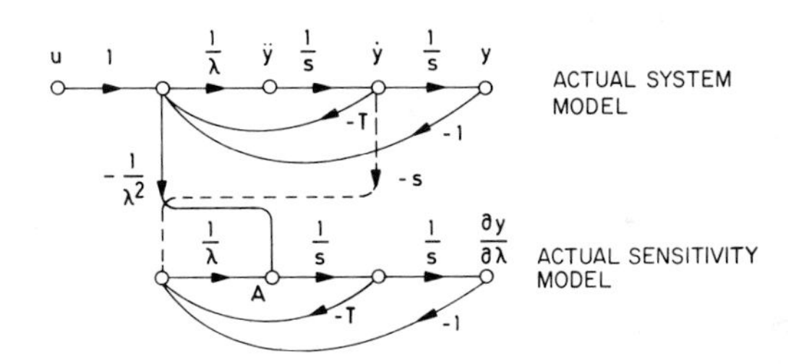

FIG. 4.5-10. Signal flow graph of the complete measuring circuit for $\partial y/\partial\lambda$.

Now we become aware of the difficulties that arise if we let λ approach zero: The transfer function of the connection $-\lambda^{-2}$ becomes $-\infty$ and the point A in the sensitivity model is no longer uniquely defined. This is due to the indefiniteness of the sensitivity equation for $\lambda = 0$.

This difficulty is avoided by moving the origin of the connection line to the right and the end to the left, as shown in Fig. 4.5–10 by the broken line. The transfer function of the resulting connection line is simply $-s$. If now the limit $\lambda \to 0$ is carried out, the transfer function of the minor control loop (in the two models) becomes

$$\lim_{\lambda\to 0} \frac{1/\lambda s}{1 + (T/\lambda s)} = \lim_{\lambda\to 0} \frac{1}{\lambda s + T} = \frac{1}{T}.$$

The resulting structural diagram for measuring $\sigma(t, 0)$ is shown in Fig. 4.5–11. Finally, to avoid the differentiation in the connection line which raises practical problems, the diagram of Fig. 4.5–11 is rearranged such as shown in Fig. 4.5–12.

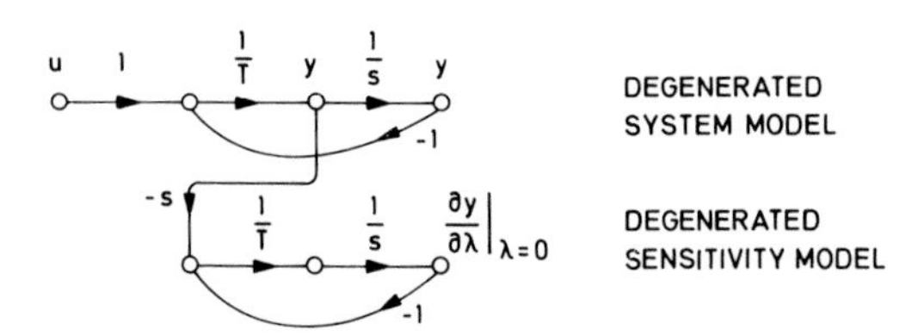

FIG. 4.5-11. Signal flow graph for measuring $\sigma(t, 0)$.

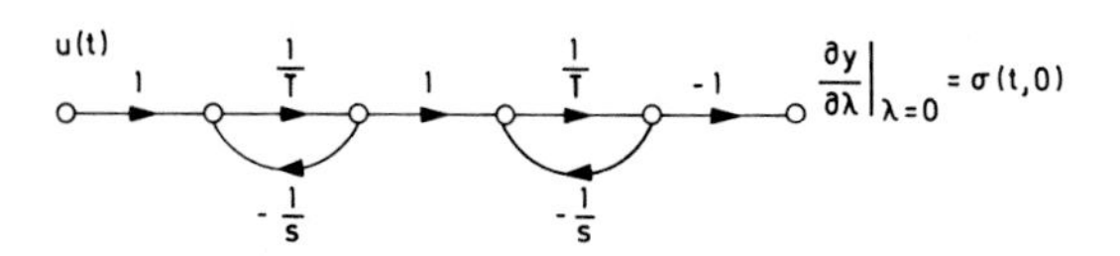

FIG. 4.5-12. Final form of the measuring circuit of Example 4.5–2.

As a main result of this procedure, it can be observed that the problem of initial conditions that causes considerable mathematical expenditure in the case of an analytical treatment is solved here automatically: For $t = +0$, the initial condition is found immediately from the above diagram (Fig. 4.5-12) by setting $s = \infty$ as

$$\sigma(+0, 0) = -u(+0)/T^2.$$

Since for a step input $u(+0) = 1$, we obtain $\sigma(+0, 0) = -1/T^2$, which is in agreement with earlier results. The solution of the initial conditions problem consists here of simple block digaram manipulations.

This procedure is evidently applicable to arbitrary changes of the order and provides substantial simplification compared with the analytical streatment of λ-variations.

In the case of *β-parameters*, the measurement circuit for the output sensitivity function can be found as follows: Consider a linear time-invariant system symbolized by Fig. 4.5–13a. Let the initial conditions $y^{(i)}(0)$ be the parameters of interest. To find the sensitivity model, we take the partial derivatives with respect to $y^{(i)}(0)$ of all signals of the original system. As a result, we see that the same system as the original system is required, since the structure is independent of $y^{(i)}(0)$. However, the derivative of the input signal $\partial u/\partial y^{(i)}(0)$ vanishes as well as those of all initial conditions except that of the initial condition under consideration. The latter becomes $\partial y^{(i)}(0)/\partial y^{(i)}(0) = 1$. Hence, the complete measuring circuit for $\sigma_i(t, \mathbf{y}_0{}^0) = (\partial y/\partial y^{(i)}(0))_{\mathbf{y}_0}$ consists simply of the sensitivity model as shown in Fig. 4.5–13b.

Example 4.5-3 Given the control loop shown in Fig. 4.5–14: Find the measurement diagram for the output sensitivity function with respect to the initial condition $y_1{}^0 = \dot{y}(0)$.

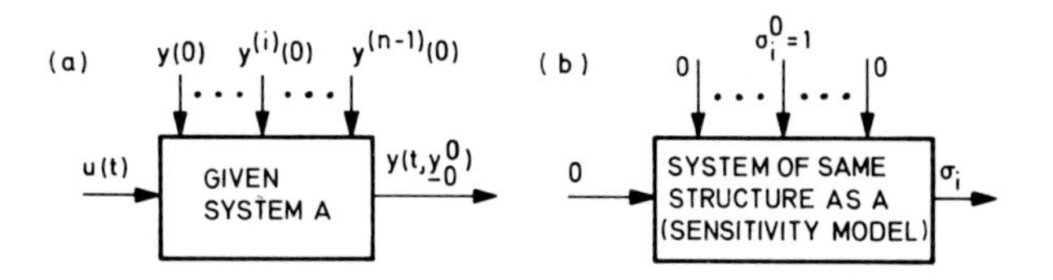

FIG. 4.5-13. (a) Given system model, (b) measuring circuit for the output sensitivity function with respect to the ith initial condition.

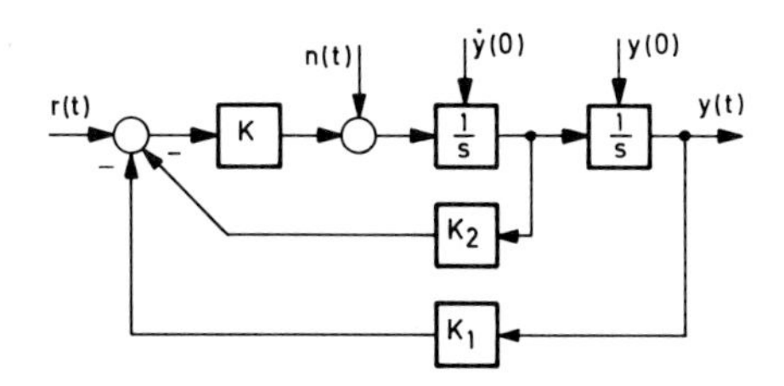

FIG. 4.5-14. Given control loop of example 4.5–3: $r(t)$, reference input; $n(t)$, disturbance.

Solution The structure required is the same as that of the original system but with the input signal and all initial conditions equal to zero except $\dot{\sigma}(0) = \partial\dot{y}(0)/\partial\dot{y}(0) = 1$. The resulting measuring circuit is shown in Fig. 4.5–15.

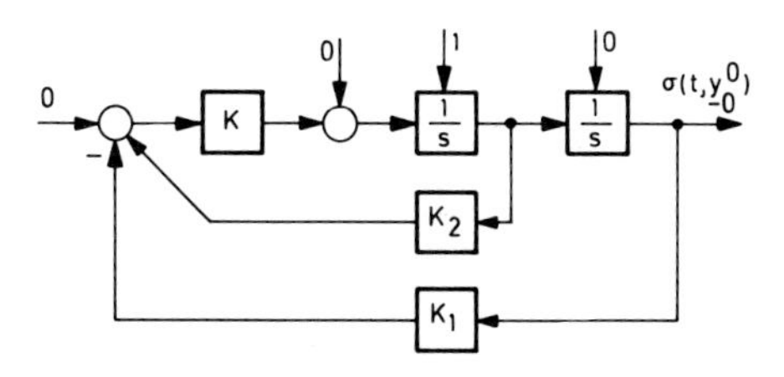

FIG. 4.5-15. Block diagram for measuring the output sensitivity function with respect to $\dot{y}(0)$ of Example 4.5–3.

4.5.2 Derivation for Nonlinear Systems

So far we have confined ourselves to linear systems. In this section, the variable component method will be extended to nonlinear systems involving linear and/or nonlinear variable components. This extension was done by Vusković and Ćircić [106].

Consider a nonlinear system that contains r linear and s nonlinear subsystems as shown in Fig. 4.5–16. The input and output signals of the linear subsystems are denoted by u_i or y_i, respectively; the corresponding signals of the *nonlinear subsystems* are denoted by $\tilde{u}_k$ or $\tilde{y}_k$, respectively. The nonlinear components are characterized by double lines.

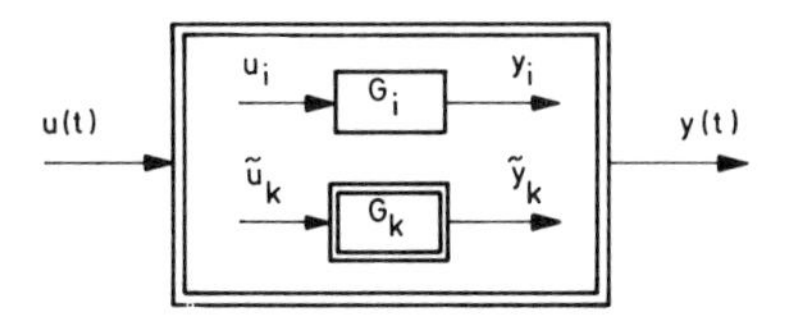

FIG. 4.5-16. Block diagram of the nonlinear system containing linear and nonlinear subsystems.

For the *linear components r* relations of the form

$$Y_i(s, \alpha) = G_i(s, \alpha) U_i(s, \alpha), \qquad i = 1, 2, \ldots, r, \tag{4.5–10}$$

hold, where $G_i(s, \alpha)$ is the transfer function of the ith component.

For the *nonlinear components*, s relations of the form

$$\tilde{y}_k(t, \alpha) = G_k(p, \alpha)\, \tilde{u}_k(t, \alpha), \qquad k = 1, 2, \ldots, s, \tag{4.5–11}$$

hold, where $G_k(p, \alpha)$ represents the nonlinear operator of the kth component applied to the corresponding input $\tilde{u}_k(t, \alpha)$.

Now suppose that the components are interconnected by the *coupling equations*

$$\sum_{i=0}^{r+1}(a_{ji}u_i + b_{ji}y_i) + \sum_{k=1}^{s}(c_{jk}\tilde{u}_k + d_{jk}\tilde{y}_k) = 0, \qquad j = 1, 2, \ldots, q, \tag{4.5–12}$$

where, formally, $y_0 \equiv 0$, $u_0 \equiv u$, $y_{r+1} \equiv y$ and $u_{r+1} \equiv 0$. The coefficients a_{ji}, b_{ji}, c_{jk}, d_{jk} may assume the values -1 or 0 or $+1$ (in accordance with Kirchhoff's network laws).

If the partial derivative of Eq. (4.5–12) with respect to the parameter of interest α is taken and α is set equal to α_0, the signals of the original system get transformed to their corresponding sensitivity functions, that is,

$$\begin{aligned}
&u_i \to \left.\frac{\partial u_i}{\partial \alpha}\right|_{\alpha 0}, \qquad && y_i \to \left.\frac{\partial y_i}{\partial \alpha}\right|_{\alpha 0}, \\
&\tilde{u}_k \to \left.\frac{\partial \tilde{u}_k}{\partial \alpha}\right|_{\alpha 0}, \qquad && \tilde{y}_k \to \left.\frac{\partial \tilde{y}_k}{\partial \alpha}\right|_{\alpha 0}, \\
&u_0 \to 0, \qquad && y_{r+1} \to \left.\frac{\partial y_{r+1}}{\partial \alpha}\right|_{\alpha 0} = \left.\frac{\partial y}{\partial \alpha}\right|_{\alpha 0} = \sigma.
\end{aligned} \tag{4.5–13}$$

How the structure changes through the differentiation depends on how the components depend on α. Let us distinguish two cases:

(1) the nonlinear components do not depend on α,
(2) even nonlinear components depend on α.

Case (1) Suppose that only the first ρ linear components depend on α whereas the reamining r-ρ linear components and the nonlinear components are independent of α. Thus for the r-ρ linear α-independent components, we have

$$\left.\frac{\partial Y_i}{\partial \alpha}\right|_{\alpha 0} = G_{i0} \left.\frac{\partial U_i}{\partial \alpha}\right|_{\alpha 0}, \qquad i = \rho + 1, \rho + 2, \ldots, r, \tag{4.5–14}$$

where the subscript 0 denotes nominal values. This indicates that the α-independent linear portion of the sensitivity model is identical with the corresponding part of the nominal original system.

For the ρ α-dependent linear components, we have

$$\left.\frac{\partial Y_i}{\partial \alpha}\right|_{\alpha 0} = U_{i0} \left.\frac{\partial G_i}{\partial \alpha}\right|_{\alpha 0} + G_{i0} \left.\frac{\partial U_i}{\partial \alpha}\right|_{\alpha 0}, \qquad i = 1, 2, \ldots, \rho. \tag{4.5–15}$$

Consequently, as above, the corresponding components of the sensitivity model are identical with those of the nominal original system. However, to the output of each component in the sensitivity model a signal has to be added that is generated from the input of the corresponding component u_i, in the original system by a block with the transfer function $(\partial G_i/\partial \alpha)_{\alpha 0}$.

So far, there is no distinction from the all linear case.

For the s nonlinear components presumed to be independent of α, we have

$$\left.\frac{\partial \tilde{y}_k}{\partial \alpha}\right|_{\alpha 0} = \left.\frac{\partial G_k}{\partial \tilde{u}_k}\right|_{\alpha 0} \left.\frac{\partial \tilde{u}_k}{\partial \alpha}\right|_{\alpha 0}, \qquad k = 1, 2, \ldots, s. \tag{4.5–16}$$

The term $(\partial G_k/\partial \tilde{u}_k)_{\alpha 0}$ represents a time-variant coefficient of $(\partial \tilde{u}_k/\partial \alpha)_{\alpha 0}$. This means that the sensitivity function $(\partial \tilde{y}_k/\partial \alpha)_{\alpha 0}$ is obtained by multiplication of the corresponding input sensitivity function $(\partial \tilde{u}_k/\partial \alpha)_{\alpha 0}$ by $(\partial G_k/\partial \tilde{u}\ )_{\alpha 0}$. Consequently, in the sensitivity model, a multiplier has to be inserted instead of the operator G_{k0}, and this multiplier has to be connected with the corresponding input signal of the nominal original system by a nonlinear block with the operator $(\partial G_k/\partial \tilde{u}_k)_{\alpha 0}$.

The resulting structure is shown in Fig. 4.5–17 (solidlines). For simplicity of representation, all subscripts 0 and α_0 have been skipped. Notice that the whole structure includes nominal parameters.

Case (2) Suppose that there are also nonlinear components that depend on α. The relations for the nonlinear components depending on α are given by

$$\tilde{y}_k(t, \alpha) = G_k(p, \alpha)\{\tilde{u}_k(t, \alpha)\}, \qquad k = 1, 2, \ldots, s,$$

that is, here the nonlinear operator is also a function of α. Therefore, partial differentiation with respect to α yields

$$\left.\frac{\partial \tilde{y}_k}{\partial \alpha}\right|_{\alpha 0} = \left.\frac{\partial G_k\{\}}{\partial \alpha}\right|_{\alpha 0} + \left.\frac{\partial G_k\{\}}{\partial \tilde{u}_k}\right|_{\alpha 0} \frac{\partial \tilde{u}_k}{\partial \alpha}. \tag{4.5–17}$$

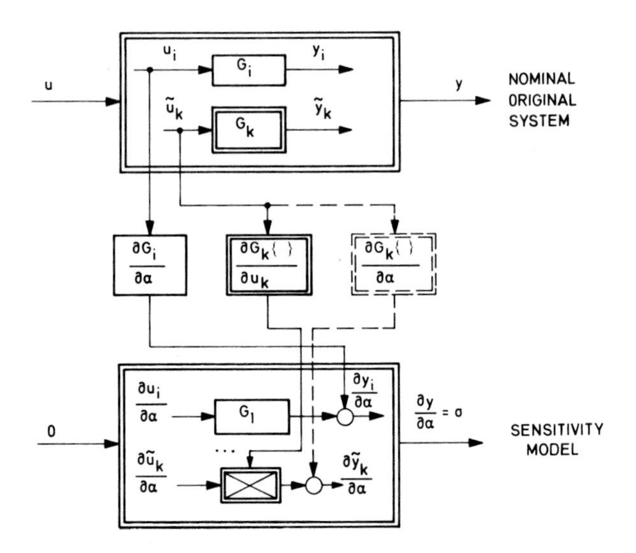

FIG. 4.5-17. Measuring circuit for the output sensitivity function in the case of nonlinear components: α-independent nonlinear components, solid lines; α-dependent nonlinear components, with broken-line appendix.

This says that, in the sensitivity model, a signal $(\partial G_k\{\,\}/\partial\alpha)_{\alpha 0}$ has to be added to the output of the corresponding multiplier. The rest of the structure remains unchanged. The corresponding necessary appendix in the measuring structure is shwon in Fig. 4.5–17 (broken lines).

Finally, consider the case in which the original system structure includes multipliers itself. Here again two cases have to be distinguished:

(1) Variables depending on α are multiplied by external inputs that do not depend on α (bilinear systems).

(2) Variables depending on α are multiplied with each other.

In the *first case*, the operation of the multiplier is of the form

$$\tilde{y}_k(t, \alpha) = \tilde{u}_j(t, \alpha)u(t), \tag{4.5–18}$$

where $\tilde{y}_k$ is the output of the multiplier, $u(t)$ is an external input, and $\tilde{u}_j$ can be any other signal of the system. Differentiating Eq. (4.5.18) with respect to α and setting $\alpha = \alpha_0$ yields

$$\left.\frac{\partial \tilde{y}_k(t, \alpha)}{\partial \alpha}\right|_{\alpha 0} = u(t)\left.\frac{\partial \tilde{u}_j(t, \alpha)}{\partial \alpha}\right|_{\alpha 0}. \tag{4.5–19}$$

Hence a multiplier has to be inserted in the sensitivity model between the input and the sensitivity function of u_j. In other words, the multiplier of the original system has to be rebuilt without any change in the sensitivity model.

In the *second case*, the operation of the multiplier is of the form

$$\tilde{y}_k(t, \alpha) = \tilde{u}_j(t, \alpha)\tilde{u}_i(t, \alpha), \tag{4.5-20}$$

where $\tilde{u}_j$ and $\tilde{u}_i$ are any two variables of the system. Partial differentiation with respect to α (and setting $\alpha = \alpha_0$) yields

$$\left.\frac{\partial \tilde{y}_k}{\partial \alpha}\right|_{\alpha_0} = \tilde{u}_{j0}\left.\frac{\partial \tilde{u}_i}{\partial \alpha}\right|_{\alpha_0} + \tilde{u}_{i0}\left.\frac{\partial \tilde{u}_j}{\partial \alpha}\right|_{\alpha_0}. \tag{4.5-21}$$

Due to this result, two multiplications have to be established in the sensitivity model or, in other words, all multipliers of the original system that multiply two inner variables have to be doubled in the sensitivity model. In general, the multiplication of n inner variables in the original system requires n multipliers in the sensitivity model.

4.5.3 General Rules to Set Up the Block Diagram

(1) The block diagram of the original system is separated into

(a) blocks that do not depend on α,

(b) blocks that depend on α.

(2) A second block diagram is drawn which differs from the original system diagram as follows:

(a) All external inputs and initial conditions are zero.

(b) The linear blocks that are independent of α are repeated with no change in wiring.

(c) To the outputs of the α-dependent linear blocks, signals are added which are generated from the inputs of the corresponding blocks of the original diagram by means of blocks with the transfer functions $(\partial G_i/\partial \alpha)_{\alpha_0}$.

(d) All nonlinear blocks are replaced by multipliers which multiply their input signals with signals generated from the inputs of the corresponding blocks of the original diagram by operators $(\partial G_k/\partial \tilde{u}_k)_{\alpha_0}$. For those components which are α-dependent, a signal has to be added, at the output of the corresponding multiplier, which is generated from the input signal of the corresponding block of the original diagram by the operator $(\partial G_k/\partial \alpha)_{\alpha_0}$.

(e) If n α-dependent variables are multiplied in the original system, then n multipliers have to be provided in the sensitivity model.

(3) If there are multiple parameters of interest, an individual sensitivity model is required for each parameter, so that there are required as many sensitivity models as there are parameters of interest.

Example 4.5-4 For the bilinear system with the differential equation $\dot{y} + ay = byu$ (y output, u input), draw a block diagram from which the output sensitivity function with respect to the parameter a can be taken.

Solution Figure 4.5-18a shows the structure of the original system. Due to this,

$$\left.\frac{\partial v}{\partial a}\right|_{a_0} = u\left.\frac{\partial y}{\partial a}\right|_{a_0},$$

$$\left.\frac{\partial z}{\partial a}\right|_{a_0} = a_0\left.\frac{\partial y}{\partial a}\right|_{a_0} + y_0.$$

All other sensitivity equations are of the same type as the original equations. Therefore, the resulting diagram is as shown in Fig. 4.5–18b.

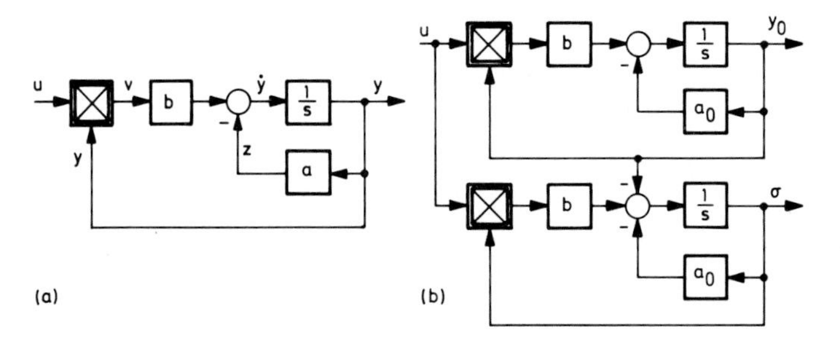

FIG. 4.5-18. Bilinear system of Example 4.5-4: (a) actual original system, (b) corresponding block diagram to measure the output sensitivity function $\sigma = (\partial y/\partial a)_{a_0}$.

Example 4.5-5 Figure 4.5-19a shows the field circuit of a separately excited dc motor. The nonlinearity of the magnetic field is taken into account by the nonlinear characteristic $\tilde{F}(\psi_f) = a\psi_f + b\psi_f^{\,3}$, where ψ_f is the magnetic flux

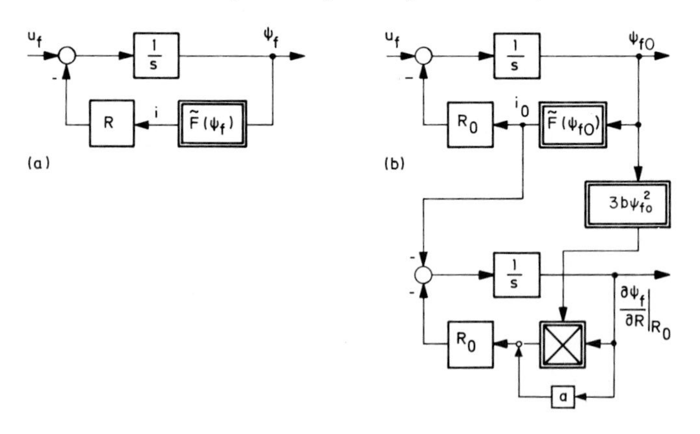

FIG. 4.5-19. Field circuit of Example 4.5-5: (a) actual diagram, (b) complete diagram to measure $\sigma = (\partial\psi_f/\partial R)_{R_0}$.

and a and b are constants. Draw the complete block diagram to measure the sensitivity function of the flux ψ_f with respect to the resistance R.

Solution From Fig. 4.5–19a we have $i_f(t, R) = \tilde{F}(\psi_f) = a\psi_f + b\psi_f^3$. Taking the partial derivative with respect to R and letting R go to R_0, we obtain

$$\left.\frac{\partial i_f}{\partial R}\right|_{R_0} = a\left.\frac{\partial \psi_f}{\partial R}\right|_{R_0} + 3b\psi_{f0}^2\left.\frac{\partial \psi_f}{\partial R}\right|_{R_0}.$$

All other equations are of the same type as the original equations. Therefore, the diagram of Fig. 4.5–19b is obtained.

Example 4.5-6 Figure 4.5–20 shows the block diagram of a separately excited dc motor where the nonlinear characteristic of the magnetic field is taken into account by $\tilde{F}(\psi_f) = a\psi_f + b\psi_f^3 + c\psi_f^5$. Here $a, b, c, R_a, L_a, R_f, J, k_1, k_2$, and k are the parameters of the motor, M_1 and M_2 are multipliers. m_L denotes the load moment, v_a the armature voltage, v_f the field voltage, and ω the angular velocity. Draw a block diagram to measure the sensitivity function of ω with respect to the constant b.

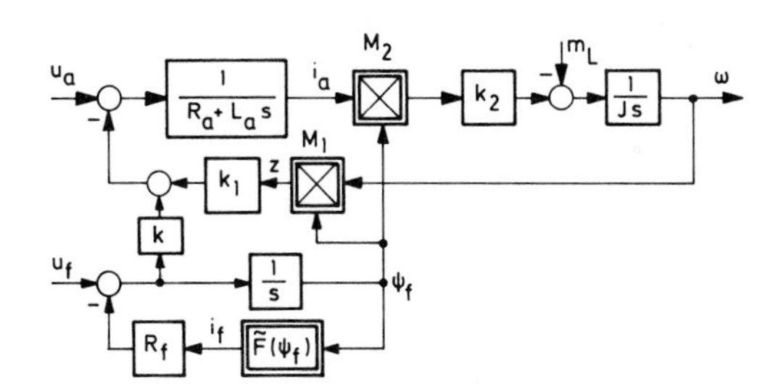

FIG. 4.5-20. Block diagram of the separately excited dc motor of Example 4.5-6.

Solution All linear relations remain unchanged in the sensitivity model. For the multipliers M_1 and M_2 we obtain

$$\left.\frac{\partial z}{\partial b}\right|_{b0} = \omega_0\left.\frac{\partial \psi_f}{\partial b}\right|_{b0} + \psi_{f0}\left.\frac{\partial \omega}{\partial b}\right|_{b0},$$

$$\left.\frac{\partial x}{\partial b}\right|_{b0} = i_{a0}\left.\frac{\partial \psi_f}{\partial b}\right|_{b0} + \psi_{f0}\left.\frac{\partial i_a}{\partial b}\right|_{b0},$$

and for the field nonlinearity we have

$$\left.\frac{\partial i_f}{\partial b}\right|_{b0} = a_0\left.\frac{\partial \psi_{f0}}{\partial b}\right|_{b0} + 3b_0\psi_{f0}\left.\frac{\partial \psi_f}{\partial b}\right|_{b0} + 5c\psi_{f0}^4\left.\frac{\partial \psi_f}{\partial b}\right|_{b0} + \psi_{f0}^3.$$

According to these relations, the diagram of Fig. 4.5-21 is obtained. For

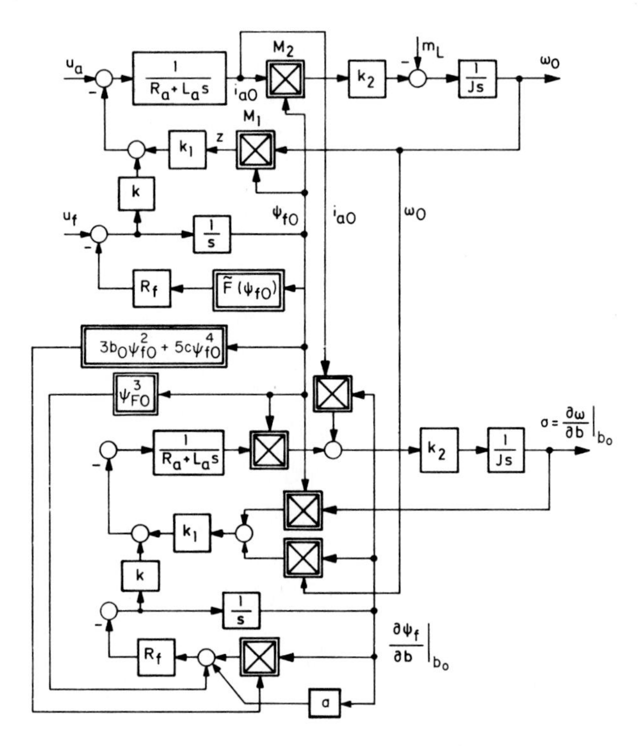

FIG. 4.5-21. Block diagram for measuring the sensitivity function of the speed ω of a dc motor with respect to the constant b of the nonlinearity in the magnetic field (Example 4.5-6).

further exercise, determine the corresponding block diagram taking the coefficient a as the parameter of interest.

4.5.4 Measurement of the Output Error Due to Finite Parameter Deviations

Now it will be shown how the parameter-induced error $\Delta y = y - y_0$ can be measured by the variable component method in the case of finite parameter deviations. Consider the measurement diagram for Δy given in Fig. 4.5-4 in the case of infinitesimal parameter changes. This diagram reflects the fact that $\Delta U_i\, \Delta G$ was neglected and, therefore, $\delta Y_i = U_i\, \Delta G$.

If the parameter changes are finite, the term $\Delta U_i\, \Delta G$ can no longer be neglected so that

$$\delta Y_i = \Delta G\, U_i + \Delta G\, \Delta U_i.$$

In terms of the block diagram, this means that an additional connection has to be installed from the input of G of the sensitivity model to the input of ΔG which is shown in broken lines in Fig. 4.5-22.

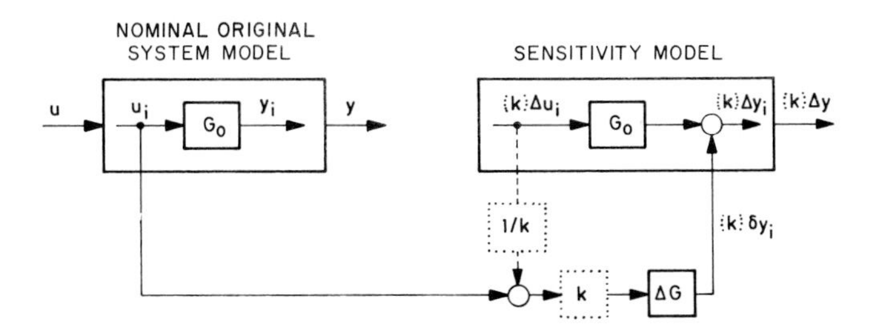

FIG. 4.5-22. Combined system in the case of finite parameter changes.

By inserting the gains k and $1/k$ (dotted lines) where k is large, Δy can be measured with increased accuracy, since the output signal will then be $k\ \Delta y$, instead of Δy.

Instead of the relation (4.5-6),

$$\Delta Y(s, \alpha) = G_1(s, \alpha_0)G_2(s, \alpha_0)\ \Delta G(s, \alpha)\ U(s),$$

we now have

$$\Delta Y(s,\alpha) = \frac{G_1(s, \alpha_0)G_2(s, \alpha_0)\ \Delta G(s, \alpha)}{1 - \Delta G(s, \alpha)\ G_3(s, \alpha_0)}\ U(s),$$

where $G_3(s, \alpha_0) = \Delta U_i(s, \alpha)/\delta Y_i(s, \alpha)$ and G_1, G_2 are as defined in Eqs. (4.5-5). As we see, the only distinction from the previous result is that the denominator now is $1 - \Delta G(s, \alpha)\ G_3(s, \alpha)$ instead of 1.

4.5.5 Limits of the Variable Component Method

The variable component method is applicable in a very wide field of problems. Nevertheless, there are some cases where the implementation of the sensitivity model may cause difficulties. Such a case occurs if the sensitivity function diverges with time which is equivalent to an unstable sensitivity model.

Consider, for example, a system that executes sustained harmonic oscillations (Example 3.2-2), so that the output is $y(t, \omega) = \sin \omega t$, and the corresponding sensitivity function with respect to ω becomes $\sigma(t, \omega_0) = t \cos \omega_0 t$. The corresponding sensitivity model is unstable. Another example is a nonlinear system with a characteristic with discontinuities, such as a relay characteristic $y = f[x(t, \alpha)] = \operatorname{sgn} x(t, \alpha)$. If we take the derivative with respect to α, we obtain $\partial y/\partial\alpha = (\partial f/\partial x)(\partial x/\partial\alpha)$. In the case of sgn $x(t, \alpha)$, $\partial f/\partial x$ is zero, except at $x = 0$ where it becomes infinite. An operation like this cannot be realized exactly in the sensitivity model. However, there are certain situations where these difficulties can be avoided [85].

4.6 THE METHOD OF SENSITIVITY POINTS

For some applications, e.g., for model-adaptive or self-optimizing systems, the sensitivity functions with respect to all parameters are required simultaneously. If the variable component method is applied for this purpose and if there are r parameters of interest, r sensitivity models would be needed. This implies a rather extensive analog computer program or an adequate amount of digital-computing time.

Kokotović has proposed a method of determining all output sensitivity functions of a system simultaneously by a single sensitivity model [55]. This method, which is also called the method of sensitivity points, is restricted to linear systems. The idea was later extended to state space considerations by Wilkie and Perkins (see Chapter 6). The idea of the method of sensitivity points shall be described in this section.

4.6.1 General Derivation via Block Diagram

Consider a linear time-invariant system with the input u, the output y, and the transfer function $G(s, \boldsymbol{\alpha})$ that contains r components with the transfer function $G_j(s, \alpha_j)$, $j = 1, 2, \ldots, r$. For the sake of mathematical simplicity, let us assume that each component G_j depends only on a single parameter α_j. The semirelative output sensitivity functions

$$\tilde{\sigma}_j(t, \boldsymbol{\alpha}_0) = \left.\frac{\partial y(t, \boldsymbol{\alpha})}{\partial \ln \alpha_j}\right|_{\boldsymbol{\alpha}_0}, \qquad j = 1, 2, \ldots, r, \tag{4.6-1}$$

are to be determined.

If the order of Laplace transformation and partial differentiation with respect to $\ln \alpha_j$ (including the limiting process $\alpha \to \alpha_0$) is reversible (see Section 4.2), then the transform of the sensitivity function may be written as

$$\tilde{\Sigma}_j(s, \boldsymbol{\alpha}_0) = \left.\frac{\partial Y(s, \boldsymbol{\alpha})}{\partial \ln \alpha_j}\right|_{\boldsymbol{\alpha}_0}. \tag{4.6-2}$$

With the aid of $Y(s, \boldsymbol{\alpha}) = G(s, \boldsymbol{\alpha})U(s)$, and by application of the chain rule (Section 2.4), Eq. (4.6-2) may be rewritten as

$$\tilde{\Sigma}_j(s, \boldsymbol{\alpha}_0) = \left[\frac{\partial G(s, \boldsymbol{\alpha})}{\partial G_j(s, \alpha_j)} \frac{\partial G_j(s, \alpha_j)}{\partial \ln \alpha_j}\right]_{\boldsymbol{\alpha}_0} U(s). \tag{4.6-3}$$

Substituting $U(s) = Y(s, \boldsymbol{\alpha})/G(s, \boldsymbol{\alpha})$ yields

$$\Sigma_j(s, \boldsymbol{\alpha}_0) = \left[\frac{\partial \ln G}{\partial \ln G_j} \frac{\partial \ln G_j}{\partial \ln \alpha_j} Y(s, \boldsymbol{\alpha})\right]_{\boldsymbol{\alpha}_0} = S_{G_j}^{G} S_{\alpha_j}^{G_j} Y(s, \boldsymbol{\alpha}_0), \tag{4.6-4}$$

where

$$S_{G_j}^{G}(s) = \left.\frac{\partial \ln G}{\partial \ln G_j}\right|_{\boldsymbol{\alpha}_0} \triangleq F_j(s) \tag{4.6-5}$$

and

$$S_{\alpha_j}^{G_j}(s) = \left.\frac{\partial \ln G_j}{\partial \ln \alpha_j}\right|_{\boldsymbol{\alpha}_0} \triangleq H_j(s) \tag{4.6-6}$$

are the Bode sensitivity functions.

In terms of a structural interpretation, the semirelative sensitivity functions σ_j can be obtained at the output of a series connection of $F_j(s)$ and $H_j(s)$ as $y_0 = y(t, \boldsymbol{\alpha}_0)$ is applied to the input. In order to determine all r sensitivity functions by this procedure, r series connections $F_j(s)H_j(s)$ would be needed, where F_j only depends on the system structure and H_j specifies the dependence of G_j on α_j only. Now it can be shown that the transfer functions F_j are already given by the nominal original system as the sections between the input and certain points, the so-called sensitivity points.

To show this, let us consider the feedback configuration depicted in Fig. 4.6.1. This configuration consists of r blocks, each of them depending on a single parameter according to $G_j = G_j(s, \alpha_j)$. This block diagram is first brought into a more condensed form by the following block diagram manipulations.

Let the part enclosed by broken lines be represented by a single block L_i, and let all the feedback paths beyond the point B_{2i-1} be combined into a single block which has the transfer function

$$M_i = \sum_{k=i+1}^{r/2} G_{2k} \prod_{\nu=1}^{k-1} G_{2\nu+1} \tag{4.6-7}$$

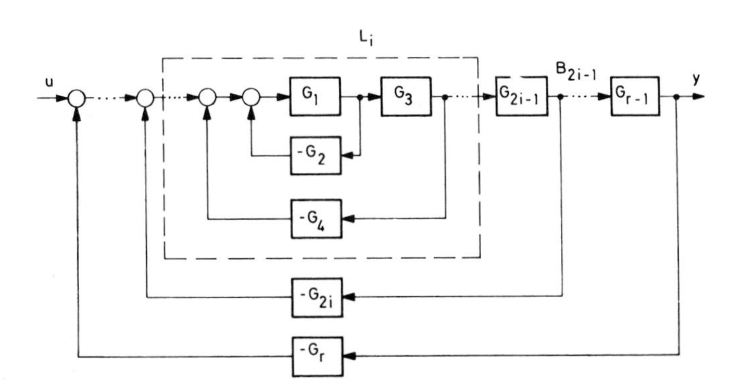

FIG. 4.6-1. Original system configuration for the derivation of the method of sensitivity points.

Similarly, combine the series connection of all components between B_{2i-1} and the output to a single block which has the transfer function

$$N_i = \prod_{k=i+1}^{r/2} G_{2k-1} = G_{2i+1} \cdots G_{r-1}. \tag{4.6-8}$$

Then the diagram of Fig. 4.6–1 can be substituted equivalently by the block diagram of Fig. 4.6-2 (solid lines).

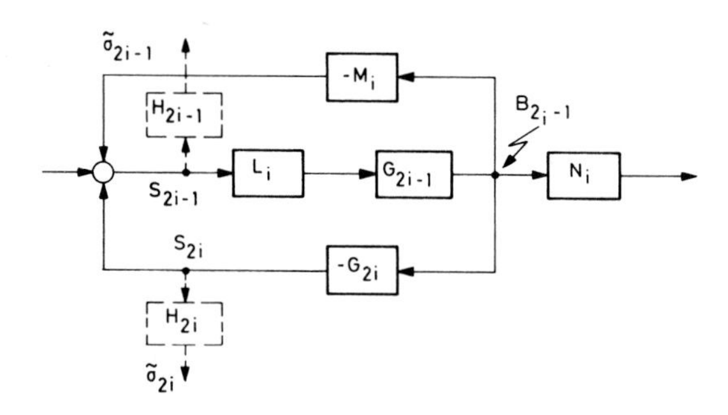

FIG. 4.6-2. Equivalent block diagram of Fig. 4.6-1.

If M_i and G_{2i} are combined to a single block with $P_i = M_i + G_{2i}$, the transfer function Y/U of the above block diagram becomes

$$G = \frac{N_i L_i G_{2i-1}}{1 + L_i P_i G_{2i-1}}. \tag{4.6-9}$$

Taking the partial derivative of G with respect to G_{2i-1} and multiplying numerator and denominator by G_{2i-1}, one obtains

$$\frac{\partial G}{\partial G_{2i-1}} = \frac{N_i L_i G_{2i-1}}{(1 + L_i P_i G_{2i-1})^2 G_{2i-1}} = \frac{1}{1 + L_i P_i G_{2i-1}} \frac{G}{G_{2i-1}}. \tag{4.6-10}$$

Hence, the Bode sensitivity function $F_{2i-1} = \partial \ln G / \partial \ln G_{2i-1}$ becomes

$$F_{2i-1} = (1 + L_i P_i F_{2i-1})^{-1}, \tag{4.6-11}$$

where the subscript 0 is skipped for notational ease. We observe that F_{2i-1} appears in the block diagram of Fig. 4.6–2 between the input and the point S_{2i-1} at the input of L_i. In other words, if the input of the above configuration is excited by the output of the nominal original system $y(t, \boldsymbol{\alpha}_0)$, rather than by $u(t)$, then the signal

$$F_{2i-1} Y(s, \boldsymbol{\alpha}_0) = \left[\frac{\partial \ln G}{\partial \ln G_{2i-1}} Y(s, \boldsymbol{\alpha})\right]_{\boldsymbol{\alpha}_0} \tag{4.6–12}$$

appears at the point S_{2i-1}. Hence, according to Eq. (4.6–4), in order to measure $\tilde{\sigma}_{2i-1} = (\partial y / \partial \ln \alpha_{2i-1})_{\boldsymbol{\alpha}0}$, an additional block with the transfer function $H_{2i-1} = (\partial \ln G_{2i-1} / \partial \ln \alpha_{2i-1})_{\boldsymbol{\alpha}0}$ has to be connected at the point

S_{2i-1}, and $\tilde{\sigma}_{2i-1}$ is then obtained at the output of H_{2i-1}. This is illustrated in Fig. 4.6–2 in broken lines.

The points S_{2i} can be found in a similar way by taking the partial derivative of G with respect to G_{2i} $(i = 1, 2, \ldots, r/2)$. For the sake of simplicity, let us denote the transfer function of the feedback loop consisting of L_i, G_{2i-1} and M_i by

$$Q_i \triangleq \frac{L_i G_{2i-1}}{1 + M_i L_i G_{2i-1}}. \tag{4.6-13}$$

Using Q_i, the overall transfer function of the system can be written as

$$G = \frac{N_i Q_i}{1 + Q_i G_{2i}}. \tag{4.6-14}$$

Partial differentiation of G with respect to G_{2i} and multiplication of numerator and denominator by G_{2i} gives

$$\frac{\partial G}{\partial G_{2i}} = \frac{-N_i Q_i^2 G_{2i}}{(1 + Q_i G_{2i})^2 G_{2i}} = -\frac{Q_i G_{2i}}{1 - Q_i G_{2i}} \frac{G}{G_{2i}}. \tag{4.6-15}$$

Thus the Bode sensitivity function $F_{2i} = \partial \ln G / \partial \ln G_{2i}$ becomes

$$F_{2i} = \frac{Q_i G_{2i}}{1 + Q_i G_{2i}}. \tag{4.6-16}$$

This transfer function appears in the block diagram between the input and the point S_{2i}. If, in accordance with Eq. (4.6-4), a block with the transfer function $H_{2i} = (\partial \ln G_{2i} / \partial \ln \alpha_{2i})_{\boldsymbol{\alpha}_0}$ is connected to S_{2i} and y_0 is applied to the input of the whole system, then $\tilde{\sigma}_{2i} = (\partial y / \partial \ln \alpha_{2i})_{\boldsymbol{\alpha}_0}$ is obtained at the output of H_{2i}. This is illustrated in Fig. 4.6–2 (broken lines).

The above holds for any $i = 1, 2, \ldots, r/2$. Therefore, there exists such a point S_j for any parameter α_j. The points S_j are called the *sensitivity points* of the system. Evidently, the sensitivity points are defined by the original system; they are an essential part of the system structure. The resulting block diagram for the simultaneous measurement of the semirelative output sensitivity functions with respect to all parameters is shown in Fig. 4.6-3. The result is summarized by the following theorem.

Theorem 4.6-1 The Bode sensitivity functions of a linear system with respect to its variable components can be obtained at the sensitivity points. The sensitivity points are essential structural properties of the system.

The above derivations, including Fig. 4.6–3, also hold in the case such that each of the components G_j depends on more than a single parameter. If, for example,

$$G_j = G_j(s, \alpha_{j1}, \alpha_{j2}),$$

then, according to Eq. (3.3-7), the signal at S_j has to be applied to the inputs

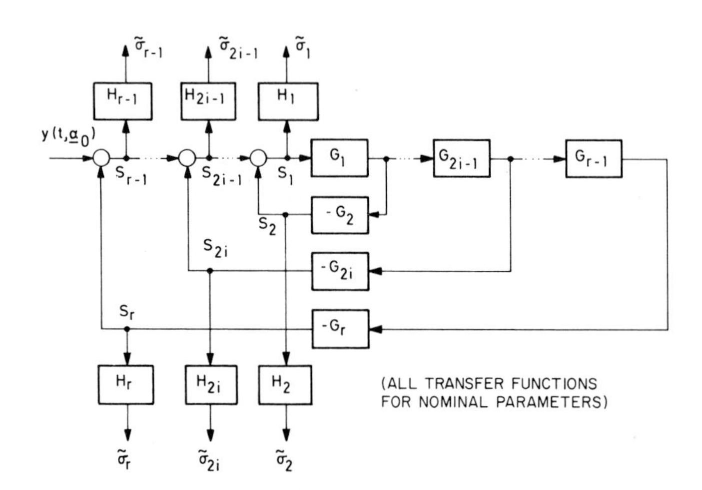

FIG. 4.6-3. Resulting block diagram for the simultaneous measurement of all semi-relative output sensitivity functions.

of two blocks with the Bode sensitivity functions H_{j1} and H_{j2}, at the outputs of which the corresponding semirelative output sensitivity functions with respect to α_{j1} and α_{j2}, respectively, can be obtained. This procedure can be extended to any number of parameter influences in G_j.

In an analogous manner, higher-order sensitivity functions can be determined by repeated application of the above procedure [18].

Finally, two cases of practical importance will be considered:

(1) If the parameters of interest are the gain factors of a system, i.e., $G_i(s, \alpha_i) = \alpha_i A_i(s)$ where $A_i(s)$ does not depend on α_i, then $H_i(s) = 1$. Hence the sensitivity functions σ_i are obtained directly at the sensitivity points.

(2) If the transfer function of a variable component is the product of two transfer functions, only one of which depending on the parameter α_i, i.e., $G_i(s, \alpha_i) = A_i(s)B_i(s, \alpha_i)$, then $H_i(s, \alpha_i) = \partial \ln B_i(s, \alpha_i)/\partial \ln \alpha_i$; or, in other words, $H_i(s, \alpha_i)$ is independent of the parameter-independent component.

In practice the measurement of $\tilde{\sigma}_i$ can be carried out in two different ways:

(1) The input $u(t)$ is given to a model of the original nominal system. The corresponding output $y_0(t)$ is measured and stored. Then $y_0(t)$ is applied as a new input to the same system, and the sensitivity functions $\tilde{\sigma}_i$ are obtained at the sensitivity points;

(2) The nominal system is modeled twice and connected in series as shown in Fig. 4.6-4.

Note that the results are valid for *linear systems* only.

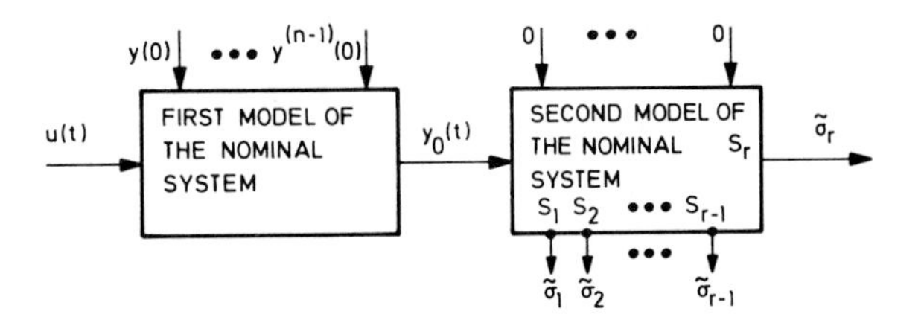

FIG. 4.6-4. Simultaneous measurement of all sensitivity functions.

Example 4.6-1 Consider the control loop depicted in Fig. 4.6-5 where the gain factors α_1 and α_2 are the parameters of interest.

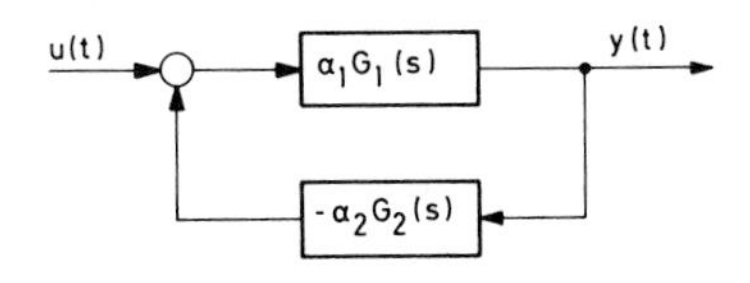

FIG. 4.6-5. Given feedback system of Example 4.6-1.

(1) Mark the sensitivity points S_1 and S_2 with respect to the parameters α_1 and α_2.

(2) Draw the complete block diagram for the simultaneous measurement of the semirelative sensitivity functions $\tilde{\sigma}_1$ and $\tilde{\sigma}_2$ from the input signal $u(t)$.

(3) Give analytical expressions for

$$\tilde{\Sigma}_1 = \mathscr{L}\{\tilde{\sigma}_1\} \quad \text{and} \quad \tilde{\Sigma}_2 = \mathscr{L}\{\tilde{\sigma}_2\}.$$

(4) How can all second-order sensitivity functions

$$\tilde{\nu}_1 = \left.\frac{\partial^2 y}{(\partial \ln \alpha_1)^2}\right|_{\boldsymbol{\alpha}_0}, \qquad \tilde{\nu}_2 = \left.\frac{\partial^2 y}{(\partial \ln \alpha_2)^2}\right|_{\boldsymbol{\alpha}_0}, \qquad \tilde{\nu}_{12} = \tilde{\nu}_{21} = \left.\frac{\partial^2 y}{\partial \ln \alpha_1 \partial \ln \alpha_2}\right|_{\boldsymbol{\alpha}_0}$$

be measured simultaneously?

Solutions (1) The sensitivity points can readily be marked: S_1 is at the input of $\alpha_1 G_1$ and S_2 is at the output of $-\alpha_2 G_2$.

(2) In the present case, the Bode sensitivity functions are $H_1(s, \alpha_1) = 1$, $H_2(s, \alpha_2) = 1$. Hence the complete block diagram for $\tilde{\sigma}_1$ and $\tilde{\sigma}_2$ is as shown in Fig. 4.6-6.

(3) From the block diagram, the transfer functions can be given immediately:

$$\tilde{\Sigma}_1(s,\boldsymbol{\alpha}_0) = \frac{Y(s)}{1 + \alpha_{10} G_1(s) \alpha_{20} G_2(s)} = \frac{\alpha_{10} G_1(s) U(s)}{[1 + \alpha_{10}\alpha_{20} G_1(s) G_2(s)]^2},$$

$$\tilde{\Sigma}_2(s,\boldsymbol{\alpha}_0) = \frac{-\alpha_{10} G_1(s) \alpha_{20} G_2(s) Y(s)}{1 + \alpha_{10}\alpha_{20} G_1(s) G_2(s)} = \frac{-\alpha_{10}^2 \alpha_{20} G_1{}^2(s) G_2(s) U(s)}{[1 + \alpha_{10}\alpha_{20} G_1(s) G_2(s)]^2}.$$

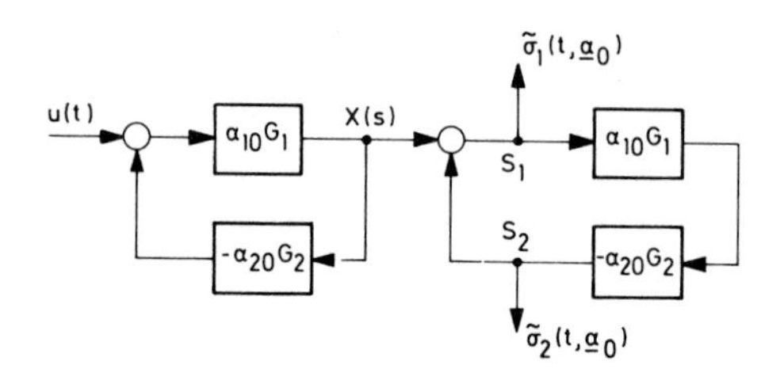

FIG. 4.6-6. Complete block diagram for measuring $\tilde{\sigma}_1$ and $\tilde{\sigma}_2$.

Notice that by application of this procedure, the derivatives need not be found analytically.

(4) The Laplace transforms of the second-order sensitivity functions $\tilde{N}_i = \mathscr{L}\{\tilde{\nu}_i\}$ can be found by taking the second derivative of Y with respect to the parameters and then multiplying by α_{i0}^2. By this procedure the first function is found as follows:

$$\begin{aligned}
\tilde{N}_1 &= \frac{\partial}{\partial\alpha_1}\left(\frac{G_1(s)U(s)}{[1+\alpha_1\alpha_{20}G_1(s)G_2(s)]^2}\right)_{\alpha_{10}}\alpha_{10}^2 \\
&= \left.\frac{\alpha_{10}^2 G_1(s)U(s)[-2\alpha_{20}G_1(s)G_2(s)(1+\alpha_1\alpha_{20}G_1(s)G_2(s))]}{[1+\alpha_1\alpha_{20}G_1(s)G_2(s)]^4}\right|_{\alpha_{10}} \\
&= \frac{-2\alpha_{10}^2\alpha_{20}G_1{}^2(s)G_2(s)U(s)}{[1+\alpha_{10}\alpha_{20}G_1(s)G_2(s)]^3} = -\frac{2\alpha_{10}\alpha_{20}G_1(s)G_2(s)\tilde{\Sigma}_1}{1+\alpha_{10}\alpha_{20}G_1(s)G_2(s)}.
\end{aligned}$$

For the remaining functions $\tilde{N}_2$ and $\tilde{N}_{1,2}$, one obtains

$$\tilde{N}_2 = \mathscr{L}\{\tilde{\nu}_2\} = -2\frac{\alpha_{10}G_1\alpha_{20}G_2}{1+\alpha_{10}\alpha_{20}G_1G_2}\tilde{\Sigma}_2,$$

$$\tilde{N}_{1,2} = \mathscr{L}\{\tilde{\nu}_{1,2}\} = 2\frac{1}{1+\alpha_{10}\alpha_{20}G_1G_2}\tilde{\Sigma}_1.$$

Hence $\frac{1}{2}\tilde{\nu}_1$ and $\frac{1}{2}\tilde{\nu}_{1,2}$ can be measured at the sensitivity points S_2 and S_1, respectively, of a second sensitivity model coupled to the sensitivity point S_1 of the first sensitivity model. $\frac{1}{2}\tilde{\nu}_2$ can be measured at S_2 of a third sensitivity model coupled to S_2 of the first sensitivity model.

4.6.2 Application to Differential Equations with No Right-Hand Derivatives

In this section we consider linear systems of nth order differential equations of the form

$$\sum_{i=0}^{n} a_i y^{(i)}(t) = u(t), \qquad a_n = 1. \tag{4.6-17}$$

Such a system can be represented by the signal flow graph shown in Fig. 4.6-7.

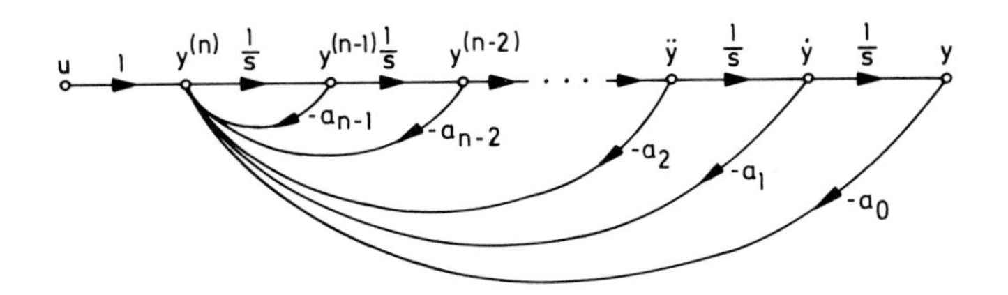

FIG. 4.6-7 Signal flow graph of a system described by a differential equation in the form of Eq. (4.6-17).

From the results of the preceding paragraph, it can be seen immediately that the semirelative sensitivity functions of the output $y(t)$ with respect to the coefficients a_i can be measured at the outputs of the corresponding variable components $-a_i$ (see Fig. 4.6-7) as the input is excited by the output $y_0(t)$ of the nominal original system.

This can also be shown directly by taking the partial derivatives of the original differential equation with respect to all $\ln a_i$. Doing this, and skipping the subscript 0 for notational ease, we obtain for $i \neq j$

$$\frac{\partial a_i y^{(i)}}{\partial \ln a_j} = a_i \tilde{\sigma}_j^{(i)} \tag{4.6-18}$$

and for $i = j$

$$\begin{aligned}\frac{\partial a_j y^{(j)}}{\partial \ln a_j} &= a_j \frac{\partial a_j y^{(j)}}{\partial a_j} = a_j \left(a_j \frac{\partial y^{(j)}}{\partial a_j} + y^{(j)} \right) \\ &= a_j \frac{\partial y^{(j)}}{\partial \ln a_j} + a_j y^{(j)} = a_j \tilde{\sigma}_j^{(j)} + a_j y^{(j)}. \end{aligned} \tag{4.6-19}$$

Thus, the differentiated differential equation has the form

$$\sum_{i=0}^{n} a_i \tilde{\sigma}_j^{(i)} = -a_j y^{(j)}, \qquad j = 0, 1, \ldots, n. \tag{4.6-20}$$

Let us now consider the differential equation of the same system with $y(t)$, i.e., the solution of Eq. (4.6-17), as the input and the variable $z(t)$ as the new output:

$$\sum_{i=0}^{n} a_i z^{(i)} = y, \qquad a_n = 1. \tag{4.6-21}$$

Differentiating this equation j times with respect to t, we obtain

$$\sum_{i=0}^{n} a_i z^{(i+j)} = y^{(j)}. \tag{4.6-22}$$

Now we define

$$\tilde{\eta}_j \triangleq -a_j z^{(j)} \tag{4.6-23}$$

and substitute

$$z^{(j)} = -\frac{1}{a_j}\tilde{\eta}_j \tag{4.6-24}$$

in Eq.(4.6-22). This yields

$$-\sum_{i=0}^{n} a_i \frac{1}{a_j}\eta_j^{(i)} = y^{(j)} \tag{4.6-25}$$

or

$$\sum_{i=0}^{n} a_i \tilde{\eta}_j^{(i)} = -a_j y^{(j)}. \tag{4.6-26}$$

Comparing this result with Eq. (4.6–20), we see that

$$\tilde{\eta}_j = \tilde{\sigma}_j, \tag{4.6-27}$$

and therefore, from Eq.(4.6–23),

$$\tilde{\sigma}_j = -a_j z^{(j)}. \tag{4.6-28}$$

From this result we may conclude that the signals at the outputs of the variable components $-a_j$ in the system excited by $y(t)$ are the corresponding semirelative sensitivity functions $\tilde{\sigma}_j$. The sensitivity model in which all sensitivity functions $\tilde{\sigma}_j$ can be measured simultaneously is therefore given by the signal flow graph shown in Fig. 4.6-8. The corresponding sensitivity points are labeled as $S_0, S_1, \ldots, S_{n-1}$.

On the other hand, by virtue of the definition

$$\sigma_j \triangleq \frac{1}{a_j}\tilde{\sigma}_j, \tag{4.6-29}$$

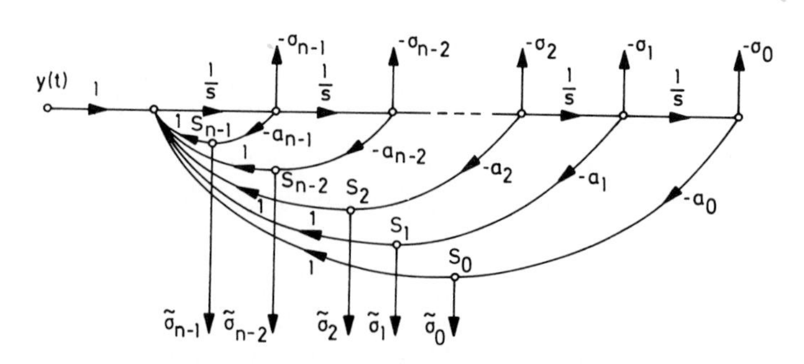

FIG. 4.6-8. Signal flow graph of the sensitivity model for simultaneous measurement of all semirelative sensitivity functions.

the negative *absolute* sensitivity functions can be measured as the phase variables of the sensitivity model. In other words, the variables $z^{(i)}$ of Eq. (4.6-21) are identical with the (negative) absolute sensitivity functions.

Example 4.6-2. Draw an analog computer diagram to simultaneously measure the semirelative sensitivity functions of y with respect to the settings of the coefficients a and b for the analog computer diagram shown in Fig. 4.6-9.

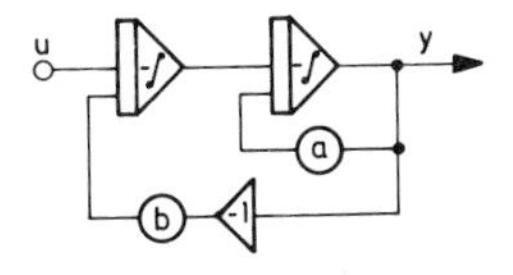

FIG. 4.6-9. Given analog computer diagram of Example 4.6-2.

Solution As a first step, the diagram is brought into an equivalent form which allows the definition of the sensitivity points. This diagram is shown in Fig. 4.6-10. Now, if this circuit is driven by the output signal y of the nominal original circuit, then the semirelative sensitivity functions $\tilde{\sigma}_a$ and $\tilde{\sigma}_b$ can be observed at the sensitivity points S_a and S_b, respectively, labeled in the above diagram. σ_b is available at the output of the inverter, and σ_a can be observed between the two integrators.

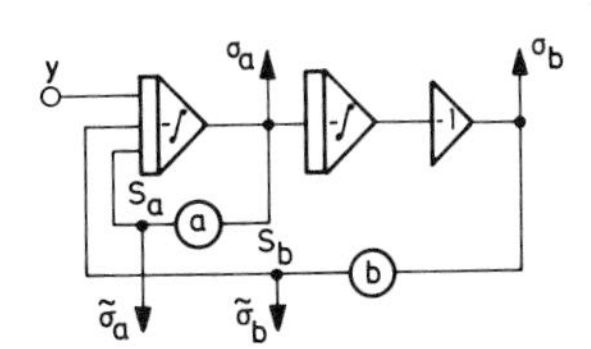

FIG. 4.6-10. Sensitivity model of the analog computer setup of Fig. 4.6-9

4.6.3 Application to General Rational Transfer Functions

Now we show how the structure for the simultaneous measurement of the semirelative sensitivity functions with respect to all coefficients a_i,b_k can be found for systems with the transfer functions of the form

$$G(s) = \frac{b_n s^{n-1} + \cdots + b_1}{s^n + a_n s^{n-1} + \cdots + a_1}, \tag{4.6-30}$$

where the coefficients a_i and b_k are real and at least one of the b_k's is nonzero.

First, let us consider the a_i's and b_k's as the independent parameters of the system. It is useful for the solution of the problem to represent the system by a signal flow graph of the first canonical form as shown in Fig. 4.6–11. Note

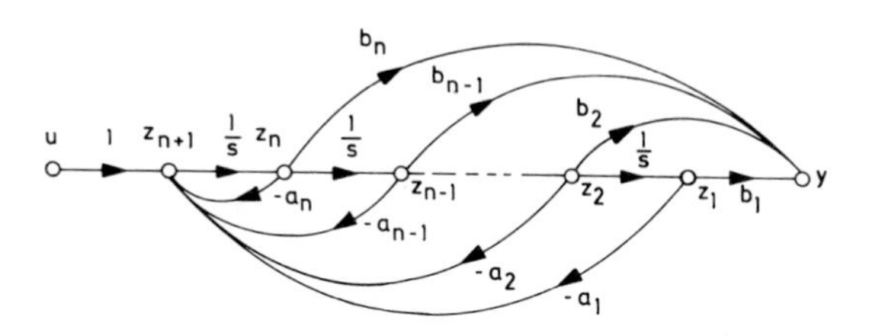

FIG. 4.6-11. Signal flow graph of the system described by Eq. (4.6-30).

that in this form of representation, the output signal y is given as the sum of the phase variables z_i weighted by the corresponding coefficients of the numerator of the transfer function, i.e.,

$$Y(s) = \sum_{i=1}^{n} Y^{(i)}(s) = \sum_{i=1}^{n} b_i Z_i(s), \qquad i = 1, 2, \ldots, n. \qquad (4.6\text{-}31)$$

Let us first determine the signal flow diagram for generating the semirelative output sensitivity functions with respect to the coefficients a_i of the *denominator.*

The result of the preceding section is not applicable here since no forward paths as present in the above configuration were taken into account. Therefore the problem will be solved by making use of the variable component method (Section 4.5). According to this, in order to generate the absolute output sensitivity function $\sigma_{a_i} \triangleq \partial y/\partial a_i$ (the subscript 0 will be dropped again), the nominal original system has to be doubled. A connection line has to be drawn from the input of the variable component, which is identical with the feedback branch $-a_i$, to the corresponding output in the second model. In the present case, the latter is identical with the input of the second system. The transfer function of the connection line is

$$\frac{\partial(-a_i)}{\partial a_i} = -1.$$

This leads to the diagram symbolized by Fig. 4.6–12.

Since the systems are linear, the order of arrangement of the two systems may be inverted without any change of the overall transfer function. This leads to the arrangement of Fig. 4.6–13 which now gives access to σ_{a_i} at the

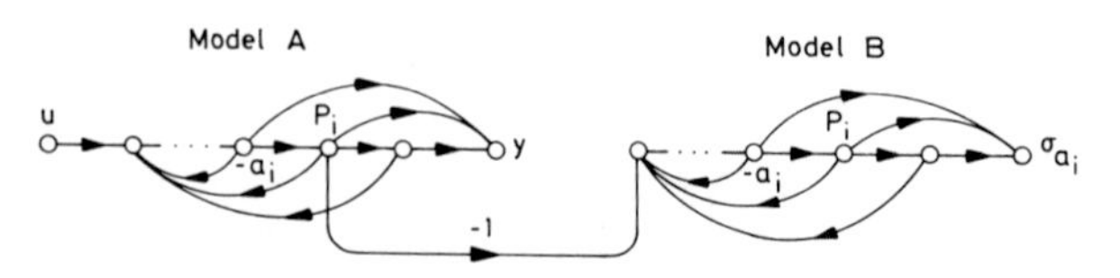

FIG. 4.6-12. Scheme for generating σ_{ai} according to the variable component method.

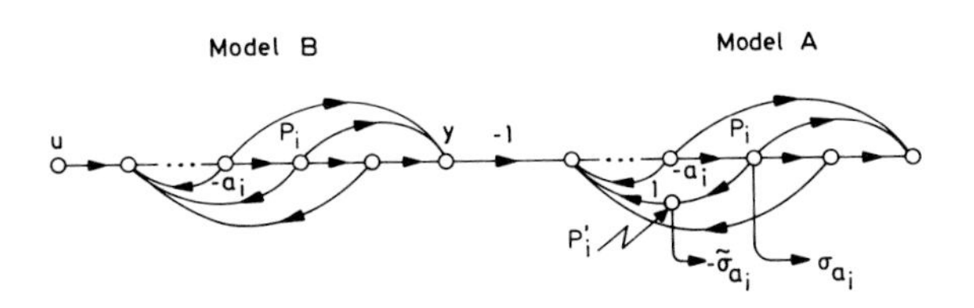

FIG. 4.6-13. Scheme for generating σ_{ai}, which is equivalent to that of Fig. 4.6-12.

point P_i of model A, that is, at the input of the variable component under consideration. Additionally, the negative semirelative sensitivity function $-\tilde{\sigma}_{a_i}$ appears at the point P_i'.

Evidently, the above scheme holds for any i; the only difference is that the pickoff points in the model A change according to the parameters under consideration. This implies that the output sensitivity functions with respect to all a_i $(i = 1, 2, \ldots, n)$ can be obtained at the sensitivity points P_i of model A. Finally, since the forward branches of model A have no influence on the signals at P_i, they may be omitted for structural simplification. The resulting sensitivity model for determining all sensitivity functions σ_{a_i} or $\tilde{\sigma}_{a_i}$ simultaneously is therefore the same as obtained in Section 4.6.2 (Fig. 4.6-8). This reveals the interesting fact that the sensitivity model A is simpler than the nominal original system model B.

In order to derive an adequate scheme for generating the sensitivity functions σ_{b_k} or $\tilde{\sigma}_{b_k}$ with respect to the zeros of $G(s)$, let us recall that by the definition of the transformed sensitivity function $\tilde{\Sigma}_{b_k}$ we have

$$\tilde{\Sigma}_{b_k} = \mathscr{L}\{\tilde{\sigma}_{b_k}\} = \frac{\partial Y}{\partial \ln b_k} = \sum_{i=1}^{n} \frac{\partial Y^{(i)}}{\partial \ln b_k} = \frac{\partial(b_k Z^{(k)})}{\partial \ln b_k} = b_k Z^{(k)} = b_k Z_k \quad (4.6\text{-}32)$$

for $k = 1, 2, \ldots, n$. This is due to the fact that, as shown above, the Z_k's are not functions of b_k. The corresponding time domain expression is therefore

$$\tilde{\sigma}_{b_k} = b_k z_k. \quad (4.6\text{-}33)$$

Note that the z_k's are the phase variables in the nominal original system. This implies that the $\tilde{\sigma}_{b_k}$ can be measured in the feedforward branches of the nominal *original* systems. The complete signal flow diagram to generate all sensitivity functions is shown in Fig. 4.6-14. This result shows again the interesting fact that the sensitivity model is simpler than the original system.

Let us finally cope with the problem of determining the sensitivity functions with respect to *arbitrary parameters* α_j, which are not identical with the coefficients a_i and b_k. In general, each of the coefficients a_i and b_k can depend

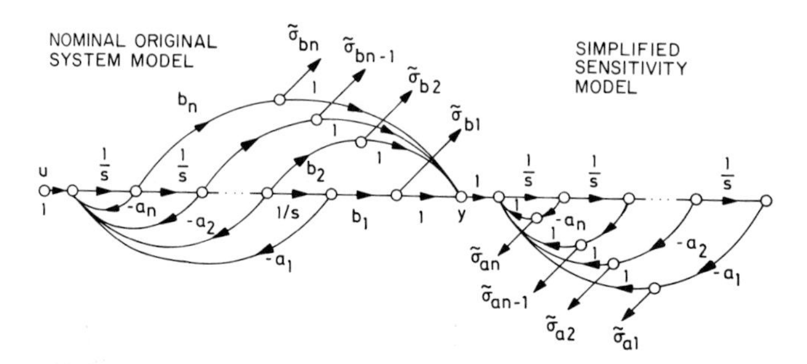

FIG. 4.6-14 Complete diagram for the simultaneous generation of the semirelative output sensitivity functions $\tilde{\sigma}$ with respect to all coefficients a_i and b_k.

on any number r of parameters, i.e., $a_i = a_i(\alpha_1, \ldots, \alpha_r)$, $b_k = b_k(\alpha_1, \ldots, \alpha_r)$. Therefore, the sensitivity function $\tilde{\sigma}_{\alpha_j}$ obeys the equation

$$\tilde{\sigma}_{\alpha j} = \sum_{j=1}^{n} \tilde{\sigma}_{a_i} S_{\alpha_j}^{a_i} + \sum_{k=1}^{n} \tilde{\sigma}_{b_k} S_{\alpha_j}^{b_k}, \tag{4.6-34}$$

where

$$S_{\alpha_j}^{a_i} \triangleq \frac{\partial \ln a_i}{\partial \ln \alpha_j}, \qquad S_{\alpha_j}^{b_k} \triangleq \frac{\partial \ln b_k}{\partial \ln \alpha_j}.$$

Due to this and in accordance with the results of Section 4.6.1, the jth semirelative sensitivity function $\tilde{\sigma}_{\alpha j}$ can be obtained by coupling systems with the transfer functions $S_{\alpha j}^{a_i}$ and $S_{\alpha_j}^{b_k}$ to the sensitivity points S_{a_i} and S_{b_k}, respectively, and then adding all their output signals. This has to be done for every parameter α_j, $j = 1, 2, \ldots, r$. Thus, r such summation circuits are required in general.

Example 4.6-3 In self-optimizing and adaptive systems or in parameter identification schemes, the principle of model adaptation is often used. This implies that a model is connected in parallel to the original system according to Fig. 4.6–15. The parameters of the model are changed until the output error is zero or, in the case of additional noise, minimal. As a suitable cri-

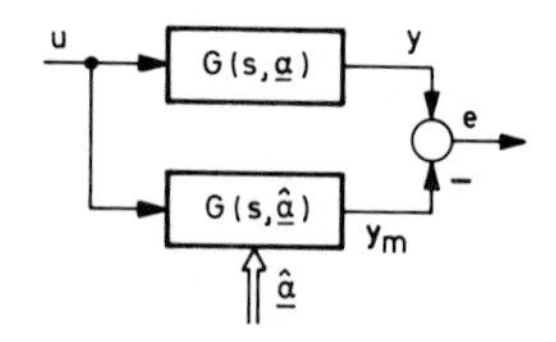

FIG. 4.6-15. Model adaptation scheme of Example 4.6-3.

terion for the quality of adaptation may serve the squared output error, which is defined as

$$J = e^2 = (y - y_m)^2 \tag{4.6-35}$$

or its integral with respect to time.

Let the parameters of the original system be denoted by $\alpha_1, \ldots, \alpha_r$ and the parameters of the model by $\hat{a}_1, \ldots, \hat{a}_r$. If we choose the parameter change $\hat{\boldsymbol{a}}$ such that the change of e^2 is always negative or maximally zero, i.e.,

$$\dot{J} = \frac{de^2}{dt} = 2e\frac{de}{dt} = 2e\frac{\partial e}{\partial \hat{\boldsymbol{\alpha}}}\hat{\boldsymbol{\alpha}} \leq 0, \tag{4.6-36}$$

the output error $e = y - y_m$ approaches zero (or the minimum) asymptotically. This can be achieved by choosing α as

$$\dot{\boldsymbol{\alpha}} = -k\left(\frac{\partial e^2}{\partial \hat{\boldsymbol{\alpha}}}\right)^{\mathrm{T}} = -k\,2e\left(\frac{\partial e}{\partial \hat{\boldsymbol{\alpha}}}\right)^{\mathrm{T}}. \tag{4.6-37}$$

k is a positive constant by which the velocity of adaptation can be controlled. Introducing Eq.(4.6–37) into Eq.(4.6–36) yields

$$\dot{J} = -4ke^2\frac{\partial e}{\partial \hat{\boldsymbol{\alpha}}}\left(\frac{\partial e}{\partial \hat{\boldsymbol{\alpha}}}\right)^{\mathrm{T}}. \tag{4.6-38}$$

This proves that $\dot{J}$ is always negative, as was anticipated.

Since $e = y - y_m$ and y is not a function of $\hat{\boldsymbol{\alpha}}$, we obtain $\partial e/\partial \hat{\boldsymbol{\alpha}} = -\partial y_m/\partial \hat{\boldsymbol{\alpha}}$. Therefore,

$$\hat{\boldsymbol{\alpha}} = +2ke\left(\frac{\partial y_m}{\partial \hat{\boldsymbol{\alpha}}}\right)^{\mathrm{T}}. \tag{4.6-39}$$

This implies that in order to generate $\hat{\boldsymbol{\alpha}}$ all output sensitivity functions of the model are needed simultaneously. A block diagram for the generation of $\boldsymbol{\alpha}$ due to Eq. (4.6-39) is shown in Fig. 4.6-16.

According to Kokotović's method of sensitivity points, a second model (sensitivity model) with the same parameter vector $\hat{\boldsymbol{\alpha}}$ is connected in series with the parallel model of the plant. The sensitivity model gives simultaneous access to all sensitivity functions $(\partial y_m/\partial \alpha_i)_{\boldsymbol{\alpha}_0}$, $i = 1, 2, \ldots, r$. Multiplication of the sensitivity vector by $2k$ and by the output error e produces the time derivative $\dot{\hat{\boldsymbol{\alpha}}}$. Hence the parameter vector $\hat{\boldsymbol{\alpha}}$ of the model and the sensitivity model is obtained by integration. Once e^2 is zero, the parameters $\hat{\boldsymbol{\alpha}}$ are equal to the parameters $\boldsymbol{\alpha}$ of the original system to be identified.

It should be noted that in the above situation the parameters under consideration are not time-invaraint as presumed in the method applied. Therefore, what is obtained at the output of the sensitivity model is the vector of the *approximate* sensitivity functions defined in Section 4.3.2. Thus, the above

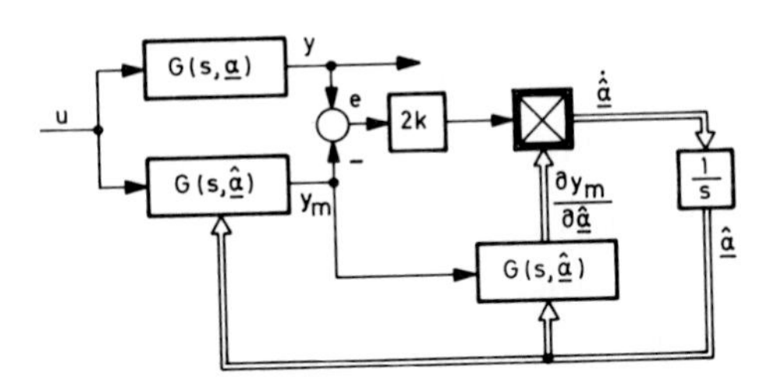

FIG. 4.6-16. Model-adaptive parameter identification scheme of Example 4.6-3.

scheme is restricted to slow changes of $\boldsymbol{\alpha}$ since then the approximate sensitivity functions are almost identical with the true sensitivity functions.

4.6.4 Generation of the Pseudosensitivity Function

The general concept of pseudosensitivity was introduced in Section 2.2. In the particular case of output sensitivity, the pseudosensitivity function defined by the identity

$$\Delta y(t,\alpha) \equiv \sigma^*(t,\alpha)\, \Delta\alpha, \tag{4.6-40}$$

where σ^* (t,α) is the output pseudosensitivity function. By the definition $\sigma^*(t, \alpha)$, the identity of Eq. (4.6–40) is satisfied for arbitrary values of $\Delta\alpha$.

As an interesting fact of great practical importance, $\sigma^*(t, \alpha)$ can be generated by the same sensitivity model as $\sigma(t, \alpha_0)$ except that the input of the sensitivity model is identical with the actual output y rather than with the nominal output y_0 of the original system. In other words, the nominal original system model in the combined system has to be replaced by an actual system model as illustrated in Fig. 4.6-17.

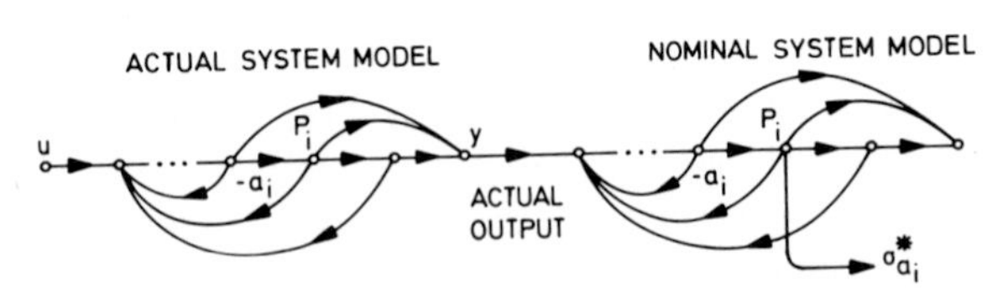

FIG. 4.6-17. Scheme for generating the pseudosensitivity function σ^*_{ai} of the output y with respect to the parameter a_i.

This can easily be shown by the aid of Fig. 4.5-22 which shows the combined system from which the output error Δy is determined in the case of arbitrary parameter variations. If the sensitvity model is divided by $\Delta\alpha$, then $\Delta y/\Delta\alpha$ instead of Δy is obtained at the output. It is easy to see that the connection from Δu_i to the input of ΔG in the sensitivity model, which was established to account for large parameter variations, could be omitted if, instead, G_0 would be replaced by the actual transfer function $G = G_0 + \Delta G$. This would imply that model B in Fig. 4.6-12 has to be replaced by an *actual* model of the original system. The inversion of the order of the two models according to Fig. 4.6-13 then yields the arrangement of Fig. 4.6–17, the output of which is $\sigma_{a_i}^* = \Delta y/\Delta a_i$, thus satisfying the relation $\Delta y \equiv \sigma_{a_i}^* \, \Delta a_i$. Hence, $\sigma_{a_i}^*$ is identical with the pseudosensitivity function.

Example 4.6-4 The employment of pseudosensitivity functions is of great importance for parameter identification, optimization, and adaptation purposes using a reference model. It is also called the pseudogradient method. The basic idea is illustrated in Fig. 4.6–18. Notice that the pseudogradient method converges faster than when using the ordinary sensitivity functions (Fig. 4.6-16).

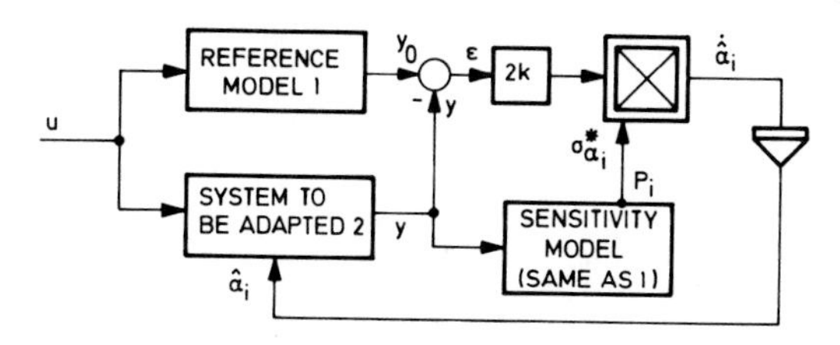

FIG. 4.6-18. Application of the pseudosensitivity function σ_{ai}^* for the adaptation of a system to a reference model (pseudogradient method).

4.6.5 The Three-Points Method

According to the sensitivity points method (Fig. 4.6-13), the output sensitivity function of a system can be viewed as the output of a series connection of two *nominal* models of the system, the first of which has the *overall* "transference" and the second has the "transference" of the section between the input and the *sensitivity point*, with the regular input signal applied to the input. In view of the special nature of this structure, it is possible to reject one of the models, i.e., the calculate the sensitivity function from the input, the output, and the sensitivity point signal of the nominal original system. This method is called the three-points method and is based on the fact that complete information about a sensitivity function of a system is contained in the input, output, and sensitivity point of the nominal system [94].

To show this, recall that, according to Fig. 4.6-13, the transformed output sensitivity function with respect to a_i is given by

$$\Sigma_{a_i} = \mathscr{L}\{\sigma_{a_i}\} = G_0 G_{i0} U, \qquad (4.6\text{-}41)$$

where $U = U(s)$ denotes the transformed input, G_0 denotes the nominal overall transfer function of the system, and G_{i0} denotes the transfer function of the section from the input to the sensitivity point P_i. In terms of the signals u, y_0, and z_i specified by Fig. 4.6-19, the above transfer functions can be written as

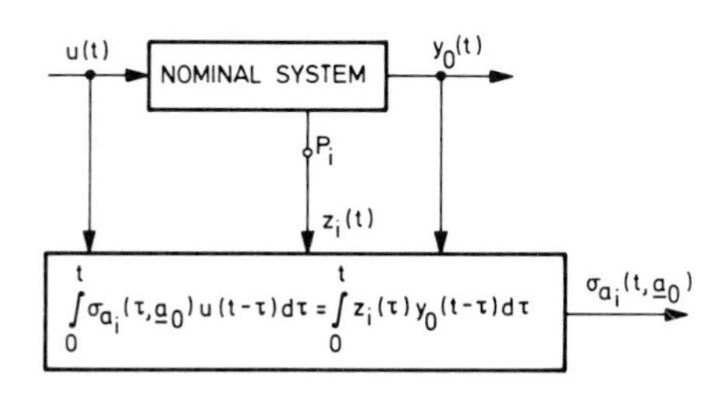

FIG. 4.6-19. Generation of the output sensitivity function $\sigma^*{}_{ai}$ by the three-points method.

$$G_0 = \frac{Y_0}{U}, \qquad G_{i0} = \frac{Z_i}{U}. \tag{4.6-42}$$

Substituting Eqs. (4.6-42) for G_0 and G_{i0} in Eq. (4.6-41) yields

$$U\Sigma_{a_i} = Y_0 Z_i \tag{4.6-43}$$

or, in the time domain,

$$\int_0^t \sigma_{a_i}(\tau, \boldsymbol{a}_0)u(t-\tau)\,d\tau = \int_0^t z_i(\tau)y_0(t-\tau)\,d\tau. \tag{4.6-44}$$

The solution of this integral equation for σ_{a_i} gives the sensitivity function in terms of the three signals $u(t)$, $y_0(t)$, and $z_i(t)$ that can be obtained at the nominal original system.

Example 4.6-5 Suppose the nominal system of Fig. 4.6-19 is excited by a δ-function and the corresponding weight function g_0 obtained at the output is continuous at $t = \tau$.

(a) Give the expression for the corresponding sensitivity function $\gamma_{a_i} = \partial g/\partial a_i|_{\boldsymbol{a}_0}$ in terms of the signals g_0 and z_i.

(b) Give the analytic expression to determine the sensitivity function σ_{a_i} in the case of an arbitrary input $u(t)$ in terms of γ_{a_i} and u.

Solution (a) Substituting $u(t-\tau) = \delta(t-\tau)$in Eq. (4.6–44), we obtain for γ_{a_i} the convolution integral

$$\gamma_{a_i} = \int_0^t z_i(\tau)g_0(t-\tau)\,d\tau.$$

(b) According to Eq. (4.6–41), we obtain in the time domain

$$\sigma_{a_i} = \int_0^t \gamma_{a_i}(\tau, \boldsymbol{a}_0)u(t-\tau)\,d\tau$$

since $\Gamma_{a_i} = \mathscr{L}\{\gamma_{a_i}\} = G_0 G_{j0}$.

4.7 SENSITIVITY ANALYSIS OF NETWORKS

In this section, methods for the determination of the sensitivity function of networks will be discussed. On one hand, these methods may serve to study the influence of tolerances of network elements on the behavior of an electrical network, on the other hand they may be used for the analysis of arbitrary systems if these systems are represented by generalized networks.

First, the *variable component* method is applied to networks. In fact, this was the way the principle of variable component was originally introduced in the literature by Bikhovski [24, 25]. Then, a method will be described which is based on the so-called adjoint system first described by Director and Rohrer [35]. Finally, two theorems on sensitivity invariants of networks will be given.

4.7.1 Variable Component Method for Networks

The objective is to find a sensitivity model for a given electrical network, the analysis of which allows the determination of the output sensitivity function. This problem may be solved either by using the balance equations, i.e., Kirchhoff's current and voltage law, or by structural considerations. Here we shall pursue the first procedure.

Consider first a *linear* network with N_v nodes and N_l meshes (Fig. 4.7-1). Nonlinear networks will be treated later. Let i_j ($j = 1, 2, \ldots, N_j$) be the currents in the kth node ($k = 1, 2, \ldots, N_k$) and u_j ($j = 1, 2, \ldots, N_j$) be the voltages of the lth mesh ($l = 1, 2, \ldots, N_l$).

Then Kirchhoff's *current law* gives

$$\sum_{j=1}^{Nj} a_{kj} i_j = 0, \qquad a_{kj} = 0 \quad \text{or} \quad 1 \quad \text{or} \quad -1, \tag{4.7-1}$$

and Kirchhoff's *voltage law* yields

$$\sum_{=1}^{Nj} b_{lj} u_j = \underset{j}{0}, \qquad b_{lj} = 0 \quad \text{or} \quad 1 \quad \text{or} \quad -1. \tag{4.7-2}$$

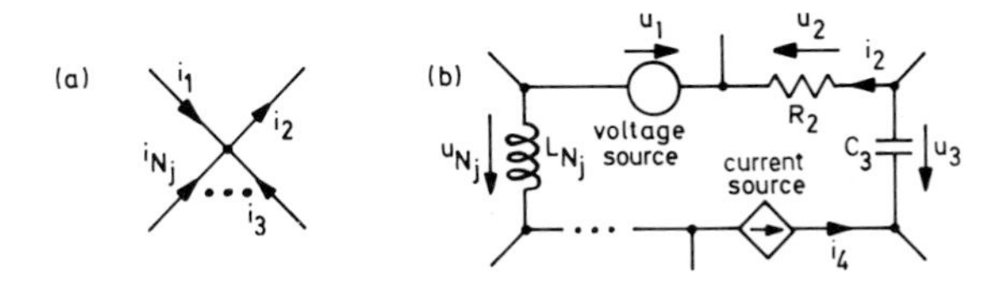

FIG. 4.7-1. Illustration of (a) kth node, (b) lth mesh of a linear network.

In addition, the constitutive relations of the elements are required. They are

$$u_j = R_j i_j \qquad \text{(at the resistance } R_j\text{)}, \tag{4.7-3}$$

$$i_k = C_k \frac{du_k}{dt} \qquad \text{(at the capacitance } C_k\text{)}, \tag{4.7-4}$$

$$u_e = L_e \frac{di_e}{dt} \qquad \text{(at the inductance } L_e\text{)}. \tag{4.7-5}$$

Finally, we take into account independent voltage and current sources defined by

$$u_j = \text{const.}, \qquad i_k = \text{const.} \tag{4.7-6}$$

Equations (4.7-1)–(4.7-6) represent the mathematical model of the network under consideration.

Now suppose that one of the elements, say R_i, is the parameter of interest. In order to find the corresponding sensitivity model of the network, Eqs. (4.7-1)–(4.7-6) have to be partially differentiated with respect to R_i. Partial differentiation of Eqs. (4.7-1) and (4.7-2) yields

$$\sum_{j=1}^{N_j} a_{kj} \left.\frac{\partial i_j}{\partial R_i}\right|_{R_{i0}} = 0, \tag{4.7-7}$$

$$\sum_{j=1}^{N_j} b_{lj} \left.\frac{\partial u_j}{\partial R_i}\right|_{R_{i0}} = 0. \tag{4.7-8}$$

Referring to Eqs. (4.7-3)-(4.7-5), we assume that the order of taking the time derivative and the partial derivative with respect to R_i may be interchanged. Hence, differentiation of Eqs. (4.7-3)–(4.7-5) yields

for $j \neq i$:

$$\left.\frac{\partial u_j}{\partial R_i}\right|_{R_{i0}} = R_j \left.\frac{\partial i_j}{\partial R_i}\right|_{R_{i0}}, \tag{4.7-9}$$

$$\left.\frac{\partial i_k}{\partial R_i}\right|_{R_{i0}} = C_k \left.\frac{d}{dt}\frac{\partial u_k}{\partial R_i}\right|_{R_{i0}}, \tag{4.7-10}$$

$$\left.\frac{\partial u_e}{\partial R_i}\right|_{R_{i0}} = L_e \left.\frac{d}{dt}\frac{\partial i_e}{\partial R_i}\right|_{R_{i0}}, \tag{4.7-11}$$

for $j = i$:

$$\left.\frac{\partial u_j}{\partial R_i}\right|_{R_{i0}} = R_{i0} \left.\frac{\partial i_i}{\partial R_i}\right|_{R_{i0}} + i_{i0}. \tag{4.7-12}$$

The partial derivatives of Eq.(4.7–6) disappear, i.e.,

$$\frac{\partial u_j}{\partial R_i} = 0, \qquad \frac{\partial i_k}{\partial R_i} = 0. \tag{4.7-13}$$

Equations (4.7-7)-(4.7-13) define the sensitivity model of the original network. The structural interpretation of the equations obtained reveals the following features of the sensitivity model:

The sensitivity equations (4.7-7) and (4.7-8) are of the same form as the original network equations except that all independent source terms are zero. This implies that the sensitivity model has the same mesh and node configuration as the original network, however with all independent current sources opened and all independent voltage sources short circuited. The voltages and currents in the sensitivity model represent the sensitivity functions of the voltages and currents in the corresponding branches of the original network.

Equations (4.7-9)-(4.7-11) verify that all elements of the original network have to be repeated in the sensitivity network. However, due to Eq.(4.7-12), a voltage source of magnitude i_{i0} has to be provided in series connection with the element R_{i0} which is the parameter of interest. This result may be summarized by the following theorem.

Theorem 4.7-1 The sensitivity model of a linear network with respect to one of its resistances R_i is of the same structure as the nominal original network except that

(1) all independent voltage sources are short circuited,
(2) all independent current sources are opened,
(3) a controlled voltage source with a voltage of the magnitude i_{i0} is connected in series with R_{i0}, i_{i0} is obtained from the nominal original network.

The voltages and currents of the sensitivity network represent the sensitivity functions of the corresponding voltages and currents, respectively, of the original network. If, for example, R_2 is the parameter of interest in Fig. 4.7-1, the corresponding kth node and lth mesh of the sensitivity model are as shown in Fig. 4.7-2. Note that the directions of the voltages and currents of the sensitivity network may be different from those of the original network.

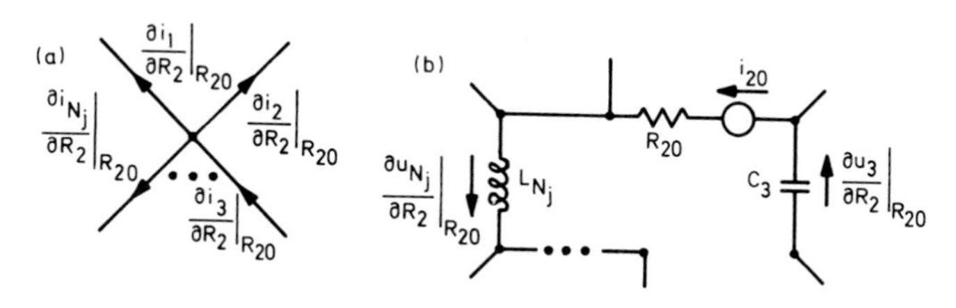

FIG. 4.7-2. Sensitivity models of the kth node (a) and lth mesh (b) of Fig. 4.7-1 with R_2 being the parameter of interest.

Example 4.7-1 For the network shown in Fig. 4.7-3, draw a sensitivity model that produces the output sensitivity function of u_2 with respect to R_1.

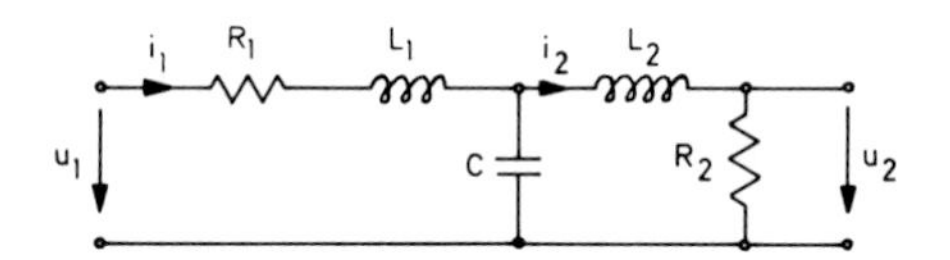

FIG. 4.7-3. Original network of Example 4.7-1.

Solution Using the rules of Theorem 4.7-1, the sensitivity network can be drawn immediately. This is shown in Fig. 4.7-4. The sensitivity network differs from the nominal original network simply by the short circuited input and by the voltage i_{10} that has to be connected in series with R_{10} according to

$$\left.\frac{\partial u_1}{\partial R_1}\right|_{R_{10}} = R_{10} \left.\frac{\partial i_1}{\partial R_1}\right|_{R_{10}} + i_{10}. \tag{4.7-14}$$

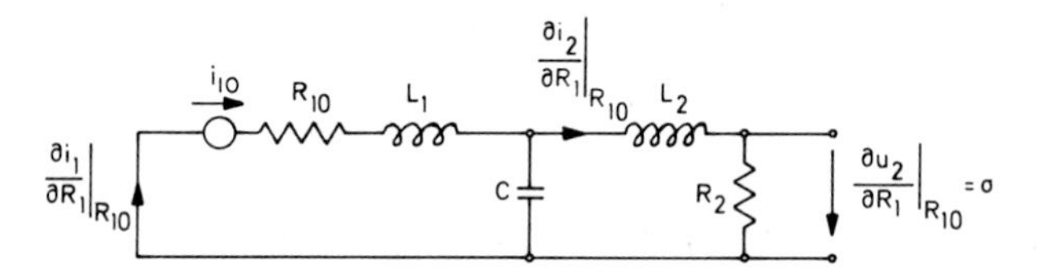

FIG. 4.7-4. Sensitivity network of Fig. 4.7-3.

The voltage i_{10} of the sensitivity network can be generated from the current i_{10} of the nominal original network by means of an amplifier that impresses a voltage proportional to i_{10}. Another object of the amplifier is to isolate both networks from each other.

Thus we see that all currents and voltages of the sensitivity network represent the sensitivity functions of the currents and voltages in the corresponding branches of the original network with the exception of $(\partial u_1/\partial R_1)_{R_{10}}$ which is given by Eq.(4.7-14). Particularly, the desired output sensitivity function σ is obtained at the output of the sensitivity model.

A generalized block diagram for the generation of the output sensitivity function of a network with respect to a resistance R_i is shown in Fig. 4.7-5. The amplifier has two functions:

(1) to generate the voltage i_{i0} of the sensitivity network by multiplying u_{i0} by $1/R_{i0}$ due to the relation $i_{i0} = u_{i0}/R_{i0}$;

(2) to isolate both networks from each other, i.e., to avoid feedback from the sensitivity network to the original network.

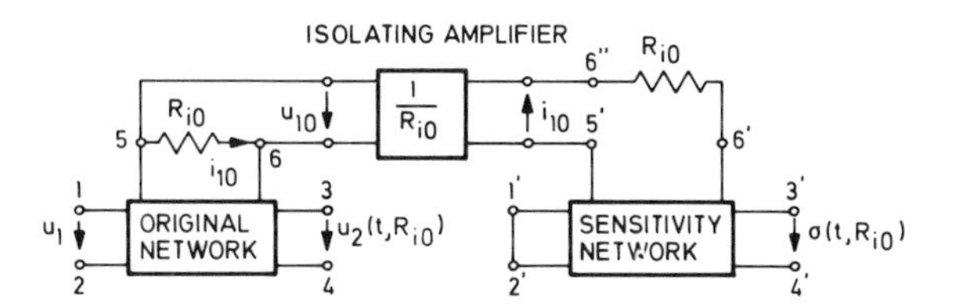

FIG. 4.7-5. Complete circuit for the generation of the output sensitivity function of a network with respect to a resistance R_i.

If the gain of the amplifier is chosen 1 instead of $1/R_{i0}$, then the signal obtained at the output of the sensitivity network is the semirelative sensitivity function

$$\tilde{\sigma}(t, R_{i0}) = R_{i0} \left.\frac{\partial u_2}{\partial R_i}\right|_{R_{i0}} = \left.\frac{\partial u_2}{\partial \ln R_i}\right|_{R_{i0}}. \tag{4.7-15}$$

By a similar procedure, the corresponding measurement circuits can be developed for an inductance or a capacitance as the parameter of interest. If L_e is the parameter of interest, then from Eq. (4.7-5) we obtain

$$\left.\frac{\partial u_e}{\partial L_e}\right|_{Le0} = L_e \frac{d}{dt} \left.\frac{\partial i_e}{\partial L_e}\right|_{Le0} + \frac{di_{e0}}{dt} = L_e \frac{d}{dt} \left.\frac{\partial i_e}{\partial L_e}\right|_{Le0} + \frac{u_{e0}}{L_{e0}}. \tag{4.7-16}$$

On the other hand, if C_k is considered as a parameter, then, from Eq. (4.7-4), we obtain

$$\left.\frac{\partial u_k}{\partial C_k}\right|_{Ck0} = \frac{1}{C_{k0}} \int \left.\frac{\partial i_k}{\partial C_k}\right|_{Ck0} dt - \frac{1}{C_{k0}^2} \int i_{k0}\, dt = \frac{1}{C_{k0}} \int \left.\frac{\partial i_k}{\partial C_k}\right|_{Ck0} dt - \frac{u_{k0}}{C_{k0}}. \tag{4.7-17}$$

These results reveal that in both cases the voltage at the concerned element in the nominal original network has to be multiplied by $1/L_{e0}$ or $-1/C_{k0}$, respectively, and the resulting voltages have to be connected in series with the corresponding element in the sensitivity network.

In order to derive analytic expressions for the output sensitivity functions, consider the general structure of Fig. 4.7-5. Defining the transfer function between the input u_1 (terminals 1-2) and the voltage u_{i0} (terminals 5-6) of the original network by $G_1 = U_{i0}/U_1$ and the transfer function between the input i_{i0} (terminals 5′-6″) and the output σ (terminals 3′-4′) of the sensitivity network by $G_2 = \Sigma/I_{i0}$, we have

$$\Sigma = \left.\frac{\partial U_2}{\partial R_i}\right|_{Ri0} = \frac{1}{R_{i0}} G_1 G_2 U_1 \tag{4.7-18}$$

or

$$\sigma = \mathscr{L}^{-1}\left\{\frac{1}{R_{i0}} G_1 G_2 U_1\right\}. \tag{4.7-19}$$

For example, if u_1 is the δ-impulse, then because $\mathscr{L}\{\delta\} = 1$, we obtain

$$\sigma = \mathscr{L}^{-1}\left\{\frac{1}{R_{i0}} G_1 G_2\right\}. \tag{4.7-20}$$

The analogous relations for inductances or capacitances as the parameters of interest are given by

$$\left.\frac{\partial U_2}{\partial L_e}\right|_{Le0} = \frac{1}{L_{e0}} G_1 G_2 U_1, \tag{4.7-21}$$

$$\left.\frac{\partial U_2}{\partial C_k}\right|_{Ck0} = -\frac{1}{C_{k0}} G_1 G_2 U_1, \tag{4.7-22}$$

and for the conductance A_{i0} as the parameter of interest, we have

$$\left.\frac{\partial U_2}{\partial A_i}\right|_{Ai0} = -\frac{1}{A_{i0}} G_1 G_2 U_1. \tag{4.7-23}$$

The proof of Eqs. (4.7-21) and (4.7-22) is analogous to that of Eqs. (4.7-16) and (4.7-17), respectively.

Now we consider *nonlinear* networks. The goal is to find a linear sensitivity network. If the original nonlinear network is given analytically by its network equations, the linear sensitivity model is obtained by partial differentiation of the network equations with respect to the parameter of interest. This procedure is always practicable and is without any difficulty.

It will be shown here how the linear sensitivity network of a nonlinear network can be found by structural considerations.

Suppose that the node and mesh configuration is as in the linear case [Eqs. (4.7-1) and (4.7-2)]. However, instead of the linear constitutive relations of the elements (Eqs. 4.7-3-4.7-5), let us assume that there are also nonlinear elements, the constitutive relations of which are

$$u_j = R_j(i_j) i_j \quad , \qquad R_j(i_j) = \frac{u(i_j)}{i_j}, \tag{4.7-24}$$

$$i_k = C_k(u_k)\frac{du_k}{dt}, \qquad C_k(u_k) \triangleq \frac{dQ(u_k)}{du_k}, \tag{4.7-25}$$

$$u_e = L_e(i_e)\frac{di_e}{dt}, \qquad L_e(i_e) \triangleq \frac{d\Phi(i_e)}{di_e}. \tag{4.7-26}$$

Here Q and Φ denote the electrical charge or the magnetic flux, respectively. It is further assumed that the functions $R_j(i_j)$, $C_k(u_k)$, and $L_e(i_e)$ are known analytically.

Since the nominal original network is given, it is possible to measure the time functions of the voltages and currents in the individual branches of this network. If these time functions are introduced into the expressions R_j, C_k, and L_e, we obtain time functions of the form

$$g_j(t) = R_j[i_j(t)], \tag{4.7-27}$$

$$g_k(t) = C_k[u_k(t)], \tag{4.7-28}$$

$$g_e(t) = L_e[i_e(t)]. \tag{4.7-29}$$

This implies that the elements are now given by their time functions, the latter being known from the nominal original system. Thus the sensitivity network reduces to a linear network with time-variant parameters. The time-variant parameters can be realized using function generators that are controlled by the currents or voltages, respectively, of the nominal original network.

Now two categories of parameter changes have to be distinguished:

(1) changes of linear elements, and
(2) changes of nonlinear elements.

Case (1) Apart from the fact that the nonlinear elements have to be replaced by function generators according to Eqs. (4.7-27)-(4.7-29), the sensitivity model is completely the same as in the case of linear networks. This is illustrated by Fig. 4.7-6 with the linear resistance R_q being the parameter of interest, where the subscript 0 denotes nominal values.

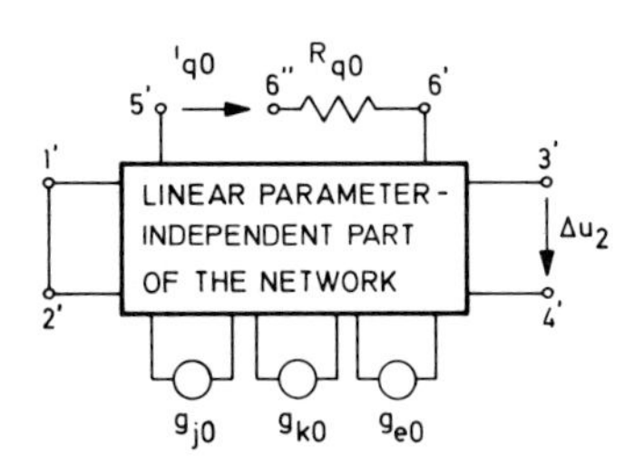

FIG. 4.7-6. Sensitivity model of a nonlinear network in which a linear resistance R_q is the parameter of interest.

Case (2) Suppose that a nonlinear resistance, say R_p, is the parameter of interest. If the deviation from the nominal value R_{p0} is known,

$$\Delta R_p(i_p) \triangleq R_p(i_p) - R_p(i_{p0}), \tag{4.7-30}$$

then, because of the knowledge of i_p as a function of time from the original system, we also know $\Delta R_p = g_p(t) - g_{p0}(t) = \Delta g_p(t)$ as a function of time.

Hence, in order to measure the corresponding deviation Δu_2 at the terminals 3′-4′ of the sensitivity model, a voltage source $\Delta g_p i_{p0}$ has to be provided additionally in series with $g_{p0}(t)$ where now both Δg_p and i_{p0} are functions of time. The rest of the sensitivity network remains the same as above. The result is illustrated in Fig. 4.7-7.

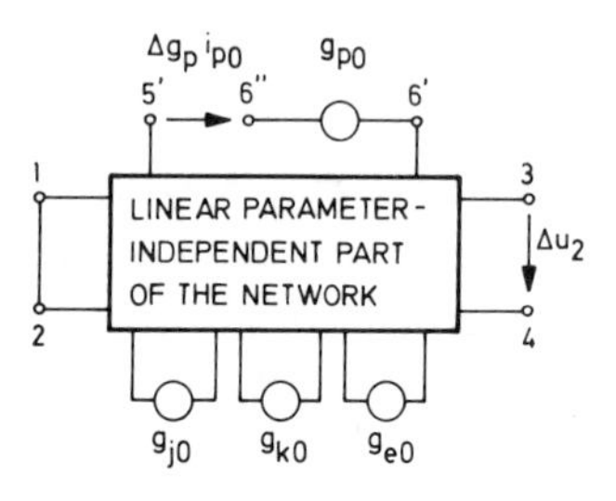

FIG. 4.7-7. Sensitivity network of a nonlinear network in which the nonlinear resistance $R_q = g_q(t)$ is the parameter of interest.

If, instead of the parameter-induced error Δu_2, the sensitivity function $\sigma_p = (\partial u_2/\partial R_q)_{Rq0}$ is to be generated, then a controlled voltage source i_{p0} instead of $\Delta g_p\, i_{p0}$ has to be provided between 5′-6″, according to the relation

$$\left.\frac{\partial u_p}{\partial R_p}\right|_{Rp0} = g_{p0} \left.\frac{\partial i_p}{\partial R_p}\right|_{Rp0} + i_{p0}. \tag{4.7-31}$$

Similar results are obtained when nonlinear inductances or capacitances are the parameters of interest.

Example 4.7-2 Consider the network of Fig. 4.7-8. Suppose the diode characteristic is approximated by

$$i_{\mathrm{D}} = \begin{cases} a u_{\mathrm{D}}^2 & \text{for} \quad u_{\mathrm{D}} > 0, \\ 0 & \text{for} \quad u_{\mathrm{D}} < 0. \end{cases} \tag{4.7-32}$$

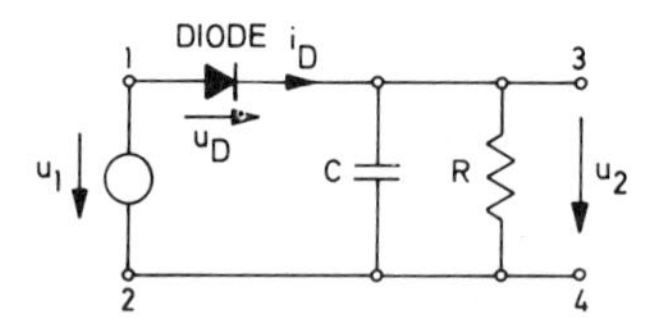

FIG. 4.7-8. Nonlinear network of Example 4.7-2.

Draw a sensitivity network for the generation of the output sensitivity function with respect to (a) the resistance R, (b) the coefficient a of the diode characteristic.

Solution (a) The nominal nonlinear resistance of the diode is

$$R_{\mathrm{D}}(i_{\mathrm{D}}) \triangleq \frac{u_{\mathrm{D}}(i_{\mathrm{D}})}{i_{\mathrm{D}}} = \begin{cases} \dfrac{1}{\sqrt{a i_{\mathrm{D}}}} & \text{for} \quad u_{\mathrm{D}} > 0, \\ \infty & \text{for} \quad u_{\mathrm{D}} < 0. \end{cases} \tag{4.7-33}$$

Its time dependence $g_D(t) = R_D[i_D(t)]$ is known since $i_D(t)$ can be obtained from the nominal original network. $g_D(t)$ has to be substituted for the diode in the sensitivity model. In addition, the voltage source u_1 has to be short circuited and a new voltage source i_{R0} has to be inserted in series with R_0. The resulting sensitivity network is shown in Fig. 4.7-9(a).

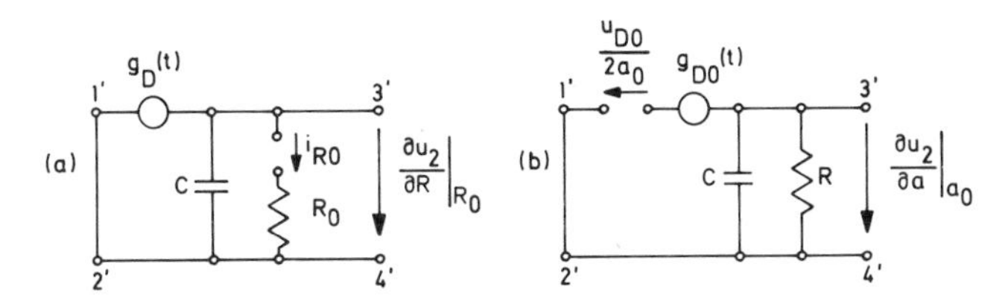

FIG. 4.7-9. Sensitivity networks of Example 4.7-2.

(b) If a is the parameter of interest with a_0 being the nominal value, then we obtain, by partially differentiating $u_D = R_D[a, i_D(a)]i_D(a)$ with respect to a and setting $a = a_0$,

$$\frac{\partial u_D}{\partial a}\Big|_{a0} = R_{D0}\frac{\partial i_D}{\partial a}\Big|_{a0} + i_{D0}\left(\frac{\partial R_D}{\partial i_D}\frac{\partial i_D}{\partial a} + \frac{\partial R_D}{\partial a}\right)_{a0}, \qquad u_{D0} > 0,$$

or, with $R_{D0} = 1/\sqrt{a_0 i_{D0}}$,

$$\frac{\partial u_D}{\partial a}\Big|_{a0} = \frac{1}{2\sqrt{a_0}\sqrt{i_{D0}}}\frac{\partial i_D}{\partial a}\Big|_{a0} - \frac{1}{2a_0\sqrt{a_0}}\sqrt{i_{D0}}, \qquad u_{D0} > 0,$$

and

$$\frac{\partial u_D}{\partial a}\Big|_{a0} = \infty, \qquad u_{D0} < 0.$$

This reveals that in the sensitivity network the diode has to be substituted by a series connection of a function generator with $g_{D0}(t) = 1/(2\sqrt{a_0}\sqrt{i_{D0}(t)})$ where $i_{D0}(t)$ is obtained from the nominal network, and a voltagesource of

$$-\frac{1}{2a_0}\frac{\sqrt{i_{D0}}}{\sqrt{a_0}} = -\frac{u_{D0}}{2a_0},$$

where u_{D0} is again obtained from the nominal network. The result is shown in Fig. 4.7-9(b).

4.7.2 Adjoint Network Method

The determination of the output sensitivity of a network due to a certain percentage change of its elements is of great practical importance in tolerance analysis in connection with network design. There is a particular procedure to solve this problem on the basis of the theorem of Tellegen. The procedure allows us to find the output sensitivity functions with respect to all network elements from the currents of a nominal network and a so-called *adjoint* network by means of simple mathematical operations. The basic idea was first proposed by Director and Rohrer [35] and was then developed to a high degree of perfection by Calahan [3]. In this section, the basic idea will be described for the special case of resistor networks.

For the purpose of derivation of the procedure, consider two linear networks N and $\hat{N}$ which have the same mesh and node configuration (i.e., the same topology) but may have different network elements between corresponding nodes. Even open and short circuits are admissible as branches between the individual nodes. Let i_ν and $\hat{i}_\nu$ be the currents and u_ν and $\hat{u}_\nu$ the voltages in the corresponding branches of N and $\hat{N}$. The arrows of current and voltage of a load are defined in the same direction, and the directions of currents $\hat{i}_\nu$ and voltages $\hat{u}_\nu$ in $\hat{N}$ are, by definition, the same as of i_ν and u_ν in N.

Then the theorem of Tellegen states that if N and $\hat{N}$ have n_z branches, the following two balance equations hold:

$$\sum_{\nu=1}^{n_z} n_\nu \hat{i}_\nu = 0 \qquad \text{and} \qquad \sum_{\nu=1}^{n_z} i_\nu \hat{u}_\nu = 0. \tag{4.7-34}$$

The proof of this theorem can be found in the literature on network theory (see, for example, Calahan [3]).

Now suppose that the currents and voltages in N are perturbed by infinitesimal amounts di_ν and du_i, respectively. Then, reusing Tellegen's theorem, we obtain

$$\sum_{\nu=1}^{n_z} (u_\nu + du_\nu)\hat{i}_\nu = 0 \qquad \text{and} \qquad \sum_{\nu=1}^{n_z} (i_\nu + di_\nu)\hat{u}_\nu = 0. \tag{4.7-35}$$

In addition, Eqs. (4.7-34) are still valid. Hence

$$\sum_{\nu=1}^{n_z} du_\nu\, \hat{i}_\nu = 0 \qquad \text{and} \qquad \sum_{\nu=1}^{n_z} di_\nu\, \hat{u}_\nu = 0 \tag{4.7-36}$$

or

$$\sum_{\nu=1}^{n_z} (du_\nu\, \hat{i}_\nu - di_\nu\, \hat{u}_\nu) = 0. \tag{4.7-37}$$

Equation (4.7-37) serves as the basis of the method described.

Let us suppose, for example, that the output sensitivity function $(\partial u_2/\partial R_k)R_{k0}$ of the network shown in Fig. 4.7-10 is to be determined. A change of R_k by dR_k is evidently associated with a change of all currents and voltages of the network. For the voltage change in the branch of the variable component du_k, we have

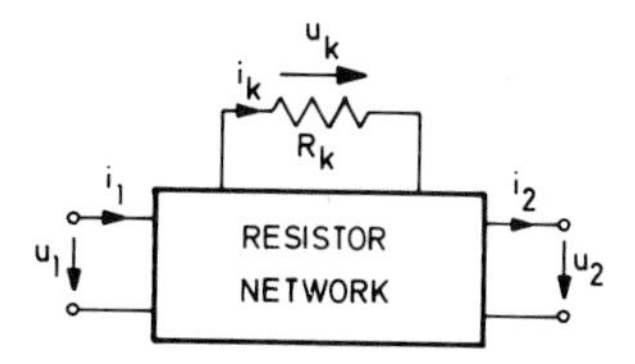

FIG. 4.7-10. Resistor network under consideration with kth branch.

$$du_k = \frac{\partial u_k}{\partial R_k}\, dR_k + \frac{\partial u_k}{\partial i_k}\, di_k = i_k\, dR_k + R_k\, di_k, \tag{4.7-38}$$

and for all others ($j \neq k$),

$$du_j = \frac{\partial u_j}{\partial i_j}\, di_j = R_j\, di_j. \tag{4.7-39}$$

If we now choose the branches of the network $\hat{N}$ such that all terms in Eq. (4.7-37) disappear except the two that contain the index k and the index 2—how this is achieved will be discussed later—then Eq. (4.7-37) reduces to the form

$$du_k\, \hat{i}_k - di_k\, \hat{u}_k + du_2\, \hat{i}_2 - di_2\, \hat{u}_2 = 0. \tag{4.7-40}$$

With the aid of Eq. (4.7-38), Eq. (4.7-40) becomes

$$i_k \hat{i}_k\, dR_k + R_k \hat{i}_k\, di_k - \hat{u}_k\, di_k + du_2 i_2 - di_2\, \hat{u}_2 = 0. \tag{4.7-41}$$

Recognize that $di_2 = 0$. Furthermore, in $\hat{N}$ we choose $\hat{i}_2 = 1$ and $\hat{R}_k = R_k$. This implies that $\hat{u}_k = R_k \hat{i}_k$. Then, from Eq. (4.7-41), $du_2 = -i_k \hat{i}_k\, dR_k$ or

$$\frac{du_2}{dR_k} = -i_k \hat{i}_k = \frac{\partial u_2}{\partial R_k}. \tag{4.7-42}$$

Thus under the quoted conditions, the desired sensitivity function of u_2 with respect to the kth element is simply given by the product of the currents in the kth branch of the nominal original network N and the corresponding branch of the network $\hat{N}$, $\hat{N}$ is also called *adjoint network* of N.

The question now is how to achieve the quoted conditions, that is, how the adjoint network can be found. In fact, the sufficient but not necessary conditions to obtain Eq. (4.7-42) presumed above may serve as a guideline to construct $\hat{N}$. For the case discussed, $\hat{N}$ has to be defined such that the individual expressions $du_\nu\, \hat{i}_\nu - di_\nu\, \hat{u}_\nu$ in Eq. (4.3-37) disappear for all ν except the two (j,k) that appear in the sensitivity function under consideration, $\partial u_j/\partial R_k$ or $\partial i_j/\partial R_k$. This requires that the above expressions disappear for all inputs (voltage and current sources), for those output voltages and currents that are not the output under consideration, and for all network elements that are not the parameters of interest.

In order to fulfill these requirements, the adjoint network $\hat{N}$ must have the following specifications:

(1) All resistances of $\hat{N}$ are the same as in the original network N; that is, $\hat{R}_\nu = R_\nu$.

(2) All independent voltage sources of N are replaced in $\hat{N}$ by short circuits; all independent current sources in N are replaced in $\hat{N}$ by open circuits.

(3) If the output under consideration is a voltage (e.g.,u_j), then the corresponding branch in $\hat{N}$ is a unit current $\hat{i}_j = 1$; conversely, if the output under consideration is a current (e.g.,i_j), then the corresponding branch in $\hat{N}$ is a unit voltage $\hat{u}_j = 1$.

Under these conditions the sufficient premises of Eq. (4.7-42) are met, and the formula (4.7-42) then holds for any resistance R_k ($k = 3, \ldots, n_z$) considered as the parameter of interest. This implies that the effect of any of the resistances upon the output of a resistor network can be determined from two networks, a nominal network and an adjoint network, as specified above. A signal analysis of N and $\hat{N}$ is necessary to find i_k and $\hat{i}_k$.

The method derived above for linear resistor networks can be extended to more general networks containing controlled sources, capacitances, inductances, and even nonlinear elements such as diodes etc. [3]. However, the complexity of the adjoint network may then increase considerably.

Example 4.7-3 Given the attenuator circuit shown in Fig. 4.7-11a. The nominal values of the resistances are $R_3 = 100$ 0hm, $R_4 = 10$ 0hm, $R_5 = 50$ 0hm, and the input voltage is $u_1 = 10$ V. Study the influence of resistor tolerances on the output voltage u_2 by determining the relative output sensitivity functions with respect to R_3, R_4, and R_5 by the aid of an adjoint network.

Solution First, the adjoint network $\hat{N}$ is set up by application of the above rules. This yields the circuit of Fig. 4.7-11b. The currents $i_3, \ldots, i_5$ and $\hat{i}_3, \ldots, \hat{i}_5$ in both circuits are now calculated by classical network analysis.

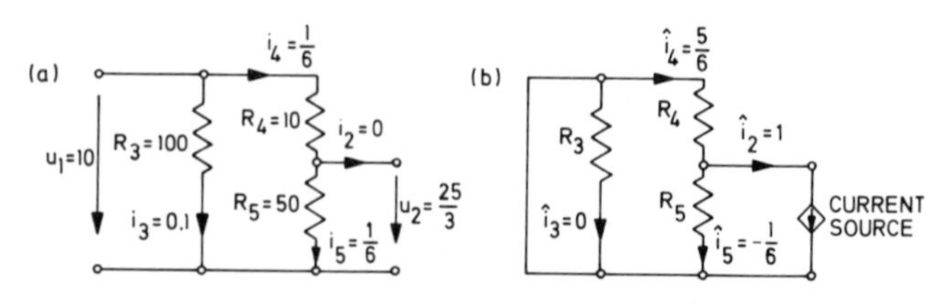

FIG. 4.7-11. (a) Given attenuator of Example 4.7-3, (b) adjoint network N.

The results are given in the diagrams. With these values the following relative output sensitivity functions are obtained:

$$\bar{\sigma}_3 \triangleq \frac{R_3}{u_2}\frac{\partial u_2}{\partial R_3} = -\frac{R_3}{u_2} i_3 \hat{i}_3 = 0,$$

$$\bar{\sigma}_4 \triangleq \frac{R_4}{u_2}\frac{\partial u_2}{\partial R_4} = -\frac{R_4}{u_2} i_4 \hat{i}_4 = -\frac{1}{6},$$

$$\bar{\sigma}_5 \triangleq \frac{R_5}{u_2}\frac{\partial u_2}{\partial R_5} = -\frac{R_5}{u_2} i_5 \hat{i}_5 = \frac{1}{6}.$$

A great advantage of this method is that a single network analysis for each of the two circuits is necessary and sufficient to calculate all sensitivity functions of the network.

4.7.3 Sensitivity Invariants of Linear Networks

The relative sensitivity functions of a linear network with respect to the network parameters R, L, $D = C^{-1}$ are not independent of each other. This will be shown in this section.

Consider a linear network of R, L, and C elements as indicated in Fig. 4.7-12. Let us use the reciprocal of C, i.e., $D = C^{-1}$ instead of C for the analytic treatment of the network.

FIG. 4.7-12. Given R, L, C network.

Assertion: Suppose that all elements R_i, ωL_i, and D_i/ω are subject to the same relative change from R_i, ωL_i, D_i/ω to λR_i, λL_i, $\lambda D_i/\omega$, where $\lambda = 1 + \alpha$ with α an infinitesimal number and ω constant. This change does not affect the transfer function $G = U_2/U_1$ of the network, that is,

$$G(\lambda R_1, \ldots, \lambda R_{NR}, \lambda L_1, \ldots, \lambda L_{NL}, \lambda D_1, \ldots, \lambda D_{ND})$$
$$= G(R_1, \ldots, R_{NR}, L_1, \ldots, L_{NL}, D_1, \ldots, D_{ND}). \tag{4.7-43}$$

Proof: Differentiating Eq. (4.7-43) with respect to λ gives, since the right-hand side does not depend on λ,

$$\sum_{i=1}^{NR} \frac{\partial G}{\partial \lambda R_i}\frac{\partial \lambda R_i}{\partial \lambda} + \sum_{i=1}^{NL} \frac{\partial G}{\partial \lambda L_i}\frac{\partial \lambda L_i}{\partial \lambda} + \sum_{i=1}^{ND} \frac{\partial G}{\partial \lambda D_i}\frac{\partial \lambda D_i}{\partial \lambda} = 0. \tag{4.7-44}$$

Now dividing by G, carrying out the differentiation, and letting λ approach 1 (i.e., $\alpha = 0$), we obtain

$$\sum_{i=1}^{NR} \frac{R_i}{G}\frac{\partial G}{\partial R_i} + \sum_{i=1}^{NL} \frac{L_i}{G}\frac{\partial G}{\partial L_i} + \sum_{i=1}^{ND} \frac{D_i}{G}\frac{\partial G}{\partial D_i} = 0 \tag{4.7-45}$$

or in terms of Bode's sensitivity function,

$$\sum_{i=1}^{N_R} S_{R_i}^G(s) + \sum_{i=1}^{N_L} S_{Li}^G(s) + \sum_{i=1}^{N_D} S_D{}^G(s) = 0. \tag{4.7-46}$$

If we designate all elements R_i, L_i, D_i, by α_i and the total number of elements $N_R + N_L + N_D$ by N, we get

$$\sum_{i=1}^{N} S_{\alpha_i}^G(s) = 0. \tag{4.7-47}$$

Now recall that $Y = GU$, so that

$$S_{\alpha_i}^G = \frac{\alpha_i}{Y}\frac{\partial Y}{\partial \alpha_i} = \bar{\Sigma}_i(s, \boldsymbol{\alpha}) = \mathscr{L}\{\bar{\sigma}_i(t, \boldsymbol{\alpha})\}.$$

Therefore, the above balance equation also holds for the relative output sensitivity functions $\bar{\sigma}_i(t, \boldsymbol{\alpha})$. That is,

$$\sum_{i=1}^{N} \bar{\sigma}_i(t, \boldsymbol{\alpha}) = 0. \tag{4.7-48}$$

This result can be summarized as follows.

Theorem 4.7-2: The sum of the relative sensitivity functions (in the frequency domain and the time domain) of a linear network with respect to the same proportional change of all its elements R_i, L_i, D_i (or their reciprocals) is for all equivalent networks constant and equal to zero.

It is easy to show [9] that similar relations hold for the sensitivity of the transfer impedance and admittance of a linear network:

Theorem 4.7-3: In linear networks the sum of the relative sensitivity functions of the transfer admittance with respect to the same relative change of all elements is equal to -1. If the transfer impedance is considered instead, the corresponding sum is $+1$.

This result shows that the sum of all Bode or relative output sensitivity functions is no proper means to specify the sensitivity of a linear network. On the other hand, only N–1 sensitivity functions need be calculated when in a linear network the influence of all N parameters is to be studied in terms of the Bode or relative output sensitivity function.

Example 4.7-4: Referring to Example 4.7–3, show that the sum of all relative output sensitivity functions is zero.

Solution: The addition of $\bar{\sigma}_3 = 0$, $\bar{\sigma}_4 = -\frac{1}{6}$ and $\bar{\sigma}_5 = \frac{1}{6}$ gives $\sum \bar{\sigma}_i = 0$.

PROBLEMS

4.1 Determine the output sensitivity equations with respect to a ($a_0 \neq 0$) of the following differential equations:

(a) $t^2\ddot{y} + t\dot{y} + (t^2 - a^2)y = 0$, $y(0) = \beta_0$, $\dot{y}(0) = \beta_1$,
(b) $t(1-t)\ddot{y} + [y - (a + b + 1)t]\dot{y} - aby = 0$, $y(0) = \beta_0$, $\dot{y}(0) = \beta_1$,
(c) $\ddot{y} + t\dot{y}^3 - e^{ay} = 0$, $y(0) = \beta_0$, $\dot{y}(0) = \beta_1$,;
(d) $\dot{y} + ay^2 - bty = u(t)$, $y(0) = \beta_0$,;
(e) $\dot{y}^2 - a^2y^2 = t^2$, $y(0) = \beta_0$;
(f) $\ddot{y} + ay\dot{y} + by = u(t)$, $y(0) = \beta_0$, $\dot{y}(0) = \beta_1$;
(g) $\dot{y}^2 + c(y^2 + a)\dot{y} - bty = 0$, $y(0) = \beta_0$;
(h) $\dot{y}^2 + bt\dot{y} - y = u(\mathrm{t})$, $\quad y(0) = \mathrm{a}$.

4.2 Determine the second-order sensitivity equation with respect to α of Van der Pol's equation $\ddot{y}$-$\alpha(1-y^2)\dot{y}+y = 0$ for (a) $\alpha_0 = 1$ and (b) $\alpha_0 = 0$.

4.3 Given the differential equation

$$a_1a_2\ddot{y} + a_1a_3\dot{y} + a_3y = 1(t), \qquad y(0) = 0, \quad \dot{y}(0) = 0, \quad a_{i0} \neq 0.$$

(a) Determine the sensitivity equations with respect to a_1 ($a_{10} \neq 0$) and to the initial condition $y(0)$.

(b) Draw a signal flow graph to measure the sensitivity function $\sigma_1 = (\partial y/\partial a_1)_{a10}$.

(c) Give the expression for the Laplace transformed sensitivity function $\boldsymbol{\Sigma}_1 = \mathscr{L}\{\sigma_1\}$.

4.4 Suppose a point mass m is falling in a vacuum by constant gravitational attraction g. The mass starts at a position $y(t_0)$ with the initial velocity $\dot{y}(t_0)$. Determine the sensitivity functions of the position $y(t)$ with respect to m, $y(t_0)$, $\dot{y}(t_0)$.

4.5 Referring to Problem 3.15, decide whether or not the direct calculation of the sensitivity function with respect to α by use of the Laplace transform is allowed.

4.6 Given the third-order differential equation

$$a_4a_3\dddot{y} + a_4a_2\ddot{y} + (a_1 + a_4)\dot{y} + y = bt, \qquad y(0) = \dot{y}(0) = \ddot{y}(0) = 0,$$

with a_1,a_2,a_3 and b nonzero constants. Suppose that the constant a_4 is small enough to be neglected. Find the output sensitivity function with respect to a_4 for $a_{40} = 0$, making use of the sensitivity equation. (Apply Chang's method or the alternative procedure given in Section 4.3.4.)

4.7 Consider a third-order system with the transfer function

$$G(s,T_1) = [(1 + T_1s)\,(1 + Ts)^2]^{-1}.$$

Suppose T_1 is small compared to T and is neglected, *i.e.*, $T_{10} = 0$. Determine the sensitivity function $\sigma_1 = (\partial y/\partial T_1)_{T10}$ for all initial conditions equal to zero and the input $u(t) = 1(t)$ (step function).

4.8 Consider the system with the transfer function

$$G(s, T_1) = [(1 + T_1 s)(1 + T_2 s)]^{-1},$$

with the nominal parameter values $T_1 = 10\text{s}$, $T_2 = 5\text{s}$ and the initial conditions equal to 0. Suppose that T_1 changes at $t_1 = 15\text{s}$ infinitesimally. Determine the corresponding output sensitivity function σ_ε for a step input $u(t) = 1(t)$.

4.9 For the feedback control system shown in Fig. 4.P-1, draw a measuring circuit for the output sensitivity function σ with respect to T (T_0 nominal value) according to the variable component method.

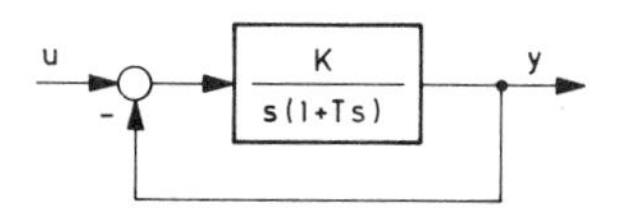

FIG. 4.P-1. Feedback control system of Problem 4.9.

4.10 Given the feedback control system of Fig. 4.P-2: Draw the block diagram to generate the output sensitivity functoin σ with respect to the dead-time T_t of the plant according to the method of variable component.

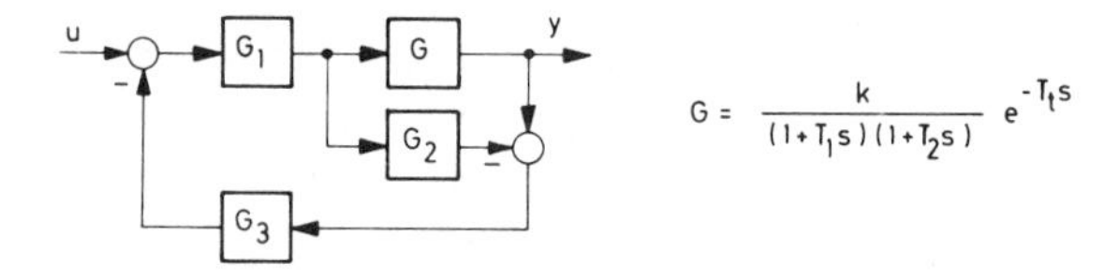

FIG. 4.P-2. Feedback control system of Problem 4.10.

4.11 Referring to Problem 4,6. draw the block diagram to measure $\sigma = (\partial y/\partial a_4)_{a40=0}$ according to the variable component method and from the resulting diagram deduce the analytic expression for $\boldsymbol{\Sigma} = \mathscr{L}\{\sigma\}$.

4.12 Consider a system with the differential equation

$$\ddot{y} + a_1 \dot{y} + a_0 y = u, \qquad y(0) = \dot{y}(0) = 0,$$

where the nominal parameter values are $a_{00} = 4, a_{10} = 0$. Determine the output sensitivity function with respect to a_1 for steplike inputs $u = 1(t)$ with the aid of the variable component method.

4.13 A system with the transfer function

$$G(s, T_1) = \frac{K}{(1 + T_1 s)^3 (1 + T_2 s)} = \frac{Y(s,T_1)}{U(s,T_1)}$$

is the forward part in a unity feedback control system. That is, $u(\imath, T_1) =$

$r(t) - y(t, T_1)$ where u and y are the input and output of $G(s, T_1)$, respectively, and r is the reference input. Suppose that T_1 is sufficiently small to be neglected in comparison with T_2. Determine the output sensitivity function $\sigma = (\partial y/\partial T_1)_0$ employing the variable component method.

4.14 For the time-variant system with the differential equation $\ddot{y} + (a + b \cos \omega t) = 0$ and the initial conditions $y(0) = \beta_0$, $\dot{y}(0) = \beta_1$,

(a) determine the output sensitivity equation with respect to a,

(b) draw an analog computer diagram to measure the output sensitivity function.

4.15 Given the block diagram of Fig. 4.P-3: Draw the block diagram to measure the output sensitivity function

(a) with respect to ω,
(b) with respect to β_0.

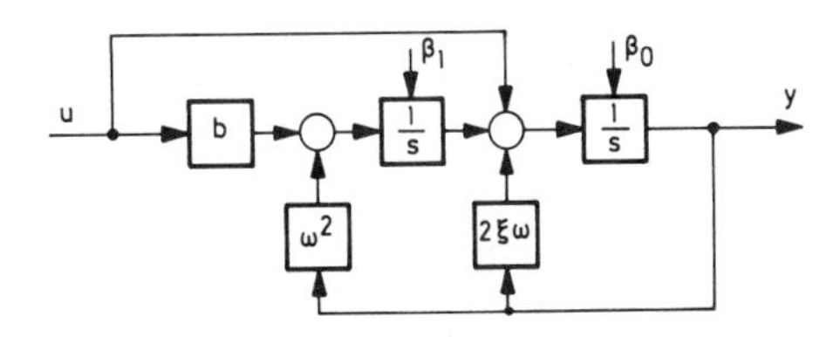

FIG. 4.P-3. Given block diagram of Problem 4.15.

4.16 Draw a block diagram to measure the output error vector $\Delta \boldsymbol{y}$ of the multivariable feedback system shown in Fig. 4. P-4 due to finite changes of the parameter vector $\boldsymbol{\alpha}$ using the variable component method.

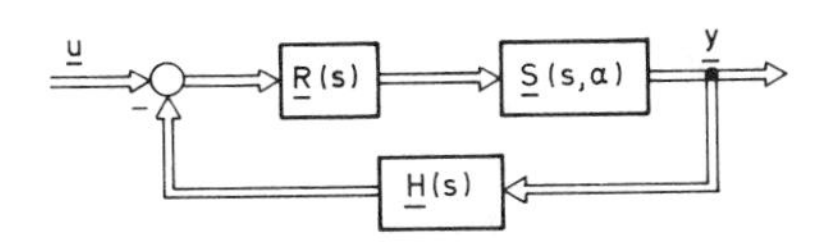

FIG. 4.P-4. Multivariable feedback system of Problem 4.16.

4.17 Referring to Example 4.5-5 (Fig. 4.5-19a), draw the block diagram to measure the output sensitivity function with respect to the nonlinear term $b\psi_f^3$ in $\tilde{F}(\psi_f)$, that is, with respect to b for $b = b_0 = 0$.

4.18 Referring to the dc motor of Example 4.5-6, determine the block diagram to measure the sensitivity functions of i_a and ω with respect to the inductance L_a for $L_a = 0$.

4.19 Draw a block diagram to measure the output sensitivity function with respect to the coefficient c of the nonlinear characteristic $x = cz^3$ in the system shown in Fig. 4.P-5.

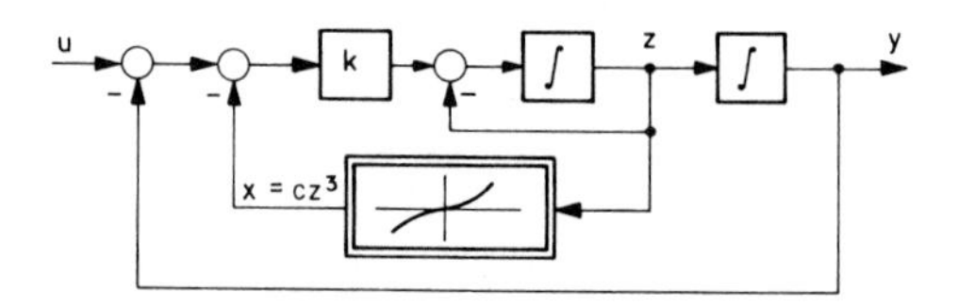

FIG. 4.P-5. Given system of Problem 4.19.

4.20 Consider the motion of a pendulum where nonlinear friction proportional to the angular velocity and to its power is taken into account. It can be described by the following differential equation

$$\ddot{\varphi} + a\dot{\varphi} + b(\operatorname{sgn}\dot{\varphi})\dot{\varphi}^2 + c\sin\varphi = 0, \qquad \varphi(0) = \beta_0, \quad \dot{\varphi}(0) = \beta_1,$$

where φ is the angle, a,b,c are constants. The coresponding analog computer diagram is shown in Fig. 4.P–6.

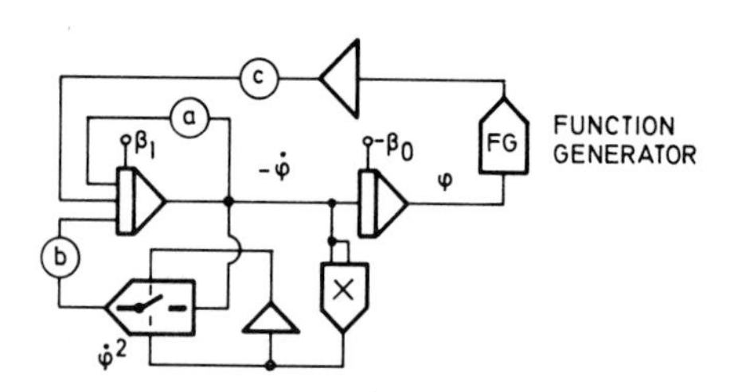

FIG. 4.P-6. Analog computer diagram of the motion of the pendulum of Problem 4 20.

(a) Determine the output sensitivity equations with respect to a,b,c.
(b) Draw an analog computer diagram to measure the sensitivity functions of φ and $\dot{\varphi}$ with respect to a,b,c.

4.21 For the system described by the differential equation

$$y^{(6)} + a_5 y^{(5)} + \cdots + a_1\dot{y} + a_0 y = 1(t), \qquad y^{(\nu)}(0) = 0, \quad \nu = 0, 1, \ldots, 5,$$

draw an analog computer diagram to measure all output sensitivity functions $\sigma_\nu = (\partial y/\partial a_\nu)_{a=0}$ simultaneously.

4.22 Referring to the feedback control system of Problem 4.10 (Fig. 4.P-2), draw the block diagram including the sensitivity point with respect to the time delay T_t.

4.23 The relation between the throttle position $u(t)$ and the vehicle acceleration y of a spacecraft is given by the differential equation

$$\ddot{y} + a_1\dot{y} + a_0 y = b_2\ddot{u} + b_1\dot{u}.$$

Draw a signal flow graph to generate the semirelative sensitivity functions of y with respect to all parameters a_0, a_1, b_1, b_2, simultaneously.

4.24 Given a system with the transfer function

$$G(s) = \frac{s + b}{(s + a_1)(s + a_2)} = \frac{Y(s)}{U(s)}.$$

Draw a signal flow graph to generate simultaneously the absolute output sensitivity functions with respect to all parameters a_1, a_2, and b.

4.25 For the given feedback control system of Fig. 4.P-7, show how the output sensitivity function with respect to the gain factor k can be calculated by means of the three-point method.

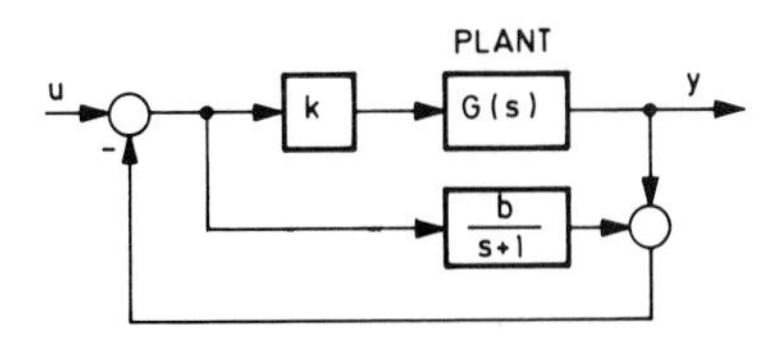

FIG. 4.P-7. Given feedback control system of Problem 4.25.

4.26 Given a system with the transfer function

$$G(s) = \frac{Y(s)}{U(s)} = \frac{K(1 + T_1 s)}{(1 + T_2 s)(1 + T_3 s)}.$$

Draw a signal flow graph which provides the simultaneous measurement of all (absolute) output sensitivity functions with respect to the parameters K, T_1, T_2, T_3.

4.27 Figure 4.P-8 shows a network in which the capacitance C is the parameter of interest. Complete this diagram such that the output sensitivity function $\sigma = (\partial u_2/\partial C)_{C_0}$ can be measured as the output voltage of the completed network.

4.28 Draw an electrical network to measure the output sensitivity function $\sigma = (\partial u_2/\partial C)_{C_0}$ of the network of Problem 3.6 (Fig. 3.5-2).

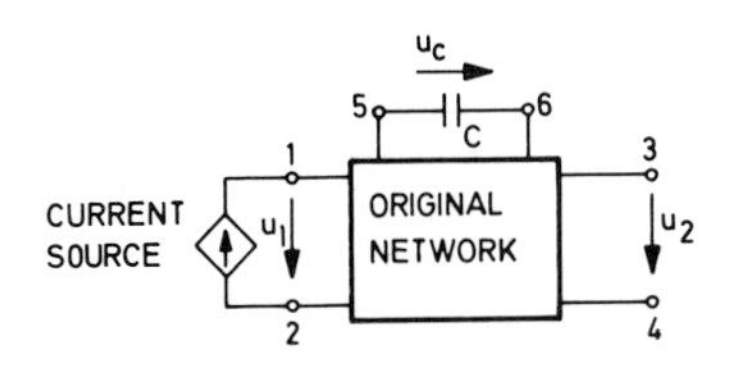

FIG. 4.P-8. Given network of Problem 4.27.

4.29 For the lattice network shown in Fig. 4.P-9, determine the relative output sensitivity functions $\bar{\sigma}_\nu = \partial \ln u_2/\partial \ln R_\nu$, $\nu = 1, 2, \ldots, 5$ using the adjoint network method. Show that the sum of all relative sensitivity functions is equal to zero.

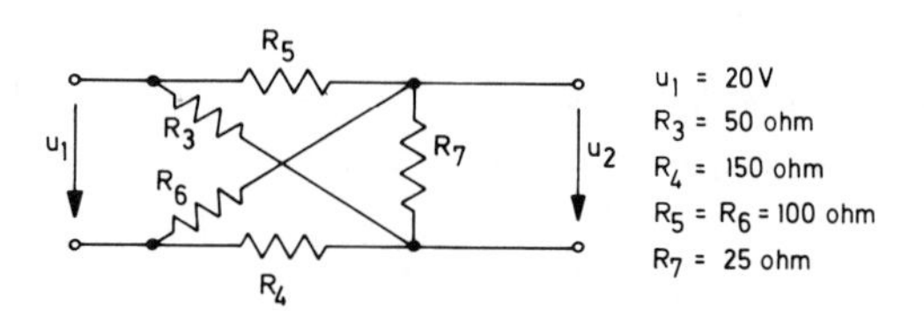

FIG. 4.P-9. Lattice network of Problem 4.29.

Chapter 5

Methods for Calculation or Measurement of the Trajectory Sensitivity Functions

5.1 INTRODUCTION

The trajectory sensitivity functions can be calculated or measured in a manner similar to the way output sensitivity functions are calculated or measured, that is, by making use of a sensitivity model. The associated sensitivity models as well as the complete measuring circuits will be derived in this chapter for linear and nonlinear continuous and discontinuous systems including sampled-data systems. One of the basic questions is that of minimal structure for the simultaneous measurement of all sensitivity functions. It turns out indeed that for the simultaneous measurement of all trajectory sensitivity functions substantial structural reductions of the complete measuring circuit are possible, since a number of trajectory sensitivity functions are identical. The theorems of Wilkie and Perkins, which give information about the maximum number of independent trajectory sensitivity functions as well as about the minimum structure necessary to measure all sensitivity functions simultaneously, are presented, proved, and illustrated by simple examples.

In sampled-data and discrete systems, besides the common parameters, the sampling period T has to be considered as an additional parameter affecting the state $\mathbf{x}$. For this case, the sensitivity equations will be derived at the sampling times as well as at the times in between. In the latter case, the deduction of the sensitivity equation is more complicated since the continuity condition is violated.

Besides this, methods will be developed for the determination of the first variation of $\mathbf{x}$ in the case of time-varying parameters. It will be shown that for the first variation of $\mathbf{x}$ in this case, the same structural rules apply as for the trajectory sensitivity functions in the case of time-invariant parameters. This implies that the former can be determined and measured by the same methods.

5.2 TRAJECTORY SENSITIVITY EQUATIONS OF CONTINUOUS SYSTEMS WITH TIME–INVARIANT PARAMETER VARIATIONS

Trajectory sensitivity equations can be found in a manner similar to the one output sensitivity equations are found, namely, either by a Taylor series expansion or by partial differentiation of the state equation with respect to the parameters of interest. To apply the latter procedure, consider a general continuous system described by the state equation (3.2-30),

$$\dot{\boldsymbol{x}} = \boldsymbol{f}(\boldsymbol{x}, t, \boldsymbol{u}, \boldsymbol{\alpha}), \qquad \boldsymbol{x}(t_0) = \boldsymbol{x}^0, \tag{5.2-1}$$

where $\boldsymbol{x}$ denotes the n-dimensional state vector, $\boldsymbol{\alpha}$ the r-dimensional parameter vector, $\boldsymbol{f}$ an n-dimensional vector function, and $\boldsymbol{u}$ the input vector independent of $\boldsymbol{\alpha}$. It is assumed that the continuity conditions of Section 3.2.3 are fulfilled and that $\boldsymbol{\alpha}$ is time-invariant. The above general form of $\boldsymbol{f}$ includes general nonlinear systems.

Now consider *α-parameters.* Taking the partial derivative of $\dot{\boldsymbol{x}}$ [(Eq. (5.2-1)] with respect to α_j, we obtain, by the application of the chain rule (Section 2.4),

$$\frac{\partial \dot{\boldsymbol{x}}}{\partial \alpha_j} = \frac{\partial \boldsymbol{f}}{\partial \boldsymbol{x}} \frac{\partial \boldsymbol{x}}{\partial \alpha_j} + \frac{\partial \boldsymbol{f}}{\partial \alpha_j}, \qquad \frac{\partial \boldsymbol{x}^0}{\partial \alpha_j} = \boldsymbol{0}, \quad j = 1, 2, \ldots, \text{r}. \tag{5.2-2}$$

Note that the derivative of the initial conditions vector $\boldsymbol{x}^0$ with respect to α_j disappears since $\boldsymbol{x}^0$ does not depend on $\boldsymbol{\alpha}$. If $\boldsymbol{\alpha}$ is r-dimensional, there are r equations of the form (5.2-2). If we now interchange the sequence of taking the derivative with respect to time t and α_j, which is admissible by virtue of the.assumptions made, and then let $\boldsymbol{\alpha}$ approach $\boldsymbol{\alpha}_0$, we obtain

$$\dot{\boldsymbol{\lambda}}_j = \left.\frac{\partial \boldsymbol{f}}{\partial \boldsymbol{x}}\right|_{\boldsymbol{\alpha}0} \boldsymbol{\lambda}_j + \left.\frac{\partial \boldsymbol{f}}{\partial \alpha_j}\right|_{\boldsymbol{\alpha}0}, \qquad \boldsymbol{\lambda}(0) = 0, \quad j = 1, 2, \ldots, \text{r}. \tag{5.2-3}$$

Here $\boldsymbol{\lambda}_j \triangleq (\partial \boldsymbol{x}/\partial \alpha_j)_{\boldsymbol{\alpha}_0}$ is the *trajectory sensitivity vector* with respect to the jth parameter as defined in Section 3.2.3. Equation (5.2-3) is called the sensitivity equation in the state space or the *trajectory sensitivity equation.*

Note that the $n \times n$ matrix $\partial \boldsymbol{f}/\partial \boldsymbol{x} = [\partial f_i/\partial x_k]$, $i, k = 1, 2, \ldots, n$, is the Jacobian matrix (Definition 2.3-9). The above shows that for α-parameters all initial conditions of the trajectory sensitivity equation are equal to zero.

Now consider the output vector equation

$$\boldsymbol{y} = \boldsymbol{g}(\boldsymbol{x}, t, \boldsymbol{u}, \boldsymbol{\alpha}). \tag{5.2-4}$$

In a procedure similar to that above we obtain the algebraic sensitivity equation

$$\boldsymbol{\sigma}_j = \left.\frac{\partial \boldsymbol{y}}{\partial \alpha_j}\right|_{\boldsymbol{\alpha}0} = \left.\frac{\partial \boldsymbol{g}}{\partial \boldsymbol{x}}\right|_{\boldsymbol{\alpha}0} \boldsymbol{\lambda}_j + \left.\frac{\partial \boldsymbol{g}}{\partial \alpha_j}\right|_{\boldsymbol{\alpha}0}, \tag{5.2-5}$$

which relates the output sensitivity vector $\boldsymbol{\sigma}_j = (\partial \boldsymbol{y}/\partial \alpha_j)_{\boldsymbol{\alpha}0}$ to the trajectory sensitivity vector $\boldsymbol{\lambda}_j$. This equation is called the *vector output sensitivity equation.*

Using the trajectory sensitivity matrix $\boldsymbol{\lambda}$ (Definition 3.2-6) and the output sensitivity matrix $\boldsymbol{\sigma}$ (Definition 3.2-2), the above result can be rewritten in the following general form

$$\dot{\boldsymbol{\lambda}} = \left.\frac{\partial \boldsymbol{f}}{\partial \boldsymbol{x}}\right|_{\boldsymbol{\alpha}0} \boldsymbol{\lambda} + \left.\frac{\partial \boldsymbol{f}}{\partial \boldsymbol{\alpha}}\right|_{\boldsymbol{\alpha}0}, \qquad \boldsymbol{\lambda}^0 = \mathbf{0},$$

$$\boldsymbol{\sigma} = \left.\frac{\partial \boldsymbol{g}}{\partial \boldsymbol{x}}\right|_{\boldsymbol{\alpha}0} \boldsymbol{\lambda} + \left.\frac{\partial \boldsymbol{g}}{\partial \boldsymbol{\alpha}}\right|_{\boldsymbol{\alpha}0}. \tag{5.2-6}$$

Equations (5.2-6) are simply called the *state sensitivity equations* of the system. It is seen that these equations are linear whether the original system is linear or not. In general, the coefficients are time-variant and the initial conditions are zero. The result may be summarized by the following theorem.

Theorem 5.2-1 The trajectory and output sensitivity functions of a continuous linear or nonlinear system with time invariant α-parameters can be determined from a system of first-order linear differential equations with, in general, time-variant coefficients and zero initial conditions.

Example 5.2-1 Determine the trajectory sensitivity equation with respect to α of Van der Pol's equation

$$\dot{x}_1 = x_2 \qquad x_1(0) = \beta_1,$$

$$\dot{x}_2 = -x + \alpha(1 - x_1{}^2)x_2, \qquad x_2(0) = \beta_2.$$

Solution Taking the partial derivatives with respect to α and letting α approach α_0, we obtain

$$\dot{\lambda}_1 = \lambda_2, \qquad \lambda_1\,(0) = 0,$$

$$\dot{\lambda}_2 = -(1 + 2\alpha_0 x_{10} x_{20})\,\lambda_1 + \alpha_0(1 - x_{10}^2)\,\lambda_2 + (1 - x_{10}^2)x_{20}, \qquad \lambda_2(0) = 0,$$

where $\lambda_1 = (\partial x_1/\partial \alpha)_{\alpha 0}$ and $\lambda_2 = (\partial x_2/\partial \alpha)_{\alpha 0}$. In terms of the trajectory sensitivity vector $\boldsymbol{\lambda} = [\lambda_1\ \lambda_2]^{\mathrm{T}}$, we have

$$\dot{\boldsymbol{\lambda}} = \begin{bmatrix} 0 & 1 \\ -(1 + 2\alpha_0 x_{10} x_{20}) & \alpha_0(1 - x_{10}^2) \end{bmatrix} \boldsymbol{\lambda} + \begin{bmatrix} 0 \\ (1 - x_{10}^2)x_{20} \end{bmatrix}, \qquad \boldsymbol{\lambda}(0) = \mathbf{0},$$

which is a linear state equation with time-variant coefficients.

If, in particular, the original system is linear, the state equations take the form

$$\dot{\boldsymbol{x}} = \boldsymbol{A}\,\boldsymbol{x} + \boldsymbol{B}\,\boldsymbol{u}, \qquad \boldsymbol{x}(t_0) = \boldsymbol{x}^0,$$

$$\boldsymbol{y} = \boldsymbol{C}\,\boldsymbol{x} + \boldsymbol{D}\,\boldsymbol{u}, \tag{5.2-7}$$

where, in general, $\boldsymbol{A} = \boldsymbol{A}(\boldsymbol{\alpha})$, $\boldsymbol{B} = \boldsymbol{B}(\boldsymbol{\alpha})$, $\mathbf{C} = \boldsymbol{C}(\boldsymbol{\alpha})$, $\boldsymbol{D} = \boldsymbol{D}(\boldsymbol{\alpha})$, and $\boldsymbol{x} = \boldsymbol{x}(t,\boldsymbol{\alpha})$,

$\mathbf{y} = \boldsymbol{y}(t, \boldsymbol{\alpha})$. Note, however, that $\boldsymbol{u}$ is not a function of $\boldsymbol{\alpha}$ if $\boldsymbol{u}$ is defined as the external input of the system.

Now taking the partial derivatives with respect to α_j, reversing the order of differentiations with respect to time and α_j, and letting $\boldsymbol{\alpha}$ approach $\boldsymbol{\alpha}_0$, the *trajectory sensitivity equation* is obtained as

$$\dot{\boldsymbol{\lambda}}_j = \boldsymbol{A}_0 \boldsymbol{\lambda}_j + \left.\frac{\partial \boldsymbol{A}}{\partial \alpha_j}\right|_{\boldsymbol{\alpha}0} \boldsymbol{x}_0 + \left.\frac{\partial \boldsymbol{B}}{\partial \alpha_j}\right|_{\boldsymbol{\alpha}0} \boldsymbol{u}(t), \qquad \boldsymbol{\lambda}_j(t_0) = \mathbf{0}, \tag{5.2-8}$$

where $\boldsymbol{A}_0 = \boldsymbol{A}(\boldsymbol{\alpha}_0)$, $\boldsymbol{x}_0 = \boldsymbol{x}(t, \boldsymbol{\alpha}_0)$, and $j = 1, 2, \ldots, r$. The initial condition vector $\boldsymbol{\lambda}_j(t_0)$ is again zero since $\boldsymbol{x}(t_0)$ does not depend on $\boldsymbol{\alpha}$.

By a similar procedure the *vector output sensitivity equation* becomes

$$\boldsymbol{\sigma}_j = \boldsymbol{C}_0 \boldsymbol{\lambda}_j + \left.\frac{\partial \boldsymbol{C}}{\partial \alpha_j}\right|_{\boldsymbol{\alpha}0} \boldsymbol{x}_0 + \left.\frac{\partial \boldsymbol{D}}{\partial \alpha_j}\right|_{\boldsymbol{\alpha}0} \boldsymbol{u}(t), \tag{5.2-9}$$

where $\boldsymbol{C}_0 = \boldsymbol{C}(\boldsymbol{\alpha}_0)$ and $j = 1, 2, \ldots, r$. The result may be summarized by the following theorem.

Theorem 5.2-2 In the case of linear systems, the vector sensitivity equations have the same $\boldsymbol{A}$ matrix as the nominal state equations and hence the same characteristic matrix $s\boldsymbol{I} - \boldsymbol{A}$. They differ from the nominal original state equations only in the driving function and the initial conditions. The latter are all zero. The driving function can be obtained by solving the nominal state equations.

A graphical interpretation of Eqs. (5.2-8) and (5.2-9) is given in Fig. 5.2-1. If the original input $\boldsymbol{u}$ is applied to this structure, the sensitivity vectors $\boldsymbol{\lambda}_j$ and $\boldsymbol{\sigma}_j$ are obtained at the points a and b, respectively.

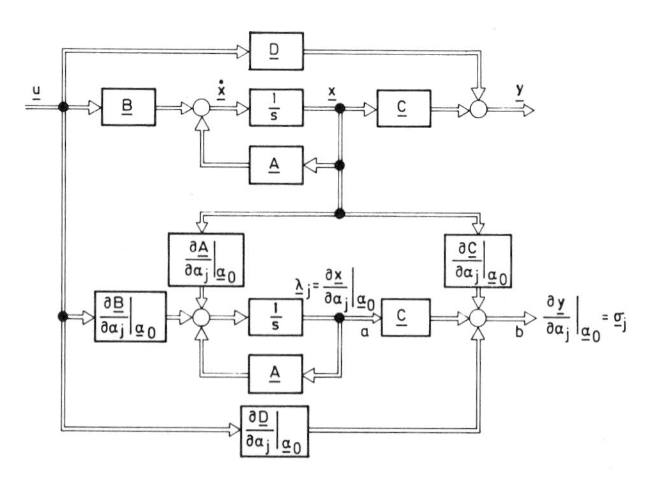

FIG. 5.2-1. Graphical interpretation of Eqs. (5.2-8) and (5.2-9) for the generation of $\boldsymbol{\lambda}_j$ and $\boldsymbol{\sigma}_j$ in the case of linear systems.

The result can be written in the general form

$$\dot{\boldsymbol{\lambda}} = \boldsymbol{A}_0 \boldsymbol{\lambda} + \left.\frac{\partial \boldsymbol{A}}{\partial \boldsymbol{\alpha}}\right|_{\boldsymbol{\alpha}0} \boldsymbol{x}_0 + \left.\frac{\partial \boldsymbol{B}}{\partial \boldsymbol{\alpha}}\right|_{\boldsymbol{\alpha}0} \boldsymbol{u}(t), \quad \boldsymbol{\lambda}(t_0) = \boldsymbol{0},$$

$$\boldsymbol{\sigma} = \boldsymbol{C}_0 \boldsymbol{\lambda} + \left.\frac{\partial \boldsymbol{C}}{\partial \boldsymbol{\alpha}}\right|_{\boldsymbol{\alpha}0} \boldsymbol{x}_0 + \left.\frac{\partial \boldsymbol{D}}{\partial \boldsymbol{\alpha}}\right|_{\boldsymbol{\alpha}0} \boldsymbol{u}(t), \tag{5.2-10}$$

where $\boldsymbol{\lambda}$ and $\boldsymbol{\sigma}$ are the trajectory and output sensitivity matrices, respectively. These are the *state sensitivity equations* of a continuous linear system with respect to α-variations.

The evaluation of the derivatives

$$\left.\frac{\partial \boldsymbol{A}}{\partial \boldsymbol{\alpha}}\right|_{\boldsymbol{\alpha}0} \boldsymbol{x}_0, \qquad \left.\frac{\partial \boldsymbol{B}}{\partial \boldsymbol{\alpha}}\right|_{\boldsymbol{\alpha}0} \boldsymbol{u}, \qquad \left.\frac{\partial \boldsymbol{C}}{\partial \boldsymbol{\alpha}}\right|_{\boldsymbol{\alpha}0} \boldsymbol{x}_0, \qquad \left.\frac{\partial \boldsymbol{D}}{\partial \boldsymbol{\alpha}}\right|_{\boldsymbol{\alpha}0} \boldsymbol{u}(t),$$

where $\boldsymbol{A}$, $\boldsymbol{B}$, $\boldsymbol{C}$, $\boldsymbol{D}$ donote matrices, may present computational difficulties. There are two ways to cope with this problem. The first is to perform the above operation in a componentwise fashion. For example, the ijth component of $(\partial \mathbf{A}/\partial \boldsymbol{\alpha})\boldsymbol{x}_0$ can be written as

$$\left[\frac{\partial \boldsymbol{A}}{\partial \boldsymbol{\alpha}} \boldsymbol{x}_0\right]_{ij} = \left[\sum_{k=1}^{n} \frac{\partial A_{ik}}{\partial \alpha_j} x_{k0}\right]_{\substack{i=1,2,\cdots,n\\ j=1,2,\cdots,r}} \tag{5.2-11}$$

where A_{ik} denotes the ikth component of $\boldsymbol{A}$ and x_{k0} the kth component of $\boldsymbol{x}_0 = \boldsymbol{x}(t, \boldsymbol{\alpha}_0)$.

An alternative procedure is the following. Consider the operation which led to the term

$$\frac{\partial(\boldsymbol{A}\,\boldsymbol{x})}{\partial \boldsymbol{\alpha}} = \frac{\partial \boldsymbol{A}}{\partial \boldsymbol{\alpha}} \boldsymbol{x} + \boldsymbol{A} \frac{\partial \boldsymbol{x}}{\partial \boldsymbol{\alpha}}. \tag{5.2-12}$$

This may be rewritten as

$$\frac{\partial \boldsymbol{A}(\boldsymbol{\alpha})}{\partial \boldsymbol{\alpha}} \boldsymbol{x}(t, \boldsymbol{\alpha}) + \boldsymbol{A}(\boldsymbol{\alpha}) \frac{\partial \boldsymbol{x}(t, \boldsymbol{\alpha})}{\partial \boldsymbol{\alpha}} = \left[\frac{\partial[\boldsymbol{A}(\boldsymbol{\alpha})\boldsymbol{x}(t, \boldsymbol{a})]}{\partial \boldsymbol{\alpha}} + \frac{\partial[\boldsymbol{A}(\boldsymbol{a})\boldsymbol{x}(t, \boldsymbol{\alpha})]}{\partial \boldsymbol{\alpha}}\right]_{\boldsymbol{a}=\boldsymbol{\alpha}}.$$

The multiplication $\boldsymbol{A}(\boldsymbol{\alpha})\boldsymbol{x}(t, \boldsymbol{a})$ in the term $[\partial[\boldsymbol{A}(\boldsymbol{\alpha})\boldsymbol{x}(t, \boldsymbol{a})]/\partial \boldsymbol{\alpha}]_{\boldsymbol{a}=\boldsymbol{\alpha}}$ results in the column vector

$$\begin{bmatrix} A_{11}(\boldsymbol{\alpha})x_1(t,\boldsymbol{a}) + \cdots + A_{1n}(\boldsymbol{\alpha})x_n(t,\boldsymbol{a}) \\ \vdots \\ A_{n1}(\boldsymbol{\alpha})x_1(t,\boldsymbol{a}) + \cdots + A_{nn}(\boldsymbol{\alpha})x_n(t,\boldsymbol{a}) \end{bmatrix}. \tag{5.2-13}$$

For this the derivative operation with respect to the row vector $\boldsymbol{\alpha}$ which has yet to be carried out, is well defined.

Example 5.2-2 Find the state sensitivity equations of a damped mass–spring system described by

$$\dot{\boldsymbol{x}} = \boldsymbol{A}\boldsymbol{x} + \boldsymbol{b}u, \qquad \dot{y} = \boldsymbol{c}\boldsymbol{x},$$

where u and y designate the force and the position of the mass, respectively, and

$$\boldsymbol{A} = \begin{bmatrix} 0 & 1 \\ \alpha_1 & \alpha_2 \end{bmatrix}, \qquad \boldsymbol{b} = \begin{bmatrix} 0 \\ \alpha_3 \end{bmatrix}, \qquad \boldsymbol{c} = [1 \quad 0], \qquad \boldsymbol{\alpha} = \begin{bmatrix} \alpha_1 \\ \alpha_2 \\ \alpha_3 \end{bmatrix}.$$

Solution Making use of Eq. (5.2-13), the state sensitivity equations (5.2-10) become

$$\dot{\boldsymbol{\lambda}} = \boldsymbol{A}_0 \boldsymbol{\lambda} + \left.\frac{\partial \boldsymbol{A}}{\partial \boldsymbol{\alpha}}\right|_{\boldsymbol{\alpha}0} \boldsymbol{x}_0 + \left.\frac{\partial \boldsymbol{b}}{\partial \boldsymbol{\alpha}}\right|_{\boldsymbol{\alpha}0} u, \qquad \boldsymbol{\sigma} = \boldsymbol{c}_0 \boldsymbol{\lambda},$$

where

$$\boldsymbol{\lambda} = \begin{bmatrix} \lambda_{11} & \lambda_{12} & \lambda_{13} \\ \lambda_{21} & \lambda_{22} & \lambda_{23} \end{bmatrix}, \qquad \lambda_{ij} = \left.\frac{\partial x_i}{\partial \alpha_j}\right|_{\boldsymbol{\alpha}0},$$

$$\boldsymbol{A}_0 = \begin{bmatrix} 0 & 1 \\ \alpha_{10} & \alpha_{20} \end{bmatrix}, \qquad \boldsymbol{c}_0 = [1 \quad 0],$$

$$\left.\frac{\partial \boldsymbol{A}}{\partial \boldsymbol{\alpha}}\right|_{\boldsymbol{\alpha}0} \boldsymbol{x}_0 = \left.\frac{\partial}{\partial \boldsymbol{\alpha}} \begin{bmatrix} x_2(t,\boldsymbol{\alpha}) \\ \alpha_1 x_1(t,\boldsymbol{\alpha}) + \alpha_2 x_2(t,\boldsymbol{\alpha}) \end{bmatrix}\right|_{\boldsymbol{\alpha}=\boldsymbol{\alpha}0} = \begin{bmatrix} 0 & 0 & 0 \\ x_{10} & x_{20} & 0 \end{bmatrix},$$

$$\left.\frac{\partial \boldsymbol{b}}{\partial \boldsymbol{\alpha}}\right|_{\boldsymbol{\alpha}0} = \begin{bmatrix} 0 & 0 & 0 \\ 0 & 0 & 1 \end{bmatrix}.$$

The correctness of this result can easily be verified by taking the partial derivatives of the individual state equations with respect to α_1, α_2, and α_3 according to Eqs. (5.2-8) and (5.2-9).

If there is an external feedback so that $\boldsymbol{u}$ is a function of $\boldsymbol{y}(t, \boldsymbol{\alpha})$ and hence depends on $\boldsymbol{\alpha}$ as well, the right-hand side of the trajectory sensitivity equation (5.2-8) will be increased by the term $\boldsymbol{\mu} = \boldsymbol{B}_0(\partial \boldsymbol{u}/\partial \alpha_j)_{\boldsymbol{\alpha}0}$ and the right-hand side of the output sensitivity equation by the term $\boldsymbol{D}_0(\partial \boldsymbol{u}/\partial \alpha_j)_{\boldsymbol{\alpha}0}$. $\boldsymbol{\mu}$ is called the "sensitivity control vector" because $\boldsymbol{\mu}$ can be employed to compensate for the effect of the parameter-induced "sensitivity disturbance vector" $(\partial \boldsymbol{A}/\partial \alpha_j)_{\boldsymbol{\alpha}0} \boldsymbol{x}_0 + (\partial \boldsymbol{B}/\partial \alpha_j)_{\boldsymbol{\alpha}0} \boldsymbol{u}$ in Eq. (5.2-8). This can be achieved by an appropriate choice of external feedback which is the basis for the design of parameter-insensitive control systems [61].

Now the trajectory sensitivity equations for *β-variations* are derived. Since $\boldsymbol{f}$ is not an explicit function of the initial conditions, the partial derivative of the state equation (5.2-1) with respect to the initial conditions $x_j^0 = \beta_j$ at nominal values becomes

$$\left.\frac{\partial \dot{\boldsymbol{x}}}{\partial \beta_j}\right|_{\beta 0} = \left.\frac{\partial \boldsymbol{f}}{\partial \boldsymbol{x}}\right|_{\beta 0} \left.\frac{\partial \boldsymbol{x}}{\partial \beta_j}\right|_{0}, \tag{5.2-14}$$

where $\boldsymbol{\beta}_0 = \boldsymbol{x}_0{}^0$ is the nominal value of the initial condition vector. Taking the derivative of the initial condition vector $\boldsymbol{x}^0 = \boldsymbol{\beta}$ with respect to β_j gives, in terms of the components,

$$\frac{\partial x_k}{\partial x_j{}^0} = \frac{\partial \beta_k}{\partial \beta_j} = \begin{cases} 1 & \text{for } k = j, \\ 0 & \text{for } k \neq j. \end{cases} \tag{5.2-15}$$

Using the definition $\boldsymbol{\lambda}_j \triangleq (\partial \boldsymbol{\lambda}/\partial \beta_j)_{\beta_{j0}}$, the result can be written as

$$\dot{\boldsymbol{\lambda}}_j = \left.\frac{\partial \boldsymbol{f}}{\partial \boldsymbol{x}}\right|_{\beta_0} \boldsymbol{\lambda}_j, \qquad \lambda_{kj}^0 = \begin{cases} 0 & \text{for } k \neq j, \\ 1 & \text{for } k = j. \end{cases}$$

Note that the trajectory sensitivity equation in the case of β-variations is homogeneous. The initial conditions are equal to zero except the one with respect to which the derivative is taken, the latter being 1.

In an analogous manner, the vector output sensitivity equation is obtained as

$$\boldsymbol{\sigma}_j = \left.\frac{\partial \boldsymbol{g}}{\partial \boldsymbol{x}}\right|_{\beta_0} \boldsymbol{\lambda}_j. \tag{5.2-16}$$

In the linear case, the trajectory sensitivity equations with respect to β-variations reduce to the following form:

$$\dot{\boldsymbol{\lambda}}_i = \boldsymbol{A}_0 \boldsymbol{\lambda}_j, \qquad \lambda_{kj}^0 = \begin{cases} 0 & \text{for } k \neq j, \\ 1 & \text{for } k = j, \end{cases} \tag{5.2-17}$$

$$\boldsymbol{\sigma}_j = \boldsymbol{C}_0 \boldsymbol{\lambda}_j. \tag{5.2-18}$$

This means that the sensitivity equations are identical to the homogeneous nominal original state equations except that the initial conditions are as shown in Eq. (5.2-17).

Let us now derive the trajectory sensitivity equation for *λ-parameters*. To avoid confusion, let us again denote the λ-parameter by α. The trajectory sensitivity equations are then found by partially differentiating Eqs. (3.2-41)-(3.2-43) with respect to α and then letting α approach zero. First, the partial differentiation of the actual state equations yields

$$\frac{\partial \dot{\boldsymbol{x}}}{\partial \alpha} = \frac{\partial \boldsymbol{f}}{\partial \boldsymbol{x}}\frac{\partial \boldsymbol{x}}{\partial \alpha} + \frac{\partial \boldsymbol{f}}{\partial \boldsymbol{z}}\frac{\partial \boldsymbol{z}}{\partial \alpha} + \frac{\partial \boldsymbol{f}}{\partial \alpha}, \qquad \frac{\partial \boldsymbol{x}^0}{\partial \alpha} = \boldsymbol{0}, \tag{5.2-19}$$

$$\alpha \frac{\partial \dot{\boldsymbol{z}}}{\partial \alpha} = \frac{\partial \boldsymbol{f}_1}{\partial \boldsymbol{x}}\frac{\partial \boldsymbol{x}}{\partial \alpha} + \frac{\partial \boldsymbol{f}_1}{\partial \boldsymbol{z}}\frac{\partial \boldsymbol{z}}{\partial \alpha} + \frac{\partial \boldsymbol{f}_1}{\partial \alpha} - \dot{\boldsymbol{z}}, \qquad \frac{\partial \boldsymbol{z}^0}{\partial \alpha} = \boldsymbol{0}, \tag{5.2-20}$$

$$\frac{\partial \boldsymbol{y}}{\partial \alpha} = \frac{\partial \boldsymbol{g}}{\partial \boldsymbol{x}}\frac{\partial \boldsymbol{x}}{\partial \alpha} + \frac{\partial \boldsymbol{g}}{\partial \boldsymbol{z}}\frac{\partial \boldsymbol{z}}{\partial \alpha} + \frac{\partial \boldsymbol{g}}{\partial \alpha}. \tag{5.2-21}$$

Note that in this stage all initial conditions are zero as in the case of α-variations.

It is easily seen that for $\alpha = 0$ the above system of equations reduces, in the same manner as the original state equations, from the order $n + r$ to the order n. Setting $\alpha = 0$ and defining $\boldsymbol{\lambda} \triangleq (\partial \mathbf{x}/\partial \alpha)_0$ and $\boldsymbol{\eta} = (\partial \mathbf{z}/\partial \alpha)_0$, we obtain the following system of differential equations;

$$\dot{\boldsymbol{\lambda}} = \left.\frac{\partial \mathbf{f}}{\partial \mathbf{x}}\right|_0 \boldsymbol{\lambda} + \left.\frac{\partial \mathbf{f}}{\partial \mathbf{z}}\right|_0 \boldsymbol{\eta} + \left.\frac{\partial \mathbf{f}}{\partial \alpha}\right|_0, \tag{5.2-22}$$

$$\boldsymbol{\eta} = \left(\frac{\partial \mathbf{f}_1}{\partial \mathbf{z}}\right)_0^{-1} \left[\dot{\mathbf{z}}_0 - \left.\frac{\partial \mathbf{f}_1}{\partial \mathbf{x}}\right|_0 \boldsymbol{\lambda} - \left.\frac{\partial \mathbf{f}_1}{\partial \alpha}\right|_0\right]. \tag{5.2-23}$$

If the expression for $\boldsymbol{\eta}$ from Eq. (5.2-23) is substituted into Eq. (5.2-22), the following reduced set of n-dimensional differential equations is obtained:

$$\dot{\boldsymbol{\lambda}} = \left[\frac{\partial \mathbf{f}}{\partial \mathbf{x}} - \frac{\partial \mathbf{f}}{\partial \mathbf{z}}\left(\frac{\partial \mathbf{f}_1}{\partial \mathbf{z}}\right)^{-1}\frac{\partial \mathbf{f}_1}{\partial \mathbf{x}}\right]_0 \boldsymbol{\lambda} + \left[\frac{\partial \mathbf{f}}{\partial \mathbf{z}}\left(\frac{\partial \mathbf{f}_1}{\partial \mathbf{z}}\right)^{-1}\right]_0 \dot{\mathbf{z}}_0$$
$$- \left[\frac{\partial \mathbf{f}}{\partial \mathbf{z}}\left(\frac{\partial \mathbf{f}_1}{\partial \mathbf{z}}\right)^{-1}\frac{\partial \mathbf{f}_1}{\partial \alpha}\right]_0 + \left.\frac{\partial \mathbf{f}}{\partial \alpha}\right|_0. \tag{5.2-24}$$

This vector differential equation constitutes the degenerate (reduced) *trajectory sensitivity equation* with respect to λ-parameters, the solution of which yields the trajectory sensitivity vector $\boldsymbol{\lambda}$. Note that $\dot{\mathbf{z}}_0$ is a given function obtained from the nominal original state equations.

The sensitivity vector $\boldsymbol{\eta}$ can be determined from the r-dimensional algebraic equation (5.2-23) as

$$\boldsymbol{\eta} = -\left[\left(\frac{\partial \mathbf{f}_1}{\partial \mathbf{z}}\right)^{-1}\frac{\partial \mathbf{f}_1}{\partial \mathbf{x}}\right]_0 \boldsymbol{\lambda} + \left(\frac{\partial \mathbf{f}_1}{\partial \mathbf{z}}\right)_0^{-1} \dot{\mathbf{z}}_0 - \left(\frac{\partial \mathbf{f}_1}{\partial \mathbf{z}}\right)_0^{-1} \left.\frac{\partial \mathbf{f}_1}{\partial \alpha}\right|_0. \tag{5.2-25}$$

In a manner analogous to Eq. (5.2-24), the *vector output sensitivity equation* is found by substituting Eq. (5.2-23) into Eq. (5.2-21) after having set $\alpha = 0$. The result is

$$\boldsymbol{\sigma} = \left[\frac{\partial \mathbf{g}}{\partial \mathbf{x}} - \frac{\partial \mathbf{g}}{\partial \mathbf{z}}\left(\frac{\partial \mathbf{f}_1}{\partial \mathbf{z}}\right)^{-1}\frac{\partial \mathbf{f}_1}{\partial \mathbf{x}}\right]_0 \boldsymbol{\lambda} + \left[\frac{\partial \mathbf{g}}{\partial \mathbf{z}}\left(\frac{\partial \mathbf{f}_1}{\partial \mathbf{z}}\right)^{-1}\right] \dot{\mathbf{z}}_0$$
$$- \left[\frac{\partial \mathbf{g}}{\partial \mathbf{z}}\left(\frac{\partial \mathbf{f}_1}{\partial \mathbf{z}}\right)^{-1}\frac{\partial \mathbf{f}_1}{\partial \alpha}\right]_0 + \left.\frac{\partial \mathbf{g}}{\partial \alpha}\right|_0. \tag{5.2-26}$$

From this equation, $\boldsymbol{\sigma}$ can be evaluated after having determined $\boldsymbol{\lambda}$ from Eq. (5.2-24).

Example 5.2-3 Let us once again consider the armature controlled dc motor treated previously in Example 3.2-4. The actual state equations are according to Eqs. (3.2-50) and (3.2-51):

$$\dot{x} = az,$$
$$L_a z = -kx - R_a z + u.$$

Taking the partial derivative with respect to L_a and assuming that the order of the derivatives with respect to time and L_a may be interchanged, we obtain the relations

$$\frac{d}{dt}\left[\frac{\partial x}{\partial L_a}\right] = a\frac{\partial z}{\partial L_a},$$

$$L_a\frac{d}{dt}\left[\frac{\partial z}{\partial L_a}\right] + \frac{dz}{dt} = -k\frac{\partial x}{\partial L_a} - R_a\frac{\partial z}{\partial L_a}.$$

Now if L_a is set equal to zero and $\partial z/\partial L_a$ is substituted by the first relation, we obtain, in terms of $\lambda \triangleq (\partial x/\partial L_a)_0$,

$$\dot{\lambda} = -\frac{ak}{R_a}\lambda - \frac{a}{R_a}\dot{z}_0.$$

This is the degenerate (reduced) sensitivity equation of the motor. Evidently, the same result is obtained when substituting the given data into the general equations.

Particular attention has to be drawn to the initial conditions $\lambda_1^0, \ldots, \lambda_n^0$ of Eq. (5.2-24). It turns out that no longer are all initial conditions zero. This is true because of the neglect of the r initial conditions $\partial z_{n+\nu}/\partial \alpha$ ($\nu = 1, 2, \ldots, r$) while passing from the nonreduced to the reduced sensitivity equation, which is equivalent to a discontinuity of the term $\dot{\boldsymbol{z}}_0$ at the initial time t_0 (see Section 4.3.4). This implies that the vector of initial conditions λ_j^0, $j = 1, 2, \ldots, r$, is discontinuous at t_0 as well. Hence the sensitivity equations are only valid for t > 0.

There are several methods of solving the problem of initial conditions. Two procedures have already been outlined in Section 4.3.4. A third one involving block diagram mainpulations has been discussed in connection with the variable component method (Example 4.5-2).

A general analytical method to determine the initial conditions at t_0^+ was proposed by Vasileva [11], which will be given in the sequel without an analytical proof. In accordance with this, the vector of initial conditions at $t_0 = t_0^+$ is found from the formula

$$\boldsymbol{\lambda}(t_0) = \int_{t_0}^{\infty} \{\boldsymbol{f}[\boldsymbol{x}(t_0), \boldsymbol{z}(\tau), t_0, \boldsymbol{u}(t_0), 0] - \boldsymbol{f}[\boldsymbol{x}(t_0), \boldsymbol{z}_0(t_0), t_0, \boldsymbol{u}(t_0), 0]\}\, d\tau, \qquad (5.2\text{-}27)$$

where $\boldsymbol{x}(t_0)$ and $\boldsymbol{u}(t_0)$ are the initial values of the nonreduced state equations, $\boldsymbol{z}_0(t_0)$ the initial conditions $\boldsymbol{z}_0^0$ of the reduced equations, and $\boldsymbol{z}(\tau)$ is the solution of the auxiliary equation

$$\frac{d\boldsymbol{z}(\tau)}{d\tau} = \boldsymbol{f}_1[\boldsymbol{x}(t_0), \boldsymbol{z}(\tau), 0, \boldsymbol{u}(t_0), t_0] \qquad (5.2\text{-}28)$$

with the initial condition $\boldsymbol{z}(\tau)|_{\tau=t_0} = \boldsymbol{z}^0$. Note that $\boldsymbol{z}^0$ is the given initial conditions vector of the nonreduced system state equations.

Example 5.2-4 Consider a series connection of two first-order lag systems as shown in Fig. 5.2-2. Suppose the time constant T_1 is significantly smaller than T_2 so that it might be neglected when treating this system mathematically. Hence $\alpha = T_1$ and $\alpha_0 = T_{10} = 0$. As can be seen, the nonreduced state equations $\dot{x} = -T_2^{-1}x + T_2^{-1}z$, $T_1\dot{z} = -z + u$ reduce for $T_1 = 0$ to $\dot{x} = -T_2^{-1}x + T_2^{-1}z$, $z = u$. In these terms,

$$f = -\frac{1}{T_2}x + \frac{1}{T_2}z,$$

$$f_1 = -z + u.$$

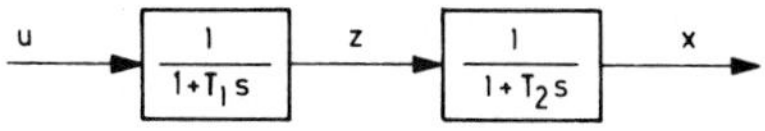

FIG. 5.2-2. Given system of Example 5.2–4.

The trajectory sensitivity functions $\lambda = (\partial x/\partial T_1)_0$ and $\eta = (\partial z/\partial T_1)_0$ are to be determined for a steplike input, that is, $u(t) = 0$ for $t \leq 0$ and $u(t) = U_0$ for $t > 0$.

Solution Differentiating the nonreduced state equation with respect to α and setting $\alpha = 0$ yields the reduced sensitivity equation

$$\dot{\lambda} = -T_2^{-1}\lambda + T_2^{-1}\eta,$$

$$0 = -\eta + \dot{z}_0.$$

Since, in this case, f_1 reduces to $-z + u$, where $u(t_0) = U_0$, the auxiliary equation reads

$$\frac{dz(\tau)}{d\tau} = -z(\tau) + U_0, \qquad z(t_0) = z^0.$$

The solution of this equation is given by

$$z(\tau) = z^0 e^{-\tau} + U_0(1 - e^{-\tau}).$$

Equation (5.2-27) becomes

$$\lambda(t_0) = \frac{1}{T_2}\int_{t_0}^{\infty} [-x^0 + z(\tau) + x^0 - z_0^{\,0}]\, d\tau,$$

where $z_0^{\,0} = U_0$. Substituting for $z(\tau)$ by the expression obtained from the auxiliary equation and carrying out the integration, one obtains

$$\lambda(t_0) = \frac{1}{T_0}(z^0 - U_0).$$

Using this initial condition and realizing that $\eta = -\dot{z}_0 = -\dot{u} = 0$, the solution of the sensitivity equation becomes

$$\lambda(t) = \frac{1}{T_2} (z^0 - U_0)\, e^{-t/T_2}.$$

The correctness of the result can easily be verified by taking the derivative of the solution $x(t,\alpha)$ of the nonreduced original state equation and then letting α approach zero.

5.3 TRAJECTORY SENSITIVITY EQUATIONS OF CONTINUOUS SYSTEMS WITH TIME-VARYING PARAMETERS

In this paragraph the sensitivity equations are derived for systems with rapidly changing parameters. It will be shown that the same sensitivity equations are obtained as for time-invariant parameter changes if the sensitivity functions are replaced by the corresponding first variations of $\boldsymbol{x}$ and $\boldsymbol{y}$, respectively. This will first be shown for systems described by general state equations. A different approach will be given for linear systems.

Consider the general actual vector state equation of a continuous, possibly nonlinear system,

$$\dot{\boldsymbol{x}} = \boldsymbol{f}(\boldsymbol{x}, \boldsymbol{\alpha}, t, \boldsymbol{u}), \tag{5.3-1}$$

where $\boldsymbol{\alpha}$ is a time-varying parameter according to $\boldsymbol{\alpha}(t) = \boldsymbol{\alpha}_0 + \varepsilon\delta\boldsymbol{\alpha}(t)$. The first variation $\delta\boldsymbol{\alpha}$ is subject to the requirements of uniform boundedness and integrability as defined in Section 3.2.1. $\boldsymbol{\alpha}_0$ and ε are constant with respect to time; ε is a small number.

It is assumed that the solution of the above state equation can be written as

$$\boldsymbol{x}(t, \varepsilon) = \boldsymbol{x}_0(t) + \varepsilon\, \delta\boldsymbol{x}(t), \tag{5.3-2}$$

where $\boldsymbol{x}_0(t)$ is the solution of the nominal vector state equation

$$\boldsymbol{x}_0 = \boldsymbol{f}(\boldsymbol{x}_0, \boldsymbol{\alpha}_0, t, \boldsymbol{u}), \qquad \boldsymbol{x}(t_0) = \boldsymbol{x}^0. \tag{5.3-3}$$

The first variation $\delta\boldsymbol{x}$ allows for the characterization of the sensitivity of the system in a manner similar to the characterization of the sensitivity function. This follows from the fact that the relation between the parameter-induced error and $\delta\boldsymbol{x}$ is given by

$$\boldsymbol{x} - \boldsymbol{x}_0 = \varepsilon\, \delta\boldsymbol{x}. \tag{5.3-4}$$

It will be shown that $\delta\boldsymbol{x}$ can be determined from sensitivity equations of the same form as for the sensitivity functions in the case of time-invariant parameters.

Using Eq. (5.3-3), Eq. (5.3-1) can be rewritten as

$$\dot{\boldsymbol{x}}_0 + \varepsilon\, \delta\dot{\boldsymbol{x}} = \boldsymbol{f}(\boldsymbol{x}_0 + \varepsilon\, \delta\boldsymbol{x}, \boldsymbol{\alpha}_0 + \varepsilon\, \delta\boldsymbol{\alpha}, t, \boldsymbol{u}). \tag{5.3-5}$$

Expanding the right-hand side of Eq. (5.3-5) into a Taylor series at the point $\boldsymbol{x}_0$, $\boldsymbol{\alpha}_0$, we obtain

$$\dot{\boldsymbol{x}}_0 + \varepsilon\,\delta\dot{\boldsymbol{x}} = \boldsymbol{f}(\boldsymbol{x}_0, \boldsymbol{\alpha}_0, t, \boldsymbol{u}) + \varepsilon \left.\frac{\partial \boldsymbol{f}(\boldsymbol{x}, \boldsymbol{\alpha}, t, \boldsymbol{u})}{\partial \boldsymbol{x}}\right|_{\boldsymbol{\alpha}_0} \delta x + \varepsilon \left.\frac{\partial \boldsymbol{f}(\boldsymbol{x}, \boldsymbol{\alpha}, t, \boldsymbol{u})}{\partial \boldsymbol{\alpha}}\right|_{\boldsymbol{\alpha}_0} \delta\boldsymbol{\alpha} + \boldsymbol{R}(\varepsilon). \tag{5.3-6}$$

Let us assume that for the term $\boldsymbol{R}(\varepsilon)$ containing all higher-order terms in ε the following limit exists:

$$\lim_{\varepsilon \to 0}\left[\boldsymbol{R}(\varepsilon)\frac{1}{\varepsilon}\right] = \boldsymbol{0}. \tag{5.3-7}$$

Introducing the nominal state equation (5.3-3) into Eq. (5.3-5), one obtains

$$\varepsilon\,\delta\dot{\boldsymbol{x}} = \varepsilon \left.\frac{\partial \boldsymbol{f}(\boldsymbol{x}, \boldsymbol{\alpha}, t, \boldsymbol{u})}{\partial \boldsymbol{x}}\right|_{\boldsymbol{\alpha}_0} \delta\boldsymbol{x} + \varepsilon \left.\frac{\partial \boldsymbol{f}(\boldsymbol{x}, \boldsymbol{\alpha}, t, \boldsymbol{u})}{\partial \boldsymbol{\alpha}}\right|_{\boldsymbol{\alpha}_0} \delta\boldsymbol{\alpha} + \boldsymbol{R}(\varepsilon).$$

If we now divide both sides by ε and then take the limit $\varepsilon = 0$, we have

$$\delta\dot{\boldsymbol{x}} = \left.\frac{\partial \boldsymbol{f}(\boldsymbol{x}, \boldsymbol{\alpha}, t, \boldsymbol{u})}{\partial \boldsymbol{x}}\right|_{\boldsymbol{\alpha}_0} \delta\boldsymbol{x} + \left.\frac{\partial \boldsymbol{f}(\boldsymbol{x}, \boldsymbol{\alpha}, t, \boldsymbol{u})}{\partial \boldsymbol{\alpha}}\right|_{\boldsymbol{\alpha}_0} \delta\boldsymbol{\alpha}. \tag{5.3-8}$$

The initial condition vector $\delta\boldsymbol{x}^0$ of the above differential equation is zero since $\boldsymbol{x}^0$ is not a function of $\boldsymbol{\alpha}$. In a more concise notation, the result may be written as

$$\delta\dot{\boldsymbol{x}} = \left.\frac{\partial \boldsymbol{f}}{\partial \boldsymbol{x}}\right|_{\boldsymbol{\alpha}_0} \delta\boldsymbol{x} + \left.\frac{\partial \boldsymbol{f}}{\partial \boldsymbol{\alpha}}\right|_{\boldsymbol{\alpha}_0} \delta\boldsymbol{\alpha}, \qquad \delta\boldsymbol{x}^0 = \boldsymbol{0}. \tag{5.3-9}$$

This is the general *sensitivity equation* of a continuous, possibly nonlinear system with *time-varying parameters*. Note that it is linear with zero-initial conditions and, in general, time-varying coefficients. The solution of this vector differential equation yields the parameter-induced first variation of the trajectory of the system $\delta\boldsymbol{x}$.

In the same way the vector output "sensitivity" equation can be found, which is a linear algebraic equation.

Comparing the results with the sensitivity equations obtained for time-invariant parameters reveals that the forms of the sensitivity equations are completely the same. The only difference is that in the case of time-varying parameters it is the first variation $\delta\boldsymbol{x}$ that is considered instead of the sensitivity function $\boldsymbol{\lambda}_j$. In addition, the form of the time function $\delta\boldsymbol{\alpha}(t)$ must be known in order to solve the sensitivity equation.

Let us now derive the sensitivity equations for *linear systems* in a different way. In this case the state equations are

$$\dot{\boldsymbol{x}}(t, \boldsymbol{\alpha}) = \boldsymbol{A}(\boldsymbol{\alpha})\boldsymbol{x}(t, \boldsymbol{\alpha}) + \boldsymbol{B}(\boldsymbol{\alpha})\boldsymbol{u}(t), \qquad \boldsymbol{x}(t_0) = \mathbf{x}^0, \tag{5.3-10}$$

$$\boldsymbol{y}(t, \boldsymbol{\alpha}) = \boldsymbol{C}(\boldsymbol{\alpha})\boldsymbol{x}(t, \boldsymbol{\alpha}) + \boldsymbol{D}(\boldsymbol{\alpha})\boldsymbol{u}(t). \tag{5.3-11}$$

Suppose the parameter vector varies again with time according to $\boldsymbol{\alpha}(t) = \boldsymbol{\alpha}_0 + \varepsilon\, \delta\boldsymbol{\alpha}(t)$, $\boldsymbol{\alpha}_0$ and ε being defined as above.

The change of $\boldsymbol{A}$ associated with a parameter change may be written as

$$\Delta \boldsymbol{A} = \boldsymbol{A}(\boldsymbol{\alpha}_0 + \varepsilon\, \delta\boldsymbol{\alpha}(t)] - \boldsymbol{A}(\boldsymbol{\alpha}_0) = \varepsilon\, \delta \boldsymbol{A} + \boldsymbol{R}(\varepsilon). \tag{5.3-12}$$

This is in agreement with the time-invariant case, where

$$\Delta \boldsymbol{A} = \frac{\partial \boldsymbol{A}}{\partial \boldsymbol{\alpha}} \Delta \boldsymbol{\alpha} + \tilde{\boldsymbol{R}}(\Delta \boldsymbol{\alpha}). \tag{5.3-13}$$

$\boldsymbol{R}$ and $\tilde{\boldsymbol{R}}$ contain the higher-order terms of the Taylor series expansion. If $\Delta \boldsymbol{\alpha}$ is replaced by $\varepsilon\, \delta\boldsymbol{\alpha}$, relation (5.3-13) is obtained. It is again assumed that the following limit exists:

$$\lim_{\varepsilon \to 0} \left[\boldsymbol{R}(\varepsilon) \frac{1}{\varepsilon} \right] = \boldsymbol{0}. \tag{5.3-14}$$

The first variations $\delta \boldsymbol{B}$, $\delta \boldsymbol{C}$, and $\delta \boldsymbol{D}$ may be defined in the same manner as $\delta \boldsymbol{A}$.

Then the state equations (5.3-10) and (5.3-11) can be rewritten as follows:

$$\frac{d}{dt}(\boldsymbol{x}_0 + \varepsilon\, \delta \boldsymbol{x}) = (\boldsymbol{A}_0 + \Delta \boldsymbol{A})\, (\boldsymbol{x}_0 + \varepsilon\, \delta \boldsymbol{x}) + (\boldsymbol{B}_0 + \Delta \boldsymbol{B}) \boldsymbol{u}, \tag{5.3-15}$$

$$\boldsymbol{y}_0 + \varepsilon\, \delta \boldsymbol{y} = (\boldsymbol{C}_0 + \Delta \boldsymbol{C})\, (\boldsymbol{x}_0 + \varepsilon\, \delta \boldsymbol{x}) + (\boldsymbol{D}_0 + \Delta \boldsymbol{D}) \boldsymbol{u}, \tag{5.3-16}$$

with the initial conditions $\boldsymbol{x}(t_0) = \boldsymbol{x}^0$, $\delta \boldsymbol{x}(t_0) = 0$.

With the aid of Eqs. (5.3-10) and (5.3-11), the above equations can be written as

$$\varepsilon\, \delta \dot{\boldsymbol{x}} = \boldsymbol{A}_0\, \varepsilon\, \delta \boldsymbol{x} + \Delta \boldsymbol{A}\, \boldsymbol{x}_0 + \Delta \boldsymbol{A}\, \varepsilon\, \delta \boldsymbol{x} + \Delta \boldsymbol{B}\, \boldsymbol{u}, \qquad \delta \boldsymbol{x}(t_0) = \boldsymbol{0},$$

$$\varepsilon\, \delta \boldsymbol{y} = \boldsymbol{C}_0\, \varepsilon\, \delta \boldsymbol{x} + \Delta \boldsymbol{C}\, \boldsymbol{x}_0 + \Delta \boldsymbol{C}\, \varepsilon\, \delta \boldsymbol{x} + \Delta \boldsymbol{D}\, \boldsymbol{u}.$$

Now dividing these equations by ε and then taking the limit $\varepsilon = 0$, we obtain the final *sensitivity equations* from which the variations $\delta \boldsymbol{x}$ and $\delta \boldsymbol{y}$ can be calculated:

$$\delta \dot{\boldsymbol{x}} = \boldsymbol{A}_0\, \delta \boldsymbol{x} + \delta \boldsymbol{A}\, \boldsymbol{x}_0 + \delta \boldsymbol{B}\, \boldsymbol{u}, \qquad \delta \boldsymbol{x}^0 = \boldsymbol{0}, \tag{5.3-17}$$

$$\delta \boldsymbol{y} = \boldsymbol{C}_0\, \delta \boldsymbol{x} + \delta \boldsymbol{C}\, \boldsymbol{x}_0 + \delta \boldsymbol{D}\, \boldsymbol{u}. \tag{5.3-18}$$

Note that $\boldsymbol{A}_0 = \boldsymbol{A}(\boldsymbol{\alpha}_0)$ and $\boldsymbol{C}_0 = \boldsymbol{C}(\boldsymbol{\alpha}_0)$ are time-invariant matrices whereas $\delta \boldsymbol{A}$, $\delta \boldsymbol{B}$, $\delta \boldsymbol{C}$, $\delta \boldsymbol{D}$ are time-variant.

Comparing the above equations with the sensitivity equations for time-invariant parameters, we again recognize the complete structural identity. Let us summarize the result by the following theorem.

Theorem 5.3-1 The first variations $\delta \boldsymbol{x}$ and $\delta \boldsymbol{y}$ in the case of time-varying parameters satisfy vector equations of the same type as the sensitivity equations in the case of time-invariant parameters.

5.4 TRAJECTORY SENSITIVITY EQUATIONS OF DISCONTINUOUS SYSTEMS

The derivation of the trajectory sensitivity equations for discontinuous systems gives rise to certain difficulties caused by the discrete nature of these systems. First, the number of parameters is increased by the sampling instants and/or the sampling periods. Second, the assumption of continuity of the sensitivity functions does not apply in general, which complicates the mathematical treatment of these systems.

In this section the trajectory sensitivity equations are first derived for sampled-data systems and then for systems with variable structure characterized by state equations with discontinuous right-hand sides.

5.4.1 Sampled-Data Systems

Consider a continuous linear system described by the state equation $\dot{\boldsymbol{x}} = \boldsymbol{A}\boldsymbol{x} + \boldsymbol{b}u$. The input u and the state vector $\boldsymbol{x}$ are sampled at equidistant instants kT ($k = 0, 1, \ldots$). The samples of u are given to a zero-order hold before entering the system. Inspecting the resulting system at the sampling times kT only, its behavior can be described by the difference equation

$$\boldsymbol{x}[(k+1)T, \boldsymbol{\alpha}] = \boldsymbol{\varphi}(T, \boldsymbol{\alpha})\boldsymbol{x}(kT, \boldsymbol{\alpha}) + \boldsymbol{h}(T, \boldsymbol{\alpha})u(kT), \qquad \boldsymbol{x}(0) = \boldsymbol{x}^0. \quad (5.4\text{-}1)$$

The symbols have the following meanings:

$\boldsymbol{\varphi}(T, \boldsymbol{\alpha}) = e^{\boldsymbol{A}T}$ is the transition matrix of the original system,

$\boldsymbol{h}(T, \boldsymbol{\alpha}) = \int_0^T e^{\boldsymbol{A}t}\boldsymbol{b}\, dt,$

T is the sampling period,

a is an $r \times 1$ parameter vector (α-parameters).

In a more concise notation, Eq. (5.4-1) can be written as

$$\boldsymbol{x}_{k+1}(T, \boldsymbol{\alpha}) = \boldsymbol{\varphi}(T, \boldsymbol{\alpha})\boldsymbol{x}_k(T, \boldsymbol{\alpha}) + \boldsymbol{h}(T, \boldsymbol{\alpha})u_k(T), \qquad \boldsymbol{x}(t_0) = \boldsymbol{x}^0. \quad (5.4\text{-}2)$$

Note that $u(kT)$, although being an external signal, depends on the sampling period T which will be important when T is considered as the parameter of interest.

In order to derive the trajectory sensitivity equation with respect to α_j, Eq. (5.4-2) has to be differentiated with respect to α_j and then $\boldsymbol{\alpha}$ has to be set equal to $\boldsymbol{\alpha}_0$. Under the same continuity conditions as in the continuous-time case, and assuming that u is not a function of $\boldsymbol{\alpha}$, we obtain

$$\boldsymbol{\lambda}_{k+1}(T, \boldsymbol{\alpha}_0) = \boldsymbol{\varphi}(T, \boldsymbol{\alpha}_0)\boldsymbol{\lambda}_{k,j}(T, \boldsymbol{\alpha}_0) + \left.\frac{\partial \boldsymbol{\varphi}(T, \boldsymbol{\alpha})}{\partial \alpha_j}\right|_{\boldsymbol{\alpha}_0} \boldsymbol{x}_k(T, \boldsymbol{\alpha}_0)$$

$$+ \left.\frac{\partial \boldsymbol{h}(T, \boldsymbol{\alpha})}{\partial \alpha_j}\right|_{\boldsymbol{\alpha}_0} u_k(T), \qquad j = 1, 2, \ldots, r, \qquad (5.4\text{-}3)$$

where $\boldsymbol{\lambda}_{k,j}$ is the trajectory sensitivity vector at the kth sampling instant $\boldsymbol{\lambda}_{k,j} \triangleq (\partial \boldsymbol{x}_k(T, \boldsymbol{\alpha})/\partial \alpha_j)_{\boldsymbol{\alpha}0}$.

Equation (5.4-3) is the *trajectory sensitivity equation* for α-parameters of the sampled-data system. For each sampling instant kT ($k = 0, 1, . . .$), there are r such sensitivity vectors $\boldsymbol{\lambda}_{k,j}$ ($j = 0, 1, . . ., r$), which can be combined to a sensitivity matrix $\boldsymbol{\lambda}_k = [\boldsymbol{\lambda}_{k,1}\ \boldsymbol{\lambda}_{k,2}\ . . .\ \boldsymbol{\lambda}_{k,r}]$. To characterize the sensitivity of the system, the sensitivity matrices for all k sampling instants are required.

The above reveals that the trajectory sensitivity equation has a similar form as in the continuous-time case: It is also linear with disappearing initial conditions. A structural interpretation is given in Fig. 5.4-1. Note that the integrator of the continuous-time case has to be replaced by a shifting operator.

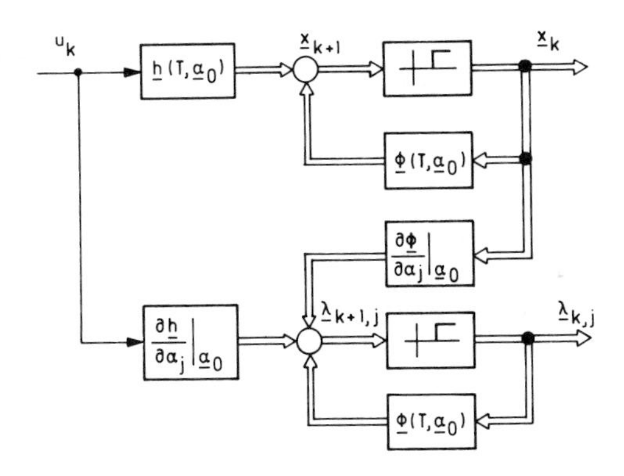

FIG. 5.4-1. Structural interpretation of the trajectory sensitivity equation (5.4-3).

The far-reaching analogy with the continuous-time case suggests that the trajectory sensitivity equations for nonlinear discrete-time systems can be determined in a way analogous to the determination of the nonlinear continuous-time case. This is why we shall not deal further with this problem.

In the case of sampled-data systems, a significant additional parameter is the *sampling period* T. It substantially affects the performance of the system. Moreover, it strongly influences other design and implementation considerations, including word length requirements and choice of quantization levels.

The sensitivity of the state vector $\boldsymbol{x}$ with respect to the sampling period T at the sampling instants is characterized by the *global trajectory sensitivity vector* (see Defirition 3.2-8). The sensitivity equation from which this vector can be determined is obtained by differentiating the state equation (5.4-1) with respect to T and then letting T approach the nominal value T_0. This differentiation does not cause any problem since $\boldsymbol{x}$ is a continuous function of

T. Dropping the α-dependence of $\boldsymbol{x}$ for notational ease without using a new symbol, and realizing that $d\boldsymbol{\varphi}/dT = \boldsymbol{A}\boldsymbol{\varphi}(T)$ and $d\boldsymbol{h}/dT = \boldsymbol{\varphi}(T)\boldsymbol{b}$, we obtain

$$\boldsymbol{\lambda}[(k+1)T_0] = \boldsymbol{\varphi}(T_0)\boldsymbol{\lambda}(kT_0) + \boldsymbol{\varphi}(T_0)\boldsymbol{A}\boldsymbol{x}(kT_0) + \boldsymbol{h}(T_0)\left.\frac{du(kT)}{dT}\right|_{T_0} + \boldsymbol{\varphi}(T_0)\boldsymbol{b}u(kT_0). \tag{5.4-4}$$

Note that a term du/dT occurs since u is a function of T. Assuming that the initial state $\boldsymbol{x}^0$ is independent of the sampling period, the initial condition of Eq. (5.4-4) becomes $\boldsymbol{\lambda}(0) = d\boldsymbol{x}^0/dT = \boldsymbol{0}$.

From Eq. (5.4-4), the sequence of sensitivity vectors $\boldsymbol{\lambda}(kT_0)$, $k = 0, 1, \ldots$, can be computed in a straightforward procedure. The order of the sensitivity equation can be reduced by the fact that the individual sensitivity vectors are related by linear transformations. A reduction scheme of the sensitivity equation is described in Perkins [79].

Between sampling instants the sensitivity to T is characterized by the *continuous sampling-period sensitivity vector* as defined by Definition 3.2-9. The state equation describing the behavior between the sampling instants is given by

$$\dot{\boldsymbol{x}}(kT+\tau) = \boldsymbol{A}\boldsymbol{x}(kT+\tau) + \boldsymbol{b}u(kT), \qquad 0 \le \tau < T; \tag{5.4-5}$$

where the initial conditions for $\tau = 0$ are obtained from Eq. (5.4-1). The above equation shows that $\dot{\boldsymbol{x}}$ is not a continuous function of T since the right-hand side of this equation is discontinuous at kT. Thus the sensitivity equation cannot be obtained by simple differentiation of Eq. (5.4-5) with respect to T. This difficulty can be avoided by excluding the discontinuities from consideration and taking the initial values at kT^+, i.e., at the right hand sides of the discontinuities. Then the derivatives exist except at the sampling instants.

With these restrictions, the sensitivity equation in the time interval between the sampling instants is given by

$$\dot{\boldsymbol{\lambda}}(kT+\tau) = \boldsymbol{A}\boldsymbol{\lambda}(kT+\tau) + \boldsymbol{b}\,\frac{du(kT)}{dT}, \tag{5.4-6}$$

with the initial conditions $\boldsymbol{\lambda}(0) = \boldsymbol{0}$ and discontinuous boundary conditions. Referring to Rozenvasser [89], and Rozenvasser and Yusupov [90], the latter are given by the relation

$$\boldsymbol{\lambda}(kT^+) = \boldsymbol{\lambda}(kT^-) - k\boldsymbol{b}[u(kT) - u((k-1)T)], \tag{5.4-7}$$

with $\boldsymbol{\lambda}(0) = \boldsymbol{0}$. This type of problem will be treated in more detail in Section 5.4.2.

Now let us turn to the general, possibly *nonlinear sampled-data system*, described by the following general vector state equation

$$\boldsymbol{x}[(k+1)T] = \boldsymbol{f}[\boldsymbol{x}(kT), u(kT), kT], \qquad \boldsymbol{x}(0) = \boldsymbol{x}^0, \qquad k = 0, 1, \ldots. \tag{5.4-8}$$

Differentiating this difference equation with respect to T and letting T approach T_0, the trajectory sensitivity equation is obtained as

$$\boldsymbol{\lambda}[(k+1)T_0] = \boldsymbol{F}(kT_0)\boldsymbol{\lambda}(kT_0) + \left.\frac{\partial \boldsymbol{f}}{\partial u}\right|_{T_0} \left.\frac{du(kT)}{dT}\right|_{T_0} + \left.\frac{\partial \boldsymbol{f}}{\partial T}\right|_{T_0}, \tag{5.4-9}$$

where

$$\boldsymbol{F}(kT_0) = \begin{bmatrix} \dfrac{\partial f_1(kT)}{\partial x_1} & \cdots & \dfrac{\partial f_1(kT)}{\partial x_n} \\ \vdots & & \vdots \\ \dfrac{\partial f_n(kT)}{\partial x_1} & \cdots & \dfrac{\partial f_n(kT)}{\partial x_n} \end{bmatrix}_{T_0}. \tag{5.4-10}$$

The initial conditions are $\boldsymbol{\lambda}(0) = \mathbf{0}$.

Equation (5.4-9) is the trajectory sensitivity equation of the general sampled-data system with respect to T at the sampling instants. Note that, as in the linear case, a term du/dT appears since the sample of u is a function of T even if there is no external feedback from $\boldsymbol{x}$ to u.

If the sampling instants are not equidistant but arbitrary, the sensitivity to sampling inaccuracies is characterized by the *local sensitivity vector* $\boldsymbol{\lambda}(t_{k0})$ according to Definition 3.2-7. In this case the general state equation of the system is given by

$$\boldsymbol{x}(t_{k+1}) = \boldsymbol{f}[\boldsymbol{x}(t_k), u(t_k), t_k]. \tag{5.4-11}$$

Differentiation with respect to t and setting $t = t_k$ gives the sensitivity equation

$$\boldsymbol{\lambda}(t_{(k+1)0}) = \boldsymbol{F}(t_{k0})\boldsymbol{\lambda}(t_{k0}) + \left.\frac{\partial \boldsymbol{f}}{\partial u}\right|_{t_{k0}} \left.\frac{du(t_k)}{dt_k}\right|_{t_{k0}} + \left.\frac{\partial \boldsymbol{f}}{\partial t}\right|_{t_{k0}}. \tag{5.4-12}$$

The initial condition of the first interval is zero. The initial conditions of the subsequent intervals follow from Eq. (5.4-12) successively. $\boldsymbol{F}$ is the Jacobian matrix defined by Eq. (5.4-10).

5.4.2 Systems with Variable Structure

Let us now derive the trajectory sensitivity equations for discrete systems whose differential equations are defined differently over different time intervals. The basic work on this subject was done by Rozenvasser [89] and Rozenvasser and Yusupov [90]. Recently Uuspää [102] has contributed to this subject by providing methods for the calculation of the sensitivity functions for a class of pulse-frequency modulated feedback control systems.

Referring to Section 3.2.5, let the general state equation of a discrete system with variable structure be defined within the ith interval by the vector differential equation

$$\dot{\boldsymbol{x}} = \boldsymbol{f}_i(\boldsymbol{x}, \boldsymbol{u}, t, \boldsymbol{\alpha}), \qquad t_{i-1} < t < t_i, \quad i = 1, 2, \ldots, \tag{5.4-13}$$

where it is assumed that the state $\boldsymbol{x}$ and the input vector $\boldsymbol{u}$ are both functions of $\boldsymbol{\alpha}$. The limits of the intervals, i.e., the switching instants, may be defined either implicitly by relations of the form

$$\theta_i(\boldsymbol{x}_1^-, t_i, \boldsymbol{\alpha}) = 0, \qquad i = 1, 2, \ldots, \tag{5.4-14}$$

or explicitly according to

$$t_i = t_{i-1} + T_{i-1}(\boldsymbol{x}_{i-1}^-, \boldsymbol{\alpha}), \tag{5.4-15}$$

as outlined in Section 3.2.5. The initial conditions are $\boldsymbol{x}(t_0, \boldsymbol{\alpha}) = \boldsymbol{x}^0$. The initial conditions of the individual intervals may be given by a so-called "reset condition"

$$\boldsymbol{x}_i^+ = \boldsymbol{\varphi}_i(\boldsymbol{x}_i^-, \mathbf{t}_i, \boldsymbol{\alpha}), \tag{5.4-16}$$

defining discontinuities of the state $\boldsymbol{x}$ at the switching instants.

In order to find the sensitivity equations, the above state equations have to be differentiated with respect to the parameter vector $\boldsymbol{\alpha}$. Since the sensitivity matrix $\boldsymbol{\lambda} = (\partial \boldsymbol{x}/\partial \boldsymbol{\alpha})_{\boldsymbol{\alpha}0}$ is, in general, discontinuous at the limits of the intervals, as was illustrated in Section 3.5.5, the derivatives can be taken only in the interior of the intervals. At the boundaries, the so-called jump conditions have to be taken into account.

Let us assume that $\boldsymbol{f}_i$ as well as θ_i, $\mathbf{t}_i$, and $\boldsymbol{\varphi}_i$ have continuous first derivatives with respect to all arguments. Then, taking the partial derivative of Eq. (5.4-13) with respect to $\boldsymbol{\alpha}$ within the interval $t_{i-1} < t < t_i$ and setting $\boldsymbol{\alpha} = \boldsymbol{\alpha}_0$, we obtain the *state sensitivity equation* of the ith interval as

$$\dot{\boldsymbol{\lambda}} = \left.\frac{\partial \boldsymbol{f}_i}{\partial \boldsymbol{x}}\right|_{\boldsymbol{\alpha}0} \boldsymbol{\lambda} + \left.\frac{\partial \boldsymbol{f}_i}{\partial \boldsymbol{u}}\frac{\partial \boldsymbol{u}}{\partial \boldsymbol{\alpha}}\right|_{\boldsymbol{\alpha}0} + \left.\frac{\partial \boldsymbol{f}_i}{\partial \boldsymbol{\alpha}}\right|_{\boldsymbol{\alpha}0}, \qquad t_{i-1} < t < t_i. \tag{5.4-17}$$

The *initial conditions* are found as follows. The general solution of the state equation (5.4-13) for the first interval is

$$\boldsymbol{x}(t, \boldsymbol{\alpha}) = \int_{t_0}^{t} \boldsymbol{f}_1(\boldsymbol{x}, t, \boldsymbol{\alpha})\, dt + \boldsymbol{x}^0, \qquad t_0 < t < t_1. \tag{5.4-18}$$

Partial differentiation of $\boldsymbol{x}(t,\boldsymbol{\alpha})$ with respect to $\boldsymbol{\alpha}$ gives

$$\frac{\partial \boldsymbol{x}}{\partial \boldsymbol{\alpha}} = -\boldsymbol{f}_1(\boldsymbol{x}^0, t_0, \boldsymbol{\alpha}) \frac{dt_0}{d\boldsymbol{\alpha}} + \int_0^t \frac{\partial \boldsymbol{f}_1}{\partial \boldsymbol{\alpha}}\, dt + \frac{d\boldsymbol{x}^0}{d\boldsymbol{\alpha}}. \tag{5.4-19}$$

For $\boldsymbol{\alpha} = \boldsymbol{\alpha}_0$ and $t = t_0$ it follows that

$$\boldsymbol{\lambda}^0 = -\boldsymbol{f}_1(\boldsymbol{x}^0, t_0, \boldsymbol{\alpha}_0) \left.\frac{dt_0}{d\boldsymbol{\alpha}}\right|_{\boldsymbol{\alpha}0} + \left.\frac{d\boldsymbol{x}^0}{d\boldsymbol{\alpha}}\right|_{\boldsymbol{\alpha}0}. \tag{5.4-20}$$

This is the vector of *initial conditions* of Eq. (5.4-17).

Let us now determine the *jump conditions* $\Delta\boldsymbol{\lambda}_i$ at the boundaries of the intervals, which provide the necessary corrections at the discontinuity points when going from one interval (i) to the next ($i + 1$). These developments will be carried out using classical mathematics. A more transparent development, using distributions, will be given later in connection with Example 5.4-2.

Denote the limit for approaching t_i from the left-hand and right-hand side by t_i^- and t_i^+, respectively, and the corresponding expressions of $\boldsymbol{x}$ and $\boldsymbol{\lambda}$ by $\boldsymbol{x}_i^- \triangleq \boldsymbol{x}(t_i^-, \boldsymbol{\alpha})$, $\boldsymbol{\lambda}_i^- \triangleq \boldsymbol{\lambda}(t_i^-, \boldsymbol{\alpha})$ and $\boldsymbol{x}_i^+ \triangleq \boldsymbol{x}(t_i^+, \boldsymbol{\alpha})$, $\boldsymbol{\lambda}_i^+ \triangleq \boldsymbol{\lambda}(t_i^+, \boldsymbol{\alpha})$, respectively. Then the required jump conditions at t_i are defined by $\Delta\boldsymbol{\lambda}_i \triangleq \Delta\boldsymbol{\lambda}_i^+ - \Delta\boldsymbol{\lambda}_i^-$.

Let us first determine $\boldsymbol{\lambda}_i^-$. Taking into account that t_i depends on $\boldsymbol{\alpha}$, the total differential quotient of $\boldsymbol{x}_i^- = \boldsymbol{x}[t_i^-(\boldsymbol{\alpha}), \boldsymbol{\alpha}]$ at t_i^- becomes

$$\frac{d\boldsymbol{x}_i^-}{d\boldsymbol{\alpha}} = \frac{\partial \boldsymbol{x}_i}{\partial t_i^-}\frac{dt_i^-}{d\boldsymbol{\alpha}} + \frac{\partial \boldsymbol{x}_i^-}{\partial \boldsymbol{\alpha}}. \tag{5.4-21}$$

However, due to Eq. (5.4-13), we have

$$\frac{\partial \boldsymbol{x}_i^-}{\partial t_i^-} = \boldsymbol{f}_i(\boldsymbol{x}_i^-, t_i^-, \boldsymbol{\alpha}) \triangleq \boldsymbol{f}_i^-. \tag{5.4-22}$$

Substituting this expression into Eq. (5.4-21), setting $\boldsymbol{\alpha} = \boldsymbol{\alpha}_0$ and solving for $\boldsymbol{\lambda}_i^- \triangleq (\partial\boldsymbol{x}_i^-/\partial\boldsymbol{\alpha})_{\boldsymbol{\alpha}0}$ gives

$$\boldsymbol{\lambda}_i^- = \frac{d\boldsymbol{x}_i^-}{d\boldsymbol{\alpha}} - \boldsymbol{f}_i^-\frac{dt_i^-}{d\boldsymbol{\alpha}} = \frac{d\boldsymbol{x}_i^-}{d\boldsymbol{\alpha}} - \boldsymbol{f}_i^-\frac{dt_i}{d\boldsymbol{\alpha}}, \qquad \boldsymbol{\alpha} = \boldsymbol{\alpha}_0. \tag{5.4-23}$$

In an analogous manner, we obtain $\boldsymbol{\lambda}_i^+$. Taking the derivative at t_i^+ gives

$$\frac{d\boldsymbol{x}_i^+}{d\boldsymbol{\alpha}} = \frac{\partial \boldsymbol{x}_i^+}{\partial t_i^+}\frac{dt_i^+}{d\boldsymbol{\alpha}} + \frac{\partial \boldsymbol{x}_i^+}{\partial \boldsymbol{\alpha}}. \tag{5.4-24}$$

Since t_i^+ already belongs to the ($i + 1$)th time interval, where

$$\dot{\boldsymbol{x}} = \boldsymbol{f}_{i+1}(\boldsymbol{x}, t, \boldsymbol{\alpha}), \tag{5.4-25}$$

the term $\partial\boldsymbol{x}_i^+/\partial t_i^+$ in Eq.(5.4-24) becomes

$$\frac{\partial \boldsymbol{x}_i^+}{\partial t_i^+} = \boldsymbol{f}_{i+1}(\boldsymbol{x}_i^+, t_i^+, \boldsymbol{\alpha}). \tag{5.4-26}$$

Defining $\boldsymbol{f}_{i+1}(\boldsymbol{x}_i^+, t_i^+, \boldsymbol{\alpha}) \triangleq \boldsymbol{f}_i^+$ and letting $\boldsymbol{\alpha}$ approach $\boldsymbol{\alpha}_0$, Eq.(5.4–20) becomes

$$\boldsymbol{\lambda}_i^+ = \frac{d\boldsymbol{x}_i^+}{d\boldsymbol{\alpha}} - \boldsymbol{f}_i^+\frac{dt_i^-}{d\boldsymbol{\alpha}} = \frac{d\boldsymbol{x}_i^+}{d\boldsymbol{\alpha}} - \boldsymbol{f}_i^+\frac{dt_i}{d\boldsymbol{\alpha}}, \qquad \boldsymbol{\alpha} = \boldsymbol{\alpha}_0. \tag{5.4-27}$$

On the other hand we have, according to Eq. (5.4–16),

$$\boldsymbol{x}_i^+ = \boldsymbol{\varphi}_i(\boldsymbol{x}_i^-, t_i, \boldsymbol{\alpha}) \triangleq \boldsymbol{\varphi}_i^-, \tag{5.4-28}$$

whence

$$\frac{d\boldsymbol{x}_i^+}{d\boldsymbol{\alpha}} = \frac{\partial \boldsymbol{\varphi}_i^-}{\partial \boldsymbol{x}_i^-}\frac{d\boldsymbol{x}_i^-}{d\boldsymbol{\alpha}} + \frac{\partial \boldsymbol{\varphi}_i^-}{\partial t_i}\frac{dt_i}{d\boldsymbol{\alpha}} + \frac{\partial \boldsymbol{\varphi}_i^-}{\partial \boldsymbol{\alpha}}. \tag{5.4-29}$$

Substituting this expression into Eq. (5.4-27) and forming $\Delta\boldsymbol{\lambda}_i = \boldsymbol{\lambda}_i^+ - \boldsymbol{\lambda}_i^-$, we obtain

$$\Delta\boldsymbol{\lambda}_i = \left(-\boldsymbol{f}_i^+ + \boldsymbol{f}_i^- + \frac{\partial\boldsymbol{\varphi}_i^-}{\partial t_i}\right)\frac{dt_i}{d\boldsymbol{\alpha}} + \left(\frac{\partial\boldsymbol{\varphi}_i^-}{\partial\boldsymbol{x}_i^-} - \boldsymbol{I}\right)\frac{d\boldsymbol{x}_i^-}{d\boldsymbol{\alpha}} + \frac{\partial\boldsymbol{\varphi}_i^-}{\partial\boldsymbol{\alpha}}. \tag{5.4-30}$$

Reusing Eq. (5.4-23) and denoting $\Delta\boldsymbol{f}_i = \boldsymbol{f}_i^+ - \boldsymbol{f}_i^-$ finally yields

$$\Delta\boldsymbol{\lambda}_i = \left[-\Delta\boldsymbol{f}_i + \left(\frac{\partial\boldsymbol{\varphi}_i^-}{\partial\boldsymbol{x}_i^-} - \boldsymbol{I}\right)\boldsymbol{f}_i^- + \frac{\partial\boldsymbol{\varphi}_i^-}{\partial t_i}\right]\frac{dt_i}{d\boldsymbol{\alpha}} + \left[\frac{\partial\boldsymbol{\varphi}_i^-}{\partial\boldsymbol{x}_i^-} - \boldsymbol{I}\right]\boldsymbol{\lambda}_i^- + \frac{\partial\boldsymbol{\varphi}_i^-}{\partial\boldsymbol{\alpha}}, \tag{5.4-31}$$

where evidently all expressions have to be taken at nominal parameter values $\boldsymbol{\alpha} = \boldsymbol{\alpha}_0$ (the subscript $\boldsymbol{\alpha}_0$ has been dropped for ease of notation). Equation (5.4-31) represents the required *jump conditions* of the trajectory sensitivity matrix $\boldsymbol{\lambda}$ at the switching instants t_i ($i = 1, 2, . . .$).

The term $dt_i/d\boldsymbol{x}$ in Eq. (5.4-31) is found by differentiating Eq. (5.4-14), $\theta_i(\boldsymbol{x}_i^-, t_i, \boldsymbol{\alpha}) = \theta_i^- = 0$, with respect to $\boldsymbol{\alpha}$:

$$\frac{\partial\theta_i^-}{\partial\boldsymbol{x}_i^-}\frac{d\boldsymbol{x}_i^-}{d\boldsymbol{\alpha}} + \frac{\partial\theta_i^-}{\partial t_i}\frac{dt_i}{d\boldsymbol{\alpha}} + \frac{\partial\theta_i^-}{\partial\boldsymbol{\alpha}} = 0. \tag{5.4-32}$$

Replacing $d\boldsymbol{x}_i^-/d\boldsymbol{\alpha}$ by Eq.(5.4-23) and solving Eq.(5.4-32) for $dt_i/d\boldsymbol{\alpha}$ gives

$$\frac{dt_i}{d\boldsymbol{\alpha}} = \frac{(\partial\theta_i^-/\partial\boldsymbol{x}_i^-)\,\boldsymbol{\lambda}_i^- + (\partial\theta_i^-/\partial\boldsymbol{\alpha})}{(\partial\theta_i^-/\partial\boldsymbol{x}_i^-)\,\boldsymbol{f}_i^- + (\partial\theta_i^-/\partial t_i)}, \tag{5.3–33}$$

Thus all expressions are known and the state sensitivity equations (5.4-17) can be solved successively for increasing $i = 1, 2, ...$

A particular case of great practical interest is that of a continuous change-over of $\boldsymbol{x}$ from one interval to the next. This implies that Eq. (5.4-14) reduces to the identity $\boldsymbol{x}_i^+ = \boldsymbol{x}_i^-$. In this case, $\partial\boldsymbol{\varphi}_i^-/\partial\boldsymbol{x}_i^- = \boldsymbol{I}$, $\partial\boldsymbol{\varphi}_i^-/\partial\boldsymbol{\alpha} = \boldsymbol{0}$, and $\partial\boldsymbol{\varphi}_i^-/\partial t = \boldsymbol{0}$ so that the jump conditions (5.4-31) reduce to

$$\Delta\boldsymbol{\lambda}_i = -\Delta\boldsymbol{f}_i\frac{dt_i}{d\boldsymbol{\alpha}}, \qquad \boldsymbol{\alpha} = \boldsymbol{\alpha}_0. \tag{5.4-34}$$

Moreover, $\partial\theta_i^-/\partial\boldsymbol{x}_i^- = \partial\theta_i^+/\partial\boldsymbol{x}_i^+ = \partial\theta_i/\partial\boldsymbol{x}_i$, $\partial\theta_i^-/\partial t = \partial\theta_i^+/\partial t = \partial\theta_i/\partial t$, and $\partial\theta_i^-/\partial\boldsymbol{\alpha} = \partial\theta_i^+/\partial\boldsymbol{\alpha} = \partial\theta_i/\partial\boldsymbol{\alpha}$. Hence Eq.(5.4–32) becomes

$$\frac{\partial\theta_i}{\partial\boldsymbol{x}_i}\frac{d\boldsymbol{x}_i}{d\boldsymbol{\alpha}} + \frac{\partial\theta_i}{\partial t_i}\frac{dt_i}{d\boldsymbol{\alpha}} + \frac{\partial\theta_i}{\partial\boldsymbol{\alpha}} = \boldsymbol{0}. \tag{5.4-35}$$

According to Eqs. (5.4-23) and (5.4-27), we have

$$\frac{d\boldsymbol{x}_i}{d\boldsymbol{\alpha}}\boldsymbol{\lambda}_i^- + \boldsymbol{f}_i^-\frac{dt_i}{d\boldsymbol{\alpha}} = \boldsymbol{\lambda}_i^+ + \boldsymbol{f}_i^+\frac{dt_i}{d\boldsymbol{\alpha}}, \qquad \boldsymbol{\alpha} = \boldsymbol{\alpha}_0. \tag{5.4-36}$$

With the aid of Eq.(5.4-36), Eq.(5.4-33) becomes

$$\frac{dt_i}{d\boldsymbol{\alpha}} = \frac{(\partial\theta_i/\partial\boldsymbol{x})\,\boldsymbol{\lambda}_i^- + (\partial\theta_i/\partial\boldsymbol{\alpha})}{(\partial\theta_i/\partial\boldsymbol{x})\,\boldsymbol{f}_i^- + (\partial\theta_i/\partial t_i)} = -\frac{(\partial\theta_i/\partial\boldsymbol{x})\,\boldsymbol{\lambda}_i^+ + (\partial\theta_i/\partial\boldsymbol{\alpha})}{(\partial\theta_i/\partial\boldsymbol{x})\,\boldsymbol{f}_i^+ + (\partial\theta_i/\partial t_i)}, \qquad \boldsymbol{\alpha} = \boldsymbol{\alpha}_0. \tag{5.4-37}$$

This is a rigorous derivation of what has been shown intuitively in Fig. 3.2-7, namely, that $\boldsymbol{\lambda}$ can be discontinuous even if the trajectory $\boldsymbol{x}$ is continuous. According to Eq. (5.4-34), $\boldsymbol{x}$ is continuous in t in either of the two cases:

(1) $\Delta\boldsymbol{f}_i = \boldsymbol{0}$, i.e., the right-hand sides of the state equations (5.4-13) are continuous at the switching instants t_i,

(2) $dt_i/d\boldsymbol{\alpha} = \boldsymbol{0}$, i.e., the switching instants do not depend on the parameters.

It is emphasized that the formulas presented above also remain valid if the system state equations are of a different order in the individual intervals. In this case the Jacobians $\partial\varphi_i/\partial\boldsymbol{x}_i$ will be rectangular and not square.

5.4.3 Pulse-Frequency Modulated Feedback Systems

Let us now apply the above procedure to pulse-frequency modulated feedback systems, a special group of discrete systems, in which the structure remains unchanged but the switching instants vary from interval to interval.

The general block diagram of a pulse-frequency modulated control system is shown in Fig. 5.4.2. One component of the state vector $\boldsymbol{x}_R$, after being measured behind the freedback element, is compared with the reference input $w(t)$. The output of the modulator provides a train of preassigned uniform pulses of the width γ. The intervals between the pulses are functions of the error $e(t)$ behind the comparator: If $e(t)$ is large, the interval is small and vice versa.

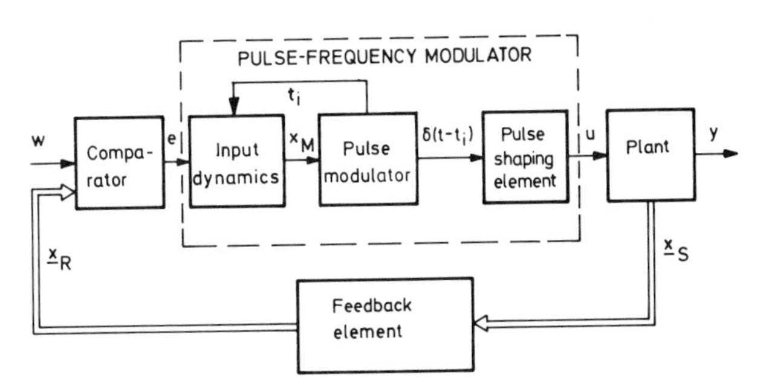

FIG. 5.4-2. Block diagram of a pulse-frequency modulated feedback control system with a type II-PFM.

Basically, two types of modulators have to be distinguished. The *type I* pulse-frequency modulator does not contain an input dynamic element. Instead, a sampler and zero-order hold is employed taking samples of $e(t)$ at t_{i-1} and holding them over the interval $t_i - t_{i-1}$. The sampling instant t_i that coincides with the beginning of the pulse is calculated in the pulse modulator as a function of $e(t_{i-1})$ by a nonlinear characteristic such as of the form of Eq. (5.4-15).

The operation of the *type II* pulse-frequency modulator can be described quite generally by the structure shown in Fig. 5.4-2. The input dynamic produces an auxiliary signal x_M from the error $e(t)$, for example, by integration (IPFM) or by a lag system (Neuro PFM). The pulse modulator provides a Dirac impulse as soon as x_M has reached a certain boundary value, thus defining the switching instant t_i. At that moment, x_M is set to a certain value, e.g., to zero again, and the operation starts again. The operation of setting x_M equal to zero is defined by a relation of the form of Eq. (5.4-16). The condition defining the coincidence of x_M with the preassigned boundary value can be expressed by a relation of the form of Eq.(5.4-14). In this case, θ_i is a nonlinear characteristic. Finally, the Dirac impulses at the output of the pulse modulator are modified by the pulse shaping element in such a way that they take a preassigned form.

In both cases the switching instants t_i are functions of the parameter vector $\boldsymbol{\alpha}$, that is, $t_i = t_i(\boldsymbol{\alpha})$. Whereas in the case of the type I modulator the trajectory is continuous at t_i, it is changed discontinuously in the case of the type II modulator. From this it follows that, according to Eq. (5.4-31), the sensitivity matrix $\boldsymbol{\lambda} = (\partial \boldsymbol{x}/\partial \boldsymbol{\alpha})_{\boldsymbol{\alpha}0}$ in the case of the type II modulator is always discontinuous. In the case of the type I modulator, $\boldsymbol{\lambda}$ is discontinuous according to Eq. (5.4-34), if the input pulses of the plant begin or end discontinuously, otherwise $\boldsymbol{\lambda}$ is continuous.

To derive the sensitivity equations, suppose the input u of the plant is a train of uniform pulses of the width γ starting at instants t_i ($i = 0, 1, \ldots$) that are functions of the state $\boldsymbol{x}_S$ of the plant and of the reference input w. Let us combine the state vectors $\boldsymbol{x}_S$, $\boldsymbol{x}_R$ and the auxiliary variable $\boldsymbol{x}_M$ to a state vector $\boldsymbol{x}$ according to

$$\boldsymbol{x} = \begin{bmatrix} \boldsymbol{x}_S \\ \boldsymbol{x}_R \\ \boldsymbol{x}_M \end{bmatrix}, \tag{5.4-38}$$

and let us combine all parameters into one parameter vector $\boldsymbol{\alpha}$. Furthermore, let us denote the time function of the pulse during the pulse-on time by u_i, so that

$$u = \begin{cases} u_i & \text{for} \quad t_i < t < t_i + \gamma, \\ 0 & \text{for} \quad t_i + \gamma < t < t_{i+1}. \end{cases} \tag{5.4-39}$$

Then the state equations of the feedback system in the ith interval are of the form

$$\dot{\boldsymbol{x}} = \boldsymbol{f}(\boldsymbol{x}, u_i, w, t, \boldsymbol{\alpha}), \qquad t_i < t < t_i + \gamma, \tag{5.4-40}$$

$$\dot{\boldsymbol{x}} = \boldsymbol{f}(\boldsymbol{x}, 0, w, t, \boldsymbol{\alpha}), \qquad t_i + \gamma < t < t_{i+1}. \tag{5.4-41}$$

$\boldsymbol{x}$ is of course a function of $\boldsymbol{\alpha}$. Note that u_i is also a function of $\boldsymbol{\alpha}$ since the parameters of the pulse shaping element are included in $\boldsymbol{\alpha}$. However, w is not a function of $\boldsymbol{\alpha}$.

Thus, taking the partial derivatives of the above state equations with respect to $\boldsymbol{\alpha}$ and setting $\boldsymbol{\alpha} = \boldsymbol{\alpha}_0$ (the subscript $\boldsymbol{\alpha}_0$ will be dropped in the sequel for ease of notation), the following *trajectory sensitivity equations* are obtained:

$$\dot{\boldsymbol{\lambda}} = \frac{\partial \boldsymbol{f}}{\partial \boldsymbol{x}} \boldsymbol{\lambda} + \frac{\partial \boldsymbol{f}}{\partial u_i} \frac{\partial u_i}{\partial \boldsymbol{\alpha}} + \frac{\partial \boldsymbol{f}}{\partial \boldsymbol{\alpha}}, \qquad t_i < t < t_i + \gamma, \tag{5.4-42}$$

$$\dot{\boldsymbol{\lambda}} = \frac{\partial \boldsymbol{f}}{\partial \boldsymbol{x}} \boldsymbol{\lambda} + \frac{\partial \boldsymbol{f}}{\partial \boldsymbol{\alpha}}, \qquad t_i + \gamma < t < t_{i+1}. \tag{5.4-43}$$

The *initial conditions* $\boldsymbol{\lambda}^0 = \boldsymbol{\lambda}(t_0, \boldsymbol{\alpha}_0)$ are found by differentiating the initial conditions $\boldsymbol{x}^0 = \boldsymbol{x}(t_0, \boldsymbol{\alpha})$ with respect to $\boldsymbol{\alpha}$ (and then setting $\boldsymbol{\alpha} = \boldsymbol{\alpha}_0$). Because $t_0 = t_0(\boldsymbol{\alpha})$, one obtains

$$\frac{d\boldsymbol{x}^0}{d\boldsymbol{\alpha}} = \frac{\partial \boldsymbol{x}^0}{\partial t_0} \frac{dt_0}{d\boldsymbol{\alpha}} + \frac{\partial \boldsymbol{x}^0}{\partial \boldsymbol{\alpha}}. \tag{5.4-44}$$

Since $\partial \boldsymbol{x}^0/\partial t_0 = \boldsymbol{f}(\boldsymbol{x}^0, t_0, \boldsymbol{\alpha})$ and $\boldsymbol{\lambda}^0 = \partial \boldsymbol{x}^0/\partial \boldsymbol{\alpha}$, Eq. (5.4-44) gives

$$\boldsymbol{\lambda}^0 = \frac{d\boldsymbol{x}^0}{d\boldsymbol{\alpha}} - \boldsymbol{f}(\boldsymbol{x}^0, t_0, \boldsymbol{\alpha}_0) \frac{dt_0}{d\boldsymbol{\alpha}}. \tag{5.4-45}$$

If the initial time is independent of $\boldsymbol{\alpha}$, which is commonly the case, Eq. (5.4-45) reduces to $\boldsymbol{\lambda}^0 = d\boldsymbol{x}^0/d\boldsymbol{\alpha}$. With knowledge of $\boldsymbol{\lambda}^0$, the sensitivity equation can be solved for the first interval $t_0 < t < t_0 + \gamma$.

To solve the sensitivity equations for the subsequent intervals, the discontinuities, called *jump conditions* of $\boldsymbol{\lambda}$ at t_i and $t_i + \gamma$, $i = 0, 1, \ldots$, must be calculated to provide the necessary corrections at pulse switching points. The following development is based on the ideas of Rozenvasser [89] and the methods developed by Uuspää [102].

At $t_i + \gamma$ $(i = 0, 1, \ldots)$ the auxiliary variable $\boldsymbol{x}_M$ will not be changed discontinuously as at t_i. Hence the state $\boldsymbol{x}$ is continuous and the discontinuities of $\boldsymbol{\lambda}$ at $t_i + \gamma$ are of the form

$$\Delta \boldsymbol{\lambda} = -\Delta \boldsymbol{f} \frac{d(t_i + \gamma)}{d\boldsymbol{\alpha}}, \qquad t = t_i + \gamma, \tag{5.4-46}$$

where $\Delta \boldsymbol{f} = \boldsymbol{f}^+ - \boldsymbol{f}^-$ designates the discontinuity of the right-hand side of the state equation at $t_i + \gamma$.

Since γ is one of the parameters included in $\boldsymbol{\alpha}$, Eq. (5.4-46) becomes

$$\Delta \boldsymbol{\lambda} = -\Delta \boldsymbol{f} \frac{dt_i}{d\boldsymbol{\alpha}} - \Delta \boldsymbol{f} \frac{d\gamma}{d\boldsymbol{\alpha}}, \qquad t = t_i + \gamma. \tag{5.4-47}$$

Note that $\Delta \boldsymbol{f}$ at $t_i + \gamma$ and $d\gamma/d\boldsymbol{\alpha}$ are known from the mathematical model of the pulse shaping element, i.e., by the form of u_i in the state equations. An expression for $dt_i/d\boldsymbol{\alpha}$ will be derived later.

To find the jump conditions *at* t_i ($i = 1, 2, \ldots$), one has to take into account that the state vector $\boldsymbol{x}$ at t_i may depend on $\boldsymbol{\alpha}$ in several ways:

(1) directly (e.g., via the plant parameters),
(2) due to the dependence of t_i on $\boldsymbol{\alpha}$,
(3) due to the reset condition, Eq. (5.4-16); this only affects $\boldsymbol{x}_i^+ = \boldsymbol{x}(t_i^+, \boldsymbol{\alpha})$, i.e., the right-hand limit of $\boldsymbol{x}$ at t_i.

The first-order approximation of the change in $\boldsymbol{x}$ due to the first two influences is

$$\Delta \boldsymbol{x}_i^- = \dot{\boldsymbol{x}}_i^- \Delta t_i + \boldsymbol{\lambda}_i^- \Delta \boldsymbol{\alpha}, \tag{5.4-48}$$

$$\Delta \boldsymbol{x}_i^+ = \dot{\boldsymbol{x}}_i^+ \Delta t_i + \boldsymbol{\lambda}_i^+ \Delta \boldsymbol{\alpha}. \tag{5.4-49}$$

Letting $\Delta\boldsymbol{\alpha}$ approach infinitesimal values $d\boldsymbol{\alpha}$ and substituting $\dot{\boldsymbol{x}}_i^-$ and $\dot{\boldsymbol{x}}_i^+$ by

$$\dot{\boldsymbol{x}}(t_i^-, \boldsymbol{\alpha}) = \boldsymbol{f}[\boldsymbol{x}(t_i^-, \boldsymbol{\alpha}), o, w(t_i), t_i(\boldsymbol{\alpha}), \boldsymbol{\alpha}] \triangleq \boldsymbol{f}_i^-, \tag{5.4-50}$$

$$\dot{\boldsymbol{x}}(t_i^+, \boldsymbol{\alpha}) = \boldsymbol{f}[\boldsymbol{x}(t_i^+, \boldsymbol{\alpha}), u_i(t_i, \boldsymbol{\alpha}), w(t_i), t_i(\boldsymbol{\alpha}), \boldsymbol{\alpha}] \triangleq \boldsymbol{f}_i^+, \tag{5.4-51}$$

we obtain

$$\frac{d\boldsymbol{x}_i^-}{d\boldsymbol{\alpha}} = \boldsymbol{f}_i^- \frac{dt_i}{d\boldsymbol{\alpha}} + \boldsymbol{\lambda}_i^-, \tag{5.4-52}$$

$$\frac{d\boldsymbol{x}_i^+}{d\boldsymbol{\alpha}} = \boldsymbol{f}_i^+ \frac{dt_i}{d\boldsymbol{\alpha}} + \boldsymbol{\lambda}_i^+. \tag{5.4-53}$$

The third kind of influence of $\boldsymbol{\alpha}$ on $\boldsymbol{x}$ is described by the reset condition, Eq. (5.4-16). Denoting $\boldsymbol{\varphi}_i^- \triangleq \boldsymbol{\varphi}(\boldsymbol{x}_i^-, t_i, \boldsymbol{\alpha})$, and taking the total derivative of Eq. (5.4-16) with respect to $\boldsymbol{\alpha}$ gives

$$\frac{d\boldsymbol{x}_i^+}{d\boldsymbol{\alpha}} = \frac{\partial \boldsymbol{\varphi}_i^-}{\partial \boldsymbol{x}_i^-}\frac{d\boldsymbol{x}_i^-}{d\boldsymbol{\alpha}} + \frac{\partial \boldsymbol{\varphi}_i^-}{\partial t_i}\frac{dt_i}{d\boldsymbol{\alpha}} + \frac{\partial \boldsymbol{\varphi}_i^-}{\partial \boldsymbol{\alpha}}. \tag{5.4-54}$$

If we substitute Eqs. (5.4-52) and (5.4-53) for $d\boldsymbol{x}_i^+/d\boldsymbol{\alpha}$ and $d\boldsymbol{x}_i^-/d\boldsymbol{\alpha}$ in Eq.(5.4-54) and form $\Delta\boldsymbol{\lambda}_i \triangleq \boldsymbol{\lambda}_i^+ - \boldsymbol{\lambda}_i^-$, we obtain

$$\Delta\boldsymbol{\lambda}_i = \left[\frac{\partial \boldsymbol{\varphi}_i^-}{\partial t_i} - \boldsymbol{f}_i^+ + \frac{\partial \boldsymbol{\varphi}_i^-}{\partial \boldsymbol{x}_i^-}\boldsymbol{f}_i^-\right]\frac{dt_i}{d\boldsymbol{\alpha}} + \left[\frac{\partial \boldsymbol{\varphi}_i^-}{\partial \boldsymbol{x}_i^-} - \boldsymbol{I}\right]\boldsymbol{\lambda}_i^- + \frac{\partial \boldsymbol{\varphi}_i^-}{\partial \boldsymbol{\alpha}}. \tag{5.4-55}$$

This is the general expression for the desired jump condition of the sensitivity matrix $\boldsymbol{\lambda}$ at t_i, $i = 1, 2, \ldots$. The result is in full agreement with Eq. (5.4-31). For further evaluation of Eq. (5.4-55), the two types of modulators have to be distinguished.

In the case of the *type II* modulator, the state $\boldsymbol{x}$ is changed discontinuously at t_i so that the above formula cannot be simplified any further. However, in the case of the type I modulator, $\boldsymbol{x}$ is continuous at t_i so that $\partial\boldsymbol{\varphi}_i^-/\partial\boldsymbol{x}_i = \boldsymbol{I}$, $\partial\boldsymbol{\varphi}_i^-/\partial t_i = \boldsymbol{0}$ and $\partial\boldsymbol{\varphi}_i^-/\partial\boldsymbol{\alpha} = \boldsymbol{0}$. Hence Eq. (5.4-55) reduces to

$$\Delta\boldsymbol{\lambda}_i = -\Delta\boldsymbol{f}_i \frac{dt_i}{d\boldsymbol{\alpha}}, \tag{5.4-56}$$

where $\Delta\boldsymbol{f}_i = \boldsymbol{f}_i^+ - \boldsymbol{f}_i^-$ is the discontinuity on the right-hand side of the state equation. Since the term $dt_i/d\boldsymbol{\alpha}$ never vanishes in the case of pulse-frequency modulated control systems, we can conclude that $\boldsymbol{\lambda}$ is discontinuous whenever the right-hand sides of the state equations are discontinuous, that is, when the pulses are discontinuous. This is in agreement with what was shown intuitively in Section 3.2–5.

Let us now focus our attention on the derivatives $dt_i/d\boldsymbol{\alpha}$ that have to be known in order to evaluate Eqs. (5.4-47), (5.4-55), and (5.4-56).

In the case of the *type I* modulator, the relation for determining t_i is given by Eq. (5.4-15). Differentiating this equation with respect to $\boldsymbol{\alpha}$ and realizing that $\boldsymbol{x}_{i-1}^- = \boldsymbol{x}_{i-1}$ because of the continuity, we have

$$\frac{dt_i}{d\boldsymbol{\alpha}} = \frac{dt_{i-1}}{d\boldsymbol{\alpha}} + \frac{\partial T_{i-1}}{\partial\boldsymbol{x}_{i-1}}\frac{d\boldsymbol{x}_{i-1}}{d\boldsymbol{\alpha}} + \frac{\partial T_{i-1}}{\partial\boldsymbol{\alpha}}. \tag{5.4-57}$$

Since in the pulse modulated feedback system $\boldsymbol{x}_{i-1}$ is a function of t_{i-1} and $\boldsymbol{\alpha}$, and $d\boldsymbol{x}_{i-1}/d\boldsymbol{\alpha}$ is discontinuous at t_{i-1}, the derivative has to be taken either to the left or to the right of the discontinuity (both are equivalent). Taking the left, we have

$$\frac{d\boldsymbol{x}_{i-1}}{d\boldsymbol{\alpha}} = \frac{d\boldsymbol{x}_{i-1}^-}{dt_{i-1}^-}\frac{dt_{i-1}}{d\boldsymbol{\alpha}} + \frac{\partial\boldsymbol{x}_{i-1}^-}{\partial\boldsymbol{\alpha}} = \boldsymbol{f}_i^- \frac{dt_{i-1}}{d\boldsymbol{\alpha}} + \boldsymbol{\lambda}_{i-1}^-. \tag{5.4-58}$$

Substituting this expression into Eq. (5.4–57) gives

$$\frac{dt_i}{d\boldsymbol{\alpha}} = \frac{dt_{i-1}}{d\boldsymbol{\alpha}} + \frac{\partial T_{i-1}}{\partial\boldsymbol{x}_{i-1}}\boldsymbol{f}_i^- \frac{dt_{i-1}}{d\boldsymbol{\alpha}} + \frac{\partial T_{i-1}}{\partial\boldsymbol{x}_{i-1}}\boldsymbol{\lambda}_{i-1}^- + \frac{\partial T_{i-1}}{\partial\boldsymbol{\alpha}} \tag{5.4-59}$$

or finally

$$\frac{dt_i}{d\boldsymbol{\alpha}} = \left(1 + \frac{\partial T_{i-1}}{\partial\boldsymbol{x}_{i-1}}\boldsymbol{f}_i^-\right)\frac{dt_{i-1}}{d\boldsymbol{\alpha}} + \frac{\partial T_{i-1}}{\partial\boldsymbol{x}_{i-1}}\boldsymbol{\lambda}_{i-1}^- + \frac{\partial T_{i-1}}{\partial\boldsymbol{\alpha}}. \tag{5.4-60}$$

This is a recursive formula for determining $dt_i/d\boldsymbol{\alpha}$ in the case of type I modulators. As the above result shows, all parameter-induced deviations of the preceding sampling instants add up, which can cause divergence of $dt_i/d\boldsymbol{\alpha}$. This then implies that $\boldsymbol{\lambda}$ will diverge too.

In the case of the type II modulator the switching instants t_i are defined by a relation of the form of Eq. (5.4-14), where $\boldsymbol{x}_i^- = \boldsymbol{x}[t_i^-(\boldsymbol{\alpha}), \boldsymbol{\alpha}]$, $t_i = t_i(\boldsymbol{\alpha})$ and $t_{i-1} = t_{i-1}(\boldsymbol{\alpha})$. By differentiating this relation with respect to $\boldsymbol{\alpha}$, one obtains

$$\frac{\partial\theta_i}{\partial\boldsymbol{x}_i^-}\frac{d\boldsymbol{x}_i^-}{d\boldsymbol{\alpha}} + \frac{\partial\theta_i}{\partial t_i}\frac{dt_i}{d\boldsymbol{\alpha}} + \frac{\partial\theta_i}{\partial t_{i-1}}\frac{dt_{i-1}}{d\boldsymbol{\alpha}} + \frac{\partial\theta_i}{\partial\boldsymbol{\alpha}} = \boldsymbol{0}, \tag{5.4-61}$$

where

$$\frac{d\boldsymbol{x}_i^-}{d\boldsymbol{\alpha}} = \frac{\partial \boldsymbol{x}_i^-}{\partial t_i}\frac{dt_i}{d\boldsymbol{\alpha}} + \frac{\partial \boldsymbol{x}_i^-}{\partial \boldsymbol{\alpha}} = \boldsymbol{f}_i^-\frac{dt_i}{d\boldsymbol{\alpha}} + \boldsymbol{\lambda}_i^-. \tag{5.4-62}$$

Substituting Eq. (5.4-62) into Eq. (5.4-61) and solving Eq. (5.4-61) for $dt_i/d\boldsymbol{\alpha}$ yields

$$\frac{dt_i}{d\boldsymbol{\alpha}} = -\frac{(\partial\theta_i/\partial t_{i-1})}{(\partial\theta_i/\partial\boldsymbol{x}_i^-)\boldsymbol{f}_i^- + (\partial\theta_i/\partial t_i)}\frac{dt_{i-1}}{d\boldsymbol{\alpha}} + \frac{(\partial\theta_i/\partial\boldsymbol{x}_i^-)\,\boldsymbol{\lambda}_i^- + (\partial\theta_i/\partial\boldsymbol{\alpha})}{(\partial\theta_i/\partial\boldsymbol{x}_i^-)\,\boldsymbol{f}_i^- + (\partial\theta_i/\partial t_i)} \tag{5.4-63}$$

This is a recursive formula for evaluating $dt_i/d\boldsymbol{\alpha}$ for a feedback system using a type II pulse-frequency modulator. The knowledge of $dt_i/d\boldsymbol{\alpha}$ enables us to calculate the desired jump conditions. An alternative, more transparent derivation of the jump conditions will be given in connection with Example 5.4-2.

Example 5.4-1 Given a single axis attitude control system of a spacecraft using a type II pulse-frequency modulator (neuro PFM). The dynamics of the control system are illustrated by Fig. 5.4-3. The state equations are

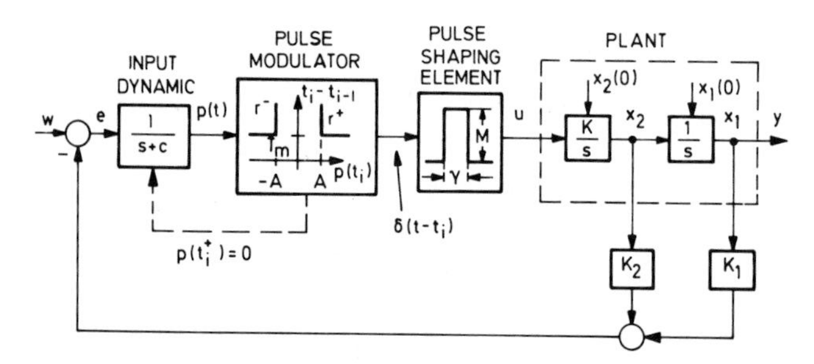

FIG. 5.4-3. Mathematical model of the given control system of Example 5.4-1.

$$\dot{x}_1 = x_2, \qquad x_1(0) = x_1^{\,0},$$

$$\dot{x}_2 = Ku = KM\sum_{i=1}^{\infty}\operatorname{sgn}[p(t_i^-)][1(t-t_i) - 1(t-t_i-\gamma)], \qquad x_2(0) = x_2^{\,0},$$

$$\dot{p} = -K_1x_1 - K_2x_2 - cp + w, \qquad p(t_0) = p_0.$$

The auxiliary variable p is set to zero at t_i^+ ($i = 1, 2, \ldots$), i.e.,

$$p(t_i^+) = 0.$$

Defining the state vector by $\boldsymbol{x} = [x_1\ x_2\ p]^{\mathrm{T}}$ and the parameter vector $\boldsymbol{\alpha}$ by

$$\boldsymbol{\alpha} = [\alpha_1 \quad \alpha_2 \quad \ldots \quad \alpha_8]^{\mathrm{T}} = [K\ K_1\ K_2\ c\ T_m\ A\ M\ \gamma]^{\mathrm{T}},$$

and taking into account that $u = M$ in the time interval $t_i < t < t_i + \gamma$ and

$u = 0$ in $t_i + \gamma < t < t_{i+1}$, the vector state equations can be written as follows:

$$\begin{aligned} \dot{\boldsymbol{x}}(t,\boldsymbol{\alpha}) &= A(\boldsymbol{\alpha})\boldsymbol{x}(t,\boldsymbol{\alpha}) + \boldsymbol{B}\,w(t) + MC(\boldsymbol{\alpha}), \qquad & t_i < t < t_i + \gamma, \\ \dot{\boldsymbol{x}}(t,\boldsymbol{\alpha}) &= A(\boldsymbol{\alpha})\boldsymbol{x}(t,\boldsymbol{\alpha}) + \boldsymbol{B}\,w(t), & t_i + \gamma < t < t_{i+1}, \end{aligned}$$

with the initial conditions $\boldsymbol{x}(t_0) = \boldsymbol{x}^0$ and the interval initial conditions defined by

$$\boldsymbol{x}(t_i^+) = \boldsymbol{\varphi}_i[\boldsymbol{x}(t_i^-, \boldsymbol{\alpha}), \boldsymbol{\alpha}, t] = \boldsymbol{D}\,\boldsymbol{x}(t_i^-,\boldsymbol{\alpha}),$$

where

$$\boldsymbol{A} = \begin{bmatrix} 0 & 1 & 0 \\ 0 & 0 & 0 \\ -K_1 & -K_2 & -c \end{bmatrix}, \qquad \boldsymbol{B} = \begin{bmatrix} 0 \\ 0 \\ 1 \end{bmatrix}, \qquad \boldsymbol{C} = \begin{bmatrix} 0 \\ K \\ 0 \end{bmatrix}, \qquad \boldsymbol{D} = \begin{bmatrix} 1 & 0 & 0 \\ 0 & 1 & 0 \\ 0 & 0 & 0 \end{bmatrix}.$$

The switching instants t_i are defined by the condition

$$(|p(t)| - A)(t - T_m - t_{i-1}) = 0 \qquad \text{if} \quad t = t_i,$$

with $p(t) \geq A$ and $t \geq t_{i-1} + T_m$. This condition will be denoted by

$$\theta_i[\boldsymbol{x}(t, \boldsymbol{\alpha}), \boldsymbol{\alpha}, t] = 0 \qquad \text{as} \quad t = t_i.$$

For this system determine the trajectory sensitivity equations in terms of $\boldsymbol{\lambda}_j = (\partial \boldsymbol{x}/\partial \alpha_j)_{\boldsymbol{\alpha}0}$, including the jump conditions at $t = t_i$, $i = 0, 1, \ldots$.

Solution By differentiating the above vector state equations with respect to α_j, setting $\boldsymbol{\alpha} = \boldsymbol{\alpha}_0$, and interchanging the order of differentiation with respect to time and α_j, one obtains the sensitivity equations

$$\begin{aligned} \dot{\boldsymbol{\lambda}}_j &= \boldsymbol{A}_0\boldsymbol{\lambda}_j + \left.\frac{\partial A}{\partial \alpha_j}\right|_{\boldsymbol{\alpha}0} \boldsymbol{x}(t, \boldsymbol{\alpha}_0) + M\left.\frac{\partial C}{\partial \alpha_j}\right|_{\boldsymbol{\alpha}0} + \boldsymbol{C}_0 \left.\frac{\partial M}{\partial \alpha_j}\right|_{\boldsymbol{\alpha}0}, \qquad & t_i < t < t_i + \gamma, \\ \dot{\boldsymbol{\lambda}}_j &= \boldsymbol{A}_0\boldsymbol{\lambda}_j + \left.\frac{\partial A}{\partial \alpha_j}\right|_{\boldsymbol{\alpha}0} \boldsymbol{x}(t, \boldsymbol{\alpha}_0), & t_i + \gamma < t < t_{i+1}. \end{aligned}$$

Employing the method of Rozenvasser, the jump conditions are found as follows: According to Eq.(5.4-55), the general formula for $\Delta\boldsymbol{\lambda}_i$ at $t = t_i$ is

$$\Delta\boldsymbol{\lambda}_i = \left[\frac{\partial \boldsymbol{\varphi}_i^-}{\partial t_i} - \boldsymbol{f}_i^+ + \frac{\partial \boldsymbol{\varphi}_i^-}{\partial \boldsymbol{x}_i^-}\boldsymbol{f}_i^-\right]\frac{dt_i}{d\boldsymbol{\alpha}} + \left[\frac{\partial \boldsymbol{\varphi}_i^-}{\partial \boldsymbol{x}_i^-} - \boldsymbol{I}\right]\boldsymbol{\lambda}_i^- + \frac{\partial \boldsymbol{\varphi}_i^-}{\partial \boldsymbol{\alpha}}$$

where

$$\begin{aligned} \boldsymbol{f}_i^- &= A(\boldsymbol{\alpha})\boldsymbol{x}(t_i^-, \boldsymbol{\alpha}) + \boldsymbol{B}\,w(t_i^-) = \boldsymbol{A}\,\boldsymbol{x}_i^- + \boldsymbol{B}\,w_i^-, \\ \boldsymbol{f}_i^+ &= A(\boldsymbol{\alpha})\boldsymbol{x}(t_i^+, \boldsymbol{\alpha}) + \boldsymbol{B}\,w(t_i^+) + MC(\boldsymbol{\alpha}) = \boldsymbol{A}\,\boldsymbol{x}_i^+ + \boldsymbol{B}\,w_i^+ + MC, \\ \boldsymbol{\varphi}_i^- &= \boldsymbol{D}\,\boldsymbol{x}(t_i^-, \boldsymbol{\alpha}) = \boldsymbol{D}\,\boldsymbol{x}_i^-, \end{aligned}$$

and

$$\frac{\partial \boldsymbol{\varphi}_i^-}{\partial \boldsymbol{x}_i^-} = \boldsymbol{D}, \qquad \frac{\partial \boldsymbol{\varphi}_i^-}{\partial t} = \boldsymbol{0}, \qquad \frac{\partial \boldsymbol{\varphi}_i^-}{\partial \boldsymbol{\alpha}} = \boldsymbol{0}.$$

Hence

$$\Delta\boldsymbol{\lambda}_i = [-\boldsymbol{A}\boldsymbol{x}_i^- - \boldsymbol{B}w_i^- + \boldsymbol{D}(\boldsymbol{A}\,\boldsymbol{x}_i^+ + \boldsymbol{B}\,w_i^+ + \boldsymbol{M}\boldsymbol{C})]\,\frac{dt_i}{d\boldsymbol{\alpha}} + (\boldsymbol{D} - \boldsymbol{I})\boldsymbol{\lambda}_i^-.$$

Analogously, we obtain for $t = t_i + \gamma$,

$$\Delta\boldsymbol{\lambda}_{i+\gamma} = [\boldsymbol{f}_{i+\gamma}^- - \boldsymbol{f}_{i+\gamma}^+]\left[\frac{dt_i}{d\boldsymbol{\alpha}} + \frac{d\gamma}{d\boldsymbol{\alpha}}\right].$$

The term $dt_i/d\boldsymbol{\alpha}$ is calculated recursively from

$$\frac{dt_i}{d\boldsymbol{\alpha}} = -\frac{(\partial\theta_i/\partial\boldsymbol{x}_i^-)\,\boldsymbol{\lambda}_i^- - (\partial\theta_i/\partial\boldsymbol{\alpha})}{(\partial\theta_i/\partial\boldsymbol{x}_i^-)\,\boldsymbol{f}_i^- + (\partial\theta_i/\partial t_i)} - \frac{(\partial\theta_i/\partial t_{i-1})}{(\partial\theta_i/\partial\boldsymbol{x}_i^-)\boldsymbol{f}_i^- + \partial\theta_i/\partial t_i)}\,\frac{dt_{i-1}}{d\boldsymbol{\alpha}}$$

where

$$\frac{\partial\theta_i}{\partial x_1^-} = \frac{\partial\theta_i}{\partial x_2^-} = 0, \qquad \frac{\partial\theta_i}{\partial p} = \operatorname{sgn}[p(t_i)](t_i - T_m - t_{i-1}),$$

$$\frac{\partial\theta_i}{\partial\alpha_m} = 0 \quad \text{for} \quad m \neq 5, 6,$$

$$\frac{\partial\theta_i}{\partial\alpha_5} = -(|p(t_i)| - A), \quad \frac{\partial\theta_i}{\partial\alpha_6} = -(t_i - T_m - t_{i-1}),$$

$$\frac{\partial\theta_i}{\partial t} = (|p(t_i)| - A), \qquad \frac{\partial\theta_i}{\partial t_{i-1}} = -(|p(t_i)| - A).$$

With the aid of the formulas obtained, all sensitivity vectors $\boldsymbol{\lambda}_j$ ($j = 1, 2, \ldots, 8$) can be calculated successively.

Example 5.4-2 Consider a single axis attitude control system of a space vehicle containing a pulse-frequency modulator type I. Details of the control system are shown in Fig. 5.4-4. The transfer function of the plant is $G(s) = K/s^2$. The input pulses are rectangular. The characteristic of the pulse modu-

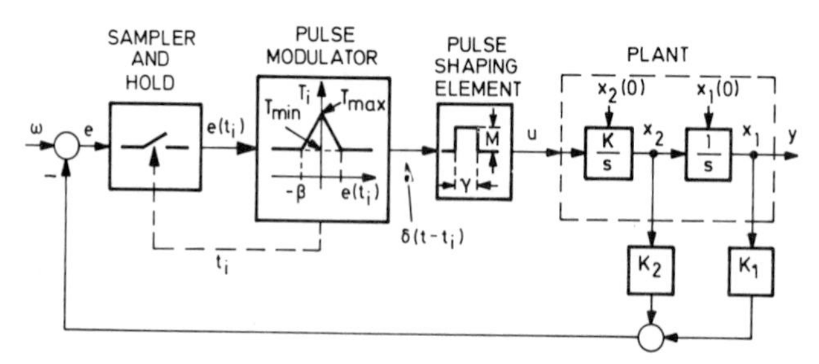

FIG. 5.4-4. Single-axis attitude control system containing a pulse-frequency modulator type I.

lator that determines the intervals between the sampling instants t_i and t_{i+1} (denoted by T_i) as a function of the error at t_i, $e(t_i)$, is given by

$$T_i = \begin{cases} T_{\max} - \dfrac{T_{\max} - T_{\min}}{\beta} |e(t_i)|, & |e| < \beta, \\ T_{\min}, & |e| > \beta. \end{cases} \tag{5.4-64}$$

The switching instants are defined as

$$t_i = t_{i-1} + T_i = t_{i-1} + T[e(t_i)]. \tag{5.4-65}$$

Let the parameter vector $\boldsymbol{\alpha} = [\alpha_1 \ldots \alpha_{10}]^{\mathrm{T}}$ be defined as

$$\boldsymbol{\alpha} = [K\ K_1\ K_2\ x_1(0)\ x_2(0)\ \beta\ T_{\max}\ T_{\min}\ \gamma\ M]^{\mathrm{T}}. \tag{5.4-66}$$

Determine the trajectory sensitivity equations of the position x_1 and the rate x_2 with respect to the above parameters.

Solution The mathematical model of the control system is given by the state equation

$$\dot{\boldsymbol{x}} = \boldsymbol{A}\boldsymbol{x} + \boldsymbol{b}u, \qquad \boldsymbol{x}(t_0) = \boldsymbol{x}^0, \tag{5.4-67}$$

where

$$\boldsymbol{x} = \begin{bmatrix} x_1 \\ x_2 \end{bmatrix}, \qquad \boldsymbol{A} = \begin{bmatrix} 0 & 1 \\ 0 & 0 \end{bmatrix}, \qquad b = \begin{bmatrix} 0 \\ K \end{bmatrix}$$

and

$$u = M \sum_{i=0}^{\infty} \operatorname{sgn}[e(t_i)][1(t - t_i) - 1(t - t_i - \gamma)], \tag{5.4-68}$$

$$t_i = t_{i-1} + T[e(t_{i-1})] = t_{i-1} + \begin{cases} T_{\max} - \dfrac{T_{\max} - T_{\min}}{\beta} |e(t_{i-1})|, & |e| < \beta, \\ T_{\min}, & |e| > \beta, \end{cases} \tag{5.4-69}$$

$$e(t_{i-1}) = w(t_{i-1}) - K_1 x_1(t_{i-1}) - K_2 x_2(t_{i-1}). \tag{5.4-70}$$

By taking the partial derivative of the state equation with respect to $\boldsymbol{\alpha}$, one obtains the state sensitivity equation

$$\dot{\boldsymbol{\lambda}} = \boldsymbol{A}\boldsymbol{\lambda} + \boldsymbol{b}\frac{\partial u}{\partial \boldsymbol{\alpha}} + \frac{\partial \boldsymbol{b}}{\partial \boldsymbol{\alpha}} u, \qquad \boldsymbol{\lambda}(t_0) = \boldsymbol{\lambda}^0. \tag{5.4-71}$$

Note that all components of $\boldsymbol{\lambda}^0$ are zero except those corresponding to the initial conditions of $\boldsymbol{x}$.

As an alternative to the method of Rozenvasser presented above, let us derive the jump conditions as an integral part of the sensitivity equations using distributions. This derivation is intuitively more satisfying and better relates the jump conditions to the originating system discontinuities. The final results are logically identical with those obtained by the method of Rozenvasser.

Suppose the initial time of each period is defined as t_i^-, so that the discontinuities of $\boldsymbol{\lambda}$ at t_i are included in the sensitivity equation. This implies that the derivatives of step functions, i.e., δ-functions (or distributions) have to be taken into consideration. Therefore, we may rewrite the above sensitivity equation in the form

$$\dot{\boldsymbol{\lambda}} = \boldsymbol{A}\boldsymbol{\lambda} + \frac{\partial \boldsymbol{b}}{\partial \boldsymbol{\alpha}} u(t, \boldsymbol{\alpha}) + \boldsymbol{b} \frac{\partial u(t, \boldsymbol{\alpha})}{\partial \boldsymbol{\alpha}}, \qquad \boldsymbol{\lambda}(t_i^-) = \boldsymbol{\lambda}_i^-, \qquad t_i \leq t \leq t_{i+1}. \tag{5.4-72}$$

The first term of Eq.(5.4-72) can be evaluated as described earlier. The second term becomes

$$\frac{\partial \boldsymbol{b}}{\partial \boldsymbol{\alpha}} u = \begin{bmatrix} 0 & 0 & 0 & 0 & 0 & 0 & 0 & 0 & 0 & 0 \\ 1 & 0 & 0 & 0 & 0 & 0 & 0 & 0 & 0 & 0 \end{bmatrix} \alpha_1 \operatorname{sgn}[e(t_i)][1(t - t_i) - 1(t - t_i - \gamma)], \tag{5.4-73}$$

where $\alpha_1 = K$ denotes the (nominal) gain of the plant.

Recalling that $t_i = t_i(\boldsymbol{\alpha})$, the third term $\partial u(t,\boldsymbol{\alpha})/\partial \boldsymbol{\alpha}$ can be evaluated as

$$\frac{\partial u(t, \boldsymbol{\alpha})}{\partial \boldsymbol{\alpha}} = \frac{\partial u(t, t_i, \boldsymbol{\alpha})}{\partial t_i} \frac{dt_i(\boldsymbol{\alpha})}{d\boldsymbol{\alpha}} + \frac{\partial u(t, t_i, \boldsymbol{\alpha})}{\partial \boldsymbol{\alpha}}, \tag{5.4-74}$$

where t_i is given by Eqs.(5.4-69) and (5.4-70), so that

$$\begin{aligned} t_i &= t_{i-1} + T_{i-1}[\boldsymbol{\alpha}, w(t_{i-1}), \boldsymbol{x}(t_{i-1},\boldsymbol{\alpha})] \\ &= t_{i-1}(\boldsymbol{\alpha}) + T_{i-1}(\boldsymbol{\alpha}, w_{i-1}, \boldsymbol{x}_{i-1}), \qquad |e| < \beta, \qquad (5.4\text{-}75a) \\ t_i &= t_{i-1}(\boldsymbol{\alpha}) + T_{\min}, \qquad |e| > \beta. \qquad (5.4\text{-}75b) \end{aligned}$$

Therefore, since w_{i-1} is not a direct function of $\boldsymbol{\alpha}$, the derivative $dt_i(\boldsymbol{\alpha})/d\boldsymbol{\alpha}$ in Eq.(5.4-74) becomes

$$\frac{dt_i(\boldsymbol{\alpha})}{d\boldsymbol{\alpha}} = \begin{cases} \dfrac{dt_{i-1}}{d\boldsymbol{\alpha}} + \dfrac{dT_{\min}}{d\boldsymbol{\alpha}}, & |e| > \beta, \\ \left[1 + \dot{w}_{i-1} \dfrac{\partial T_{i-1}}{\partial w_{i-1}} + \dot{\boldsymbol{x}}_{i-1} \dfrac{\partial T_{i-1}}{\partial \boldsymbol{x}_{i-1}}\right] \dfrac{dt_{i-1}}{d\boldsymbol{\alpha}} & \\ \quad + \dfrac{\partial T_{i-1}}{\partial \boldsymbol{\alpha}} + \boldsymbol{\lambda}_{i-1} \dfrac{\partial T_{i-1}}{\partial \boldsymbol{x}_{i-1}}, & |e| < \beta. \end{cases} \tag{5.4-76}$$

This is a recursive formula for calculating $dt_i/d\boldsymbol{\alpha}$.

Using the relation (5.4-68) of u and defining $\boldsymbol{c} \triangleq \operatorname{sgn}[e(t_i)]$, Eq.(5.4-74) can be evaluated to be

$$\frac{\partial u(t, \boldsymbol{\alpha})}{\partial \boldsymbol{\alpha}} = \alpha_1 \boldsymbol{c} \frac{\partial[1(t - t_i) - 1(t - t_i - \gamma)]}{\partial t_i} \frac{dt_i(\boldsymbol{\alpha})}{d\boldsymbol{\alpha}}$$

$$+ \alpha_1 \boldsymbol{c} \frac{\partial[-1(t - t_i - \gamma)]}{\partial \gamma} \frac{d\gamma}{d\boldsymbol{\alpha}}$$

$$+ \boldsymbol{c} \frac{\partial \alpha_1}{\partial \boldsymbol{\alpha}} [1(t - t_i) - 1(t - t_i - \gamma)], \tag{5.4-77}$$

where $e(t_i) = 0$ has been excluded from consideration. Recalling that the derivatives of the step function are the delta function, and slightly rearranging the terms of Eq.(5.4-77), we finally obtain

$$\frac{\partial u(t, \boldsymbol{\alpha})}{\partial \boldsymbol{\alpha}} = K \operatorname{sgn}[e(t_i)]\boldsymbol{b}\, \delta(t - t_i - \gamma) \left[\frac{dt_i(\boldsymbol{\alpha})}{d\boldsymbol{\alpha}} + \frac{d\gamma}{d\boldsymbol{\alpha}}\right]$$

$$- K \operatorname{sgn}[e(t_i)]\boldsymbol{b}\, \delta(t - t_i) \frac{dt_i}{d\boldsymbol{\alpha}}$$

$$+ \frac{dK}{d\boldsymbol{\alpha}} \operatorname{sgn}[e(t_i)][1(t - t_i) - 1(t - t_i - \gamma)], \tag{5.4-78}$$

where $dt_i/d\boldsymbol{\alpha}$ is given by Eq.(5.4-76).

If the resulting sensitivity equations are integrated, the above δ-functions yield step functions of $\boldsymbol{\lambda}$ at t_i and $t_i + \gamma$, being identical with the jump conditions. Thus the jump conditions have been arrived at by a straightforward derivation of the sensitivity equations in the interval $t_i \leq t \leq t_{i+1}$, where the discontinuous input u has been described in terms of step functions.

5.5 SOLUTION OF THE TRAJECTORY SENSITIVITY EQUATION

There are two basically different ways of obtaining the solution of the trajectory sensitivity equation:

(1) the analytical method, including numerical calculation (i.e., digital simulation) if necessary;

(2) the structural method, i.e., the measurement of the sensitivity functions by means of analog simulation.

The basic ideas of both methods will be outlined in this section.

The fact that the sensitivity equations are always linear regardless of the nature of the original system is a great advantage of the method especially in the case of linear systems since then the system equations and the sensitivity equations are of the same type. However, if the original system equation is nonlinear and a digital computer is employed for its evaluation, the advantage of the sensitivity equation being linear is somewhat questionable because a new computer program has to be written for the sensitivity equation, which is different from that of the original system equation. In this case it might be preferable to adopt the elementary approach of two computer runs, one for nominal and one for slightly changed parameters.

5.5.1 Analytical (Numerical) Solution

For *continuous* systems and time-invariant parameter variations, the general form of the state sensitivity equation is, according to Eq.(5.2–6),

$$\dot{\boldsymbol{\lambda}} = \left.\frac{\partial \boldsymbol{f}}{\partial \boldsymbol{x}}\right|_{\boldsymbol{\alpha}_0} \boldsymbol{\lambda} + \left.\frac{\partial \boldsymbol{f}}{\partial \boldsymbol{\alpha}}\right|_{\boldsymbol{\alpha}_0}, \qquad \boldsymbol{\lambda}(t_0) = \boldsymbol{\lambda}^0, \tag{5.5-1}$$

where $\boldsymbol{\lambda} = [\boldsymbol{\lambda}_1 \mid \ldots \mid \boldsymbol{\lambda}_r]$ represents the sensitivity matrix. This is a linear matrix differential equation in $\boldsymbol{\lambda}$ where the matrix of driving functions $(\partial \boldsymbol{f}/\partial \boldsymbol{\alpha})_{\boldsymbol{\alpha}_0}$ is obtained from the nominal original state equation. The initial conditions are either zero (α-variations) or partially zero (β- or λ-variations).

It is well known from state space theory that the above linear vector equation has the general solution

$$\boldsymbol{\lambda}(t, \boldsymbol{\alpha}_0) = \left[\boldsymbol{\varphi}(t, t_0)\boldsymbol{\lambda}(t_0, \boldsymbol{\alpha}_0) + \int_{t_0}^{t} \boldsymbol{\varphi}(t,\tau)\frac{\partial \boldsymbol{f}[\boldsymbol{x}(\tau, \boldsymbol{\alpha}), \boldsymbol{u}(\tau), \boldsymbol{\alpha}, \tau]}{\partial \boldsymbol{\alpha}}\, d\tau\right]_{\boldsymbol{\alpha}_0}, \tag{5.5-2}$$

where $\boldsymbol{\varphi}$ denotes the transition matrix satisfying the following differential equation:

$$\frac{\partial \boldsymbol{\varphi}(t, \tau)}{\partial t} = \frac{\partial \boldsymbol{f}[\boldsymbol{x}, \boldsymbol{u}, \boldsymbol{\alpha}, t]}{\partial \boldsymbol{x}}\, \boldsymbol{\varphi}(t, \tau), \tag{5.5-3}$$

with $\boldsymbol{\varphi}(t, t) = \boldsymbol{I}$.

In the case of α-variations where all initial conditions are equal to zero, i.e., $\boldsymbol{\lambda}(t_0) = \boldsymbol{0}$, the above expression reduces to

$$\boldsymbol{\lambda}(t,\boldsymbol{\alpha}_0) = \int_{t_0}^{t} \boldsymbol{\varphi}(t,\tau) \left.\frac{\partial \boldsymbol{f}[\boldsymbol{x}(\tau, \boldsymbol{\alpha}), \boldsymbol{u}(\tau), \boldsymbol{\alpha}, \tau]}{\partial \boldsymbol{\alpha}}\right|_{\boldsymbol{\alpha}_0} d\tau. \tag{5.5-4}$$

In the case of *β-variations*, $\partial \boldsymbol{f}/\partial \boldsymbol{\alpha} = \boldsymbol{0}$ and hence Eq.(5.5–2) reduces to

$$\boldsymbol{\lambda}(t, \boldsymbol{\beta}_0) = \boldsymbol{\varphi}(t, t_0)\, \boldsymbol{\lambda}(t_0, \boldsymbol{\beta}_0). \tag{5.5-5}$$

For *λ-variations*, none of the right-hand terms in Eq.(5.5–2) can be neglected.

Once $\boldsymbol{\lambda}$ is found, the output sensitivity matrix $\boldsymbol{\sigma}$ as defined by Definition 3.2-2 can be determined from the algebraic equation

$$\boldsymbol{\sigma} = \left.\frac{\partial \boldsymbol{g}}{\partial \boldsymbol{x}}\right|_{\boldsymbol{\alpha}_0} \boldsymbol{\lambda} + \left.\frac{\partial \boldsymbol{g}}{\partial \boldsymbol{\alpha}}\right|_{\boldsymbol{\alpha}_0} \tag{5.5-6}$$

Note that similar formulas in terms of the first variations of $\boldsymbol{x}$ and $\boldsymbol{y}$, respectively, hold in the case of systems with time-variant parameter variations.

Let us now focus our attention on the particular case of *linear time-invariant* systems with *α-variations* as described by Eq.(5.2-7). In this case the sensitivity equation is also time-invariant, since

$$\frac{\partial \boldsymbol{f}[\boldsymbol{x}, \boldsymbol{u}, \boldsymbol{\alpha}, t]}{\partial \boldsymbol{x}} = \boldsymbol{A}(\boldsymbol{\alpha}), \tag{5.5-7}$$

where $\boldsymbol{A}(\boldsymbol{\alpha})$ represents the system matrix that is time-invariant by the above assumption. The transition matrix becomes, in this particular case,

$$\varphi(t,\tau) = \varphi(t - \tau) = \exp[\boldsymbol{A}(\boldsymbol{\alpha}) \cdot (t - \tau)]. \tag{5.5-8}$$

Furthermore, if there is no feedback from $\boldsymbol{y}$ or $\boldsymbol{x}$ to $\boldsymbol{u}$, $\partial \boldsymbol{f}/\partial \boldsymbol{\alpha}$ takes the form

$$\frac{\partial \boldsymbol{f}[\boldsymbol{x}, \boldsymbol{u}, \boldsymbol{\alpha}, \mathrm{t}]}{\partial \boldsymbol{\alpha}} = \frac{\partial \boldsymbol{A}}{\partial \boldsymbol{\alpha}} \boldsymbol{x} + \frac{\partial \boldsymbol{B}}{\partial \boldsymbol{\alpha}} u. \tag{5.5-9}$$

Hence the solution of the sensitivity quation is

$$\boldsymbol{\lambda}(t,\boldsymbol{\alpha}_0) = \int_{t_0}^{t} \exp\,[\boldsymbol{A}(\boldsymbol{\alpha}_0)(t - \tau)] \left[\frac{\partial \boldsymbol{A}(\boldsymbol{\alpha})}{\partial \boldsymbol{\alpha}} \boldsymbol{x}(\tau,\boldsymbol{\alpha}) + \frac{\partial \boldsymbol{B}(\boldsymbol{\alpha})}{\partial \boldsymbol{\alpha}} \boldsymbol{u}(\tau)\right]_{\boldsymbol{\alpha}_0} d\tau. \tag{5.5-10}$$

With the aid of this expression, the output sensitivity matrix becomes

$$\begin{aligned}\boldsymbol{\sigma}(t,\boldsymbol{\alpha}_0) = \boldsymbol{C}(\boldsymbol{\alpha}_0) \int_{t_0}^{t} [\exp \boldsymbol{A}(\boldsymbol{\alpha}_0)(t - \tau)] \left[\left.\frac{\partial \boldsymbol{A}(\boldsymbol{\alpha})}{\partial \boldsymbol{\alpha}}\right|_{\boldsymbol{\alpha}_0} \boldsymbol{x}(\tau, \boldsymbol{\alpha}_0) + \left.\frac{\partial \boldsymbol{B}(\boldsymbol{\alpha})}{\partial \boldsymbol{\alpha}}\right|_{\boldsymbol{\alpha}_0} \boldsymbol{u}(\tau)\right] d\tau \\ + \left.\frac{\partial \boldsymbol{C}(\boldsymbol{\alpha})}{\partial \boldsymbol{\alpha}}\right|_{\boldsymbol{\alpha}_0} \boldsymbol{x}(t,\boldsymbol{\alpha}_0) + \left.\frac{\partial \boldsymbol{D}(\boldsymbol{\alpha})}{\partial \boldsymbol{\alpha}}\right|_{\boldsymbol{\alpha}_0} \boldsymbol{u}(t).\end{aligned} \tag{5.5-11}$$

When the system matrix $\boldsymbol{A}$ is constant, the transition matrix can always be found in closed form, although it may be tedious to do so for high-order systems. In the nonlinear and time-varying case it will be necessary to find numerical solutions or approximations of both the system equations and the sensitivity equations on the digital computer. The discrete solution of the sensitivity equation for the linear case is given by

$$\boldsymbol{\lambda}[(t + \Delta t)\boldsymbol{\alpha}_0) = e^{A\Delta t}\boldsymbol{\lambda}(t,\boldsymbol{\alpha}_0) + \int_{0}^{\Delta t} e^{A\tau} \left[\frac{\partial \boldsymbol{A}}{\partial \boldsymbol{\alpha}} \boldsymbol{x} + \frac{\partial \boldsymbol{B}}{\partial \boldsymbol{\alpha}} \boldsymbol{u}\right]_{\boldsymbol{\alpha}_0} d\tau. \tag{5.5-12}$$

In the same manner the nominal original system equation has to be solved simultaneously.

Whereas in the case of nonlinear systems different programs have to be set up for the original and the sensitivity equations, this can be avoided in the case of linear systems. Here the sensitivity equations can be joined to the system equations, forming the so-called *combined system*

$$\begin{bmatrix} \dot{\boldsymbol{x}}_0 \\ \dot{\boldsymbol{\lambda}} \end{bmatrix} = \begin{bmatrix} \boldsymbol{A}_0 & \boldsymbol{0} \\ \left.\dfrac{\partial \boldsymbol{A}}{\partial \alpha_j}\right|_{\boldsymbol{\alpha}0} & \boldsymbol{A}_0 \end{bmatrix} \begin{bmatrix} \boldsymbol{x}_0 \\ \boldsymbol{\lambda}_j \end{bmatrix} + \begin{bmatrix} \boldsymbol{B}_0 \\ \left.\dfrac{\partial \boldsymbol{B}}{\partial \alpha_j}\right|_{\boldsymbol{\alpha}0} \end{bmatrix} \boldsymbol{u}(\mathrm{t}), \tag{5.5-13}$$

$$\begin{bmatrix} \boldsymbol{y}_0 \\ \boldsymbol{\sigma}_j \end{bmatrix} = \begin{bmatrix} \boldsymbol{C}_0 & \boldsymbol{0} \\ \left.\dfrac{\partial \boldsymbol{C}}{\partial \alpha_j}\right|_{\boldsymbol{\alpha}_0} & \boldsymbol{C}_0 \end{bmatrix} \begin{bmatrix} \boldsymbol{x}_0 \\ \boldsymbol{\lambda}_j \end{bmatrix} + \begin{bmatrix} \boldsymbol{D}_0 \\ \left.\dfrac{\partial \boldsymbol{D}}{\partial \alpha_j}\right|_{\boldsymbol{\alpha}0} \end{bmatrix} \boldsymbol{u}(t). \tag{5.5-14}$$

The initial conditions are $\boldsymbol{x}(t_0) = \boldsymbol{x}^0$ and $\boldsymbol{\lambda}_j(t_0) = \boldsymbol{\lambda}_j^0$. The simultaneous solution of these differential equations using the same standard procedures for each of the above matrix equations yields the output vector, the state vector, and the corresponding output and trajectory sensitivity vectors $\boldsymbol{\sigma}_j$ and $\boldsymbol{\lambda}_j$. If there are r parameter variations, r systems of equations of the above type have to be solved. Since the homogeneous part of the original differential equation is identical with the homogenous part of the sensitivity equations with respect to all parameters, it is possible in certain situations to calculate all sensitivity functions simultaneously by a linear combination of the solution of a single set of equations. This clearly indicates the advantage of treating the problem in terms of the sensitivity equations.

In the case of sampled-data systems the sensitivity equation is given a priori as a difference equation and can be evaluated on the digital computer without any further modification.

Let us finally consider *systems with variable structure*. Referring to Section 5.4.2, the Eqs.(5.4-17), (5.4-20), (5.4-31) [or (5.4-34)], and (5.4-33) [or (5.4-37)] have to be solved successively and in parallel with the nominal state equations to find the course of the trajectory sensitivity matrix over the entire time domain. This calculation cannot be done without a digital computer. The sensitivity equations on the time intervals between the discontinuities are of the same form as for continuous systems. Therefore the formula for the discrete solution is as given in Eq.(5.5-12); however it is valid only for a single interval $t_i < t < t_{i+1}$. At the switching instants, jump conditions have to be added. The calculation of both the sensitivity equations and the jump conditions require the knowledge of the corresponding values of the state $\boldsymbol{x}$ of the original system. Hence the system equations and the sensitivity equations must be solved simultaneously in an iterative manner from interval to interval.

The procedure may be discussed in terms of *pulse-frequency modulated systems*. Referring to Section 5.4.3, Eqs.(5.4-42), (5.4-43), (5.4-45), (5.4-47) [or (5.4-55)], and (5.4-60) [or (5.4-63)] have to be calculated iteratively. If the control input u_i is considered "on" in the interval $t_i < t < t_i + \gamma$ and "off" in the interval $t_i + \gamma < t < t_{i+1}$, one sampling period T_i may be divided for processing on the computer into four separate parts, each requiring different calculations to be made.

(1) The *sample time* t_i. At this time the error $e(t_i)$ must be computed and in the case of a type I modulator the time T_i to the next sample time is calculated. In the case of a type II modulator the auxiliary variable is set to zero. The control input u_i is applied with the polarity determined by the sign of $e(t_i)$, and the jump conditions for t_i are added to the sensitivity functions.

(2) The *pulse-on interval* $t_i < t < t_i + \gamma$. During this interval the system and sensitivity equations must be solved with the control input applied. This may be done by using discrete solutions calculated at intervals or by using numerical integration to integrate equations [Eq. (5.5-12)]. At the end of this interval the step size Δt is adjusted to bring the system exactly to the end time $t_i + \gamma$.

(3) The *pulse end time* $t_i + \gamma$. At this time the control input is set to zero and the jump conditions for $t_i + \gamma$ are added to the sensitivity functions.

(4) The *pulse-off interval* $t_i + \gamma < t < t_{i+1}$. On this interval the system and sensitivity equations must be solved with the control input set to zero. In the case of a type II modulator the next sample time t_{i+1} has to be calculated from the system equations. The step size at the end of this interval is adjusted to bring the system exactly to the next sample time t_{i+1}.

At t_{i+1} the entire sequence has to be repeated. This concept is illustrated by the flow chart of Fig. 5.5-1. The four steps ae carried out using certain subroutines that depend on the particular nature of the modulator used. Notice that the performance of the individual steps includes the simultaneous calculation of the sensitivity and nominal state vectors.

Example 5.5-1 For the attitude control system of Example 5.4-1 containing a neuro pulse-frequency modulator, determine the trajectory sensitivity functions of x_1 and x_2 with respect to the gain factor K of the plant for step and sine wave reference inputs $w(t)$.

Solution The state equation of the control system within $t_i < t < t_{i+i}$ is

$$\dot{\boldsymbol{x}} = \boldsymbol{A}\boldsymbol{x} + \boldsymbol{B}w + \boldsymbol{C}u, \qquad \boldsymbol{x}(t_0) = \boldsymbol{x}^0,$$

where

$$\boldsymbol{A} = \begin{bmatrix} 0 & 1 & 0 \\ 0 & 0 & 0 \\ -K_1 & -K_2 & -c \end{bmatrix}, \qquad \boldsymbol{B} = \begin{bmatrix} 0 \\ 0 \\ 1 \end{bmatrix}, \qquad \boldsymbol{C} = \begin{bmatrix} 0 \\ K \\ 0 \end{bmatrix},$$

and

$$p(t_i^{+}) = 0.$$

Here u represents a full period (on and off portion) of the pulse train applied to the plant. Since neither u nor the reference input w depend directly on K, i.e., $\partial u/\partial K = \partial w/\partial K = 0$, and

$$\frac{\partial \boldsymbol{A}}{\partial K} = \boldsymbol{0}, \qquad \frac{\partial \boldsymbol{C}}{\partial K} = \begin{bmatrix} 0 \\ 1 \\ 0 \end{bmatrix},$$

the trajectory sensitivity equation within $t_i < t < t_{i+1}$ becomes

$$\dot{\boldsymbol{\lambda}} = \begin{bmatrix} 0 & 1 & 0 \\ 0 & 0 & 0 \\ -K_1 & -K_2 & -c \end{bmatrix} \boldsymbol{\lambda} + \begin{bmatrix} 0 \\ 1 \\ 0 \end{bmatrix} u, \qquad \boldsymbol{\lambda}(t_0) = \boldsymbol{0}.$$

At $t = t_i$ the jump conditions are

$$\Delta\boldsymbol{\lambda}(t_i) \triangleq \boldsymbol{\lambda}_i^+ - \boldsymbol{\lambda}_i^- = [-\boldsymbol{f}_i^+ + \boldsymbol{D}\boldsymbol{f}_i^-] \frac{dt_i}{dK} + [\boldsymbol{D} - \boldsymbol{I}]\boldsymbol{\lambda}_i^-,$$

where

$$\boldsymbol{f}_i^+ = \boldsymbol{A}\boldsymbol{x}_i^+ + \boldsymbol{B}w_i^+ + \boldsymbol{C}u_i^+,$$

$$\boldsymbol{f}_i^- = \boldsymbol{A}\boldsymbol{x}_i^- + \boldsymbol{B}w_i^-,$$

$$\boldsymbol{x}_i^+ = \boldsymbol{D}\boldsymbol{x}_i^-, \qquad \boldsymbol{D} = \begin{bmatrix} 1 & 0 & 0 \\ 0 & 1 & 0 \\ 0 & 0 & 0 \end{bmatrix}.$$

At $t = t_i + \gamma$ we have (since $d\gamma/dK = 0$):

$$\Delta\lambda(t_i + \gamma) = \boldsymbol{CM} \operatorname{sgn}[p(t_i^-)] \frac{dt_i}{dK}.$$

The expression for dt_i/dK is, in general,

$$\frac{dt_i}{dK} = -\frac{(\partial\theta_i/\partial\boldsymbol{x}^-)\,\boldsymbol{\lambda}_i^- + (\partial\theta_i/\partial K)}{(\partial\theta_i/\partial\boldsymbol{x}^-)\boldsymbol{f}_i^- + (\partial\theta_i/\partial t_i)} - \frac{(\partial\theta_i/\partial t_{i-1})}{(\partial\theta_i/\partial\boldsymbol{x}_i^-)\boldsymbol{f}_i^- + (\partial\theta_i/\partial t_i)} \frac{dt_{i-1}}{dK}.$$

Since $\theta_i = (|p(t_i^-)| - A)(t_i - T_m - T_{i-1}) = (|p_i^-| - A)(t_i - T_m - t_{i-1})$, and therefore

$$\frac{\partial\theta_i}{\partial x_{i,1}^-} = \frac{\partial\theta_i}{\partial x_{2,i}^-} = 0, \qquad \frac{\partial\theta_i}{\partial p_i^-} = \operatorname{sgn}[p_i^-](t_i - T_m - t_{i-1}),$$

$$\frac{\partial\theta_i}{\partial K} = 0, \qquad \frac{\partial\theta_i}{\partial t_i} = (|p_i^-| - A), \qquad \frac{\partial\theta_i}{\partial t_{i-1}} = -(|p_i^-| - A),$$

the above expression reduces to

$$\frac{dt_i}{dK} = -\frac{\operatorname{sgn}[p(t_i^-)](t_i - T_m - t_{i-1})(\partial p(t_i^-)/\partial K)}{\operatorname{sgn}[p(t_i^-)](t_i - T_m - t_{i-1})\dot{p}(t_i^-) + (|p(t_i^-)| - A)} \frac{dt_{i-1}}{dK}.$$

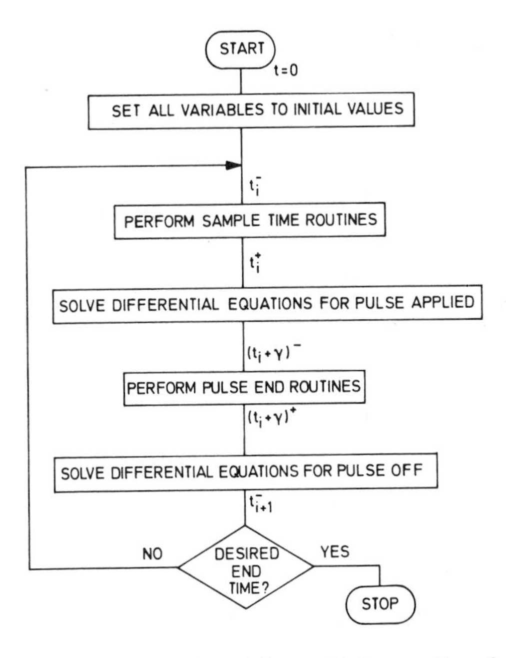

FIG. 5.5-1. Flow chart for the solution of the sensitivity equations for pulse-frequency modulated control systems.

For further evaluation of dt_i/dK the following two cases have to be distinguished:

(1) If $t_i - T_m - t_{i+1} > 0$, $(|p(t_i^-)| - A) = 0$, then we have

$$\frac{dt_i}{dK} = -\frac{1}{\dot{p}(t_i^-)} \frac{\partial p(t_i^-)}{\partial K}.$$

(2) If $t_i - T_m - t_{i-1} = 0$, $(|p(t_i^-)| - A) > 0$, then we have

$$\frac{dt_i}{dK} = \frac{dt_{i-1}}{dK}.$$

Using the above results, the nominal state $\boldsymbol{x}_0$ and the trajectory sensitivity vector $\boldsymbol{\lambda}_K = (\partial \boldsymbol{x}/\partial K)_{K_0}$ have to be computed simultaneously step by step. Notice that in this case the state $\boldsymbol{x}_0$ enters the sensitivity functions only through the jump conditions.

Plots of the time histories of x_{10}, x_{20}, and $x_{30} = p_0$ and their semirelative sensitivity functions $\tilde{\lambda}_{1k} = (\partial x_1/\partial \ln K)_{K_0}$ up to $\tilde{\lambda}_{3k} = (\partial x_3/\partial \ln K)_{K_0}$ are shown in Figs. 5.5-2-5.5-4 [112]. In all cases the initial conditions $x_1(t_0)$, $x_2(t_0)$, $p(t_0)$ are zero. In order to study the character of the sensitivity functions for different kinds of limit cycling—slow and fast—two sets of parameter settings are considered. In Figs. 5.5-2 and 5.5-3, the parameters are chosen as $K_0 = 1$, $c_0 = 0.1$, $T_{\min} = 0.2$, $K_{10} = 1$, $K_{20} = 2$, $\gamma_0 = 0.2$, $A_0 = 0.2$, $M_0 = 1$, so that no limit cycles occur during the observation time. In Fig. 5.5-2 the reference input is a step function ($w(t) = 4 \cdot 1(t) - 8 \cdot 1(t - t_s)$ with $t_s = 15$), and in Fig. 5.5-3 it is a sine wave ($w(t) = 4 \sin 0.21t$). In Fig. 5.5-4 the parameters K_{20}, M_0, and $T_{\min, 0}$ are changed into 1, 2, and 0.3, respectively, all others keeping constant, so that the system limit cycles considerably. The reference input is a single step $w(t) = 5 \cdot 1(t)$.

The results verify the discrete nature of $(\partial x_2/\partial \ln K)_{K_0}$. Furthermore they show that the sensitivity functions diverge with time when the state executes limit cycles. In the first two cases the divergence of $(\partial x_2/\partial \ln K)_{K_0}$ does not

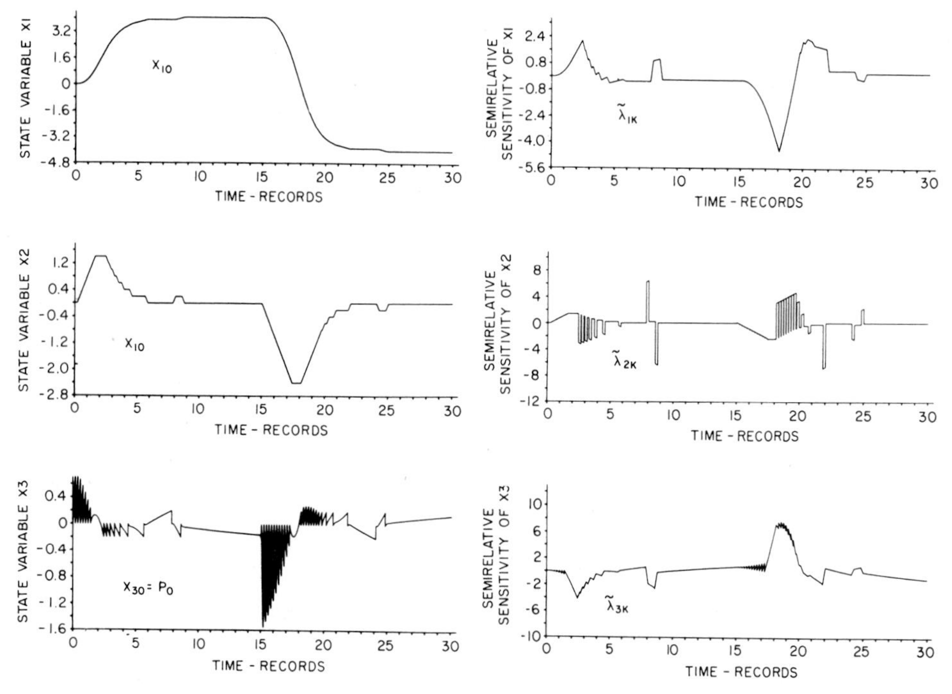

FIG. 5.5-2. Plots of $x_{10}, x_{20}, x_{30} = p_0$ and the corresponding semirelative sensitivity functions $\tilde{\lambda}_{1k}$, $\tilde{\lambda}_{2k}$, and $\tilde{\lambda}_{3k}$ of the control system of Example 5.4-1 with K the parameter of interest; where $w(t) = 4 \cdot 1(t) - 8 \cdot 1(t - t_3)$, $t_s = 15$; nominal parameters: $K_0 = 1$, $c_0 = 0.1$, $T_{m0} = 0.2$, $A_0 = 0.2$, $M_0 = 1$, $\gamma_0 = 0.2$; initial conditions zero; $K_{10} = 1$, $K_{20} = 2$ (slow limit cycles).

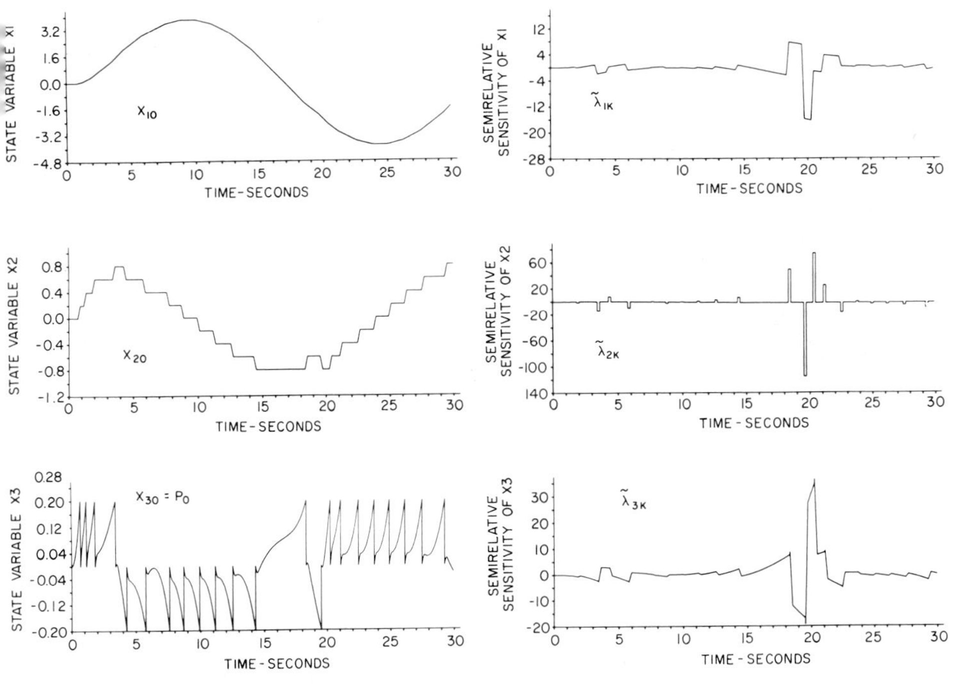

FIG. 5.5-3. Same plots as in Fig. 5.5-2, however, for a sine wave input $w(t) = 4 \sin 0.21t$.

become apparent during the time interval under consideration because of the slow limit cycling of the system.

Example 5.5-2 Referring to Example 5.4-2, determine the trajectory sensitivity equations of x_1 and x_2 with respect to the gain K of the plant and give the formulas for the trajectory sensitivity function λ_1 and λ_2. Discuss the question of the continuity of λ_1 and λ_2 with respect to time.

Solution Defining $\lambda_1 \triangleq (\partial x_1/\partial K)_{\boldsymbol{\alpha}0}$ and $\lambda_2 \triangleq (\partial x_2(\partial K)_{\boldsymbol{\alpha}0}$, and excluding the case $e(t_i) = 0$, one obtains for $\boldsymbol{\alpha} = \boldsymbol{\alpha}_0$,

$$\dot{\lambda}_1 = \lambda_2,$$

$$\begin{aligned}\dot{\lambda}_2 &= u_0 + K_0 \left.\frac{\partial u}{\partial K}\right|_{\boldsymbol{\alpha}0} \\ &= M_0 \operatorname{sgn}[e(t_{i0})][1(t - t_{i0}) - 1(t - t_{i0} - \gamma_0)] \\ &\quad + K_0 M_0 \operatorname{sgn}[e(t_{i0})][-\delta(t - t_{i0}) + \delta(t - t_{i0} - \gamma_0)] \left.\frac{dt_i}{dK}\right|_{\boldsymbol{\alpha}0},\end{aligned}$$

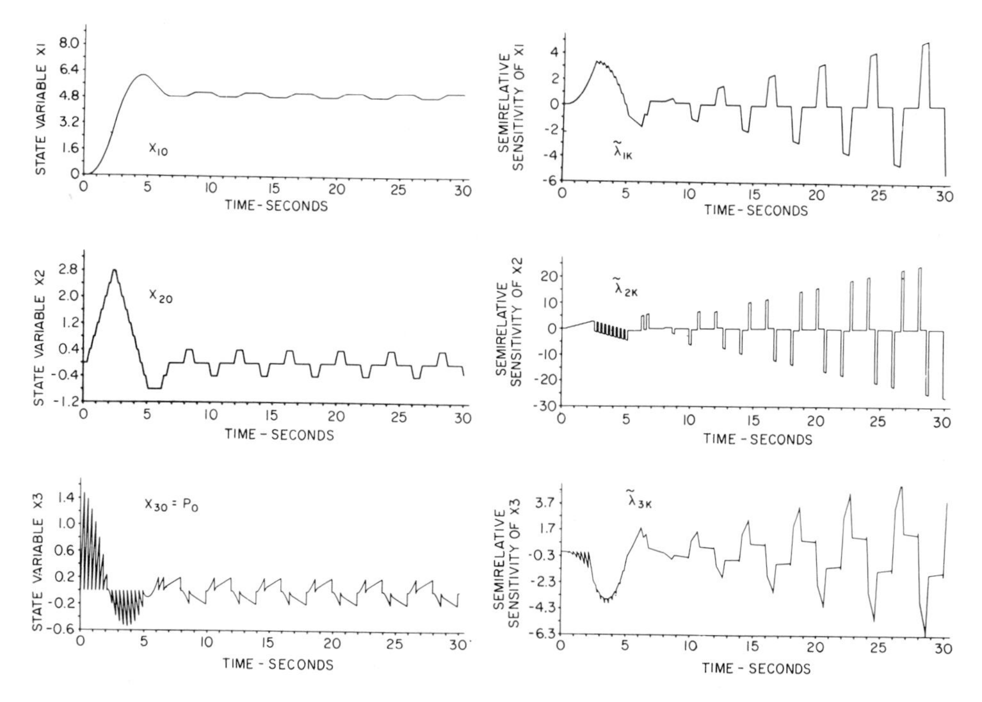

FIG. 5.5-4. Same plots as in Fig. 5.5-2, however for $K_{20} = 1$, $M_0 = 2$, $T_{\min} = 0.3$ (fast limit cycles), reference input $w(t) = 5 \cdot 1(t)$.

where (using the notation $\dot{x}_{1,i-1} = \partial x_1(t_{i-1}, \boldsymbol{\alpha})/\partial t_{i-1}$, etc.)

$$\frac{dt_i}{dK} = \frac{dt_i}{dK} - \frac{T_{\max} - T_{\min}}{\beta} \operatorname{sgn}[e(t_{i-1})] \Big\{ [\dot{w}_{i-1} - K_1 \dot{x}_{1,i-1}$$

$$- K_2 \dot{x}_{2,i-1}] \frac{dt_{i-1}}{dK} - K_1 \lambda_{1,i-1} - K_2 \lambda_{2,i-1} \Big\}, \qquad |e| < \beta,$$

$$\frac{dt_i}{dK} = \frac{dt_{i-1}}{dK}, \qquad |e| > \beta.$$

Integrating the above sensitivity equations yields, on the interval $t_i^- \leq t \leq t_{i+1}^-$,

$$\lambda_1 = \int \lambda_2 \, dt + \lambda_1(t_{i0}^-),$$

$$\lambda_2 = M_0 \operatorname{sgn}[e(t_{i0})] \{(t - t_{i0}) 1(t - t_{i0}) - (t - t_{i0} - \gamma_0) 1(t - t_{i0} - \gamma_0)\}$$

$$+ K_0 [1(t - t_{i0}) - 1(t - t_{i0} - \gamma_0)] \left.\frac{dt_i}{dK}\right|_{\boldsymbol{\alpha}0} + \lambda_2(t_i^-),$$

where dt_i/dK is as above. This result shows that λ_2 is discontinuous at t_i and $t_i + \gamma$, whereas λ_2 has corners at these instants.

Figures 5.5-5 and 5.5-6 show plots of sensitivity functions with respect to K obtained by the procedure described above [111]. The nominal parameters of the control system are $K_0 = 1$, $K_{10} = 1$, $K_{20} = 1$, $M_0 = 5$, $\gamma_0 = 0.2$, $T_{max,0} = 1$, $T_{min,0} = 0.2$, and $\beta_0 = 8$. The figures contain the time histories of the states x_1 and x_2 and of the corresponding sensitivity functions λ_1 and λ_2 for different modes of operation. In Fig. 5.5-5 the input w is zero, and the initial conditions are $x_1^0 = -10.066$, $x_2^0 = 6.9$. In Fig. 5.5-6 the initial conditions are zero and a step of $w = 20$ units is applied to the input.

The results verify the discontinuous nature of the sensitivity function of x_2. They also show that the sensitivity functions are divergent with time. This is due to the fact that the trajectory performs limit cycles. The divergence of the sensitivity functions implies that even a very small parameter variation can eventually have a significant effect on the system trajectory. For more details see Litty [111].

5.5.2 Structural Method

The basis for the *measurement* of the trajectory sensitivity functions is the

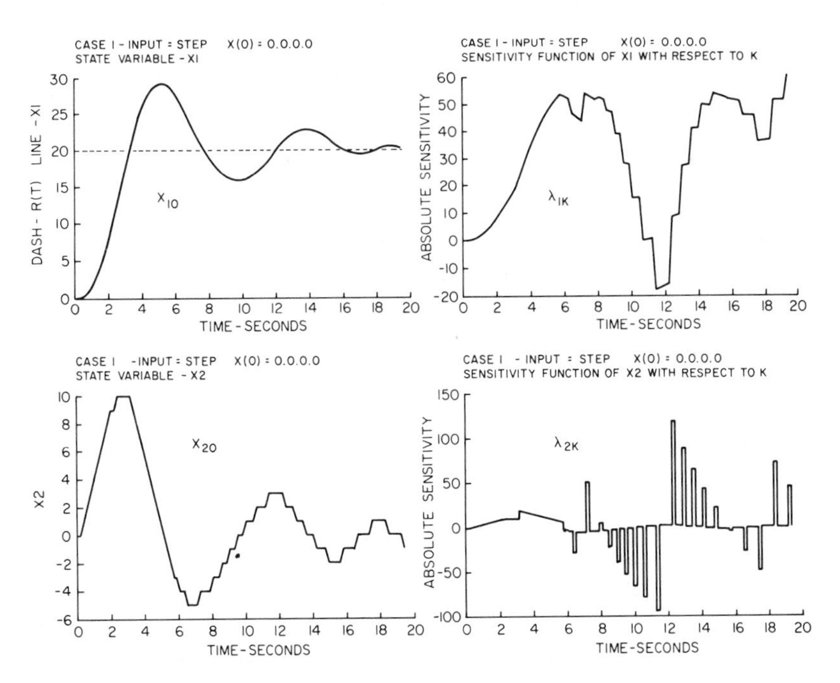

FIG. 5.5-5. Plots of x_{10}, x_{20} and the corresponding sensitivity function λ_{1k}, λ_{2k} for the control system of Example 5.4-2 with K the parameter of interest, reference input $w(t) = 0$, initial conditions $x_1(0) = -10.066$, $x_2(0) = 6.9$.

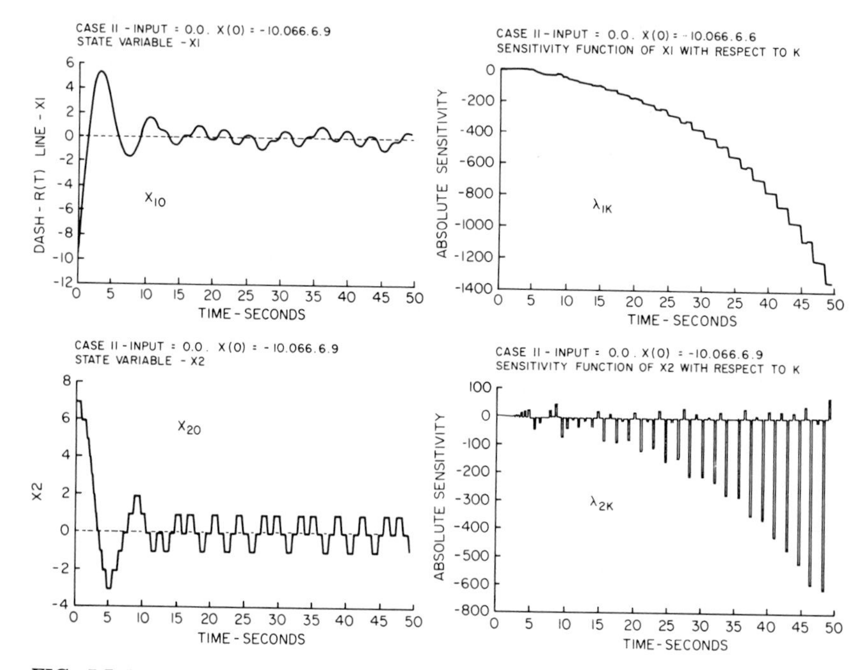

FIG. 5.5-6. Plots of x_{10}, x_{20} and λ_{1k}, λ_{2k} as in Fig. 5.5-5, however for $w(t) = 20 \cdot 1(t)$, $x_1(0) = 0$, $x_2(0) = 0$.

structural interpretation of the trajectory sensitivity equation leading to the following definition:

Definition 5.5-1 Trajectory sensitivity model. The physical system represented by the trajectory sensitivity equations is called the trajectory sensitivity model or the sensitivity model in the state space.

From the results of Section 5.2 the following theorem can be stated:

Theorem 5.5-1 Regardless of the nature of the original system the sensitivity model in the state space is always linear. In the case of a linear original system it has the same structure as the original system, merely the inputs and initial conditions are different depending on the parameter of interest.

Since the driving function of the sensitivity model contains the nominal state, the measuring circuit for the trajectory sensitivity functions consists of a connection of the nominal original system and the sensitivity model as illustrated by Fig. 5.2-1. If the actual input $\boldsymbol{u}$ is applied to this structure, the trajectory and output sensitivity vectors $\boldsymbol{\lambda}_j$ and $\boldsymbol{\sigma}_j$ can be measured at the points a and b, respectively.

In order to measure all r sensitivity vectors simultaneously, r sensitivity models are required. This may be demonstrated by the following example.

Example 5.5-3 For a system with the state equations

$$\begin{aligned}
\dot{x}_1 &= x_2, & x_1(0) &= 0,\\
\dot{x}_2 &= x_3, & x_2(0) &= 0,\\
\dot{x}_3 &= -a_0x_1 - a_1x_2 - a_2x_3 + u, & x_3(0) &= 0,\\
y &= x_1,
\end{aligned}$$

draw a signal flow diagram to measure simultaneously all trajectory sensitivity functions with respect to a_0, a_1, a_2.

Solution In vector notation, the original system equations are

$$\dot{\boldsymbol{x}} = \boldsymbol{A}\boldsymbol{x} + \boldsymbol{b}u, \qquad \boldsymbol{x}(0) = \boldsymbol{0},$$

$$y = \boldsymbol{c}\boldsymbol{x} + du,$$

where

$$\boldsymbol{A} = \begin{bmatrix} 0 & 1 & 0 \\ 0 & 0 & 1 \\ -a_0 & -a_1 & -a_2 \end{bmatrix}, \qquad \boldsymbol{b} = \begin{bmatrix} 0 \\ 0 \\ 1 \end{bmatrix}, \qquad \boldsymbol{c} = [1\ 0\ 0], \qquad d = 0.$$

To find the sensitivity equations, the following expressions have to be formed:

$$\frac{\partial \boldsymbol{A}}{\partial \alpha_0} = \begin{bmatrix} 0 & 0 & 0 \\ 0 & 0 & 0 \\ -1 & 0 & 0 \end{bmatrix}, \qquad \frac{\partial \boldsymbol{A}}{\partial \alpha_1} = \begin{bmatrix} 0 & 0 & 0 \\ 0 & 0 & 0 \\ 0 & -1 & 0 \end{bmatrix}, \qquad \frac{\partial \boldsymbol{A}}{\partial \alpha_2} = \begin{bmatrix} 0 & 0 & 0 \\ 0 & 0 & 0 \\ 0 & 0 & -1 \end{bmatrix},$$

$$\frac{\partial \boldsymbol{b}}{\partial \alpha_j} = \begin{bmatrix} 0 \\ 0 \\ 0 \end{bmatrix}, \qquad \frac{\partial \boldsymbol{c}}{\partial \alpha_j} = [0\quad 0\quad 0], \qquad \frac{\partial d}{\partial \alpha_j} = 0, \quad j = 0, 1, 2.$$

By substituting these expressions into the sensitivity equations (5.2-8) and (5.2-9), the trajectory sensitivity equations become

$$\dot{\boldsymbol{\lambda}}_j = \boldsymbol{A}_0\boldsymbol{\lambda}_j + \left.\frac{\partial \boldsymbol{A}}{\partial \alpha_j}\right|_{\boldsymbol{\alpha}0} \boldsymbol{x}_0, \qquad j = 0, 1, 2,$$

$$\sigma_j = \boldsymbol{c}_0\boldsymbol{\lambda}_j, \qquad j = 0, 1, 2,$$

or, in terms of the components $\lambda_{ij} = (\partial x_i/\partial \alpha_j)_{\boldsymbol{\alpha}0}$,

$$\begin{aligned}
\dot{\lambda}_{1j} &= \lambda_{2j},\\
\dot{\lambda}_{2j} &= \lambda_{3j},\\
\dot{\lambda}_{3j} &= -a_{00}\lambda_{1j} - a_{10}\lambda_{2j} - a_{20}\lambda_{3j} - x_{j+1,0},\\
\sigma_j &= \lambda_{1j}, \qquad j = 0, 1, 2,
\end{aligned}$$

with all initial conditions equal to zero. The corresponding structural diagram to measure all sensitivity functions simultaneously is shown in Fig. 5.5-7.

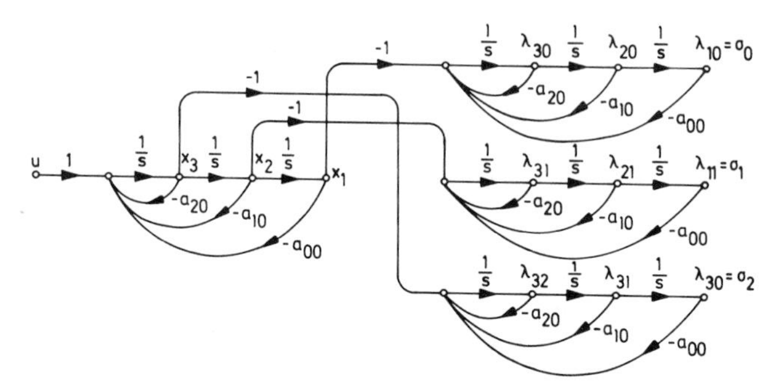

FIG. 5.5-7. Signal flow diagram to measure simultaneously all trajectory sensitivity functions of Example 5.5-3.

The above procedure may serve as a guiding rule for the construction of a measuring circuit for $\boldsymbol{\lambda}_j$ without knowing the trajectory sensitivity equations. In addition to the nominal model, other r models of the same kind have to be drawn. Then connections are established from the inputs of the variable components in the original model to the outputs of the corresponding components in the individual sensitivity models. The transfer functions of the connection lines are the derivatives of the transfer functions of the corresponding variable components with respect to the parameters under consideration. In the particular case above, these transfer function are -1. If the actual input u is then applied to the original system, the state variables of the sensitivity models represent the desired trajectory sensitivity functions.

From this procedure the great similarity to the variable component method discussed in Section 4.5 becomes apparent. Analogously, in the case of linear systems, a structural reduction is possible, generalizing the method of sensitivity points discussed in Section 4.6. This will be shown in the following section.

5.6 THE THEOREMS OF WILKIE AND PERKINS

In general, if a system has r parameters, r sensitivity models are needed in order to measure all trajectory sensitivity functions simultaneously according to the above procedure. The question of minimal structure necessary for the simultaneous measurement of all trajectory sensitivity functions was first treated by Wilkie and Perkins [107, 108]. They have shown that all nr trajectory sensitivity functions of a linear time-invariant nth order system with r parameters of interest ($r \gtreqless n$) can be obtained simultaneously by a single nth order sensitivity model in combination with the original system model. The procedure is based on two theorems stated and proved by Wilkie and Perkins [107] that will be given below.

It is well known from the theory of state space that any controllable system given by an arbitrary state equation

$$\dot{\boldsymbol{x}} = \tilde{\boldsymbol{A}}\boldsymbol{x} + \tilde{\boldsymbol{b}}u, \qquad \boldsymbol{x}(0) = \boldsymbol{0}, \tag{5.6-1}$$

$$y = \tilde{\boldsymbol{c}}\boldsymbol{x} \tag{5.6-2}$$

can be brought to the Frobenius canonical (companion) form

$$\dot{\boldsymbol{z}} = \boldsymbol{A}\boldsymbol{z} + \boldsymbol{b}u, \qquad \boldsymbol{z}(0) = \boldsymbol{0}, \tag{5.6-3}$$

$$y = \boldsymbol{c}\boldsymbol{z} \tag{5.6-4}$$

by a nonsingular transformation

$$\boldsymbol{x} = \boldsymbol{T}\boldsymbol{z}, \tag{5.6-5}$$

where

$$\boldsymbol{A} = \boldsymbol{T}^{-1}\tilde{\boldsymbol{A}}\boldsymbol{T} = \begin{bmatrix} 0 & 1 & & 0 \\ & & \ddots & \\ & & & 1 \\ -\alpha_1 & -\alpha_2 & \cdots & -\alpha_n \end{bmatrix}, \tag{5.6-6}$$

$$\boldsymbol{b} = \boldsymbol{T}^{-1}\tilde{\boldsymbol{b}} = \begin{bmatrix} 0 \\ \vdots \\ 0 \\ 1 \end{bmatrix}. \tag{5.6-7}$$

The z_i are the phase variables of the system and the elements $\alpha_1 \alpha_2, \ldots, \alpha_n$ of the Frobenius matrix $\boldsymbol{A}$ are the coefficients of the characteristic equation

$$s^n + \alpha_n s^{n-1} + \cdots + \alpha_2 s + \alpha_1 = 0. \tag{5.6-8}$$

For a system in Frobenius canonical form the following two theorems hold:

Theorem 5.6-1 (*Total symmetry property*) Define the sensitivity vector as

$$\boldsymbol{\zeta}_j \triangleq \left.\frac{\partial \boldsymbol{z}}{\partial \alpha_j}\right|_{\boldsymbol{\alpha}0}, \tag{5.6-9}$$

i.e.,

$$\zeta_{ij} = \left.\frac{\partial z_i}{\partial \alpha_j}\right|_{\boldsymbol{\alpha}0}, \qquad i, j = 1, 2, \ldots, n,$$

and the sensitivity matrix as

$$\boldsymbol{\zeta} = [\zeta_{ij}] = [\boldsymbol{\zeta}_1 \vdots \boldsymbol{\zeta}_2 \vdots \cdots \vdots \boldsymbol{\zeta}_n]. \tag{5.6-10}$$

Then the sensitivity matrix has the following property, called the *total symmetry property*:

$$\zeta_{ij} = \zeta_{i+1, j-1}, \qquad i, j = 2, 3, \ldots, n, \tag{5.6-11}$$

i.e., all the elements of the "antidiagonals" are identical:

$$\boldsymbol{\zeta} = \begin{bmatrix} \zeta_{11} & \zeta_{12} & \zeta_{13} & \cdots & \zeta_{1,n-1} & \zeta_{1n} \\ \zeta_{12} & \zeta_{13} & & & & \zeta_{2n} \\ \zeta_{13} & & & & & \vdots \\ \vdots & & & & & \\ & & & & & \zeta_{n-1,n} \\ \zeta_{1n} & \zeta_{2n} & & \cdots & \zeta_{n-1,n} & \zeta_{nn} \end{bmatrix}. \tag{5.6-12}$$

Hence there are only $n + n - 1 = 2n - 1$ independent trajectory sensitivity functions.

Proof This theorem can readily be proven in a straightforward way by differentiation of the state equations with respect to α_j. Assuming again that the sequence of differentiation with respect to time and α_j can be interchanged, we obtain the sensitivity equations (dropping the index $\boldsymbol{\alpha}_0$ for notational ease)

$$\begin{aligned} \frac{\partial \dot{z}_1}{\partial \alpha_j} &= \frac{\partial z_2}{\partial \alpha_j} \\ &\ \vdots \\ \frac{\partial \dot{z}_{n-1}}{\partial \alpha_j} &= \frac{\partial z_n}{\partial \alpha_j}, \\ \frac{\partial \dot{z}_n}{\partial \alpha_j} &= -\sum_{i=1}^{n} \alpha_i \frac{\partial z_i}{\partial \alpha_j} - z_j, \qquad \frac{\partial \mathbf{z}(0)}{\partial \alpha_j} = \mathbf{0}. \end{aligned} \tag{5.6-13}$$

On the other hand, if we take the derivative with respect to α_{j+1}, we obtain

$$\begin{aligned} \frac{\partial \dot{z}_1}{\partial \alpha_{j+1}} &= \frac{\partial z_2}{\partial \alpha_{j+1}}, \\ &\ \vdots \\ \frac{\partial \dot{z}_{n-1}}{\partial \alpha_{j+1}} &= \frac{\partial z_n}{\partial \alpha_{j+1}}, \\ \frac{\partial \dot{z}_n}{\partial \alpha_{j+1}} &= -\sum_{i=1}^{n} \alpha_i \frac{\partial z_i}{\partial \alpha_{j+1}} - z_{j+1}, \qquad \frac{\partial \mathbf{z}(0)}{\partial \alpha_{j+1}} = \mathbf{0}. \end{aligned} \tag{5.6-14}$$

Integration of the last equation over time t gives

$$\int_0^t \frac{\partial \dot{z}_n}{\partial \alpha_{j+1}} \, d\tau = \frac{\partial z_n}{\partial \alpha_{j+1}} = -\sum_{i=1}^{n} \alpha_i \int_0^t \frac{\partial z_i}{\partial \alpha_{j+1}} \, d\tau - \int_0^t z_{j+1} \, d\tau. \tag{5.6-15}$$

If we now introduce the $n - 1$ equations of (5.6-14) into the above equation, we obtain

$$\frac{\partial \dot{z}_{n-1}}{\partial \alpha_{j+1}} = -\alpha_1 \int_0^t \frac{\partial z_1}{\partial \alpha_{j+1}} \, d\tau - \sum_{i=2}^{n} \alpha_i \frac{\partial z_{i-1}}{\partial \alpha_{j+1}} - z_j. \tag{5.6-16}$$

Now we define

$$
\begin{aligned}
w_1(t) &\triangleq \int_0^t \frac{\partial z_1}{\partial \alpha_{j+1}}\, d\tau, \\
w_2(t) &\triangleq \dot{w}_1(t) = \frac{\partial z_1}{\partial \alpha_{j+1}}, \\
&\;\vdots \\
w_n(t) &\triangleq \frac{\partial z_{n-1}}{\partial \alpha_{j+1}}.
\end{aligned}
\tag{5.6-17}
$$

Use of the above substitutions in Eqs.(5.6–14) and replacement of the last equation of Eqs.(5.6-14) by Eq.(5.6-16) yields

$$
\begin{aligned}
\dot{w}_1 &= w_2, \\
\dot{w}_2 &= w_3, \\
&\;\vdots \\
\dot{w}_{n-1} &= w_n, \\
\dot{w}_n &= -\sum_{i=1}^{n} \alpha_i w_i - z_j, \qquad \boldsymbol{w}(0) = \mathbf{0}.
\end{aligned}
\tag{5.6-18}
$$

By comparison of Eqs.(5.6-18) and (5.6-13) and recalling the uniqueness theorem of differential equations, we obtain

$$
\boldsymbol{w} = \frac{\partial \boldsymbol{z}}{\partial \alpha_j} \tag{5.6-19}
$$

which is equivalent to

$$
w_1 = \int_0^t \frac{\partial z_1}{\partial \alpha_{j+1}}\, d\tau = \frac{\partial z_1}{\partial \alpha_j} \tag{5.6-20}
$$

and

$$
w_k = \frac{\partial z_{k-1}}{\partial \alpha_{j+1}} = \frac{\partial z_k}{\partial \alpha_j}, \qquad \begin{aligned} k &= 2, 3, \ldots, n, \\ j &= 1, 2, \ldots, n-1. \end{aligned} \tag{5.6-21}
$$

This proves the above theorem.

The relationship with Kokotović's method of sensitivity points becomes quite apparent if we write Eq.(5.6-20) in the form

$$
\frac{\partial z_1}{\partial \alpha_{j+1}} = \frac{\partial \dot{z}_1}{\partial \alpha_j}, \qquad j = 1, 2, \ldots, n-1. \tag{5.6-22}
$$

When setting $j = 1$, Eqs.(5.6-13) constitute Kokotović's result. Hence the method of sensitivity points can be considered as a special case of the first theorem of Wilkie and Perkins.

Theorem 5.6-2 (*Complete simultaneity property*) All the sensitivity functions

$$
\zeta_{ij} = \left.\frac{\partial z_i}{\partial \alpha_j}\right|_{\boldsymbol{\alpha}0}, \qquad i, j = 1, 2, \ldots, n, \tag{5.6-23}
$$

of the system in Frobenius canonical form can be obtained as linear algebraic combinations of state variables of a single sensitivity model and the nominal original system model.

Proof In the proof of Theorem 5.6-1 it was established that all the n sensitivity functions

$$\zeta_{1j} = \left.\frac{\partial z_1}{\partial \alpha_j}\right|_{\boldsymbol{\alpha}0}, \qquad j = 1, 2, \ldots, n,$$

can be obtained by a single sensitivity model, as shown in the signal flow diagram of Fig. 5.6-1.

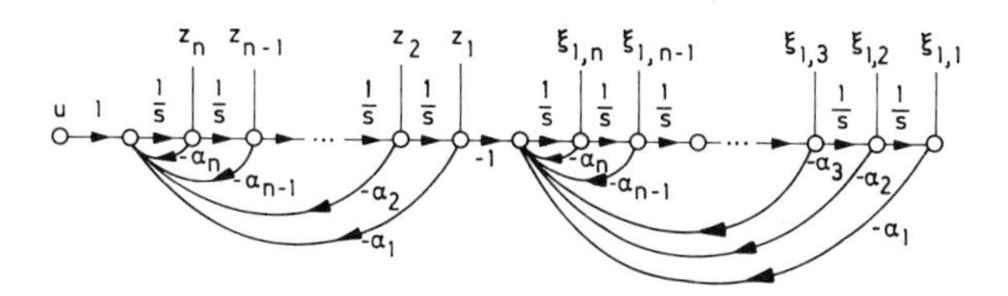

FIG. 5.6-1. Structural interpretation of Theorem 5.6–1.

In the terminology of Theorem 5.6-1, this means that all the sensitivity functions of the upper left triangle of the sensitivity matrix can be found from the above diagram. The remaining $n - 1$ independent sensitivity functions,

$$\zeta_{kn} = \left.\frac{\partial z_k}{\partial \alpha_n}\right|_{\boldsymbol{\alpha}0}, \qquad k = 2, 3, \ldots, n,$$

are obtained as follows:

Recall that the signal at node k of the sensitivity model is

$$\frac{\partial z_1}{\partial \alpha_k}, \qquad k = 1, 2, \ldots, n.$$

On the other hand from Eq.(5.6-3),

$$\frac{\partial \dot{z}_1}{\partial \alpha_n} = \frac{\partial z_2}{\partial \alpha_n}, \tag{5.6-24a}$$

and by virtue of Theorem 5.6-1,

$$\frac{\partial z_1}{\partial \alpha_n} = \frac{\partial z_n}{\partial \alpha_1}. \tag{5.6-24b}$$

Therefore,

$$\frac{\partial z_2}{\partial \alpha_n} = \frac{\partial \dot{z}_n}{\partial \alpha_1}, \tag{5.6-25}$$

and from Eqs. (5.6-13), we obtain

$$\frac{\partial z_2}{\partial \alpha_n} = -z_1 - \sum_{i=1}^{n} \alpha_i \frac{\partial z_i}{\partial \alpha_1} = -z_1 - \sum_{i=1}^{n} \alpha_i \frac{\partial z_1}{\partial \alpha_i} \tag{5.6-26}$$

Making use of this result we find

$$\frac{\partial z_3}{\partial \alpha_n} = \frac{\partial \dot{z}_2}{\partial \alpha_n} = -\dot{z}_1 - \sum_{i=1}^{n} \alpha_i \frac{\partial \dot{z}_1}{\partial \alpha_i}, \tag{5.6-27}$$

and, with the aid of Eqs.(5.6-14), we get

$$\frac{\partial z_3}{\partial \alpha_n} = -z_2 - \sum_{i=1}^{n-1} \alpha_i \frac{\partial z_{i+1}}{\partial \alpha_1} - \alpha_n \frac{\partial z_2}{\partial \alpha_n}. \tag{5.6-28}$$

Analogously, we find

$$\frac{\partial z_4}{\partial \alpha_n} = \frac{\partial \dot{z}_3}{\partial \alpha_n} = -\dot{z}_2 - \sum_{i=1}^{n-1} \alpha_i \frac{\partial \dot{z}_{i+1}}{\partial \alpha_1} - \alpha_n \frac{\partial \dot{z}_2}{\partial \alpha_n}$$

$$= -z_3 - \sum_{i=1}^{n-2} \alpha_i \frac{\partial z_{i+2}}{\partial \alpha_i} - \alpha_{n-1} \frac{\partial z_2}{\partial \alpha_n} - \alpha_n \frac{\partial z_3}{\partial \alpha_n}. \tag{5.6-29}$$

Continuation of this procedure yields all the remaining sensitivity functions:

$$\frac{\partial z_n}{\partial \alpha_n} = -z_{n-1} - \sum_{i=1}^{2} \alpha_i \frac{\partial z_{i+n-2}}{\partial \alpha_1} - \alpha_3 \frac{\partial z_2}{\partial \alpha_n}$$

$$- \alpha_4 \frac{\partial z_3}{\partial \alpha_n} - \cdots - \alpha_{n-1} \frac{\partial z_{n-2}}{\partial \alpha_n} - \alpha_n \frac{\partial z_{n-1}}{\partial \alpha_n}. \tag{5.6-30}$$

In general, we have

$$\frac{\partial z_k}{\partial \alpha_n} = -z_{k-1} - \sum_{i=1}^{n+2-k} \alpha_i \frac{\partial z_{i+k-2}}{\partial \alpha_1} - \sum_{i=n+3-k}^{n} \alpha_i \frac{\partial z_{i-1-n+k}}{\partial \alpha_n}. \tag{5.6-31}$$

Thus the sensitivity functions $\partial z_2/\partial \alpha_n, \ldots, \partial z_n/\partial \alpha_n$ can be found by linear combinations of the sensitivity functions $\partial z_1/\partial \alpha_1, \ldots, \partial z_1/\partial \alpha_n$ appearing at the nodes of the sensitivity model, with nominal state variables $z_1, \ldots, z_{n-1}$ appearing at the nodes of the nominal original system. This proves Theorem 5.6-2.

Note that $\partial z_2/\partial \alpha_n$, in particular, is obtained at the input of the first integrator of the sensitivity model of Fig. 5.6-1. For the remaining sensitivity functions $\partial z_k/\partial \alpha_n$, $k = 3, \ldots, n$, linear combinations of variables have to be made.

The two theorems demonstrate that, at most, 2n-1 *independent signals* are needed to specify all the elements of the sensitivity matrix, and that all of them can be found by measuring the signals at the nodes of the original system model and the sensitivity model. This is a generalization of Kokotović's method of sensitivity points.

Example 5.6-1 For the system of Example 5.5-3 described by the state equations

$$\begin{aligned}
\dot{z}_1 &= z_2, & z_1(0) &= 0,\\
\dot{z}_2 &= z_3, & z_2(0) &= 0,\\
\dot{z}_3 &= -\alpha_1 z_1 - \alpha_2 z_2 - \alpha_3 z_3 + u, & z_3(0) &= 0,\\
y &= z_1,
\end{aligned} \tag{5.6-32}$$

the signal flow diagram is to be drawn from which all the trajectory sensitivity functions can be obtained.

Solution Taking the partial derivatives with respect to one of the coefficients, say α_1, we obtain

$$\begin{aligned}
\frac{\partial \dot{z}_1}{\partial \alpha_1} &= \frac{\partial z_2}{\partial \alpha_1},\\
\frac{\partial \dot{z}_2}{\partial \alpha_1} &= \frac{\partial z_3}{\partial \alpha_1},\\
\frac{\partial \dot{z}_3}{\partial \alpha_1} &= -\alpha_1 \frac{\partial z_1}{\partial \alpha_1} - \alpha_2 \frac{\partial z_2}{\partial \alpha_1} - \alpha_3 \frac{\partial z_3}{\partial \alpha_1} - z_1.
\end{aligned} \tag{5.6-33}$$

According to this result, the sensitivity functions

$$\left.\frac{\partial z_1}{\partial \alpha_1}\right|_{\alpha 0}, \qquad \left.\frac{\partial z_2}{\partial \alpha_1}\right|_{\alpha 0}, \qquad \left.\frac{\partial z_3}{\partial \alpha_1}\right|_{\alpha 0}$$

can be obtained at the nodes of the signal flow diagram shown in Fig. 5.6-2.

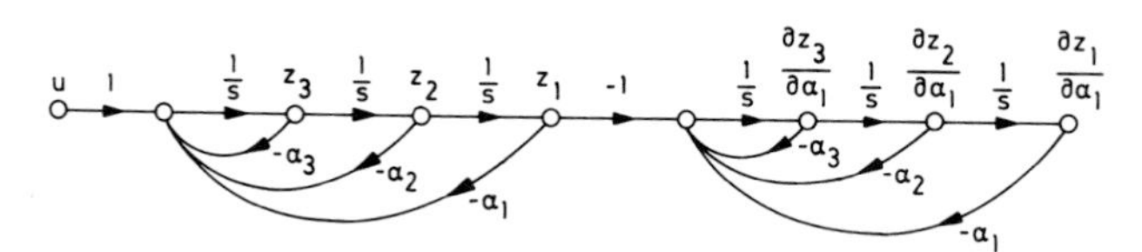

FIG. 5.6-2. Signal flow diagram to determine the trajectory sensitivity functions with respect to α_1.

By virtue of Theorems 5.6-1 and 5.6-2, it should be possible to determine the remaining six trajectory sensitivity functions (with respect to α_2 and α_3) from the same diagram without inserting additional dynamic elements. If we take the derivatives with respect to α_2, we obtain

$$\begin{aligned}
\frac{\partial \dot{z}_1}{\partial \alpha_2} &= \frac{\partial z_2}{\partial \alpha_2},\\
\frac{\partial \dot{z}_2}{\partial \alpha_2} &= \frac{\partial z_3}{\partial \alpha_2},\\
\frac{\partial \dot{z}_3}{\partial \alpha_2} &= -\alpha_1 \frac{\partial z_1}{\partial \alpha_2} - \alpha_2 \frac{\partial z_2}{\partial \alpha_2} - \alpha_3 \frac{\partial z_3}{\partial \alpha_2} - z_2.
\end{aligned} \tag{5.6-34}$$

The only difference between this sensitivity model and the one obtained earlier is the input signal, which is now z_2 instead of z_1. However, since $z_2 = \dot{z}_1$, the signals of the latter sensitivity model are the derivatives of the corresponding signals of the former. Hence

$$\frac{\partial z_1}{\partial \alpha_2} = \frac{\partial \dot{z}_1}{\partial \alpha_1} = \frac{\partial z_2}{\partial \alpha_1},$$
$$\frac{\partial z_2}{\partial \alpha_2} = \frac{\partial \dot{z}_2}{\partial \alpha_1} = \frac{\partial z_3}{\partial \alpha_1}, \tag{5.6-35}$$
$$\frac{\partial z_3}{\partial \alpha_2} = \frac{\partial \dot{z}_3}{\partial \alpha_1}.$$

Analogously, one obtains for the derivatives with respect to α_3,

$$\frac{\partial \dot{z}_1}{\partial \alpha_3} = \frac{\partial z_2}{\partial \alpha_3},$$
$$\frac{\partial \dot{z}_2}{\partial \alpha_3} = \frac{\partial z_3}{\partial \alpha_3}, \tag{5.6-36}$$
$$\frac{\partial \dot{z}_3}{\partial \alpha_3} = -\alpha_1 \frac{\partial z_1}{\partial \alpha_3} - \alpha_2 \frac{\partial z_2}{\partial \alpha_3} - \alpha_3 \frac{\partial z_3}{\partial \alpha_3} - z_3.$$

In this case the input signal is $z_3 = \ddot{z}_1$, which implies that the signals of this sensitivity model are the second derivatives of the corresponding signals of the first sensitivity model, or the first derivatives of those of the second model. Hence

$$\frac{\partial z_1}{\partial \alpha_3} = \frac{\partial \dot{z}_1}{\partial \alpha_2} = \frac{\partial z_2}{\partial \alpha_2} = \frac{\partial z_3}{\partial \alpha_1}, \tag{5.6-37}$$

$$\frac{\partial z_2}{\partial \alpha_3} = \frac{\partial \dot{z}_2}{\partial \alpha_2} = \frac{\partial z_3}{\partial \alpha_2}. \tag{5.6-38}$$

Finally, for the last function $\partial z_3/\partial \alpha_3$ we obtain

$$\frac{\partial z_3}{\partial \alpha_3} = \frac{\partial \dot{z}_3}{\partial \alpha_2} = -\alpha_1 \frac{\partial z_1}{\partial \alpha_2} - \alpha_2 \frac{\partial z_2}{\partial \alpha_2} - \alpha_3 \frac{\partial z_3}{\partial \alpha_2} - z_2. \tag{5.6-39}$$

Thus all sensitivity functions except the last one can be obtained directly as the signals in the diagram of Fig. 5.6-2. This is illustrated by Fig. 5.6-3. From the same diagram, the last sensitivity function is obtained according to Eq.(5.6-31), as shown in Fig. 5.6-4.

The results of this section can be summarized as follows: For a system in Frobenius canonical form and zero initial conditions, all n^2 trajectory sensitivity functions of the sensitivity matrix can be expressed by $2n - 1$ independent functions. These functions can be obtained simultaneously from a nominal system model and a single sensitivity model. Of these signals, $n + 1$ can be obtained directly at the nodes of the sensitivity model. The remaining $n - 2$

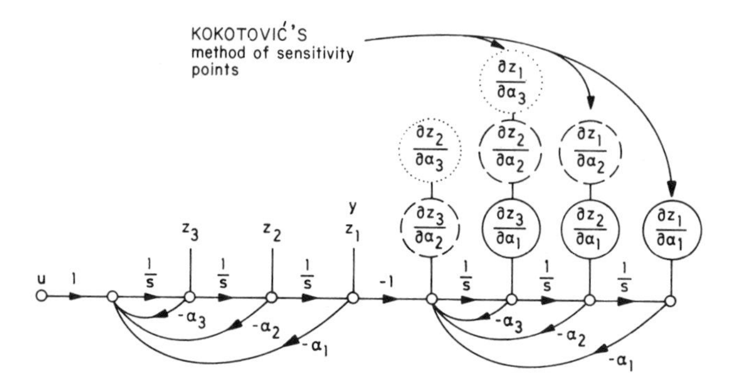

FIG. 5.6-3. Illustration of the result of Example 5.6-1.

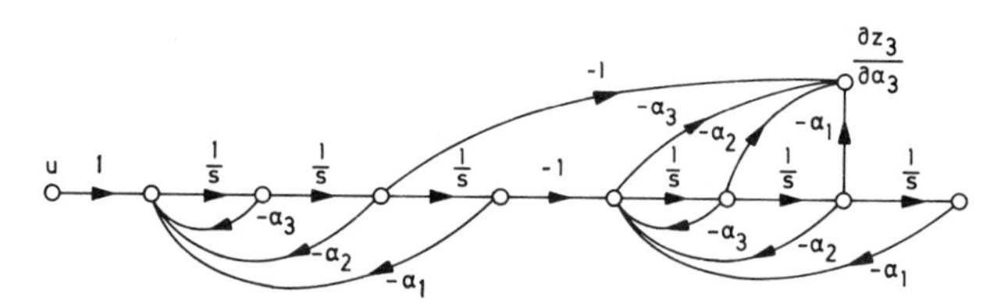

FIG. 5.6-4. Generation of $\partial z_3/\partial \alpha_3$ of Example 5.6-1.

signals are composed of the directly measurable sensitivity functions and the state variables of the nominal original system according to Eq. (5.6-31). It was shown by Guardabassi *et al.* [41] that the Wilkie-Perkins sensitivity model is of minimal order under the given circumstances. They have also proved that this sensitivity model is uncontrollable [42, 43].

In the next section, it will be shown how this procedure can be extended to single input linear time-invariant controllable systems with zero initial conditions in arbitrary state representation.

5.7 SIMULTANEOUS MEASUREMENT OF ALL TRAJECTORY SENSITIVITY FUNCTIONS OF GENERAL LINEAR SYSTEMS

In this section the methods developed in Section 5.6 for single-input linear time-invariant systems in companion form are generalized to systems described by a set of arbitrary state equations.

Suppose that the state equations of a single-input linear time-invariant controllable system are given by

$$\dot{\boldsymbol{x}} = \tilde{\boldsymbol{A}}\boldsymbol{x} + \tilde{\boldsymbol{b}}u, \qquad \boldsymbol{x}(0) = \boldsymbol{0}, \tag{5.7-1}$$
$$y = \tilde{\boldsymbol{c}}\boldsymbol{x},$$

where the matrix $\tilde{A}$ is not in Frobenius canonical form, i.e., the x_i's may be any set of state variables. The matrices $\tilde{A}, \tilde{b}, \tilde{c}$ may depend on a general parameter vector $a = [a_1\ a_2\ .\ .\ .\ a_r]^T$. Then the transform T that brings the above state equation into the companion form is also a function of a, i.e.,

$$T = T(a). \tag{5.7-2}$$

Furthermore, the elements $\alpha_1, \alpha_2, \ldots, \alpha_n$ of the A matrix of the companion form obtained from the transformation

$$A = T^{-1}\tilde{A}T \tag{5.7-3}$$

are also functions of a, that is,

$$\alpha(a) = [\alpha_1(a) \quad \alpha_2(a) \quad \cdots \quad \alpha_n(a)]^T. \tag{5.7-4}$$

This vector represents the n dimensional parameter vector after the state equations have been brought into the companion form. Whatever the dimension of the parameter vector a is, the parameter vector affecting the state z of the companion form is always given by α. Therefore the α_i's are called the *essential parameters*. Now the following theorem holds:

Theorem 5.7-1 The sensitivity functions of all the state variables x_i of Eqs. (5.7-1) with respect to all r parameters ($r = n$) can be obtained as linear combinations of the sensitivity functions $\partial z_i/\partial\alpha_j (i, j = 1, 2, \ldots, n)$ of the state equations in companion form and the state variables x_i.

Proof Since $x = Tz$ and $T = T(a)$, for any parameter a_j,

$$\begin{aligned} \frac{\partial x}{\partial a_j} &= \frac{\partial T(a)}{\partial a_j} z + T(a)\frac{\partial z}{\partial \alpha}\frac{\partial \alpha}{\partial a_j} \\ &= \frac{\partial T(a)}{\partial a_j} T^{-1}(a)x + T(a)\frac{\partial z}{\partial \alpha}\frac{\partial \alpha}{\partial a_j}. \end{aligned} \tag{5.7-5}$$

This proves the theorem.

If Theorem 5.7-1 is combined with the theorems of Wilkie and Perkins, the following theorem can be formulated:

Theorem 5.7-2 For a linear time-invariant single-input single-output system in arbitrary state description, all trajectory sensitivity functions with respect to all r parameters can be obtained by an algebraic combination of signals generated simultaneously by a nominal system model and a single sensitivity model in companion form.

This is illustrated in Fig. 5.7-1. To evaluate the above equations on a digital computer, the expressions $T(a)$, $\partial T/\partial a_j$ and $\partial\alpha_k/\partial a_j$, $k = 1, 2, \ldots, n$, $j = 1, 2\ \ldots, r$, must be obtained. The utility of this method highly depends on whether the above calculations can be carried out without finding T and the α_k as analytical expressions of a. It was shown by Wilkie and Perkins that

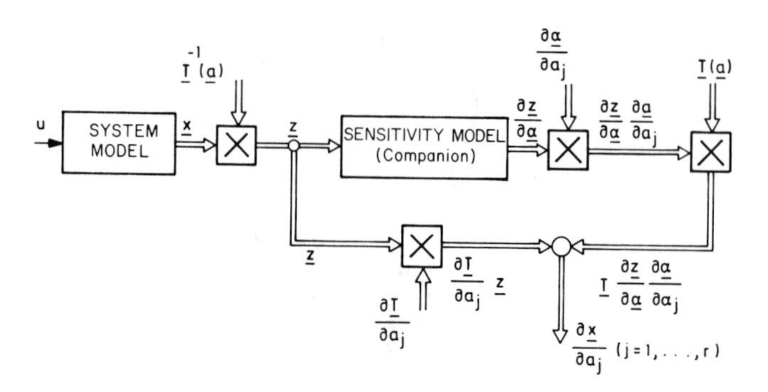

FIG. 5.7-1. Structural interpretation of Theorem 5.7-2.

the expressions $\partial \boldsymbol{T}/\partial a_j$ and $\partial \alpha_k/\partial a_j$ for any set of parameters $\boldsymbol{a}$ can be obtained recursively by an extension of the Leverrier algorithm.

As is well known from the theory of state space, the transformation $\boldsymbol{T}$ can be calculated iteratively by the relation

$$\boldsymbol{T} = [\boldsymbol{t}_1 \vdots \boldsymbol{t}_2 \vdots \ldots \vdots \boldsymbol{t}_n], \tag{5.7-6}$$

where

$$\begin{aligned}
\boldsymbol{t}_n &= \boldsymbol{b}, \\
\boldsymbol{t}_{n-1} &= \boldsymbol{A}\boldsymbol{t}_n + \alpha_n \boldsymbol{t}_n, \\
\boldsymbol{t}_{n-2} &= \boldsymbol{A}\boldsymbol{t}_{n-1} + \alpha_{n-1} \boldsymbol{t}_n, \\
&\vdots \\
\boldsymbol{t}_1 &= \boldsymbol{A}\boldsymbol{t}_2 + \alpha_2 \boldsymbol{t}_n.
\end{aligned}$$

The coefficients $\alpha_2, \ldots, \alpha_n$ required in the above procedure can be obtained by Leverrier's algorithm:

$$\begin{aligned}
\alpha_{n+1} &= 1, \\
\boldsymbol{S}_{n+1} &= \boldsymbol{I} \qquad (n \times n \text{ identity matrix}), \\
\alpha_{n-i+1} &= -\frac{1}{i} \operatorname{tr}(\boldsymbol{A}\boldsymbol{S}_{n-i+2}), \\
\boldsymbol{S}_{n-i+1} &= \alpha_{n-i+1} \boldsymbol{I} + \boldsymbol{A}\boldsymbol{S}_{n-i+2}.
\end{aligned} \tag{5.7-7}$$

Note that $\boldsymbol{S}_1 = \boldsymbol{0}$. This can be used as a check on the correctness of the calculations. Using this procedure, the transformation to the companion form can easily be accomplished on a digital computer. Taking the partial derivatives of Eqs. (5.7-6) with respect to $\boldsymbol{a}_j$, one obtains

$$\frac{\partial \boldsymbol{T}}{\partial a_j} = \left[\frac{\partial \boldsymbol{t}_1}{\partial a_j} \vdots \frac{\partial \boldsymbol{t}_2}{\partial a_j} \vdots \ldots \vdots \frac{\partial \boldsymbol{t}_n}{\partial a_j}\right],$$

$$
\begin{aligned}
\frac{\partial \boldsymbol{t}_n}{\partial a_j} &= \frac{\partial \boldsymbol{b}}{\partial a_j}, \\
\frac{\partial \boldsymbol{t}_{n-1}}{\partial a_j} &= \frac{\partial \boldsymbol{A}}{\partial a_j}\boldsymbol{t}_n + \boldsymbol{A}\frac{\partial \boldsymbol{t}_n}{\partial a_j} + \frac{\partial \alpha_n}{\partial a_j}\boldsymbol{b} + \alpha_n\frac{\partial \boldsymbol{b}}{\partial a_j}, \\
\frac{\partial \boldsymbol{t}_{n-2}}{\partial a_j} &= \frac{\partial \boldsymbol{A}}{\partial a_j}\boldsymbol{t}_{n-1} + \boldsymbol{A}\frac{\partial \boldsymbol{t}_{n-1}}{\partial a_j} + \frac{\partial \alpha_{n-1}}{\partial a_j}\boldsymbol{b} + \alpha_{n-1}\frac{\partial \boldsymbol{b}}{\partial a_j}, \\
&\vdots \\
\frac{\partial \boldsymbol{t}_1}{\partial a_j} &= \frac{\partial \boldsymbol{A}}{\partial a_j}\boldsymbol{t}_2 + \boldsymbol{A}\frac{\partial \boldsymbol{t}_2}{\partial a_j} + \frac{\partial \alpha_2}{\partial a_j}\boldsymbol{b} + \alpha_2\frac{\partial \boldsymbol{b}}{\partial a_j}.
\end{aligned}
\tag{5.7-8}
$$

Taking corresponding partial derivatives of the Leverrier algorithm, Eq. (5.7-7), one obtains

$$
\begin{aligned}
\frac{\partial \alpha_{n+1}}{\partial a_j} &= 0, \\
\frac{\partial \boldsymbol{S}_{n+1}}{\partial a_j} &= \boldsymbol{0}, \\
\frac{\partial \alpha_{n-i+1}}{\partial a_j} &= -\frac{1}{i}\operatorname{tr}\left[\frac{\partial \boldsymbol{A}}{\partial a_j}\boldsymbol{S}_{n-i+2} + \boldsymbol{A}\frac{\partial \boldsymbol{S}_{n-i+2}}{\partial a_j}\right], \\
\frac{\partial \boldsymbol{S}_{n-i+1}}{\partial a_j} &= \frac{\partial \alpha_{n-i+1}}{\partial a_j}\boldsymbol{I} + \frac{\partial \boldsymbol{A}}{\partial a_j}\boldsymbol{S}_{n-i+2} + \boldsymbol{A}\frac{\partial \boldsymbol{S}_{n-i+2}}{\partial a_j}.
\end{aligned}
\tag{5.7-9}
$$

Thus if the dependence of the matrices $\boldsymbol{A}$ and $\boldsymbol{b}$ on $\boldsymbol{a}$ is known, so that $\partial \boldsymbol{A}/\partial a_j$ and $\partial \boldsymbol{b}/\partial a_j$ can be determined, only $\partial \alpha_k/\partial a_j$ $(k = 2, 3, \ldots, n)$ need be found to calculate the expressions $\partial \boldsymbol{T}/\partial a_j$ by Eqs. (5.7-8) as easily as $\boldsymbol{T}$. The expressions $\partial \alpha_k/\partial a_j$, however, can be calculated by the extended Leverrier algorithm according to Eq. (5.7-9). Thus by using Eqs. (5.7-6)–(5.7-9), all the functions $\boldsymbol{T}(\boldsymbol{a})$, $\partial \boldsymbol{T}/\partial a_j$, and $\partial \alpha_k/\partial a_j$ $(k = 1, 2, \ldots, n, j = 1, 2, \ldots, r)$ required in the general structural diagram for the calculation of the sensitivity vectors $\boldsymbol{\lambda}_j$ can be calculated without knowing the analytical dependence of $\boldsymbol{T}$ on $\boldsymbol{a}$.

This method of calculating the sensitivity vectors $\boldsymbol{\lambda}_j$ has also been extended to linear time-invariant normal (i.e., controllable from any input) multi-input systems as well as to more general systems and systems with dead time [34, 52, 73, 107]. The benefit of this method is a substantial reduction in the complexity of the structure (in the case of an analog computer application) or in the number of necessary computations (in the case of a digital computer application); this becomes important in the study of high-order systems.

5.8 DETERMINATION OF EIGENVALUE SENSITIVITY

The methods commonly used for system representation can be unsatisfactory or even useless for the representation of large multiple-loop systems. In this case the description in terms of eigenvalues has proven to be a most useful approach. Evidently, the sensitivity of the system should then be expressed in terms of the eigenvalue or eigenvector sensitivity defined in Section 3.2.8.

Eigenvalue sensitivity can be regarded from two points of view, namely of time or frequency domain, depending on whether the mathematical model is given in the time or frequency domain. In the latter case the term "root sensitivity" will be preferred. In this section, formulas for the determination of the eigenvalue sensitivity are developed with respect to parameters of the state space representation. This problem was recently treated by several authors [63, 64, 65, 70, 82, 88, 103].

Let us first derive a relation for the semirelative eigenvalue sensitivity in terms of eigenvectors. Referring to Section 3.2.8, the free motion of a system can be described by the vector state equation

$$\dot{\boldsymbol{x}} = \boldsymbol{A}\boldsymbol{x}. \tag{5.8-1}$$

If the matrix $\boldsymbol{A}$ has distinct eigenvalues λ_i, the solution is

$$\boldsymbol{x} = C\boldsymbol{v}_i\, e^{\lambda_i t}, \tag{5.8-2}$$

where C is a scalar and $\boldsymbol{v}_i$ is an eigenvector of $\boldsymbol{A}$. Substituting the solution into the original state equation, we have

$$\boldsymbol{A}\boldsymbol{v}_i = \lambda_i \boldsymbol{v}_i. \tag{5.8-3}$$

Analogously, denoting the eigenvector of the transposed matrix $\boldsymbol{A}^{\mathrm{T}}$ by $\boldsymbol{w}_k$ (*adjoint eigenvector*), we also have

$$\boldsymbol{w}_k^{\mathrm{T}}\boldsymbol{A} = \lambda_k \boldsymbol{w}_k^{\mathrm{T}}. \tag{5.8-4}$$

The matrix $\boldsymbol{v}$ of eigenvectors and the matrix $\boldsymbol{w}$ of adjoint eigenvectors are related through $(\boldsymbol{v}^{\mathrm{T}})^{-1} = \boldsymbol{w}$, and the relation $\boldsymbol{v}_i^{\mathrm{T}}\boldsymbol{w}_j = \boldsymbol{w}_j^{\mathrm{T}}\boldsymbol{v}_i = \delta_{ij}$ holds, where δ_{ij} is the Kronecker delta, and $i, j = 1, 2, \ldots, n$.

If now any set of parameters $\boldsymbol{\alpha}$, in Eq. (5.8-1) changes, this affects the generic elements a_j of $\boldsymbol{A}$, the eigenvalues λ_i, and the eigenvectors $\boldsymbol{v}_i$ and $\boldsymbol{w}_k$ as well. Thus partial differentiation of Eq. (5.8-3) with respect to the parameters α_j $(j = 1, 2, \ldots, r)$ gives

$$\frac{\partial \boldsymbol{A}}{\partial \alpha_j}\boldsymbol{v}_i + \boldsymbol{A}\frac{\partial \boldsymbol{v}_i}{\partial \alpha_j} = \frac{\partial \lambda_i}{\partial \alpha_j}\boldsymbol{v}_i + \lambda_i \frac{\partial \boldsymbol{v}_i}{\partial \alpha_j}. \tag{5.8-5}$$

Premultiplication of Eq. (5.8–5) by $\boldsymbol{w}_i^{\mathrm{T}}$ yields

$$\boldsymbol{w}_i^{\mathrm{T}}\frac{\partial \boldsymbol{A}}{\partial \alpha_j}\boldsymbol{v}_i + \boldsymbol{w}_i^{\mathrm{T}}\boldsymbol{A}\frac{\partial \boldsymbol{v}_i}{\partial \alpha_j} = \boldsymbol{w}_i^{\mathrm{T}}\frac{\partial \lambda_i}{\partial \alpha_j}\boldsymbol{v}_i + \boldsymbol{w}_i^{\mathrm{T}}\lambda_i \frac{\partial \boldsymbol{v}_i}{\partial \alpha_j}. \tag{5.8-6}$$

Using Eq. (5.8–4) for $k = i$ and multiplying by the nominal parameter value α_{j0}, we have

$$\frac{\partial \lambda_i}{\partial \ln \alpha_j} = \frac{\boldsymbol{w}_i^{\mathrm{T}}(\partial \boldsymbol{A}/\partial \alpha_j)\,\boldsymbol{v}_i}{\boldsymbol{w}_i^{\mathrm{T}}\boldsymbol{v}_i}\,\alpha_{j0}, \qquad \begin{array}{l} i = 1, 2, \ldots, n, \\ j = 1, 2, \ldots, r. \end{array} \tag{5.8-7}$$

This formula has proved to be very useful for finding the semirelative eigenvalue sensitivities with respect to variations of any set of parameters in the state equations. In the particular case that the a_{jl} themselves are the parameters of interest, the above formula reduces to

$$\frac{\partial \lambda_i}{\partial \ln a_{jl}} = w_{ij} v_{il} a_{jl0}, \qquad i, k, l = 1, 2, \ldots, n, \tag{5.8-8}$$

where w_{ij} and v_{il} are, respectively, the jth component of w_i and the lth element of v_i and the a_{jl0} are the nominal values of a_{jl}. The $\partial\lambda_i/\partial a_{jl}$ are the desired semirelative eigenvalue sensitivities relating the changes in the eigenvalues λ_i to percentage changes in the elements a_{jl} of the system matrix A. Note that the above formulas have to be evaluated for nominal parameter values. Once the eigenvector v_i and the adjoint eigenvector w_i are known, the sensitivities are readily determined from the above formulas.

The corresponding absolute eigenvalue sensitivities may be viewed as the elements of a set of n eigenvalue sensitivity matrices

$$S_i = \left[\frac{\partial \lambda_i}{\partial a_{jl}}\right] = w_i v_i^{\mathrm{T}}, \qquad i, j, l = 1, 2, \ldots, n. \tag{5.8-9}$$

It can be shown [29] that

$$S_i S_k = \delta_{ik} S_i, \qquad (i, k = 1, 2, \ldots, n), \tag{5.8-10}$$

where δ_{ik} is the Kronecker delta, and

$$\sum_{i=1}^{n} S_i = I.$$

These properties of S_i may be useful both in theory and in checking numerical calculations.

A more suitable form for computing the eigenvalue sensitivities is obtained from Eq. (5.8-7) by a series of matrix operations [70]:

$$\frac{\partial \lambda_i}{\partial \ln \alpha_j} = \alpha_{j0} \left. \frac{\operatorname{tr}[\operatorname{Adj}(A - \lambda I)]\, \partial A/\partial \alpha_j}{\operatorname{tr}[\operatorname{Adj}(A - \lambda I)]} \right|_{\substack{\alpha=\alpha_0 \\ \lambda=\lambda_i}}. \tag{5.8-11}$$

Whereas Eq. (5.8–7) is most useful for calculating the eigenvalue sensitivities for the state equations in canonical form, Eq. (5.8-11) should be preferred in the case of arbitrary state equations. Note that the matrix $\operatorname{Adj}(A - \lambda I)$ can be efficiently computed by means of the Leverrier algorithm.

The above formula has been used by Lee and Chen [63] to show that the Jordan canonical form of the state equations is least sensitive with respect to parameter variations and is therefore the best for analog simulation of a system.

Another formula for the calculation of the semirelative eigenvalue sensitivities $\partial\lambda_i/\partial \ln a_{jk}$, where a_{jk} are the generic elements of the system matrix A, will be developed now. Recall that λ_i is a root of the characteristic equation (3.2–9),

$$f(\lambda, A) = \det[\lambda I - A] = 0. \tag{5.8-12}$$

Other useful forms of this equation are

$$f(\lambda, A) = \prod_{\nu=1}^{n} [\lambda - \lambda_\nu(A)] = \lambda^n + \sum_{\mu=1}^{n} \alpha_\mu(A)\lambda^{\mu-1}. \tag{5.8-13}$$

Taking the partial derivative of f with respect to $\ln a_{jk}$ gives

$$\frac{\partial f}{\partial \lambda}\frac{\partial \lambda}{\partial \ln a_{jk}} + \frac{\partial f}{\partial \ln a_{jk}} = 0, \qquad \lambda = \lambda_i, \tag{5.8-14}$$

or

$$\frac{\partial \lambda_i}{\partial \ln a_{jk}} = -\left.\frac{\partial f/\partial \ln a_{jk}}{\partial f/\partial \lambda}\right|_{\lambda=\lambda_i}. \tag{5.8-15}$$

Using Eq. (5.8-13), we obtain

$$\frac{\partial \lambda_i}{\partial \ln a_{jk}} = -\frac{\sum_{\mu=1}^{n} (\partial \alpha_\mu/\partial \ln a_{jk})\, \lambda_i^{\mu-1}}{\prod_{\nu=1,\nu\neq i}^{n} (\lambda_i - \lambda_\nu)}. \tag{5.8-16}$$

This is another formula to calculate the change of the ith eigenvalue due to a percentage change of one of the elements of $\boldsymbol{A}$, avoiding the knowledge of the eigenvectors and adjoint eigenvectors. Notice that all equations have to be evaluated for nominal parameter values.

As a special case, consider the system matrix as being in companion form, that is,

$$\boldsymbol{A} = \begin{bmatrix} 0 & 1 & \cdots & 0 \\ \vdots & & \ddots & \vdots \\ 0 & & & 1 \\ -\alpha_1 & -\alpha_2 & & -\alpha_n \end{bmatrix}. \tag{5.8-17}$$

Provided that all zero and unity elements are invariant, the only elements of $\boldsymbol{A}$ that can vary are the α_μ. Therefore the formula for the eigenvector sensitivity [Eq. (5.8–16)] reduces to

$$\frac{\partial \lambda_i}{\partial \ln \alpha_\mu} = -\frac{\lambda_i^{\mu-1}\alpha_\mu}{\prod_{\nu=1,\nu\neq i}^{n}(\lambda_i - \lambda_\nu)} \triangleq \frac{-\lambda_i^{\mu-1}\alpha_\mu}{f_\lambda(\lambda_i)}, \qquad \mu = 1, 2, \ldots, n. \tag{5.8-18}$$

Using this, the summed semirelative eigenvalue sensitivity, defined by Definition 3.2–14, becomes

$$\tilde{S}_A^{\lambda_i} = \lim_{\Delta\to 0} \frac{1}{\Delta} \operatorname*{supc}_{|\Delta \ln \alpha_\mu| \leq \Delta} \frac{1}{f_\lambda(\lambda_i)} \sum_{\mu=0}^{n-1} (-\lambda_i^\mu \, \Delta\alpha_{\mu+1}). \tag{5.8-19}$$

When λ_i is real, we have

$$\tilde{S}_A^{\lambda_i} = \frac{1}{|f_\lambda(\lambda_i)|} \sum_{\mu=0}^{n-1} |\lambda_i^\mu \alpha_{\mu+1}|. \tag{5.8-20}$$

For the corresponding absolute sensitivity we can write

$$S_A^{\lambda_i} = \frac{1 - |\lambda_i|^n}{(1 - |\lambda_i|)\Pi_{\nu=1,\nu\neq i}^{n}|(\lambda_i - \lambda_\nu)|}. \tag{5.8-21}$$

For the extension of the above sensitivity measure to complex eigenvalues see Singer [97]. An application of eigenvector sensitivity to the determination of the least sensitive mathematical model for the simulation on an analog computer is given by Lee and Chen [63].

Example 5.8-1 For the system of Example 3.2-8 with the state equation

$$\dot{\boldsymbol{x}} = \begin{bmatrix} 0 & 1 \\ -\alpha_1 & -\alpha_2 \end{bmatrix} \boldsymbol{x}, \qquad \alpha_{10} = 3, \quad \alpha_{20} = 4,$$

determine the semirelative eigenvalue sensitivities and the summed semirelative eigenvalue sensitivities using the above formulas.

Solution The (nominal) eigenvalues are $\lambda_{10} = -1$, $\lambda_{20} = -3$. Substituting these values into Eq. (5.8-18) yields the required eigenvalue sensitivities

$$\left.\frac{\partial \lambda_1}{\partial \ln \alpha_1}\right|_{\boldsymbol{\alpha}0} = -\frac{\lambda_{10}^0 \alpha_{10}}{(\lambda_{10} - \lambda_{20})} = -\frac{3}{2},$$

$$\left.\frac{\partial \lambda_1}{\partial \ln \alpha_2}\right|_{\boldsymbol{\alpha}0} = -\frac{\lambda_{10} \alpha_{20}}{(\lambda_{10} - \lambda_{20})} = -\frac{-1 \times 4}{2} = 2,$$

$$\left.\frac{\partial \lambda_2}{\partial \ln \alpha_1}\right|_{\boldsymbol{\alpha}0} = -\frac{\lambda_{20}^0 \alpha_{10}}{(\lambda_{20} - \lambda_{10})} = -\frac{3}{-2} = \frac{3}{2},$$

$$\left.\frac{\partial \lambda_2}{\partial \ln \alpha_2}\right|_{\boldsymbol{\alpha}0} = -\frac{\lambda_{20} \alpha_{20}}{(\lambda_{20} - \lambda_{10})} = -\frac{-3 \times 4}{-2} = -6.$$

Using Eq. (5.8-20), the summed semirelative eigenvalue sensitivities become

$$\tilde{S}_A^{\lambda_1} = \frac{|\alpha_{10}| + |\lambda_{10}\alpha_{20}|}{|\lambda_{10} - \lambda_{20}|} = \frac{7}{2},$$

$$\tilde{S}_A^{\lambda_2} = \frac{|\alpha_{10}| + |\lambda_{20}\alpha_{20}|}{|\lambda_{20} - \lambda_{10}|} = \frac{15}{2}.$$

The results are, of course, in agreement with the solutions of Example 3.2-8.

PROBLEMS

5.1 Consider a point mass m falling in a vacuum by constant gravitational attraction g. Let $\boldsymbol{x} = [y\ \dot{y}]^{\mathrm{T}}$ be considered as the state of the system. Determine the trajectory sensitivity vectors with respect to m, $\beta_1 = y(t_0)$, $\beta_2 = \dot{y}(t_0)$ assuming that the nominal values of m, β_1, β_2 are nonzero.

5.2 The state equations describing a single-axis reaction-wheel attitude control system are

$$\dot{\theta} = \omega,$$

$$\dot{\omega} = \frac{1}{J_F}\left[D + \frac{K_1^{\,2}}{R_a}\right]\Omega - \frac{K_1}{J_F R_a} v_a,$$

$$\dot{\Omega} = -\left[\frac{1}{J_F} + \frac{1}{J_w}\right]\left[\frac{K_1^{\,2}}{R_a} + D\right]\Omega + \frac{K_1}{R_a}\left[\frac{1}{J_F} + \frac{1}{J_w}\right] v_a,$$

where θ is the attitude angle of the space vehicle, ω is the angle velocity, Ω is the relative speed between the reaction-wheel and the vehicle, and v_a is the voltage of the motor of the reaction-wheel (input signal of the control system). The remaining symbols represent constants of the system. Determine the trajectory sensitivity equations of the system with respect to F_v and draw a block diagram to measure $\lambda_1 = (\partial\theta/\partial J_v)_{J_{F0}}$.

5.3 Consider the circuit shown in Fig. 5.P-1. Let the state of the circuit be defined by the voltages of the two energy storage elements $\boldsymbol{x} = [x_1\ x_2]^{\mathrm{T}}$; con-

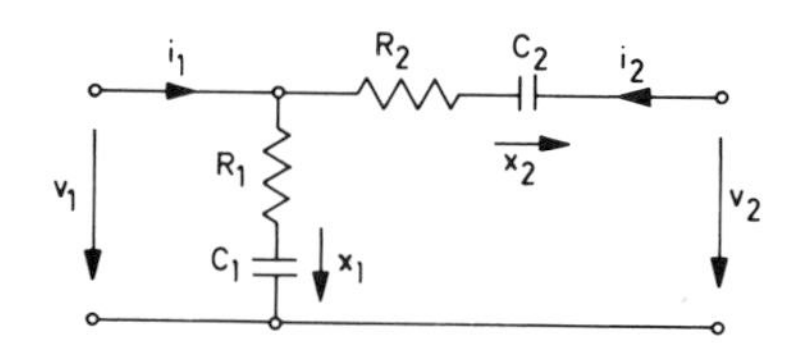

FIG. 5.P-1. Given network of Problem 5.3.

sider $\boldsymbol{v} = [v_1\ v_2]^{\mathrm{T}}$ as the input vector and $\boldsymbol{i} = [i_1\ i_2]^{\mathrm{T}}$ as the output vector of the system. In these terms, the state equations read

$$\dot{\boldsymbol{x}} = \boldsymbol{A}\boldsymbol{x} + \boldsymbol{B}\boldsymbol{v},$$

$$\boldsymbol{i} = \boldsymbol{C}\boldsymbol{x} + \boldsymbol{D}\boldsymbol{v},$$

where

$$\boldsymbol{A} = \begin{bmatrix} -\dfrac{1}{R_1C_1} & 0 \\ 0 & -\dfrac{1}{R_2C_2} \end{bmatrix}, \qquad \boldsymbol{B} = \begin{bmatrix} \dfrac{1}{R_1C_1} & 0 \\ \dfrac{1}{R_2C_2} & -\dfrac{1}{R_2C_2} \end{bmatrix}$$

$$\boldsymbol{C} = \begin{bmatrix} -\dfrac{1}{R_1} & -\dfrac{1}{R_2} \\ 0 & \dfrac{1}{R_2} \end{bmatrix}, \qquad \boldsymbol{D} = \begin{bmatrix} \dfrac{1}{R_1} + \dfrac{1}{R_2} & -\dfrac{1}{R_2} \\ -\dfrac{1}{R_2} & \dfrac{1}{R_2} \end{bmatrix}.$$

Give the state sensitivity equations with respect to

$$\boldsymbol{a} = [R_1 \quad C_1 \quad R_2 \quad C_2]^{\mathrm{T}}.$$

5.4 Given the third-order system shown in Fig. 5.P-2, where a, b, T, and k are constants. Suppose the constant T is small compared to $1/b$, so as to be neglected, and that the gain k is sufficiently high so that we can assume $k = \infty$.

(a) Draw the reduced block diagram for $T = 0$ and $k = \infty$.

(b) Write the state equations for the nonreduced and the reduced system using $1/\Delta\alpha$ instead of k and $\Delta\alpha$ instead of T with the nominal value of $\Delta\alpha$ being zero.

(c) Give the trajectory sensitivity equation with respect to $\Delta\alpha$.

(d) Determine the trajectory sensitivity functions $\lambda_1 = (\partial x_1/\partial\Delta\alpha)_0$, $\eta_1 = (\partial x_2/\partial\Delta\alpha)_0$, and $\eta_2 = (\partial x_3/\partial\Delta\alpha)_0$ for a unit step input (using the variable component method).

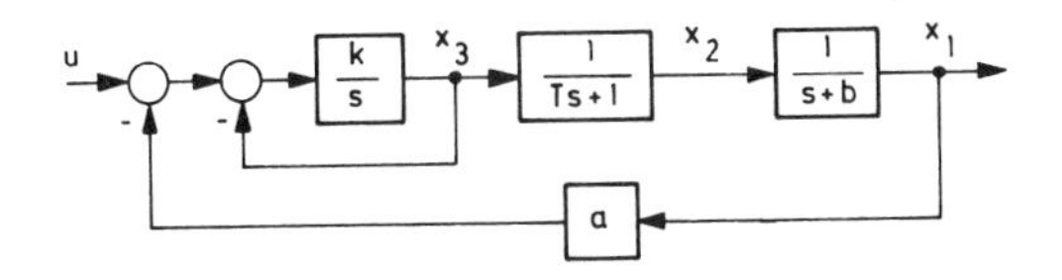

FIG. 5.P-2. Given system of Problem 5.4.

5.5 A nonlinear system with time-invariant parameters is described by the differential equations

$$\dot{x}_1 = x_2, \qquad x_1(t_0) = \beta_1,$$
$$\dot{x}_2 = -(1 - \alpha x_1^{\,2})x_2 - x_1 + u, \qquad x_2(t_0) = \beta_2,$$

where x_1, x_2 are the state variables and u is the input variable.

(a) Determine the trajectory sensitivity equation for $\boldsymbol{\lambda} = (\partial \mathbf{x}/\partial\alpha)_{\alpha_0 \neq 0}$.

(b) Draw a structural diagram for measuring $\lambda_1 = (\partial x_1/\partial\alpha)_{\alpha_0}$.

(c) Determine the sensitivity equation for the case in which the initial condition β_1 is the parameter of interest, i.e., for $\boldsymbol{\lambda} = (\partial \mathbf{x}/\partial\beta_1)_{\beta_{10}}$.

5.6 For the system described by the state equations

$$\dot{x} = z, \qquad x(0) = 0,$$
$$\alpha\dot{z} = -Tz - x + 1(t), \qquad z(0) = 0,$$

derive the trajectory sensitivity equation with respect to α with $\alpha_0 = 0$, using Vasileva's method.

5.7 For the armature controlled dc motor described by

$$\dot{x} = z, \qquad x(o) = 2,$$
$$\alpha\dot{z} = -x - z + u, \; z(0) = 2,$$

determine the trajectory sensitivity equation with respect to α for $\alpha_0 = 0$. Solve the sensitivity equation for a step input $u(t) = 1(t)$ using Vasileva's method.

5.8 Given an nth order single-input single-output process described by the state equations

$$\dot{\boldsymbol{x}} = \boldsymbol{A}\boldsymbol{x} + \boldsymbol{b}u, \qquad \boldsymbol{x}(0) = \boldsymbol{x}^0,$$

$$y = \boldsymbol{C}\boldsymbol{x} + du,$$

where $\boldsymbol{A} = \boldsymbol{A}(\boldsymbol{\alpha})$, $\boldsymbol{b} = \boldsymbol{b}(\boldsymbol{\alpha})$, $\boldsymbol{C} = \boldsymbol{C}(\boldsymbol{\alpha})$, $d = d(\boldsymbol{\alpha})$ are functions of a parameter vector $\boldsymbol{\alpha}$. Suppose that the state $\boldsymbol{x}$ is fed back such that the input obeys the law $u = r - \boldsymbol{K}\boldsymbol{x}$, where r is a reference input and $K = [k_1\ k_2\ .\ .\ .\ k_n]$ is a constant $1 \times n$ matrix ("linear state feedback").

(a) Determine the trajectory and vector output sensitivity equation with respect to one of the parameters α_j.

(b) Give a graphical representation of the result.

5.9 For a system described by the state equations

$$\dot{z}_1 = z_2, \qquad \dot{z}_2 = z_e, \qquad \dot{z}_3 = z_4,$$

$$\dot{z}_4 = -\alpha_1 z_1 - \alpha_2 z_2 - \alpha_3 z_3 - \alpha_4 z_4 + u,$$

with all initial conditions equal to zero, draw a complete signal flow graph for the simultaneous measurement of all trajectory sensitivity functions $\zeta_{ij} = (\partial z_i/\partial \alpha_j)_{\boldsymbol{\alpha}0}$.

5.10 For the system governed by the state equation $\dot{\boldsymbol{x}} = \boldsymbol{A}\boldsymbol{x} + \boldsymbol{b}u$ where

$$\boldsymbol{A} = \begin{bmatrix} 0 & 1 & 0 \\ 0 & 0 & 1 \\ -\alpha_1 & -\alpha_2 & -\alpha_3 \end{bmatrix}, \qquad \boldsymbol{b} = \begin{bmatrix} 0 \\ 0 \\ 1 \end{bmatrix},$$

determine the semirelative eigenvalue and summed eigenvalue sensitivities with respect to α_1,α_2,α_3. The nominal parameter values are $\alpha_{10} = 6$, $\alpha_{20} = 11$, $\alpha_{30} = 6$. Find the most sensitive eigenvalue to a given tolerance of α_1, α_2, α_3.

5.11 For the attitude control system of Example 5.4-1 containing a neuro pulse-frequency modulator, determine

(a) the trajectory sensitivity equations of $\boldsymbol{x} = [x_1, x_2\ p]^{\mathrm{T}}$ with respect to the feedback gains K_1, K_2 between switching instants.

(b) dt_i/dk_1 and dt_i/dk_2 for the cases (1) $p(t_i) > A$, $t_i - t_{i-1} = T_m$ and (2) $p(t_i) = A$, $t_i - t_{i-1} > T_m$;

(c) the jump conditions at $t = t_i$ and $t = t_i + \gamma$.

5.12 Given a position control system including a step motor whose block diagram is shown in Fig. 5.P–3.

(a) Derive the trajectory sensitivity equations of the state $\boldsymbol{x} = [x_1\ x_2\ x_3\ p]^T$ with respect to the gain factor K between the switching instants.

(b) Determine dt_i/dK for the following two cases: (1) $p(t_i) > A$, $t_i - t_{i-1} = T_m$, and (2) $p(t_i) = A$, $t_i - t_{i-1} > T_m$.

(c) Give the formulas for the jump conditions at t_i and $t_i + \gamma$.

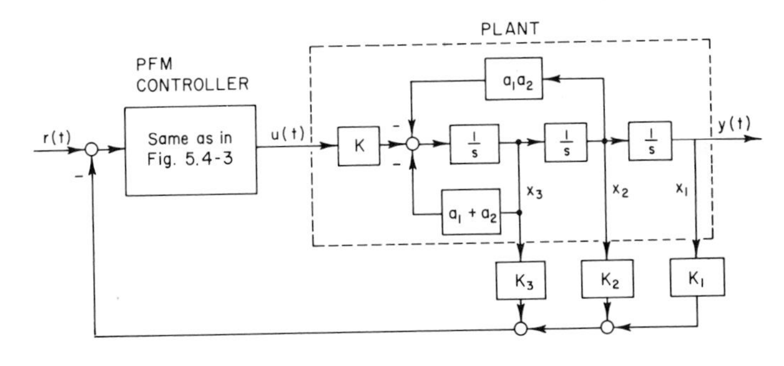

FIG. 5.P-3. Block diagram of the control loop of Problem 5.12.

Chapter 6

Determination of the Sensitivity Functions of the Frequency Domain

6.1 INTRODUCTION

This chapter presents methods for the determination of the sensitivity functions and measures of the frequency domain, such as the Bode, Horowitz, and comparison sensitivity functions and the root sensitivities.

In Section 3.3 we saw that there is a close connection between the Bode, Horowitz, and comparison sensitivity functions, the comparison sensitivity function being the most general of all. Therefore we will limit our consideration to the comparison sensitivity function with small time-invariant parameter variations.

In addition, some formulas for the determination of the root sensitivities of open- and closed-loop systems are presented. There are, of course, certain overlappings with the eigenvalue sensitivities which were treated at the end of Chapter 5. Whereas eigenvalue sensitivity was primarily discussed in view of the parameters of the state equations, root sensitivities will be discussed in terms of the parameters of the system representation in the frequency domain.

6.2 COMPARISON SENSITIVITY FUNCTION

Let us first derive a simulation scheme for the comparison sensitivity function for the linear case where the system under consideration is represented in terms of a transfer function or matrix. It was shown in Section 3.3 that the comparison sensitivity function of a closed-loop system of the form of Fig. 3.3-6 is given by

$$S_p(s, \boldsymbol{\alpha}) = [1 + P(s, \boldsymbol{\alpha})R(s)H(s)]^{-1}. \tag{6.2-1}$$

Since this expression is identical with the transference of a signal disturbance from the plant output to the output of the entire control loop, the comparison sensitivity function may be obtained from input (a) to output (b) of a nominal model of the closed-loop system as shown in Fig. 6.2-1.

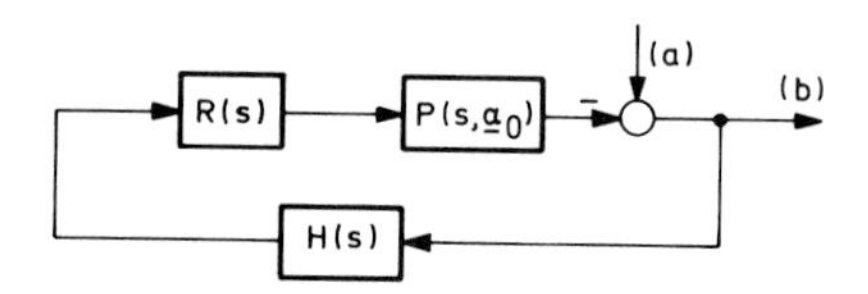

FIG. 6.2-1. Measurement of the comparison sensitivity function in the case of linear systems.

From Eq. (3.3-35) we observe that the closed-loop output error Δy_R is obtained at the output (b) in Fig. 6.2-1 as the output error of the nominally equivalent open-loop system Δy_S is applied to (a). Δy_S can be generated by the nominal system according to the relation $\Delta Y_S = \Delta P\ U_S$ where, in terms of Fig. 3.3-6,

$$U_S = \frac{Y_0}{P_0} = \frac{R}{1 + P_0 RH} W. \tag{6.2-2}$$

This leads to the complete block diagram for obtaining Δy_S and Δy_R simultaneously as shown in Fig. 6.2-2. Note that this structure is the same as that obtained by using the variable component method (Section 4.5).

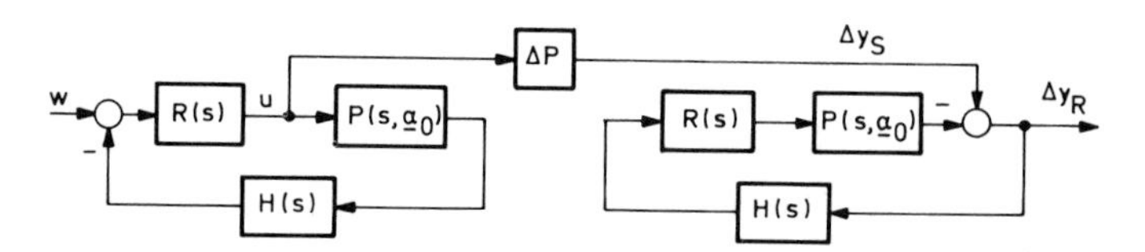

FIG. 6.2-2. Complete block diagram for obtaining Δy_S and Δy_R.

So far we have limited our consideration to linear time-invariant systems. Now consider the more general nonlinear time-varying systems represented by the state equations

$$\dot{\boldsymbol{x}} = \boldsymbol{f}(\boldsymbol{x}, \boldsymbol{u}, \boldsymbol{\alpha}, t), \tag{6.2-3}$$

$$\boldsymbol{y} = \boldsymbol{g}(\boldsymbol{x}, \boldsymbol{w}, \boldsymbol{\alpha}, t), \tag{6.2-4}$$

where $\boldsymbol{x}$ is the state, $\boldsymbol{w}$ the system input, $\boldsymbol{u}$ the plant input, and $\boldsymbol{\alpha}$ the parameter vector of interest. For nominal parameter values $\boldsymbol{\alpha}_0$ it is required that the open-loop and corresponding closed-loop systems are equivalent, that is, $\boldsymbol{u}_R(t, \boldsymbol{\alpha}_0) = \boldsymbol{u}_S(t, \boldsymbol{\alpha}_0)$ as the same input $\boldsymbol{w}$ is applied.

Let the control vector of the closed-loop system be given by

$$\boldsymbol{u}_R(t, \boldsymbol{\alpha}) = \boldsymbol{h}_R(\boldsymbol{y}, \boldsymbol{w}, t) \tag{6.2-5}$$

and of the open-loop system by

$$\boldsymbol{u}_S(t, \boldsymbol{\alpha}) = \boldsymbol{h}_S(\boldsymbol{w}, t). \tag{6.2-6}$$

After differentiation of Eqs. (6.2-3) and (6.2-4) with respect to one of the components of $\boldsymbol{\alpha}$, α_j, at nominal parameters $\boldsymbol{\alpha}_0$, we have for the closed-loop system

$$\frac{d}{dt}\frac{\partial \boldsymbol{x}_R}{\partial \alpha_j}\bigg|_{\boldsymbol{\alpha}_0} = \frac{\partial \boldsymbol{f}}{\partial \boldsymbol{x}}\bigg|_{\boldsymbol{\alpha}_0}\frac{\partial \boldsymbol{x}_R}{\partial \alpha_j}\bigg|_{\boldsymbol{\alpha}_0} + \frac{\partial \boldsymbol{f}}{\partial \boldsymbol{u}}\bigg|_{\boldsymbol{\alpha}_0}\frac{\partial \boldsymbol{h}_R}{\partial \boldsymbol{y}}\bigg|_{\boldsymbol{\alpha}_0}\frac{\partial \boldsymbol{y}_R}{\partial \alpha_j}\bigg|_{\boldsymbol{\alpha}_0} + \frac{\partial \boldsymbol{f}}{\partial \alpha_j}\bigg|_{\boldsymbol{\alpha}_0}, \tag{6.2-7}$$

$$\frac{\partial \boldsymbol{y}_R}{\partial \alpha_j}\bigg|_{\boldsymbol{\alpha}_0} = \frac{\partial \boldsymbol{g}}{\partial \boldsymbol{x}}\bigg|_{\boldsymbol{\alpha}_0}\frac{\partial \boldsymbol{x}_R}{\partial \alpha_j}\bigg|_{\boldsymbol{\alpha}_0} + \frac{\partial \boldsymbol{g}}{\partial \alpha_j}\bigg|_{\boldsymbol{\alpha}_0}, \tag{6.2-8}$$

and for the open-loop system

$$\frac{d}{dt}\frac{\partial \boldsymbol{x}_S}{\partial \alpha_j}\bigg|_{\boldsymbol{\alpha}_0} = \frac{\partial \boldsymbol{f}}{\partial \boldsymbol{x}}\bigg|_{\boldsymbol{\alpha}_0}\frac{\partial \boldsymbol{x}_S}{\partial \alpha_j}\bigg|_{\boldsymbol{\alpha}_0} + \frac{\partial \boldsymbol{f}}{\partial \alpha_j}\bigg|_{\boldsymbol{\alpha}_0}, \tag{6.2-9}$$

$$\frac{\partial \boldsymbol{y}_S}{\partial \alpha_j}\bigg|_{\boldsymbol{\alpha}_0} = \frac{\partial \boldsymbol{g}}{\partial \boldsymbol{x}}\bigg|_{\boldsymbol{\alpha}_0}\frac{\partial \boldsymbol{x}_S}{\partial \alpha_j}\bigg|_{\boldsymbol{\alpha}_0} + \frac{\partial \boldsymbol{g}}{\partial \alpha_j}\bigg|_{\boldsymbol{\alpha}_0}. \tag{6.2-10}$$

From Eqs. (6.2-7)–(6.2-10) it follows that

$$\dot{\boldsymbol{\lambda}}_R - \dot{\boldsymbol{\lambda}}_S = \frac{\partial \boldsymbol{f}}{\partial \boldsymbol{x}}\bigg|_{\boldsymbol{\alpha}_0}(\boldsymbol{\lambda}_R - \boldsymbol{\lambda}_S) + \frac{\partial \boldsymbol{f}}{\partial \boldsymbol{u}}\bigg|_{\boldsymbol{\alpha}_0}\frac{\partial \boldsymbol{h}_R}{\partial \boldsymbol{y}}\bigg|_{\boldsymbol{\alpha}_0}\boldsymbol{\sigma}_R, \tag{6.2-11}$$

$$\boldsymbol{\sigma}_R - \boldsymbol{\sigma}_S = \frac{\partial \boldsymbol{g}}{\partial \boldsymbol{x}}\bigg|_{\boldsymbol{\alpha}_0}(\boldsymbol{\lambda}_R - \boldsymbol{\lambda}_S), \tag{6.2-12}$$

where $\boldsymbol{\lambda}$ and $\boldsymbol{\sigma}$ are the trajectory and output sensitivity vectors, respectively.

Equations (6.2-11) and (6.2-12) implicitly relate $\boldsymbol{\sigma}_R$ and $\boldsymbol{\sigma}_S$. Hence one may introduce a linear time-variable operator $\boldsymbol{S}_p$ such that

$$\boldsymbol{\sigma}_S = \boldsymbol{S}_p\{\boldsymbol{\sigma}_S\}. \tag{6.2-13}$$

Note that this holds for all α_j, $j = 1, 2, \ldots, r$. Denoting $\boldsymbol{\sigma}_R$ and $\boldsymbol{\sigma}_S$ for a certain α_j by $\boldsymbol{\sigma}_{R,j}$ and $\boldsymbol{\sigma}_{S,j}$, respectively, we may also write

$$\sum_{j=1}^{r} \boldsymbol{\sigma}_{R,j}\, d\alpha_j = \boldsymbol{S}_p\{\sum_{j=1}^{r} \boldsymbol{\sigma}_{S,j}\, d\alpha_j\} \qquad \text{or} \qquad \Delta \boldsymbol{y}_R = \boldsymbol{S}_p\{\Delta \boldsymbol{y}_S\}.$$

Thus the operator $\boldsymbol{S}_p$ may be simulated by the block diagram shown in Fig. 6.2-3. The operator $\boldsymbol{S}_p$ is given by the transmission from $\boldsymbol{\sigma}_S$ (or $\Delta \boldsymbol{y}_S$) to $\boldsymbol{\sigma}_R$ (or $\Delta \boldsymbol{y}_R$). The input $\boldsymbol{\sigma}_S$ (or $\Delta \boldsymbol{y}_S$) can be obtained from the nominal original system as was similarly done in Fig. 6.2-2.

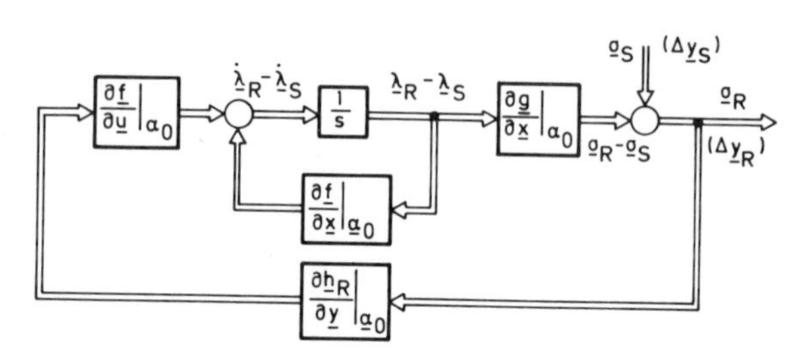

FIG. 6.2-3. Simulation of the comparison sensitivity operator of a nonlinear system.

Example 6.2-1 Consider a linear time-varying multivariable system described by

$$\dot{\boldsymbol{x}} = \boldsymbol{A}\boldsymbol{x} + \boldsymbol{B}\boldsymbol{u},$$
$$\mathbf{y} = \boldsymbol{C}\boldsymbol{x},$$

where $\boldsymbol{A} = \boldsymbol{A}(t, \boldsymbol{\alpha})$, $\boldsymbol{B} = \boldsymbol{B}(t, \boldsymbol{\alpha})$, $\boldsymbol{C} = \boldsymbol{C}(t, \boldsymbol{\alpha})$. Let the closed-loop control be given by $\boldsymbol{u}_R = \boldsymbol{H}(t)\boldsymbol{y}_R + \boldsymbol{R}(t)\boldsymbol{w}$ and the open-loop control by $\boldsymbol{u}_S = \boldsymbol{R}_S(t)\boldsymbol{w}$.

(a) Draw the simulation diagram for the corresponding comparison sensitivity operator $\boldsymbol{S}_p$.

(b) Give the complete diagram to measure $\boldsymbol{\sigma}_R$ and $\boldsymbol{\sigma}_S$ simultaneously.

Solution (a) Equations (6.2-11) and (6.2-12) become

$$\dot{\boldsymbol{\lambda}}_R - \dot{\boldsymbol{\lambda}}_S = \boldsymbol{A}(\boldsymbol{\lambda}_R - \boldsymbol{\lambda}_S) + \boldsymbol{B}\boldsymbol{H}\boldsymbol{\sigma}_R,$$
$$\boldsymbol{\sigma}_R - \boldsymbol{\sigma}_S = \boldsymbol{C}(\boldsymbol{\lambda}_R - \boldsymbol{\lambda}_S).$$

Thus the resulting simulation diagram for $\boldsymbol{S}_p$ is as shown in Fig. 6.2-4.

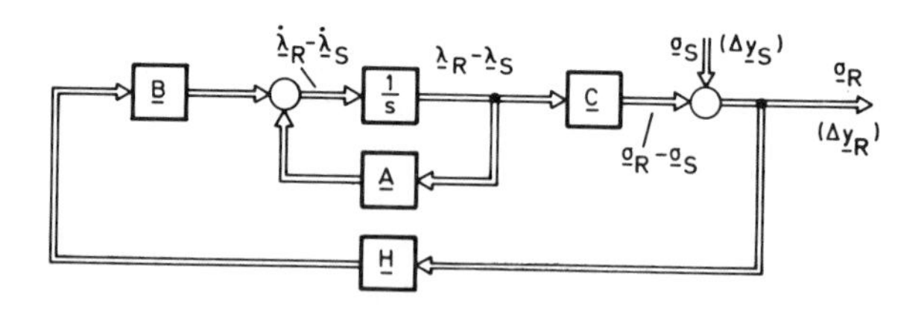

FIG. 6.2-4. Simulation diagram of the comparison sensitivity operator $\boldsymbol{S}_p$ of Example 6.2-1.

(b) The open-loop output sensitivity vector $\boldsymbol{\sigma}_S$ needed as an input in Fig. 6.2-4 is obtained from a sensitivity model which is of the same form as the one in Fig. 5.2-1, but with $\boldsymbol{D} = \mathbf{0}$ and $(\partial \boldsymbol{D}/\partial \alpha_j)_{\boldsymbol{\alpha}_0} = \mathbf{0}$. Thus, as a whole, three nominal models of the control system are needed to measure $\boldsymbol{\sigma}_S$ and $\boldsymbol{\sigma}_R$ simultaneously.

6.3 ROOT SENSITIVITY IN TERMS OF CHANGES OF THE COEFFICIENTS OF THE CHARACTERISTIC POLYNOMIAL

In this section it is shown how the relative and semirelative root sensitivities defined by Definition 3.3-6 can be found in terms of the coefficients of the characteristic polynomial or, in other words, of the transfer matrix of the system. Referring to Eqs. (3.3-47) and (3.3-49), consider the characteristic equation

$$\sum_{k=0}^{n} a_k s_i^k = 0. \tag{6.3-1}$$

Let us first assume that the characteristic coefficients a_k are continuous functions of a parameter vector $\boldsymbol{\alpha} = [\alpha_1\ \alpha_2\ \cdots\ \alpha_r]^T$ and that the roots $s = s_i$ of Eq. (6.3-1) are real or complex but simple.

Taking the partial derivative of Eq. (6.3-1) for the ith root s_i with respect to one of the parameters α_j gives

$$\sum_{k=0}^{n} \frac{\partial a_k}{\partial \alpha_j} s_i^k + \sum_{k=0}^{n} a_k k\, s_i^{k-1} \frac{\partial s_i}{\partial \alpha_j} = 0, \tag{6.3-2}$$

$$\frac{\partial s_i}{\partial \alpha_j} = -\frac{\sum_{k=0}^{n} (\partial a_k/\partial \alpha_j)\, s_i^k}{\sum_{k=0}^{n} a_k k s_i^{k-1}}. \tag{6.3-3}$$

From this we obtain the following formula for the semirelative root sensitivity:

$$\tilde{S}_{\alpha_j}^{s_i} = \left.\frac{\partial s_i}{\partial \alpha_j}\right|_{\boldsymbol{\alpha}_0} \alpha_{j0} = -\alpha_{j0} \frac{\sum_{k=0}^{n} (\partial a_k/\partial \alpha_j)|_{\boldsymbol{\alpha}_0} s_{i0}^k}{\sum_{k=1}^{n} k a_{k0} s_{i0}^{k-1}} \tag{6.3-4}$$

which is valid for simple poles, whether they are real or complex conjugate.

Example 6.3-1 Determine the semirelative root sensitivities of Example 3.3-12 with respect to a_1 using Eq. (6.3-4).

Solution The roots were found to be

$$s_{20} = -1 + j2, \qquad s_{30} = -1 - j2.$$

The nominal parameter values are $\alpha_{10} = a_{00} = 5$, $\alpha_{20} = a_{10} = 2$, and $a_{20} = 1$. Thus Eq. (6.3-4) yields

$$\tilde{S}_{\alpha_1}^{s_2} = \frac{-a_{10} s_{20}^1}{a_{10} s_{20}^0 + 2a_{20} s_{20}^1} = \frac{-2(-1+j2)}{2 + 2(-1+j2)} = -1 - j\frac{1}{2},$$

$$\tilde{S}_{\alpha_1}^{s_3} = \frac{-a_{10} s_{30}^1}{a_{10} s_{30}^0 + 2a_{20} s_{30}^1} = \frac{-2(-1-j2)}{2 + 2(-1-j2)} = -1 + j\frac{1}{2}.$$

The results are, of course, identical with the expressions obtained in Example 3.3-12.

The advantage of the above formula lies in the fact that the dependence of the roots upon the parameters need not be known analytically. This simplifies the determination of the root sensitivities since numerical methods for the determination of the roots can be employed.

The above formula can also be used as a basis for the calculation of the *relative* root sensitivities. However, in this case, we have to distinguish between real and complex conjugate (simple) roots. In the case of *real* roots, the relative root sensitivity is simply given by normalization of $\tilde{S}_{\alpha_j}^{s_i}$ upon s_{i0}, so that we have

$$\bar{S}_{\alpha_j}^{s_i} = \frac{\partial s_i}{\partial \alpha_j}\bigg|_{\boldsymbol{\alpha}_0} \frac{\alpha_{j0}}{s_{i0}} = -\alpha_{j0} \frac{\sum_{k=0}^{n} (\partial a_k / \partial \alpha_j)|_{\boldsymbol{\alpha}_0} s_{i0}^k}{\sum_{k=0}^{n} k a_{k0} s_{i0}^k}. \tag{6.3-5}$$

This is not possible in the case of *complex* roots. Here we have to normalize the real and imaginary part of $\tilde{S}_{\alpha_j}^{s_i}$ separately upon σ_{i0} and ω_{i0}, respectively. Thus we obtain

$$\bar{S}_{\alpha_j}^{s_i} = \frac{1}{\sigma_{i0}} \operatorname{Re}\{\tilde{S}_{\alpha_j}^{s_i}\} + j\frac{1}{\omega_{i0}} \operatorname{Im}\{\tilde{S}_{\alpha_j}^{s_i}\},$$

where $\tilde{S}_{\alpha_j}^{s_i}$ can again be calculated from formula (6.3-4).

Example 6.3-2 Referring to Example 6.3–1, determine the corresponding relative root sensitivities $\bar{S}_{a_1}^{s_2}$, $\bar{S}_{a_1}^{s_3}$.

Solution Dividing the real parts of $\tilde{S}_{a_1}^{s_2}$ and $\tilde{S}_{a_1}^{s_3}$ of Example 6.3-1 by $\sigma_{20} = \sigma_{30} = -1$ and the imaginary parts by $\omega_{20} = 2$ and $\omega_{30} = -2$, we obtain

$$\bar{S}_{a_1}^{s_2} = 1 - j\tfrac{1}{4}, \qquad \bar{S}_{a_1}^{s_3} = 1 - j\tfrac{1}{4}.$$

It is emphasized that the above formulas are valid only if the roots are simple. For *multiple* roots the denominators of Eqs. (6.3-4) and (6.3-5) go to zero and, consequently, the value of the root sensitivity becomes infinite. This is due to the fact that by an incremental change in the parameters the multiple real root can split up into pairs of complex roots since such a root belongs to a breakaway point of the root locus.

In order to avoid this difficulty, the sensitivity of a *double*-real root s_i might better be discussed in terms of the sensitivities of the damping ratio ξ_i and the natural frequency ω_{in} to parameter changes (Definitions 3.3-9 and 3.3-10), where the nominal values of ξ_i and ω_{in} are set to 1 or $-s_i$, respectively. Then for small variations of the parameters, the double-real root becomes two distinct real roots if ξ_i increases with the change of the parameters; two complex roots if ξ_i decreases; or they remain double-real roots if ξ_i remains constant. Therefore the sign of the sensitivity $\bar{S}_{\alpha_j}^{\xi_i}$ indicates all three cases.

The sensitivities $\bar{S}_{\alpha_j}^{\xi_i}$ can be determined as follows. Denoting $s_i^k = X_{ik} + jY_{ik}$, the characteristic equation $\sum_{k=0}^{n} a_{ik} s_i^k = 0$ can be rewritten as

$$\sum_{k=0}^{n} (a_{ik} X_{ik} + j a_{ik} Y_{ik}) = 0, \tag{6.3-6}$$

whence

$$\sum_{k=0}^{n} a_{ik} Y_{ik} = 0. \tag{6.3-7}$$

For double-real roots, the following equation must hold as well:

$$\sum_{k=0}^{n} a_{ik} Y_{i,k-1} = 0. \tag{6.3-8}$$

If Y_{ik} is expressed in terms of ξ_i and ω_{in}, then the partial derivatives of Eqs. (6.2-7) and (6.3-8) with respect to α_j give

$$\begin{aligned} &\frac{\partial \ln \omega_{in}}{\partial \ln \alpha_j} \frac{\omega_{in}}{\alpha_j} \sum_{k=0}^{n} a_{ik} \frac{\partial Y_{ik}}{\partial \omega_{in}} + \frac{\partial \ln \xi_i}{\partial \ln \alpha_j} \frac{\xi_i}{\alpha_j} \sum_{k=0}^{n} a_{ik} \frac{\partial Y_{ik}}{\partial \xi_i} \\ &\quad + \sum_{k=0}^{n} \frac{\partial a_{ik}}{\partial \alpha_j} Y_{ik} = 0, \\ &\frac{\partial \ln \omega_{in}}{\partial \ln \alpha_j} \frac{\omega_{in}}{\alpha_j} \sum_{k=0}^{n} a_{ik} \frac{\partial Y_{i,k-1}}{\partial \omega_{in}} + \frac{\partial \ln \xi_i}{\partial \ln \alpha_j} \frac{\xi_i}{\alpha_j} \sum_{k=0}^{n} a_{ik} \frac{\partial Y_{i,k-1}}{\partial \xi_i} \\ &\quad + \sum_{k=0}^{n} \frac{\partial a_{ik}}{\partial \alpha_j} Y_{i,k-1} = 0. \end{aligned} \tag{6.3-9}$$

It can further be shown [16] that

$$\frac{\partial Y_{ik}}{\partial \omega_{in}} = \frac{k-1}{\omega_{in}} Y_{ik} \tag{6.3-10}$$

and

$$(k-1)\frac{\partial Y_{i,k-1}}{\partial \xi_i} + 2\xi_i \omega_{in} k \frac{\partial Y_{ik}}{\partial \xi_i} + \omega_{in}^2 (k+1) \frac{\partial Y_{i,k-1}}{\partial \xi_i} = 0, \tag{6.3-11}$$

with $\partial Y_{i0}/\partial \xi_i \equiv \partial Y_{i1}/\partial \xi_i \equiv 0$ and $\partial Y_{i2}/\partial \zeta_i = -2\omega_{in}$. With Definitions 3.3-9 and 3.3-10, Eq. (6.3-9) can be written as

$$\begin{aligned} P_{i1} \bar{S}_{\alpha_j}^{\omega_{in}} + Q_{i1} \bar{S}_{\alpha_j}^{\xi_i} + R_{i1} = 0, \\ P_{i2} \bar{S}_{\alpha_j}^{\omega_{in}} + Q_{i2} \bar{S}_{\alpha_j}^{\xi_i} + R_{i2} = 0, \end{aligned} \tag{6.3-12}$$

where

$$\begin{aligned} P_{i1} &= \sum_{k=0}^{n} (k-2)\, a_{ik} Y_{i,k-1}, & P_{i2} &= \sum_{k=0}^{n} (k-1)\, a_{ik} Y_{ik}, \\ Q_{i1} &= \xi_i \sum_{k=0}^{n} \frac{\partial Y_{i,k-1}}{\partial \xi_i}, & Q_{i2} &= \xi_i \sum_{k=0}^{n} a_{ik} \frac{\partial Y_{ik}}{\partial \xi_i}, \\ R_{i1} &= \alpha_j \sum_{k=0}^{n} \frac{\partial a_{ik}}{\partial \alpha_j} Y_{i,k-1}, & R_{i2} &= \alpha_j \sum_{k=0}^{n} \frac{\partial a_{ik}}{\partial \alpha_j} Y_{ik}. \end{aligned} \tag{6.3-13}$$

All the equations have to be evaluated for nominal parameters.

Solving Eqs. (6.3-12) for $\bar{S}_{\alpha_j}^{\omega_{in}}$ and $\bar{S}_{\alpha_j}^{\xi_i}$ gives

$$\bar{S}_{\alpha_j}^{\omega_{in}} = \frac{Q_{i1}R_{i2} - Q_{i2}R_{i1}}{P_{i1}Q_{i2} - P_{i2}Q_{i1}}, \qquad \bar{S}_{\alpha_j}^{\xi_i} = \frac{P_{i2}R_{i1} - P_{i1}R_{i2}}{P_{i1}Q_{i2} - P_{i2}Q_{i1}}. \tag{6.3-14}$$

Equations (6.3-14) are the desired formulas to calculate the $\bar{S}_{\alpha_j}^{\omega_{in}}$ and $\bar{S}_{\alpha_j}^{\xi_i}$, $j = 1, 2, \ldots, r$.

Example 6.3-3 Determine $\bar{S}_{\alpha_1}^{\omega_{1n}}$, $\bar{S}_{\alpha_1}^{\xi_1}$, $\bar{S}_{\alpha_2}^{\omega_{1n}}$, $\bar{S}_{\alpha_2}^{\xi_1}$ of the double-real root of the characteristic equation

$$s^3 + \alpha_2 s^2 + \alpha_1 s + 1 = 0,$$

where $\alpha_{10} = 5$, $\alpha_{20} = 4.25$.

Solution The double-real root is $s_{10} = -2$. Thus Eqs. (6.3-13) and (6.3-14) have to be evaluated for $\xi_i = \xi_1 = 1$ and $\omega_{in} = \omega_{1n} = 2$; this yields:

$$\bar{S}_{\alpha_1}^{\omega_{1n}} = 0.83, \qquad \bar{S}_{\alpha_1}^{\xi_1} = -0.73.$$

The negative sign of $\bar{S}_{\alpha_1}^{\xi_1}$ indicates that for a small increase of α_1 from its nominal value $\alpha_{10} = 5$ (while $\alpha_2 = \alpha_{20} = 4.25$), the double-real root $s_{10} = -2$ is split up into two complex conjugate roots. Analogously we obtain

$$\bar{S}_{\alpha_2}^{\omega_{1n}} = -0.163, \qquad \bar{S}_{\alpha_2}^{\xi_1} = 1.21,$$

where the positive sign of $\bar{S}_{\alpha_2}^{\xi_1}$ indicates that for a small increase of the parameter α_2 from its nominal value $\alpha_{20} = 4.25$, the double-real roots s_{10} remain two distinct roots.

In the case of a *multiple* root, the sensitivity has to be defined in a slightly different manner. Consider an ith order root s_i which may be real or complex conjugate. Suppose that the characteristic equation is given by

$$F(s, \boldsymbol{\alpha}) = 0, \tag{6.3-15}$$

where $s = s_i$ is the (real or complex) root and $\boldsymbol{\alpha}$ is the parameter vector with nominal value $\boldsymbol{\alpha}_0$.

If s_i is a νth order root so that $(s\text{-}s_i)^\nu$ is a factor of F, the first $\nu - 1$ derivatives of F with respect to s_i are zero. Therefore the total differential dF should be expanded to the first nonzero order term, whence

$$dF = \frac{1}{\nu!}\frac{\partial^\nu F}{\partial s_i^\nu}(ds_i)^\nu + \sum_{j=1}^{r} \left.\frac{\partial F}{\partial \alpha_j}\right|_{\boldsymbol{\alpha}_0} d\alpha_j = 0. \tag{6.3-16}$$

Solving Eq. (6.3-16) for ds_i gives

$$ds_i = \left[-\frac{\nu!}{(\partial^\nu F/\partial s^\nu)} \sum_{j=1}^{r} \frac{\partial F}{\partial \alpha_j} d\boldsymbol{\alpha}_j\right]_{\substack{s=s_i\\ \boldsymbol{\alpha}=\boldsymbol{\alpha}_0}}^{1/\nu}. \tag{6.3-17}$$

This relates infinitesimal changes in the roots s_i to infinitesimal changes in the parameters α_j, $j = 1, 2, \ldots, r$.

Now we define the absolute root sensitivity by

$$S_{\alpha j}^{s_i} \triangleq \left.\frac{(\partial s_i)^\nu}{\partial \alpha_j}\right|_{\boldsymbol{\alpha}_0} = -\nu! \left[\frac{\partial F/\partial \alpha_j}{\partial^\nu F/\partial s^\nu}\right]_{\substack{s=s_i \\ \boldsymbol{\alpha}=\boldsymbol{\alpha}_0}}. \tag{6.3-18}$$

Then the change of the ith root s_i due to small changes of the parameters α_j, $j = 1, 2, \ldots, r$, becomes approximately

$$\Delta s_i = \left[\sum_{j=1}^{r} S_{\alpha j}^{s_i} \Delta\alpha_j\right]^{1/\nu}, \tag{6.3-19}$$

where ν is the order of the root s_i.

Similarly, the semirelative root sensitivity can be defined as follows:

$$\tilde{S}_{\alpha j}^{s_i} \triangleq \left.\frac{(\partial s_i)^\nu}{\partial \alpha_j}\right|_{\boldsymbol{\alpha}_0} \alpha_{j0} = -\nu!\, \alpha_{j0} \left[\frac{\partial F/\partial \alpha_j}{\partial^\nu F/\partial s^\nu}\right]_{\substack{s=s_i \\ \boldsymbol{\alpha}=\boldsymbol{\alpha}_0}}. \tag{6.3-20}$$

This is the general formula to evaluate the semirelative sensitivity of *a* νth order roots of a polynomial F with respect to the parameter α_j.

Frequently, the polynomial F can be brought into the form

$$F(s, \alpha) = F_1(s) + \alpha F_2(s), \tag{6.3-21}$$

where $F_1(s)$ and $F_2(s)$ are parameter-independent polynomials. Therefore $\partial F/\partial\alpha = F_2(s)$, and since $F_1(s) + F_2(s) = 0$, we can also write $\partial F/\partial\alpha = -F_1(s)/\alpha$. Introducing this expression into Eq. (6.3-20) yields

$$\tilde{S}_{\alpha}^{s_i} = \nu! \left[\frac{F_1(s)}{\partial^\nu F/\partial s^\nu}\right]_{\substack{s=s_i \\ \boldsymbol{\alpha}=\boldsymbol{\alpha}_0}}. \tag{6.3-22}$$

This is a very useful formula to evaluate the semirelative sensitivity of the roots of a polynomial F in the form of Eq. (6.3-21) with respect to the parameter α.

Example 6.3-4 (a) Determine the semirelative sensitivity of the root s_1 with respect to the parameter a_1 of the characteristic equation $F(s, a_1) = s^3 + a_2 s^2 + a_1 s + a_0 = 0$, where the nominal parameter values are $a_{00} = 8$, $a_{10} = 12$, $a_{20} = 6$.

(b) Give the change in s_1 due to a 10% change of a_1.

Solution (a) The characteristic equation has a real thirdorder root with the nominal value $s_{10} = -2$. Thus evaluating Eq. (6.3-20) gives (with $\nu = 3$, $\alpha_j = a_1$)

$$\tilde{S}_{a_1}^{s_1} = -3! \cdot 12 \frac{s_{10}}{6} = 24.$$

The same result is of course obtained using Eq. (6.3-22).

(b) The parameter-induced change of s_1 along the real axis obtained from Eq. (6.3-17) is

$$\Delta s_1 = \left[-\frac{3!}{6} s_{10} \cdot 12 \cdot 0.1 \right]^{1/3} = \sqrt[3]{2.4} = 1.339.$$

6.4 ROOT SENSITIVITIES OF FEEDBACK SYSTEMS

One important field of application of root sensitivities is the analysis and design of linear feedback systems [11]. To characterize the relative stability of a feedback system it is important to know how the poles of its transfer function or matrix are affected by the parameters of the components, i.e., how rapidly they move along the root loci due to the parameter changes. Parameters of particular interest are the loop gain k, and the coefficients, poles, and zeros of the transfer functions of the components, especially of the plant.

Methods of calculating the closed-loop root sensitivities have been developed mainly by Horowitz [11] and Chang [4] (see also [47], [64], [81]. In this section the basic relations will be presented.

6.4.1 Root Sensitivities to Parameter Changes in the Plant

Consider the single-input single-output feedback system shown in Fig. 3.3-12.

The transfer function of the closed loop $G(s, \boldsymbol{\alpha})$ can be written as

$$G(s, \boldsymbol{\alpha}) = \frac{P(s, \boldsymbol{\alpha})R(s)}{1 + P(s, \boldsymbol{\alpha})R(s)H(s)} = K(\boldsymbol{\alpha}) \frac{\prod_\mu (s - s_\mu(\boldsymbol{\alpha}))}{\prod_\nu (s - s_\nu(\boldsymbol{\alpha}))}. \tag{6.4-1}$$

where, in general, $K(\boldsymbol{\alpha})$, $s_\mu(\boldsymbol{\alpha})$, and $s_\nu(\boldsymbol{\alpha})$ are functions of the parameter vector $\boldsymbol{\alpha}$. We assume that the poles s_ν and the zeros s_μ are distinct. The goal of this section is to derive a formula for determining the semirelative real-root sensitivities $\tilde{S}_{\alpha_j}^{s_\nu} = (\partial s_\nu / \partial \ln \alpha_j)_{\boldsymbol{\alpha}_0}$.

Instead of $G(s, \boldsymbol{\alpha})$, let us consider $G^{-1}(s, \boldsymbol{\alpha})$, which can be written as

$$G^{-1}(s, \boldsymbol{\alpha}) = (P(s, \boldsymbol{\alpha})R(s, \boldsymbol{\alpha}))^{-1} + H(s, \boldsymbol{\alpha}) = L^{-1}(s, \boldsymbol{\alpha}) + H(s, \boldsymbol{\alpha}), \tag{6.4-2}$$

where $L = PR$. Taking the logarithm of G^{-1} and then differentiating with respect to $\ln \alpha_j$ gives

$$-\frac{\partial \ln G}{\partial \ln \alpha_j} = \alpha_j G \left[\frac{\partial L^{-1}}{\partial \alpha_j} + \frac{\partial H}{\partial \alpha_j} \right], \qquad j = 1, 2, \ldots, r. \tag{6.4-3}$$

If we now set $\boldsymbol{\alpha} = \boldsymbol{\alpha}_0$ and use Eq. (3.3-70), we obtain

$$-\sum_\nu \frac{1}{s - s_{\nu 0}} \tilde{S}_{\alpha_j}^{s_\nu} + \sum \frac{1}{s - s_{\mu 0}} \tilde{S}_{\alpha_j}^{s_\mu} - S_{\alpha_j}^{K} = \alpha_{j0} G_0 \left[\frac{\partial L^{-1}}{\partial \alpha_j} + \frac{\partial H}{\partial \alpha_j} \right]_{\boldsymbol{\alpha}_0}, \tag{6.4-4}$$

where the subscript 0 designates nominal values.

Both sides of Eq. (6.4-4) are now multiplied by $-(s - s_\nu)$ and then s is chosen to be s_ν. Thus we get

$$\tilde{S}_{\alpha j}^{s_\nu} = -\alpha_{j0}[(s - s_{\nu 0})G(s,\boldsymbol{\alpha}_0)]_{s_{\nu 0}} \left[\frac{\partial L^{-1}(s_\nu,\boldsymbol{\alpha})}{\partial \alpha_j} - \frac{\partial H(s_\nu,\boldsymbol{\alpha})}{\partial \alpha_j}\right]_{\boldsymbol{\alpha}_0}. \tag{6.4-5}$$

This is the general formula for determining the pole sensitivities from the transfer functions of the components of the feedback system.

In particular, if R and H are independent of $\boldsymbol{\alpha}$, which is often the case in studying the effects of plant parameters only, then the above formula reduces to

$$\tilde{S}_{\alpha j}^{s_\nu} = -\alpha_{j0}[(s - s_{\nu 0})G(s, \boldsymbol{\alpha}_0)]_{s_{\nu 0}} \left.\frac{\partial L^{-1}(s_\nu, \boldsymbol{\alpha})}{\partial \alpha_j}\right|_{\boldsymbol{\alpha}_0}. \tag{6.4-6}$$

Note that the expression in brackets is the residue of G_0 with respect to $s_{\nu 0}$, denoted by $\text{Res}_\nu(G_0)$. Moreover, since $s_{\nu 0}$ is a pole of $G(s, \boldsymbol{\alpha}_0)$, it follows from Eq. (6.4-2) that $L^{-1}(s_{\nu 0}, \boldsymbol{\alpha}_0) = -H$. Hence, for $H = 1$,

$$\tilde{S}_{\alpha j}^{s_\nu} = \text{Res}_\nu(G_0) \left.\frac{\partial \ln L^{-1}(s_\nu, \boldsymbol{\alpha})}{\partial \ln \alpha_j}\right|_{\boldsymbol{\alpha}_0}. \tag{6.4-7}$$

This result indicates that for finding the relative pole sensitivity of a feedback system to plant parameter variations, the partial derivative of $(P_0R_0)^{-1}$ with respect to α and the residue of G_0 have to be determined.

Example 6.4-1 For the problem of Example 3.3-14 where $H = 1$, determine the semirelative pole sensitivities of the closed-loop transfer function to variations of k, T_1, and T_2.

Solution Since $H = 1$, Eq. (6.4-7) applies. The nominal transfer function of the closed loop is

$$G(s,\alpha_0) = \frac{100}{s^2 + 10.1s + 101} = \frac{100}{(s - s_{10})(s - s_{20})},$$

where

$$s_{10} = -5.05 + j8.68, \qquad s_{20} = -5.05 - j8.68.$$

Now consider the first pole s_{10}. The corresponding residue is

$$\text{Res}_1(G_0) = \frac{100}{s_{10} - s_{20}} = -j5.75.$$

The terms $\partial[\ln L^{-1}(s_\nu, \boldsymbol{\alpha})]/\partial \ln \alpha_j$ become

$$\left.\frac{\partial \ln L^{-1}}{\partial \ln k}\right|_{\boldsymbol{\alpha}_0} = -1, \qquad \left.\frac{\partial \ln L^{-1}}{\partial \ln T_{10}}\right|_{\boldsymbol{\alpha}_0} = \frac{T_{10}s_{10}}{1 + T_{10}s_{10}} = 1.07 + j0.205,$$

$$\left.\frac{\partial \ln L^{-1}}{\partial \ln T_{20}}\right|_{\boldsymbol{\alpha}_0} = \frac{T_{20}s_{10}}{1 + T_{20}s_{10}} = 1.044 + j0.095.$$

Using these results in Eq. (6.4-7), we finally obtain

$$\tilde{S}_K^{s_1} = j5.75 = 5.75\underline{/90},$$
$$\tilde{S}_{T_1}^{s_1} = 1.17 - j6.15 = 6.26\underline{/-79.2},$$
$$\tilde{S}_{T_2}^{s_1} = 0.544 - j6 = 6.02\underline{/-84.8}.$$

In a similar way, the sensitivities of s_2 to the various parameters are determined. The results are symmetrical to the ones above. The above results show that the poles are almost equally sensitive to all parameters of interest.

It is also possible to interpret the pole sensitivity of a feedback system in terms of the residue of the Bode sensitivity function of the closed loop to the plant. This was first shown by Perlis [81]. It results in a formula for the calculation of the pole sensitivities that is very easy to evaluate.

In order to derive this formula, consider again a feedback configuration of the form of Fig. 3.3-12. Let the poles of the transfer functions $P(s, \boldsymbol{\alpha})$, $R(s)$, $H(s)$ be denoted by $s = s_i$, $s = s_k$, $s = s_e$, respectively, so that the open-loop transference $M(s, \boldsymbol{\alpha}) = P(s, \boldsymbol{\alpha})R(s)H(s)$ can be written as

$$M(s, \boldsymbol{\alpha}) = \frac{kN_M(s)}{\prod_{i=1}^{nP}(s - s_i)\prod_{k=1}^{nR}(s - s_k)\prod_{e=1}^{nH}(s - s_e)}. \tag{6.4-8}$$

Here $N_M(s)$ represents a polynomial in s of no further interest.

In these terms, the closed-loop transfer function $G(s)$ and the Bode sensitivity function $S_P{}^G(s)$ take the form

$$G(s) = \frac{P(s, \boldsymbol{\alpha})R(s)}{1 + M(s, \boldsymbol{\alpha})} = \frac{KN_G(s)}{\prod_{\nu=1}^{nG}(s - s_\nu)}, \tag{6.4-9}$$

$$S^G{}_P(s) = \frac{1}{1 + M(s, \boldsymbol{\alpha}_0)} = \frac{\prod_{i=1}^{nP}(s - s_i)\prod_{k=1}^{nR}(s - s_k)\prod_{e=1}^{nH}(s - s_e)}{\prod_{\nu=1}^{nG}(s - s_\nu)}, \tag{6.4-10}$$

where $n_G = n_P + n_R + n_H$ and s_ν are the poles of the closed loop. Note that these are the same as the poles of S_P^G.

Let us now take the double logarithmic derivative of the Bode sensitivity function in the two forms of Eq. (6.4-10) with respect to one of the nominal plant parameters α_{j0}. This gives

$$-S_P{}^G(s)\frac{\partial M(s, \boldsymbol{\alpha}_0)}{\partial \ln \alpha_{j0}} = -\sum_{i=1}^{np}\frac{\tilde{S}_{\alpha j}^{s_i}}{s - s_{i0}} - \sum_{k=1}^{nR}\frac{\tilde{S}_{\alpha j}^{s_k}}{s - s_{k0}}$$
$$-\sum_{e=1}^{nH}\frac{\tilde{S}_{sj}^{s_e}}{s - s_{e0}} + \sum_{\nu=1}^{nG}\frac{\tilde{S}_{\alpha j}^{s_\nu}}{s - s_{\nu 0}}. \tag{6.4-11}$$

The result shows that the semirelative root sensitivity $\tilde{S}_{\alpha j}^{s_\nu}$ of the closed-loop pole s_ν with respect to α_j can be interpreted in terms of the residue of $S_P{}^G(s)$:

$$\tilde{S}_{\alpha j}^{s_\nu} = -\text{Res}[S_P{}^G(s)]_{s=s_\nu}\left.\frac{\partial M(s, \boldsymbol{\alpha})}{\partial \ln \alpha_j}\right|_{\substack{s=s_\nu\\ \alpha=\alpha 0}}. \tag{6.4-12}$$

Since $R(s)$ and $H(s)$ are not functions of $\boldsymbol{\alpha}$ and since $M(s_{\nu 0}, \boldsymbol{\alpha}_0) = -1$, the partial derivative of $M(s, \boldsymbol{\alpha})$ with respect to $\ln \boldsymbol{\alpha}_j$ at $s = s_\nu$ and $\boldsymbol{\alpha} = \boldsymbol{\alpha}_0$ becomes

$$\left.\frac{\partial M(s, \boldsymbol{\alpha})}{\partial \ln \alpha_j}\right|_{\substack{s=s_\nu \\ \boldsymbol{\alpha}=\boldsymbol{\alpha}_0}} = -S_{\alpha_j}^P(s_\nu). \tag{6.4-13}$$

Thus

$$\tilde{S}_{\alpha_j}^{s_\nu} = \mathrm{Res}[S_P{}^G(s)]_{s=s_\nu} S_{\alpha_j}^P(s_\nu). \tag{6.4-14}$$

$\mathrm{Res}\,[S_P{}^G(s)]_{s=s_\nu}$ is very simple to evaluate in the s plane: The magnitude is given by

$$|\mathrm{Res}\,[S_P{}^G(s)]_{s_\nu}| = \frac{\text{Product of distances from each pole of } P, R, H \text{ to } s_\nu}{\text{Product of distances from all other closed-loop poles to } s_\nu}.$$

The phase is equal to the sum of the angles of P, R, H to s_ν minus the sum of the angles of all other poles of $S_P{}^G$ to s_ν. The Bode sensitivity function of P with respect to α_j often takes simple forms.

Example 6.4-2 Given a feedback configuration of the form of Fig. 3.3-12 with $R(s) = A$, $H(s) = 1 + Bs$, and $P(s, k) = k/(s(s + a))$.

Then $G(s) = \dfrac{kA}{s^2(a + kAB)s + kA}$ and

$$S_P{}^G(s) = \frac{s(s + a_0)}{s^2 + (a_0 + k_0A_0B_0)s + k_0A_0}.$$

Let the nominal values be $k_0 = 10$, $a_0 = 5$, $A_0 = 10$, $B_0 = 0.05$. Determine the semirelative sensitivity of one of the poles of $G(s)$ with respect to the gain k of the plant. Give the change of the pole due to a 10% change of k.

Solution The Bode sensitivity function $S_K{}^P$ is 1. The nominal poles of $G(s)$ and $S_P{}^G(s)$ are found to be

$$s_{10} = -5 + j5\sqrt{3}, \qquad s_{20} = -5 - j5\sqrt{3},$$

and the poles of $P(s, k)$ are $s_{1_p} = 0$, $s_{2_p} = -5$. Thus

$$\tilde{S}_k^{s_1} = \frac{\sqrt{25 + 75}\,(5\sqrt{3})}{10\sqrt{3}} \underline{/120^\circ + 90^\circ - 90^\circ} = 5\underline{/120^\circ}.$$

Hence if k changes by 10%, the pole s_1 changes by

$$\Delta s_1 = 5 \times 0.1\underline{/120^\circ} = 0.5\underline{/120^\circ}.$$

Note that in the derivations of the above formulas for the closed-loop pole sensitivities no further specifications of the parameters of interest were made. Consequently these formulas can be applied to all kinds of parameters such as the open-loop gain or the poles and zeros or the coefficients of the plant transfer function. For the individual cases more elaborate formulas will be given in the following three sections.

6.4.2 Sensitivity of Closed-Loop Poles to Open-Loop Gain

Very often in feedback systems the parameter of interest is the gain factor of the open loop, i.e., one wants to know how rapidly and in which direction the roots of the closed-loop transfer function move due to changes of the open-loop gain factor. Formulas for the calculation of the root sensitivity of that case will be given in this section.

Denoting the open-loop transfer function by

$$M(s,k) = kB(s) = k\frac{N(s)}{D(s)}, \tag{6.4-15}$$

the closed-loop transfer function, Eq.(6.4-1), can be written as

$$\begin{aligned} G(s,k) &= \frac{kB(s)H^{-1}(s)}{1 + kB(s)} = -\frac{D(s)H^{-1}(s)}{D(s) + kN(s)} \\ &= K(k)\frac{\prod_\mu (s - s_\mu)}{\prod_\nu [s - s_\nu(k)]}. \end{aligned} \tag{6.4-16}$$

This shows that k only affects the gain factor K and the poles s_ν of the closed-loop transfer function while having no effect on the zeros s_μ. Therefore in the case of simple poles the Bode sensitivity function of the closed loop with respect to k reduces to

$$S_k{}^G(s) = S_k{}^K + \sum_{\nu=1}^{nG} \frac{1}{s - s_{\nu 0}} \tilde{S}_k^{s_\nu}, \tag{6.4-17}$$

where $\tilde{S}_k^{s_\nu}$ is the semirelative pole sensitivity with respect to k, i.e., the sensitivity of the roots of $1 + kB(s) = 0$ or of $F(s,k) = D(s) + kN(s) = 0$.

It is easy to see that $\tilde{S}_k^{s_\nu}$ means the residue of the Bode sensitivity functions $S_P{}^G(s_\nu)$ or $S_K{}^G(s_\nu)$ or the product of $-H(s_\nu)$ with the residue of the closed-loop transfer function $G(s,k)$ at $s = s_\nu$ and $k = k_0$. The former follows immediately from Eq.(6.4-17) for $s = s_{\nu 0}$ or from Eq.(6.4-14) by realizing that in this case $S_K{}^P(s_\nu) = 1$. The latter follows for simple poles by taking the partial derivative of $F(s,k) = D(s) + kN(s) = 0$ with respect to $\ln k$:

$$\frac{\partial F}{\partial s}\frac{\partial s}{\partial \ln k} + \frac{\partial F}{\partial \ln k} = 0. \tag{6.4-18}$$

By letting $s \to s_\nu$ and solving for $\partial s_\nu / \partial \ln k$, one obtains

$$\tilde{S}_k^{s_\nu} = -k\left.\frac{\partial F/\partial k}{\partial F/\partial s}\right|_{\substack{s=s_\nu\\k=k_0}} = -k\left.\frac{N(s)}{\partial[D(s) + kN(s)]/\partial s}\right|_{\substack{s=s_\nu\\k=k_0}}. \tag{6.4-19}$$

With the aid of $D(s) + kN(s) = 0$, i.e., $N(s) = -k^{-1}D(s)$, we obtain

$$\tilde{S}_k^{s_\nu} = \left.\frac{D(s)}{F'(s)}\right|_{s=s_\nu} = \left.\frac{D(s)}{D'(s) + kN'(s)}\right|_{s=s_\nu} = \text{Res}\left[\frac{D(s)}{F(s)}\right]_{s=s_\nu}. \quad (6.4\text{-}20)$$

The comparison with Eq.(6.4-16) shows that this is $-H(s_\nu)\ \text{Res}_\nu(G_0)$. The same result is obtained by evaluating Eq.(6.4-6) for the present situation. Thus

$$\tilde{S}_k^{s_\nu} = -H(s_\nu)\ \text{Res}[G(s, k_0)]_{s=s_{\nu 0}} = \text{Res}[S_k{}^G(s_\nu)] = \text{Res}[S_P{}^G(s_\nu)]. \quad (6.4\text{-}21)$$

In the case of *multiple* poles one obtains, according to Eq.(6.3-22),

$$\tilde{S}_k^{s_\nu} = i!\left.\frac{D(s)}{F^{(i)}(s)}\right|_{\substack{s=s_\nu \\ k=k_0}} = \left.\frac{i!\ D(s)}{D^{(i)}(s) + k_0 N^{(i)}(s)}\right|_{s=s_\nu}, \quad (6.4\text{-}22)$$

where i is the order of the root s_ν.

In the above formulas, $D(s)$, $N(s)$, and $F(s)$ may be given either in the form of polynomials or in terms of their roots. Thus $\tilde{S}_k^{s_\nu}$ can be calculated either in terms of the coefficients of the open-loop transfer function or in terms of the open-loop and closed-loop poles.

Example 6.4-3 Solve Example 6.4-2, using formula (6.4-22) for $\tilde{S}_k^{s_1}$.

Solution The open-loop transfer function is $M(s,k) = [kA(1 + Bs)]/[s(s + a)]$. Thus using the nominal values,

$$N(s) = A(1 + Bs) = 10(1 + 0.05s), \qquad D(s) = s(s + a) = s(s + 5),$$
$$F(s) = D(s) + k_0 N(s) = s^2 + 10s + 100,$$

and from Eq.(6.4-20), we obtain

$$\tilde{S}_k^{s_1} = \text{Res}\left[\frac{s(s + 5)}{s^2 + 10s + 100}\right]_{s=-5+i5\sqrt{3}} = 5\underline{/120^\circ},$$

which is the same result as in Example 6.4-2.

6.4.3 Sensitivity of Closed-Loop Poles to Coefficients of the Plant Transfer Function

Suppose the open-loop transfer function is in the form

$$M(s,a) = \frac{N(s)}{P(s) + aQ(s)}, \quad (6.4\text{-}23)$$

where $N(s)$, $P(s)$, and $Q(s)$ are polynomials in s not depending on a, and the sensitivities of the closed-loop poles due to the parameter a are to be deter- mined. Then by slightly changing Eq.(6.4-20), this formula can be used to find the sensitivity functions in this case as well.

In terms of Eq.(6.4-23), the transfer function of the closed loop (Fig. 3.3-12) reads

$$G(s) = \frac{H^{-1}(s)N(s)}{N(s) + P(s) + aQ(s)} = K\frac{\prod_\mu (s - s_\mu)}{\prod_\nu (s - s_\nu)}, \tag{6.4-24}$$

where the poles of $G(s)$ are the roots of

$$N(s) + P(s) + aQ(s) = Z(s) + aQ(s) = F(s) = 0. \tag{6.4-25}$$

Assuming simple poles, $\tilde{S}_a^{s_\nu}$ can be obtained by a procedure similar to that for Eqs.(6.8-18)–(6.8-20) as

$$\tilde{S}_a^{s_\nu} = \left.\frac{Z(s)}{F'(s)}\right|_{s=s_\nu} = \left.\frac{N(s) + P(s)}{N'(s) + P'(s) + aQ'(s)}\right|_{s=s_\nu}, \tag{6.4–26}$$

In other words,

$$\tilde{S}_a^{s_\nu} = \text{Res}\left[\frac{N(s) + P(s)}{N(s) + P(s) + a_0 Q(s)}\right]_{s=s_\nu}. \tag{6.4-27}$$

In the case of multiple poles of the ith order we have

$$\tilde{S}_a^{s_\nu} = \left.\frac{i!(N(s) + P(s))}{N^{(i)}(s) + P^{(i)}(s) + a_0 Q^{(i)}(s)}\right|_{s=s_\nu} \tag{6.4-28}$$

These formulas apply whether the transfer function of the open loop is given in terms of poles and zeros or in the form of polynomials.

Example 6.4-4 Given a control loop of the form of Fig. 3.3-12 with $H(s) = 1$ and the open-loop transfer function

$$M(s) = R(s)P(s,b) = \frac{k(s + a)}{(s + b)(s + c)}.$$

The nominal values of the parameters are $k_0 = 2$, $a_0 = 5$, $b_0 = 2$, $c_0 = 1$. Determine the semirelative root sensitivity of one of the closed-loop poles with respect to b.

Solution The roots of the closed-loop are found from the characteristic equation

$$s^2 + (c_0 + k_0)s + a_0 k_0 + b_0(s + c_0) = 0.$$

This gives

$$s_{10} = -2.5 + j2.4, \qquad s_{20} = -2.5 - j2.4.$$

Denoting

$$N(s) + P(s) = s^2 + (c + k)s + ak,$$

the desired root sensitivity $\tilde{S}_b^{s_1}$ becomes, due to Eq.(6.4-27),

$$\tilde{S}_b^{s_1} = \text{Res}\left[\frac{s^2 + (c_0 + k_0)s + a_0 k_0}{s^2 + (c_0 + k_0)s + a_0 k_0 + b_0(s + c_0)}\right]_{s_1} = \text{Res}\left[\frac{s^2 + 3s + 10}{s^2 + 5s + 12}\right]_{s_1}$$

which is

$$\tilde{S}_b^{s_1} = -1 - j0.62 = 1.18\underline{/211.7^\circ}.$$

6.4.4 Sensitivity of Closed-Loop Roots to Open-Loop Poles or Zeros

In this section it will be shown how the sensitivity of closed-loop poles to open-loop poles or zeros can be calculated.

Suppose the open-loop transfer function has a zero s_N so that it can be written as

$$M(s, s_N) = (s - s_N)M_1(s) = (s - s_N)\frac{N(s)}{D(s)}. \tag{6.4-29}$$

Then the closed-loop transfer function is

$$G(s) = \frac{KN_G(s)}{D(s) + (s - s_N)N(s)} = \frac{K\prod_\mu(s - s_\mu)}{\prod_\nu(s - s_\nu)}. \tag{6.4-30}$$

The poles of $G(s)$ are the roots of

$$F(s,s_N) \triangleq D(s) + (s - s_N)N(s) = 0. \tag{6.4-31}$$

Let us assume that the roots of G are of order i. Then applying the same procedure as in Eqs.(6.3-16)–(6.3–18), we obtain

$$\frac{\partial F}{\partial s_N} + \frac{1}{i!}\frac{\partial^i F}{\partial s^i}\frac{\partial s^i}{\partial s_N} = -N(s) + \frac{1}{i!}F^{(i)}\frac{\partial s^i}{\partial s_N} = 0. \tag{6.4-32}$$

Reusing Eq.(6.4-31), which gives $N(s) = -(s - s_N)^{-1} D(s)$, and setting $s = s_\nu$ and $s_N = s_{N0}$, we obtain the semirelative sensitivity

$$\tilde{S}_{s_N}^{s_\nu} = \left.\frac{\partial s_\nu^i}{\partial \ln s_N}\right|_{s_{N0}} = -\frac{s_{N0}}{s_\nu - s_{N0}}\frac{i!D(s_\nu)}{F^{(i)}(s_\nu)} = S_{N0}\frac{i!F(s_\nu)}{F^{(i)}(s_\nu)}. \tag{6.4-33}$$

In the same manner, the sensitivity of an ith order closed-loop pole to a pole s_D of the open-loop transfer function can be found.

Suppose that the open-loop transfer function has the pole s_D so that it can be written as

$$M(s,s_D) = \frac{1}{s - s_D}\frac{N(s)}{D(s)}. \tag{6.4-34}$$

Then the closed-loop transfer function is

$$G(s) = \frac{KN_G(s)}{(s - s_D)D(s) + N(s)} = \frac{K\prod_\mu(s - s_\mu)}{\prod_\nu(s - s_\nu)}. \tag{6.4-35}$$

The poles of $G(s)$ are the roots of

$$F(s) = (s - s_D)D(s) + N(s) = 0. \tag{6.4-36}$$

If the roots of $G(s)$ are of order i, then by applying the same procedure as in the above case we obtain

$$\tilde{S}_{s_D}^{s_\nu} = \left.\frac{\partial s_\nu{}^i}{\partial \ln s_D}\right|_{s_{D0}} = -\frac{1}{s_\nu - s_{D0}} \frac{i!\,N(s_\nu)}{F^{(i)}(s_\nu)} = \frac{i!D(s_\nu)}{F^{(i)}(s_\nu)} \tag{6.4-37}$$

These formulas give the semirelative sensitivity of the closed-loop poles with respect to open-loop zeros and poles in terms of the open- and closed-loop poles. It is clear that the absolute and relative sensitivities can easily be obtained from these results by suitable normalizations.

PROBLEMS

6.1 For the transfer function

$$G(s) = \frac{s + a_1}{s^2 + a_1 a_2 s + a_1 a_3},$$

with the nominal values $a_{10} = 1$, $a_{20} = 5$, $a_{30} = 6$, give the semirelative root sensitivities with respect to a_1.

6.2 Given the transfer function

$$G(s) = \frac{a_1(s + a_3)}{s^3 + (a_2 + a_4)s^2 + (a_1 + a_2 a_4)s + a_1 a_3},$$

with the nominal parameter values $a_{10} = 10$, $a_{20} = 0.5$, $a_{30} = 1$, $a_{40} = 5$: Find the relative pole sensitivities with respect to a_1.

6.3 Consider the transfer function

$$G(s) = \frac{b_1 s + b_0}{a_4 s^4 + a_3 s^3 + a_2 s^2 + a_1 s + a_0},$$

with the nominal parameter values $a_{00} = 7.2$, $a_{10} = 10.2$, $a_{20} = 5.3$, $a_{30} = 1.2$, $a_{40} = 0.1$, $b_{00} = 1$, $b_{10} = 2$: Determine the semirelative sensitivities of the roots with respect to the coefficient a_2.

6.4 Given the transfer function

$$G(s) = (s^5 + a_4 s^4 + a_3 s^3 + a_2 s^2 + a_1 s + a_0)^{-1},$$

with the nominal parameter values $a_{00} = 676$, $a_{10} = 585$, $a_{20} = 272$, $a_{30} = 74$, $a_{40} = 12$: Determine the semirelative root sensitivities with respect to a_3.

6.5 Given a unity feedback system with the transfer function of the forward path

$$M(s,K) = \frac{K(s + a)}{(s + b)(s + c)}.$$

The nominal parameter values are $K_0 = 2, a_0 = 5, b_0 = 2, c_0 = 1$. Determine the simirelative sensitivities of the closed-loop poles with respect to the gain factor K.

6.6 Consider the servo system shown in Fig. 6.P-1. The nominal parameter values are $K_{10} = 9.2$, $K_{20} = 10$, $K_{30} = 100$, $a_0 = 10$, $B_0 = 100$. Determine the semirelative pole sensitivities of the closed loop with respect to K_2 and a.

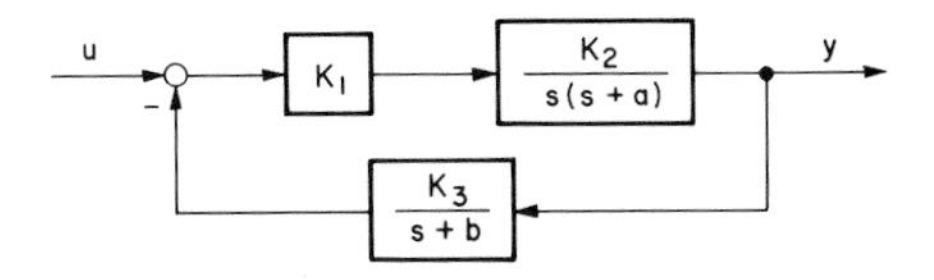

FIG. 6.P-1. Block diagram of the servo system of Problem 6.6.

6.7 Given the open-loop transfer function

$$M(s, \boldsymbol{\alpha}) = \frac{K}{(s + a)(s + b)(s^2 + cs + d)},$$

where $H(s) = 1$ and the nominal parameter values are $s_0 = 1, b_0 = 3, c_0 = 4, d_0 = 5$: Find the semirelative sensitivity function of the pole $s_1 = -1 + j$ of the closed loop with respect to the open-loop pole at $a = -1$.

Chapter 7

Sensitivity Comparison of Open-Loop and Closed-Loop Systems

7.1 INTRODUCTION

In technical feedback control systems parameter tolerances or parameter deviations during operation, especially in the plant, are basically unavoidable. Therefore the question of parameter sensitivity represents one of the fundamental problems of technical control, without whose clarification the applicability of the control remains questionable.

The two most important aims of automatic control are

(1) the reduction of errors due to external disturbing signals,
(2) the reduction of errors due to the change of system parameters.

This chapter deals with the comparison of parameter sensitivity of open-loop and closed-loop systems. The widespread view, originated from amplifier technology which asserts that feedback reduces the parameter sensitivity in any case, turns out to be incorrect. Therefore the primary question to be discussed here is whether and under what conditions a feedback control system will be less parameter-sensitive than an equivalent open-loop control system.

The first contributions to this problem were made by Bode [1]. Later, Horowitz [11], Perkins and Cruz [13], 77], Cruz [5], Cruz and Pertins [30, 31], Kreindler [58, 59] (and in Cruz [5]), and others [20, 21, 49, 92] dealt more fully with this topic. The sensitivity functions of the frequency domain are of particular importance. Bode has already succeeded in making a remarkable statement concerning the reducibility of parameter sensitivity in a closed-loop control. His so-called absolute-value integral theorem will be given and discussed in Section 7.2.3.

Besides the classical sensitivity methods of the frequency domain, in the formuluation of which we limit ourselves primarily to linear single-loop systems, the more recent methods of the time domain will be introduced in Section 7.3. In Section 7.4, we shall enter in more detail into the sensitivity comparison of optimal feedback controls.

7.2 SENSITIVITY COMPARISON IN THE FREQUENCY DOMAIN

7.2.1 THE BODE AND COMPARISON SENSITIVITY FUNCTIONS FOR OPEN-LOOP SYSTEMS

It was pointed out in Section 3.3.3 that for small parameter deviations the comparison sensitivity function S_p reduces to Bode's sensitivity function $S_P{}^G$. Let us first calculate the Bode sensitivity function for an open control chain of elements $R_1(s), \ldots, R_n(s)$, $P(s, \alpha)$, as shown in Fig. 7.2-1. It is assumed that only the plant transfer function $P(s,\alpha)$ is parameter-dependent. Determining Bode's sensitivity function with respect to P gives

$$S_P{}^{R_1 \ldots R_n P} = \frac{(R_{10} \ldots R_{no} P_0)}{P_0} \frac{P_0}{R_{10} \ldots R_{no} P_0} = 1. \tag{7.2-1}$$

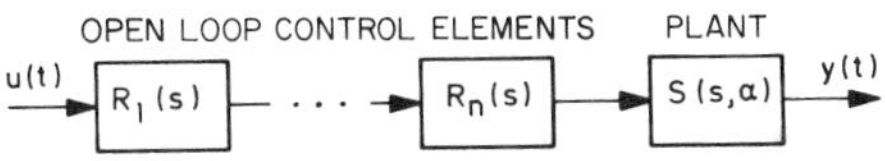

FIG. 7.2-1. For the calculation of Bode's sensitivity function of control chains.

It is seen that the Bode sensitivity function is equal to 1 no matter what the transfer functions of $R_r \ldots R_n$ are like. According to Section 3.3.1, the relative change of the output due to a parameter change $\Delta\alpha$ is given by

$$\frac{\Delta Y}{Y_0} = S_P{}^G S_\alpha{}^P \frac{\Delta\alpha}{\alpha_0} \tag{7.2-2}$$

where $S_\alpha{}^G$ is the Bode sensitivity function of the plant with respect to α. With $S_P{}^G = 1$ it follows that for an open chain,

$$\frac{\Delta Y}{Y_0} = S_\alpha{}^P \frac{\Delta\alpha}{\alpha_0}. \tag{7.2-3}$$

This says that the parameter sensitivity cannot be affected by circuit manipulations as long as the open chain structure is maintained. An alteration is only possible by introducing feedback.

7.2.2 THE SENSITIVITY FUNCTIONS FOR SINGLE-LOOP SYSTEMS

The basic idea of the reduction of parameter sensitivity by means of feedback is illustrated by the following simple example.

Example 7.2-1 Let K in the *block diagram* of Fig. 7.2-2 represent a gain

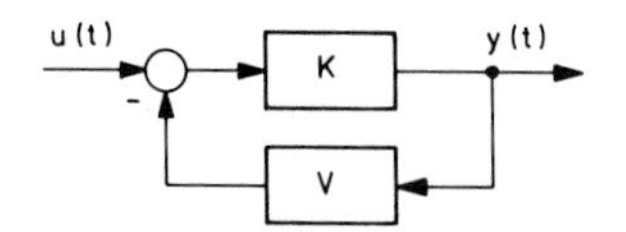

FIG. 7.2-2. Block diagram of a simple negative feedback configuration.

factor inaccurately determined or slowly varying with time, V is a constant. Determine the sensitivity of the overall transfer function $F = Y/U$ to infinitesimal deviations ΔK from K_0.

Solution The transfer function of the actual overall system is

$$G = \frac{K}{1 + KV} \tag{7.2-4}$$

and that of the nominal system is

$$G_0 = \frac{K_0}{1 + K_0 V}. \tag{7.2-5}$$

Thus the Bode sensitivity function reads

$$S_K{}^G = \left.\frac{\partial G/G}{\partial K/K}\right|_{K_0} = \frac{K_0(1 + K_0 V)}{K_0} \frac{\partial}{\partial K_0}\left[\frac{K_0}{1 + K_0 V}\right]$$
$$= \frac{(1 + K_0 V)(1 + K_0 V - K_0 V)}{(1 + K_0 V)^2}$$

or

$$S_K{}^G = \frac{1}{1 + K_0 V}. \tag{7.2-6}$$

The parameter-induced relative change $\Delta G/G_0$ then becomes

$$\frac{\Delta G}{G_0} = S_K{}^G \frac{\Delta K}{K_0} = \frac{1}{1 + K_0 V} \frac{\Delta K}{K_0}. \tag{7.2-7}$$

If there were no feedback, $S_K{}^G = 1$ and $\Delta G/G_0 = \Delta K/K_0$. This is also true if an element with the transfer function $1/(1 + K_0 V)$ is placed before the amplifier so that the open-loop system is nominally equivalent to the feedback system.

As can be seen from Eq.(7.2-7), the effect of $\Delta K/K_0$ is reduced by the factor $1/(1 + K_0 V)$ by the introduction of feedback. For $K_0 V > 0$, $1/(1 + K_0 V)$ will always be less than one and therefore the sensitivity in this example can never be greater than the sensitivity without feedback.

In order to generalize this result, let us now consider the "classical" unity feedback control system of Fig. 7.2-3. As was proved by Horowitz [11] this configuration represents all control loops with one degree of freedom. This means that for such control loops only one specification can be satisfied.

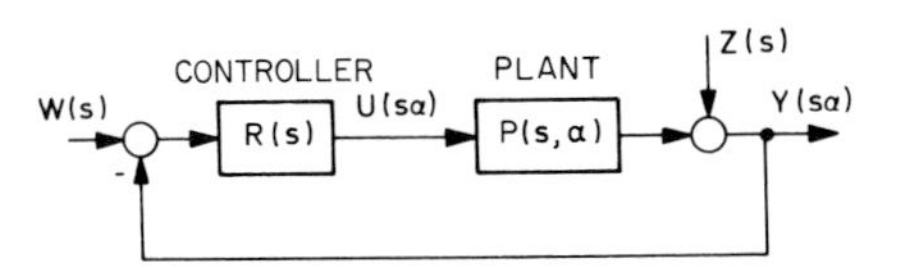

FIG. 7.2-3. Classical control loop with unity feedback and *one* degree of freedom.

Let the plant be parameter-dependent, its nominal transfer function being P_0. Then the nominal overall transfer function of the feedback control system $G_0 = Y_0/W$ becomes

$$G_0 = \frac{P_0 R}{1 + P_0 R} = \frac{F_0}{1 + F_0}, \tag{7.2-8}$$

where $F_0 = P_0 R$ represents the nominal open-loop transfer function. Then we obtain the Bode sensitivity function as

$$S_P{}^G = \frac{\partial G}{\partial P}\frac{P}{G}\bigg|_{P_0} = \frac{P_0 R(1 + P_0 R)}{(1 + P_0 R)^2 P_0 R} = \frac{1}{1 + P_0 R} = 1 - G_0. \tag{7.2-9}$$

The same result can be obtained if we proceed as did Horowitz (see Section 3.3.2):

$$\begin{aligned}\frac{\Delta G/G}{\Delta P/P} &= \frac{P(G - G_0)}{G(P - P_0)} = \frac{P}{P - P_0}\,\frac{(1 + F)[F/(1 + F) - F_0/(1 + F_0)]}{F}\\ &= \frac{P}{F(P - P_0)}\left(F - \frac{F_0(1 + F)}{1 + F_0}\right)\\ &= \frac{P(F - F_0)}{F(P - P_0)(1 + F_0)} = \frac{1}{1 + F_0} = \frac{1}{1 + P_0 R} = 1 - G_0.\end{aligned} \tag{7.2-10}$$

In order to interpret the results, let us consider the parameter-induced relative deviation $\Delta G/G_0$. If we multiply both the numerator and the denominator by W, we get

$$\frac{\Delta G\, W}{G_0 W} = \frac{1}{1 + P_0 R}\frac{\Delta P}{P_0} = \frac{\Delta Y}{Y_0}. \tag{7.2-11}$$

from which we obtain

$$\frac{\Delta Y/Y_0}{\Delta P/P_0} = \frac{1}{1 + P_0 R} = 1 - G_0. \tag{7.2-12}$$

This relation describes the relative change of the controlled variable due to a relative change of the plant transfer function.

On the other hand, we note that the Laplace transform of the control error due to inputs W and Z gives the same expression, namely,

$$E = W - Y_0 = W - \frac{P_0 R}{1 + P_0 R} W - \frac{1}{1 + P_0 R} Z = \frac{1}{1 + P_0 R}(W - Z)$$

or

$$\frac{E}{W - Z} = \frac{1}{1 + P_0 R}. \tag{7.2-13}$$

Thus comparison of Eqs.(7.2-13), (7.2-12), and (7.2-9) shows that

$$\frac{E}{W - Z} = \frac{\Delta Y/Y_0}{\Delta P/P_0} = S_P{}^G = \frac{1}{1 + P_0 R} = 1 - G_0. \tag{7.2-14}$$

It was further shown in Section 3.3.3 that the comparison sensitivity of Perkins and Cruz, which is in this case identical with $S_P{}^G$, relates the parameter-induced closed-loop error E_R to the error E_S of the nominally equivalent open loop according to the equation

$$S_P{}^G = S_p = \frac{E_R}{E_S} = \frac{\Delta Y_R}{\Delta Y_S}. \tag{7.2–15}$$

Since infinitesimal deviations are considered in the above case, we can use ∂Y for ΔY. Dividing both the numerator and denominator on the right-hand side by $\partial \alpha$, yields, with $\alpha = \alpha_0$ and $\boldsymbol{\Sigma} \triangleq \mathscr{L}\{\sigma\} = \mathscr{L}\{(\partial Y/\partial \alpha)_{\alpha 0}\}$, the relation $S_P{}^G = \Sigma_R/\Sigma_S$. This result can be summarized as follows:

Theorem 7.2-1 The Bode and the Horowitz sensitivity functions of the output of the classical control loop which are identical with the comparison sensitivity for infinitesimal deviations are equal to the inverse "return difference" $(1 + P_0 R)^{-1}$. They therefore relate the control error E to the input signal W–Z in the same fashion as the output sensitivity function $\boldsymbol{\Sigma}_R$ of the closed loop to Σ_S of the nominally equivalent open loop.

This means that small relative changes of the transfer function of the plant $\Delta P/P_0$ affect the relative change of the controlled variable $\Delta Y/Y_0$ in the same way as the external influence W–Z affects the control error E. Or in other words: If a control loop with *one* degree of freedom is designed so as to achieve a certain behavior with respect to the command and/or disturbance response, then it possesses the same behavior with respect to any parameter changes of the plant. If for example the controller contains an integrator in order to avoid a permanent control error due to any disturbing signal, this will also be true for parameter changes in the plant.

As a result, for a control loop with one degree of freedom, once the command and/or disturbance response is fixed, the parameter sensitivity will also be fixed. There is no way to alter one without the other. According to Eqs. (3.3-5) and (7.2-9), the output sensitivity function of the control loop in the frequency domain becomes,

$$\Sigma(s, \alpha_0) = S_P{}^G S_\alpha{}^P \frac{Y_0}{\alpha_0} = \frac{Y_0}{1 + P_0 R} S_\alpha{}^P \frac{1}{\alpha_0}. \tag{7.2-16}$$

In general, S_α^P is fixed for a given plant. The sensitivity can therefore only be influenced by the controller R.

7.2.3 Sensitivity Comparison of Single-Input Single-Output Control Loops

We now proceed to show how the sensitivity of a single input closed-loop control can be compared with that of a nominally equivalent open-loop control by means of the Bode sensitivity function.

For real frequencies $s = j\omega$ and $P_0(j\omega) \triangleq P(j\omega, \boldsymbol{\alpha}_0)$, the magnitude of the Bode sensitivity function of the control loop of Fig. 7.2-3 becomes

$$|S_P^G(j\omega)| = |1 + P_0(j\omega)R(j\omega)|^{-1}. \tag{7.2-17}$$

In contrast to the example of the amplifier considered earlier in this chapter, the sensitivity function of a general control loop is also a function of the frequency ω. Depending on the frequency, it can take on any value between zero and infinity. Since the sensitivity function of any control chain is always equal to one (whether or not it is nominally equivalent to the closed-loop control), S_P^G can serve as a criterion for the comparison of the sensitivity of the control loop with any control chain containing the same plant. For frequencies ω, at which $|S_P^G(j\omega)| < 1$, the parameter sensitivity of the control loop is smaller than that of the plant or of the control chain. For frequencies, at which $|S_P^G(j\omega)| > 1$, the parameter sensitivity of the control loop is larger than that of the open control chain, i.e., the parameter sensitivity is increased by introducing feedback.

Let us now illustrate the results obtained above in terms of the Nyquist locus. Suppose the Nyquist locus of the open loop $F_0(j\omega)$ is as shown in Fig. 7.2-4. The vector from the point -1 to the point $F_0(j\omega)$ represents the denominator of $S_P^G(j\omega)$. Thus

$$Q(j\omega) = 1 + F_0(j\omega). \tag{7.2-18}$$

This implies

$$|S_P^G(j\omega)| = |Q(j\omega)|^{-1}. \tag{7.2-19}$$

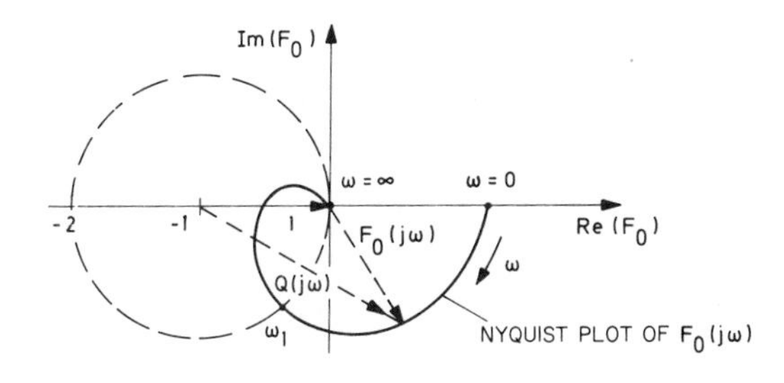

FIG. 7.2-4. Evaluating the sensitivity from the Nyquist diagram.

We can therefore make the following observation from Fig. 7.2-4: For those frequencies at which the locus $F_0(j\omega)$ cuts the unit circle centered at -1, the sensitivities of the closed-loop and open-loop systems are the same. For all those frequencies at which the locus $F_0(j\omega)$ lies outside the unit circle, $|Q^{-1}(j\omega)| < 1$ holds and the closed-loop system is less sensitive than the open-loop system. For all frequencies $\omega > \omega_1$ at which the locus $F_0(j\omega)$ lies inside the unit circle, $|S_P{}^G| = |Q^{-1}| > 1$ holds and the closed-loop system is more sensitive than the open-loop one.

From above the following can be seen: For a control system $F_0(j\omega)$ which has at least two more poles than zeros, there always exists a frequency ω_1 such that

$$|S_P{}^G(j\omega)| > 1 \qquad \text{for} \quad \omega > \omega_1. \tag{7.2-20}$$

In other words, there always exist frequencies at which the feedback control system is more sensitive than the uncontrolled one.

The frequency range $\omega < \omega_1$ for which $S_P{}^G < 1$ holds can be influenced in a wide range by a proper choice of $R(j\omega)$. This fact was already recognized by Bode and can be expressed by the following theorem:

Theorem 7.2-2 (Absolute-value-integral theorem of Bode) If the transfer function $F_0(s)$ of the open loop does not contain poles and $1 + F_0(s)$ does not contain zeros in the right half of the s-plane (poles and zeros on the imaginary axis are allowed) and if the number of poles of $F_0(s)$ exceeds the number of zeros at least by 2, then the following equality holds:

$$\int_0^\infty \ln |1 + F_0(j\omega)| \, d\omega = -\int_0^\infty \ln |S_P{}^G|(j\omega)| \, d\omega = 0. \tag{7.2-21}$$

In words: For a stable control system with a pole excess of at least 2, the logarithm of the magnitude of the sensitivity function is on the average equal to zero. This means that if the (logarithmic) Bode diagram of $|S_P{}^G|$ is drawn, the area enclosed with the 0–dB line for the region $|S_P{}^G| > 1$ is exactly the same as that for the region $|S_P{}^G| < 1$. The frequency ω_1 at which $|S_P{}^G| = 1$ can be specified by suitably choosing the transfer function $R(s)$ of the controller. For example, one may choose ω_1 such that $|S_P{}^G| < 1$ holds for the frequency range of interest, i.e., in which all the important spectral contributions of the signal lie. However, due to the above theorem, what is gained in sensitivity in the frequency range $\omega < \omega_1$ is lost again in the frequency range $\omega > \omega_1$.

The condition that the pole excess must be at least equal to 2 is almost always fulfilled in practice for systems with output feedback. In this case the above condition will surely be met if the plant possesses a low-pass of order higher than one because no pure differentiation in the controller is possible, and F_0, as a result, has at least the same pole excess as P_0. However this is not true for state feedback systems which will be treated later.

It is often desirable in connection with control-loop synthesis that the control quality be influenced without affecting the sensitivity. As we saw, the frequency ω_1 is fixed in a one-degree-of-freedom system as soon as the controller transfer function $R(s)$ is chosen. Moreover, by the choice of $R(s)$ the overall transfer function and thus the control quality will be fixed. In order to be able to influence the sensitivity and the control quality seperately, an additional degree of freedom has to be established. One may for example connect an additional element in front of the control loop. In this way, the control quality can be fixed independently of the sensitivity [11].

Example 7.2-2 The open-loop control system (a) and the closed-loop control system (b) of Fig. 7.2-5 have the same plant transfer function and the same overall transfer function $G(s) = Y(s)/W(s)$ for $K = 2$:

$$G_a(s) = \frac{K}{s^2 + 4s + 5}, \qquad G_b(s) = \frac{K}{s^2 + 4s + 3 + K}, \tag{7.2-22}$$

$$G_a(s)|_{K=2} = G_b(s)|_{K=2} = \frac{2}{s^2 + 4s + 5}, \tag{7.2-23}$$

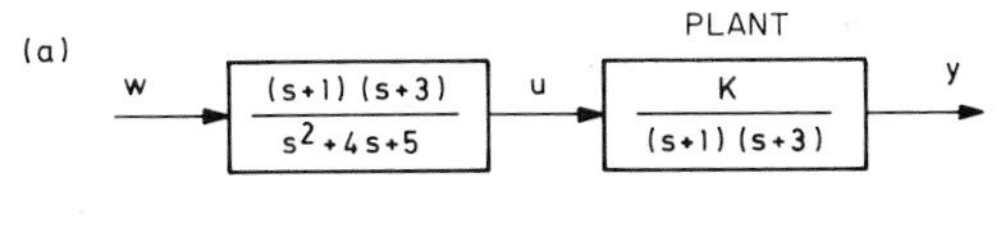

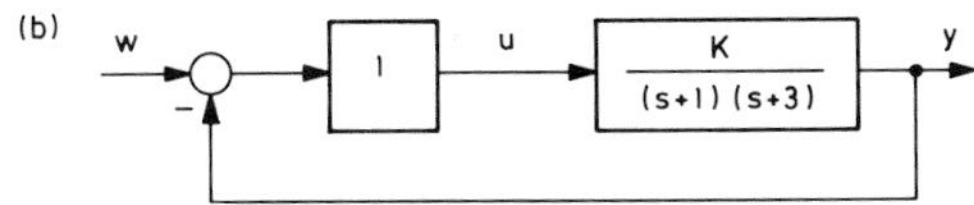

FIG. 7.2-5. Open-loop control system (a) and nominally equivalent closed-loop control system (b) of Example 7.2-2.

This means that they are nominally equivalent (for the nominal value $K = 2$). For deviations of the amplification factor K from the value 2, their properties will be considerably different. This is to be shown by means of the sensitivity functions with respect to K.

For *the open-loop system* (a), the rule [Eq. (3.3-20)] derived in Section 3.3.1 yields, with $G_1 = 0$, $G_2 = 1$, $G_3 = s^2 + 4s + 5$, and $G_4 = 0$, the same result for the Bode sensitivity function $S_K{}^G$ and the comparison sensitivity function $S_{p,a}$, namely,

$$S_K^{G_a}(s) = S_{P,a}(s) = \frac{K(s^2 + 4s + 5)}{(s^2 + 4s + 5)K} = 1. \tag{7.2-24}$$

For *the closed-loop system* (b) with $G_1 = 0$, $G_2 = 1$, $G_3 = s^2 + 4s + 3$, and $G_4 = 1$, the comparison sensitivity function $S_{p,b}$ becomes

$$S_{p,b}(s) = \frac{K(s^2 + 4s + 3)}{(s^2 + 4s + 3 + K)K} = \frac{1}{1 + K/(s^2 + 4s + 3)}, \qquad (7.2\text{-}25)$$

and with $K = 2$, the Bode sensitivity function is

$$S_K^{G_b}(s) = [1 + 2/(s^2 + 4s + 3)]^{-1}. \qquad (7.2\text{-}26)$$

Note that for the open-loop system (a), $S_K{}^G = S_p = 1$ for all values K and all frequencies. For the closed-loop system (b), the comparison sensitivity function is a function of K as well as a function of s and the Bode sensitivity function is a function of s only.

In order to determine the frequency ω_1 at which the open-loop and closed-loop controls are equally sensitive, we set

$$|S_K^{G_b}(j\omega)| = \frac{|(j\omega)^2 + 4j\omega + 3|}{|(j\omega)^2 + 4j\omega + 5|} \doteq 1. \qquad (7.2\text{-}27)$$

Solving for $\omega = \omega_1$, we obtain $\omega_1 = 2$, i.e., for frequencies $\omega < 2$ the closed-loop control is less sensitive; for frequencies in the range $2 < \omega < \infty$ the closed-loop control is more sensitive than the open-loop one. In the case of the closed-loop control the limit ω_1 can be shifted by varying K or the amplification factor of the controller element, which has so far been taken to be 1. The dependence of ω_1 on K is for $K \neq 0$ expressed by

$$\omega_1 = \sqrt{3 + \tfrac{1}{2} K}. \qquad (7.2\text{-}28)$$

It is important to note that the limitations of the absolute-value integral theorem of Bode are not concerned with *state feedback*. In this case the assumption that the pole excess must be larger than 1 is, in general, not satisfied. This will be shown next by transforming a linear single-variable control loop with complete proportional state feedback to the classical output feedback configuration [8].

Figure 7.2-6 shows the general scheme of state feedback. Let the plant functions $P_\nu(s, \boldsymbol{\alpha})$ be of the form:

$$P_\nu(s, \boldsymbol{\alpha}) = \frac{a_\nu(\boldsymbol{\alpha}) + b_\nu(\boldsymbol{\alpha})s}{c_\nu(\boldsymbol{\alpha}) + d_\nu(\boldsymbol{\alpha})s}, \qquad d_\nu \neq 0, \quad \nu = 1, 2, \ldots, n. \qquad (7.2\text{-}29)$$

If the different tappings of the plant are transferred to the plant output, the control loop takes the form of an output feedback configuration of the type (b) in Fig. 3.3-6. The transfer function of the equivalent feedback element H is now given by

$$H(s, \boldsymbol{\alpha}) = k_1 + k_2 P_1^{-1}(s, \boldsymbol{\alpha}) + \cdots + k_n \prod_{\nu=1}^{n-1} P_\nu^{-1}(s, \boldsymbol{\alpha}). \qquad (7.2\text{-}30)$$

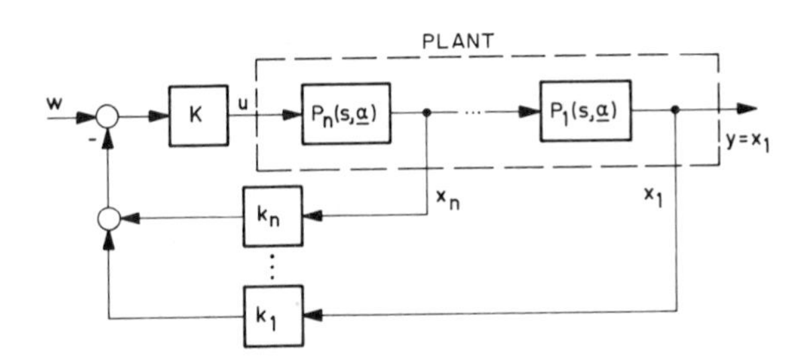

FIG. 7.2-6. Proportional state feedback.

When the Bode sensitivity function of the control loop G with respect to $\boldsymbol{\alpha}$ is calculated, it should be noted that H is also dependent on $\boldsymbol{\alpha}$. By the chain rule,

$$S_{\boldsymbol{\alpha}}{}^{G} = S_P{}^G S_{\boldsymbol{\alpha}}{}^P + S_H{}^G S_{\boldsymbol{\alpha}}{}^H$$

with $S_P{}^G = 1/(1 + H_0KP_0)$ and $S_H{}^G = -H_0KP_0/(1 + H_0KP_0)$. H_0KP_0 is given by

$$H_0KP_0 = K[k_1 \prod_{\nu=1}^{n} P_\nu(s, \boldsymbol{\alpha}) + \cdots + k_n \prod_{\nu=n}^{n} P_\nu(s, \boldsymbol{\alpha})]. \tag{7.2-31}$$

Substituting the expressions of Eq. (7.2-29) into (7.2-31), we see that H_0KP_0 i.e., the transfer function of the open loop, cannot have a pole excess greater than one. Therefore for complete state feedback the closed-loop control can be less sensitive than the open-loop control for all frequencies. Note however that this applies only to *complete* state feedback.

Example 7.2-3 Consider the state feedback of a double integration system (e.g., single-axis attitude control of a space vehicle) shown in Fig. 7.2-7. Here, $H(s) = 1 + s$ and hence

$$F_0(s) = \frac{K(1 + s)}{s^2}. \tag{7.2-32}$$

The Bode sensitivity function is then given by

$$S_P{}^G(s) = \frac{1}{1 + F_0(s)} = \frac{s^2}{s^2 + Ks + K}. \tag{7.2-33}$$

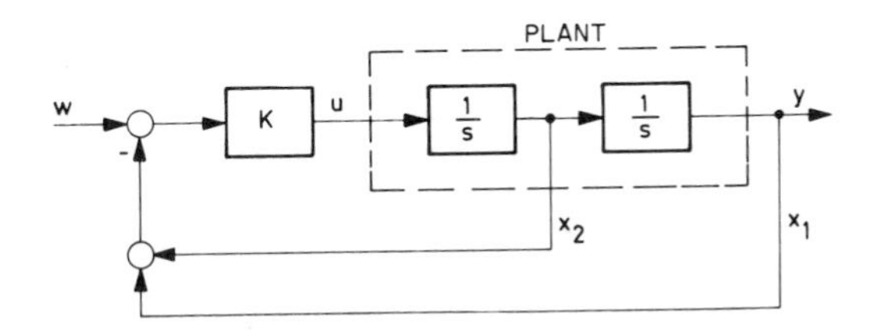

FIG. 7.2-7. State feedback of Example 7.2-3.

With $s = j\omega$, its magnitude is

$$|S_P{}^G(j\omega)| = \frac{\omega^2}{\sqrt{(K-\omega^2)^2 + K^2\omega^2}}. \tag{7.2-34}$$

As can be seen, $|S_P{}^G(j\omega)| \leq 1$ for $K = 2$ and all ω, i.e., for $K = 2$, the closed-loop control is less sensitive than the nominally equivalent open-loop control over the whole frequency range. That this happens to be true for $K = 2$ lies in the fact that this control loop is optimal according to the criterion

$$I = \int_0^\infty (w - y)^2 + \tfrac{1}{4}u^2]\, dt.$$

As will be more generally shown in Section 7.4, a control loop optimized according to such a criterion will be least sensitive as well.

7.2.4 Sensitivity Comparison for Linear Multivariable Control Systems

The results obtained so far can be directly extended to multivariable, nonlinear, and time-variant systems [5, 30–32, 77, 78].

Consider first a multivariable system with an output feedback of the type shown in Fig. 7.2-3. Here, the signals W, E, V, Z, Y are to be conceived as vectors $\boldsymbol{W}$, $\boldsymbol{E}$, $\boldsymbol{V}$, $\boldsymbol{Z}$, $\boldsymbol{Y}$ and the transfer functions R, P as transfer matrices $\boldsymbol{R}$, $\boldsymbol{P}$. According to Section 3.3.3, the parameter-induced deviations of the output vector $\Delta\boldsymbol{Y}$ for multivariable closed- and open-loop controls are related by $\Delta\boldsymbol{Y}_R = \boldsymbol{S}_p\, \Delta\boldsymbol{Y}_s$, where the comparison sensitivity $\boldsymbol{S}_p$ is given by the quadratic matrix

$$\boldsymbol{S}_p = (I + \boldsymbol{F})^{-1} = (\boldsymbol{I} + \boldsymbol{P}\,\boldsymbol{R})^{-1}. \tag{7.2-35}$$

The expression $\boldsymbol{I} + \boldsymbol{F}$ represents the generalized return difference, $\boldsymbol{F}$ the actual transfer matrix of the open loop. $\boldsymbol{S}_p$ is always quadratic. If the transfer matrix of the closed loop

$$\boldsymbol{G} = (\boldsymbol{I} + \boldsymbol{P}\boldsymbol{R})^{-1}\boldsymbol{P}\boldsymbol{R} \tag{7.2-36}$$

is quadratic and not singular for all ω, and $\boldsymbol{P}_0$, $\boldsymbol{G}_0$ denote the nominal and $\boldsymbol{P}$, $\boldsymbol{G}$ the actual matrices, as well as $\Delta\boldsymbol{P} = \boldsymbol{P}_0 - \boldsymbol{P}$ and $\Delta\boldsymbol{G} = \boldsymbol{G}_0 - \boldsymbol{G}$ the finite deviations, we can derive from Eq. (7.2-35) the following expression for $\boldsymbol{S}_p$ [30]:

$$\boldsymbol{S}_p = \Delta\boldsymbol{G}\,\boldsymbol{G}_0{}^{-1}\,\boldsymbol{P}_0\Delta\boldsymbol{P}^{-1}. \tag{7.2-37}$$

This is a direct extension of the Horowitz sensitivity function to the multivariable case. For small parameter deviations it reduces to the Bode sensitivity definition for the multivariable system. For infinitesimal parameter variations the sensitivity functions are given by the inverse return difference $(\boldsymbol{I} + \boldsymbol{F}_0)^{-1}$ at nominal plant parameter values.

If it is required that the closed-loop control should be less sensitive than the open-loop control, then, for the whole frequency range of interest, the following inequality must be satisfied:

$$\boldsymbol{S}_P{}^T(-j\omega)\boldsymbol{S}_p(j\omega) \leq \boldsymbol{I}. \tag{7.2-38}$$

Therefore, in the multivariable case, there must be an analogous limitation as in the single-variable case which may be expressed by a corresponding generalized absolute-value integral theorem.

Let us now determine the expression for the comparison sensitivity for *state feedback*. Suppose the plant is given by the actual state equation

$$\dot{\boldsymbol{x}}(t, \boldsymbol{\alpha}) = \boldsymbol{A}(\boldsymbol{\alpha})\boldsymbol{x}(t, \boldsymbol{\alpha}) + \boldsymbol{B}(\boldsymbol{\alpha})\boldsymbol{u}(t, \boldsymbol{\alpha}), \qquad \boldsymbol{x}(0) = \boldsymbol{x}^0. \tag{7.2-39}$$

It is assumed that the parameter deviations $\Delta\boldsymbol{\alpha} = \boldsymbol{\alpha}_0 - \boldsymbol{\alpha}$ are finite (instead of infinitesimal) and time-invariant.

Consider first the *open-loop case*. Denoting the parameter-induced trajectory deviation by

$$\Delta\boldsymbol{x}_S(t) \triangleq \boldsymbol{x}_S(t, \boldsymbol{\alpha}_0) - \boldsymbol{x}_S(t, \boldsymbol{\alpha}) = \boldsymbol{x}_{S0} - \boldsymbol{x}_S, \tag{7.2-40}$$

we obtain from Eq. (7.2-39), with $\boldsymbol{A} = \boldsymbol{A}_0 - \Delta\boldsymbol{A}$, $\boldsymbol{B} = \mathbf{B}_0 - \Delta\boldsymbol{B}$, the following sensitivity equation:

$$\Delta\dot{\boldsymbol{x}}_S = \boldsymbol{A}\,\Delta\boldsymbol{x}_S + \Delta\boldsymbol{A}\,\boldsymbol{x}_{S0} + \Delta\boldsymbol{B}\,\boldsymbol{u}_{S0}. \tag{7.2-41}$$

Laplace transformation yields

$$\Delta\boldsymbol{X}_S(s) = (s\boldsymbol{I} - \boldsymbol{A})^{-1}[\Delta\boldsymbol{A}\,\boldsymbol{X}_{S0}(s) + \Delta\boldsymbol{B}\,\boldsymbol{U}_{S0}(s)]. \tag{7.2-42}$$

Next consider the *closed-loop case*. Define the parameter-induced trajectory deviation correspondingly as

$$\Delta\boldsymbol{x}_R(t) \triangleq \boldsymbol{x}_R(t, \boldsymbol{\alpha}_0) - \boldsymbol{x}_R(t, \boldsymbol{\alpha}) = \boldsymbol{x}_{R0} - \boldsymbol{x}_R. \tag{7.2-43}$$

Let the state $\boldsymbol{x}_R$ be proportionally fed back according to the linear control low $-\boldsymbol{K}\boldsymbol{x}_R$. This signal is added to the reference variable vector $\boldsymbol{w}(t)$ such that

$$\boldsymbol{u}_R = \boldsymbol{w} - \boldsymbol{K}\boldsymbol{x}_R. \tag{7.2-44}$$

Then the following sensitivity equation is obtained:

$$\Delta\dot{\boldsymbol{x}}_R = \boldsymbol{A}\,\Delta\boldsymbol{x}_R + \Delta\boldsymbol{A}\,\boldsymbol{x}_{R0} + \Delta\boldsymbol{B}\,\boldsymbol{u}_{R0} - \boldsymbol{B}\boldsymbol{K}\,\Delta\boldsymbol{x}_R. \tag{7.2-45}$$

Applying Laplace transformation and solving for $\Delta\boldsymbol{X}_R(s)$, we obtain

$$\Delta\boldsymbol{X}_R(s) = (s\boldsymbol{I} - \boldsymbol{A} + \boldsymbol{B}\boldsymbol{K})^{-1}\,[\Delta\boldsymbol{A}\,\boldsymbol{X}_{R0}(s) + \Delta\boldsymbol{B}\,\boldsymbol{U}_{R0}(s)]. \tag{7.2-46}$$

For nominal equivalence, the following equalities must hold:

$$\boldsymbol{U}_{R0} = \boldsymbol{U}_{S0} \qquad \text{and} \qquad \boldsymbol{X}_{R0} = \boldsymbol{X}_{S0}. \tag{7.2-47}$$

Thus from Eqs. (7.2-42) and (7.2-46),

$$\Delta\boldsymbol{X}_R(s) = (s\boldsymbol{I} - \boldsymbol{A} + \boldsymbol{B}\boldsymbol{K})^{-1}(s\boldsymbol{I} - \boldsymbol{A})\,\Delta\boldsymbol{X}_S(s). \tag{7.2-48}$$

The comparison sensitivity matrix which relates the trajectory deviations ΔX for closed- and open-loop control, becomes

$$\tilde{S}_P = (s\boldsymbol{I} - \boldsymbol{A} + \boldsymbol{BK})^{-1}(s\boldsymbol{I} - \boldsymbol{A}). \tag{7.2-49}$$

Since the Laplace transform of the transition matrix of the actual open-loop system is

$$\phi_S(s) = (s\boldsymbol{I} - \boldsymbol{A})^{-1} \tag{7.2-50}$$

and that of the actual closed-loop system is

$$\phi_R(s) = (s\boldsymbol{I} - \boldsymbol{A} + \boldsymbol{BK})^{-1}, \tag{7.2-51}$$

$\tilde{S}_p$ can be written as

$$\tilde{S}_p(s) = \phi_R(s)\phi_S^{-1}(s). \tag{7.2-52}$$

Finally, by means of a simple transformation of Eq. (7.2-49), $\tilde{S}_p$ can also be written as

$$\tilde{S}_p(s) = [\boldsymbol{I} + \phi_S(s)\boldsymbol{BK}]^{-1}. \tag{7.2-53}$$

$\tilde{S}_p$ relates the trajectory deviation $\Delta \boldsymbol{x}_R$ of the state feedback to $\Delta \boldsymbol{x}_S$ of the nominally equivalent open-loop control chain. As can be easily seen in Fig. 7.2-8, the expression $\phi_S(s)\boldsymbol{BK} \triangleq \tilde{\boldsymbol{F}}(t)$ is the transfer matrix of the open control loop obtained by cutting off the connection at the state $\boldsymbol{x}$. $(\boldsymbol{I} + \tilde{\boldsymbol{F}})$ represents, therefore, the "return difference" of the control loop. If the trajectory of the state feedback should be less sensitive than that of the open-loop control, it must be required that the inequality

$$|\boldsymbol{I} + \phi_S(j\omega)\boldsymbol{BK}| \geq \boldsymbol{I} \tag{7.2-54}$$

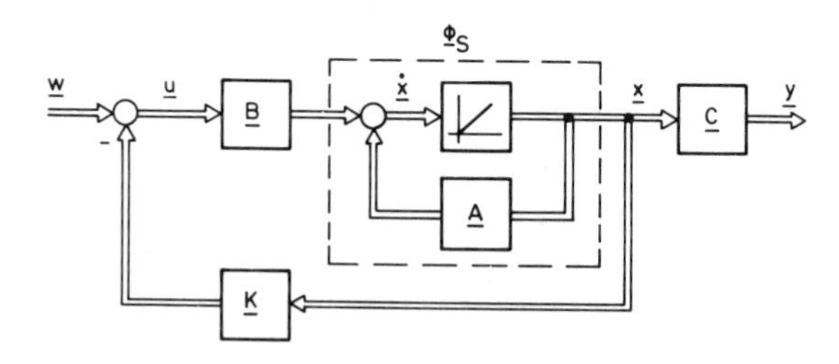

FIG. 7.2-8. General block diagram of the linear state feedback.

holds for the frequency range of interest. Note that this sensitivity measure says nothing about the sensitivity of the output vector $\boldsymbol{y} = \boldsymbol{Cx}$ to the parameters. A corresponding simple expression $\boldsymbol{S}_p$ for the output deviation ΔY, as in the case of the output feedback, does not exist here (see Kreindler [58] and Section 7.3). This means a certain limitation of the above sensitivity measure since in control systems it is often the output that is of primary interest.

7.3 SENSITIVITY COMPARISON IN THE TIME DOMAIN

It is often desirable to discuss the reduction of the sensitivity of a control loop in the time domain, i.e., in terms of output or state variables. As an appropriate measure that, on the whole, takes all contributions into consideration, a norm of the parameter-induced deviation or of the corresponding sensitivity function is used. Such a norm can be the L_2-norm, i.e., the integral of a quadratic form. This can also be regarded as a generalization of the system error integral of Mazer [66] defined in Section 3.2.7 [Eq. (3.2-81)]:

$$I_M = \int_0^{t_1} \boldsymbol{e}^{\mathrm{T}}(\mathbf{t}) \boldsymbol{Z}\, \boldsymbol{e}(t)\, dt. \tag{7.3-1}$$

When the parameters are finite, time-variant, or vectorial, $\boldsymbol{e}$ is to be set equal to the parameter-induced trajectory or output deviation; thus

$$\boldsymbol{e}(t) = \Delta \boldsymbol{y}(t) = \boldsymbol{y}(t, \boldsymbol{\alpha}_0) - \boldsymbol{y}(t, \boldsymbol{\alpha}). \tag{7.3-2}$$

However when the parameters are infinitesimal, time-invariant, or scalar, then the corresponding sensitivity vectors $\boldsymbol{\lambda}(t, \boldsymbol{\alpha}_0)$ or $\boldsymbol{\sigma}(t, \boldsymbol{\alpha}_0)$ can be used, i.e.,

$$I_M = \int_0^{t_1} \boldsymbol{\sigma}^{\mathrm{T}}(t, \alpha_0) \boldsymbol{Z}\, \boldsymbol{\sigma}(t, \alpha_0)\, dt. \tag{7.3-3}$$

The matrix $\boldsymbol{Z}$ in Eqs. (7.3-1) and (7.3-3) is a symmetric positive semidefinite matrix. By means of this matrix the different components of $\boldsymbol{e}^{\mathrm{T}} \boldsymbol{Z}\, \boldsymbol{e}$ or $\boldsymbol{\sigma}^{\mathrm{T}} \boldsymbol{Z}\, \boldsymbol{\sigma}$ in I_m can be individually weighted or some of them can even be disregarded. The constant $t_1 < 0$ represents the time elapse of interest.

Denoting the closed-loop control with the subscript R and the open-loop control with the subscript S, we now define:

Definition 7.3-1 A closed-loop control is called less parameter-sensitive than its nominally equivalent open-loop control in the sense of the criterion I_M if the inequality

$$\int_0^{t_1} \boldsymbol{e}_R{}^{\mathrm{T}}(t) \boldsymbol{Z}\, \boldsymbol{e}_R(t)\, dt \leq \int_0^{t_1} \boldsymbol{e}_S{}^{\mathrm{T}}(t) \boldsymbol{Z}\, \boldsymbol{e}_S(t)\, dt \tag{7.3-4}$$

holds for all t_1 in the interval $0 < t_1 \leq \infty$ and for a bounded reference variable $\boldsymbol{w}(t)$, that is,

$$\int_0^{t_1} \boldsymbol{w}^{\mathrm{T}}(t) \boldsymbol{Z}\, \boldsymbol{w}(t)\, dt < \infty.$$

As has already been shown by Perkins and Cruz and others [10, 30, 31, 53, 77, 78], this condition in the time domain can be expressed by a corresponding condition in terms of the comparison sensitivity. The latter condition will now be derived.

We consider first the simple special case of a linear single-variable plant with output feedback and $t_1 = \infty$ in the integral of Eq. (7.3-4). Then we generalize the result to arbitrary $t_1 > 0$ and arbitrary plants with output or state feedback. In this connection we will first consider the output sensitivity for output feedback, then the state sensitivity for state feedback, and finally the output sensitivity for state feedback.

In Section 7.2.3, we used the inequality

$$|S_P{}^G(j\omega)| \leq 1 \tag{7.3-5}$$

as a criterion for the reduction of the output sensitivity achieved by output feedback. Let us interpret this criterion in terms of the time-domain criterion (7.3-1). In the present case the latter takes the form

$$I_M = \int_0^\infty e_R{}^2(t)\, dt, \tag{7.3-6}$$

where t_1 has been set equal to ∞. Using Parseval's theorem and with $E(s) \triangleq \mathscr{L}\{e(t)\}$,

$$I_M = \int_{-\infty}^{\infty} |E_R(j\omega)|^2\, d\omega. \tag{7.3-7}$$

According to Theorem 7.2-1, E_R is related to the corresponding E_S of the nominally equivalent open-loop control by

$$E_R(s) = S_P{}^G(s)\, E_S(s). \tag{7.3-8}$$

Thus Eq. (7.3-7) can be written as

$$I_M = \int_{-\infty}^{\infty} |E_R(j\omega)|^2\, d\omega - \int_{-\infty}^{\infty} |S_P{}^G(j\omega)|^2 |E_S(j\omega)|^2\, d\omega. \tag{7.3-9}$$

Now, if $|S_P{}^G(j\omega)| \leq 1$ for all ω, then by Eq. (7.3-9),

$$\int_{-\infty}^{\infty} |E_R(j\omega)|^2\, d\omega \leq \int_{-\infty}^{\infty} |E_S(j\omega)|^2\, d\omega \tag{7.3-10}$$

and hence

$$\int_0^\infty e_R{}^2(t)\, dt \leq \int_0^\infty e_S{}^2(t)\, dt. \tag{7.3-11}$$

This result shows that condition (7.3-5) expresses the fact that the parameter-induced output error $e_R = \Delta y_R$ of the closed-loop control is, on the quadratic average, smaller than that of its nominally equivalent open-loop control e_S. This is an interesting time-domain interpretation of condition (7.3-5) which has been independently obtained by frequency domain considerations.

The above result will now be extended to $t \neq \infty$ and arbitrary output feedbacks. That means a condition for the comparison sensitivity function $\boldsymbol{S}_p$ will be given which is equivalent to the inequality (7.3-4)

$$\int_0^{t_1} \boldsymbol{e}_R{}^{\mathrm{T}}(t) \boldsymbol{Z}\, \boldsymbol{e}_R(t)\, dt \leq \int_0^{t_1} \boldsymbol{e}_S{}^{\mathrm{T}}(t) \boldsymbol{Z}\, \boldsymbol{e}_S(t)\, dt.$$

In the following derivation, we first confine ourselves to linear time-invariant multivariable systems which may have finite parameter deviations.

For the case $t_1 = \infty$, the application of the Parseval theorem enables us to write Eq. (7.3-4) as

$$\int_{-\infty}^{\infty} \boldsymbol{E}_R^{\mathrm{T}}(j\omega)\boldsymbol{Z}\,\boldsymbol{E}_R(j\omega)\,d\omega \leq \int_{-\infty}^{\infty} \boldsymbol{E}_S^{\mathrm{T}}(-j\omega)\boldsymbol{Z}\boldsymbol{E}_S(j\omega)\,d\omega, \qquad (7.3\text{-}12)$$

where $\boldsymbol{E}(s) = \mathscr{L}\{\boldsymbol{e}(t)\}$. In view of

$$\boldsymbol{E}_R(s) = \boldsymbol{S}_P(s)\boldsymbol{E}_S(s), \qquad (7.3\text{-}13)$$

we obtain

$$\int_{-\infty}^{\infty} \boldsymbol{E}_S^{\mathrm{T}}(-j\omega)[\boldsymbol{S}_P^{\mathrm{T}}(-j\omega)\boldsymbol{Z}\,\boldsymbol{S}_P(j\omega) - \boldsymbol{Z}]\boldsymbol{E}_S(j\omega)\,d\omega \leq 0. \qquad (7.3\text{-}14)$$

Obviously, a sufficient condition in order to satisfy this inequality is that for all ω,

$$\boldsymbol{S}_P^{\mathrm{T}}(-j\omega)\boldsymbol{Z}\boldsymbol{S}_p(j\omega) - \boldsymbol{Z} \leq 0, \qquad (7.3\text{-}15)$$

where the equality sign holds only for some values of ω, i.e., $\boldsymbol{S}_P^{\mathrm{T}}(-j\omega)\boldsymbol{Z} \cdot \boldsymbol{S}(j\omega) - \boldsymbol{Z}$ is negative semidefinite.

In order to consider the case $t_1 \neq \infty$, let us introduce two auxiliary functions $\tilde{\boldsymbol{e}}_S(t)$ and $\tilde{\boldsymbol{e}}_R(t)$ defined as

$$\tilde{\boldsymbol{e}}_S(t) = \begin{cases} \boldsymbol{e}_S(t) & \text{for} \quad t \leq t_1, \\ \boldsymbol{0} & \text{for} \quad t > t_1, \end{cases} \qquad (7.3\text{-}16)$$

$$\tilde{\boldsymbol{E}}_R(s) = \boldsymbol{S}_p(s)\tilde{\boldsymbol{E}}_S(s). \qquad (7.3\text{-}17)$$

Due to the causality of $\boldsymbol{S}_p$, $\tilde{\boldsymbol{e}}_R = \boldsymbol{e}_R$ for $t \leq t_1$. Furthermore, by the definition of $\tilde{\boldsymbol{e}}_S$,

$$\int_0^{\infty} \tilde{\boldsymbol{e}}_S^{\mathrm{T}}(t)\boldsymbol{Z}\,\tilde{\boldsymbol{e}}_S(t)\,dt = \int_0^{t_1} \boldsymbol{e}_S^{\mathrm{T}}(t)\boldsymbol{Z}\,\boldsymbol{e}_S(t)\,dt. \qquad (7.3\text{-}18)$$

In a manner similar to that for the case $t_1 = \infty$, we can show that Eq. (7.3-15) is a sufficient condition for

$$\int_0^{\infty} \tilde{\boldsymbol{e}}_R^{\mathrm{T}}\boldsymbol{Z}\,\tilde{\boldsymbol{e}}(t)\,dt \leq \int_0^{\infty} \tilde{\boldsymbol{e}}_S^{\mathrm{T}}(t)\boldsymbol{Z}\,\tilde{\boldsymbol{e}}_S(t)\,dt. \qquad (7.3\text{-}19)$$

Substituting Eq. (7.3-18) into the right-hand side of (7.3-19) and noting that $\tilde{\boldsymbol{e}}_R = \boldsymbol{e}_R$ for $t \leq t_1$, we obtain

$$\int_0^{t_1} \boldsymbol{e}_S^{\mathrm{T}}(t)\boldsymbol{Z}\,\boldsymbol{e}_S(t)\,dt \geq \int_0^{t_1} \boldsymbol{e}_R^{\mathrm{T}}(t)\boldsymbol{Z}\,\boldsymbol{e}_R(t)\,dt + \int_{t_1}^{\infty} \tilde{\boldsymbol{e}}_R^{\mathrm{T}}(t)\boldsymbol{Z}\,\tilde{\boldsymbol{e}}_R(t)\,dt. \qquad (7.3\text{-}20)$$

By neglecting the second integral on the right-hand side, which is permissible since it is nonnegative, Eq. (7.3-20) is reduced to Eq. (7.3-4). Hence Eq. (7.3-15) is also a sufficient condition for Eq. (7.3-4).

In the special case of a single-variable control loop, where $\boldsymbol{S}_p = S_P^G$, condition (7.3-15) reduces to $|S_P^G(j\omega)| \leq 1$. This condition can therefore be regarded as an equivalent to the time domain condition

$$\int_0^{t_1} e_R^2(t)\,dt \leq \int_0^{t_1} e_S^2(t)\,dt \qquad \text{for all} \quad t_1 > 0. \tag{7.3-21}$$

Note that the above results obtained for linear time-invariant multivariable systems is valid for finite deviations as well. Furthermore they can be generalized in an analogous manner to nonlinear time-variant systems as shown by Cruz [5], Cruz and Perkins [32], and Kreindler [58]. Here the condition (7.3-15) on the comparison sensitivity is also valid. The only difference is that in this case the comparison sensitivity is defined as an operator.

We now summarize the results obtained in the following:

Theorem 7.3-1 A sufficient but in general not necessary condition such that a closed-loop control is less sensitive (in the sense of the quadratic criterion I_M) than its nominally equivalent open-loop control is

$$\boldsymbol{S}_p^{\mathrm{T}}(-j\omega)\,\boldsymbol{Z}\boldsymbol{S}_p(j\omega) \leq \boldsymbol{Z} \tag{7.3-22}$$

for all ω, whereas the sign of equality does not hold for all ω.

Condition (7.3-22) will also be necessary when $\boldsymbol{e}_S(t)$ is an arbitrary square integrable vector. For example, this is the case when the reference vector $\boldsymbol{w}(t)$ is arbitrary and the parameter-induced deviation of the transfer matrix of the closed-loop (and of the nominally equivalent open-loop) is invertible, i.e., when $\Delta\boldsymbol{G}^{-1}$ exists, where $\Delta\boldsymbol{G} = \boldsymbol{G}(s, \boldsymbol{\alpha}_0) - \boldsymbol{G}(s, \boldsymbol{\alpha})$.

For the application of condition (7.3-22) to practical cases, it is often convenient to make the following substitutions:

$$\boldsymbol{S}_p^{-1} = \boldsymbol{I} + \boldsymbol{F}_0, \tag{7.3-23}$$

where $\boldsymbol{F}_0 = \boldsymbol{P}_0\boldsymbol{R}$ represents the transfer operator of the open-loop (in analogy to Fig. 7.2-3). Equation (7.3-22) can be rewritten as

$$[\boldsymbol{S}_p^{-1}(-j\omega)]^{\mathrm{T}}\boldsymbol{Z}\,\boldsymbol{S}_p^{-1}(j\omega) \geq \boldsymbol{Z}, \tag{7.3-24}$$

which, by substituting $\boldsymbol{I} + \boldsymbol{F}_0$ for $\boldsymbol{S}_p^{-1}$, becomes

$$[\boldsymbol{I} + \boldsymbol{F}_0(-j\omega)]^{\mathrm{T}}\boldsymbol{Z}[\boldsymbol{I} + \boldsymbol{F}_0(j\omega)] \geq \boldsymbol{Z}. \tag{7.3-25}$$

If this condition is satisfied for all ω, whereas the sign of equality does not hold for all ω, then the closed-loop control will be less sensitive than its nominally equivalent open-loop control in the sense of the criterion $\boldsymbol{I}_M$ for all $t_1 > 0$.

In an analogous manner, we find the condition such that the *trajectory sensitivity* or the *trajectory deviation* of a system with *state feedback* is smaller than that of the corresponding open-loop chain in the sense of the I_M criterion. This is

$$\tilde{\boldsymbol{S}}_p^{\mathrm{T}}(-j\omega)\boldsymbol{Z}\,\tilde{\boldsymbol{S}}_p(j\omega) \leq \boldsymbol{Z}, \tag{7.3-26}$$

with $\tilde{S}_p$ given by Eq. (7.2-53).

Finally, if it is required that the *output sensitivity* or the *output deviation* of a *state-feedback system* should be smaller than that of the open-loop system in the sense of the following criterion:

$$\int_0^{t_1} e_R^{\mathrm{T}}(t)\tilde{Z}\, e_R(t)\, dt \leq \int_0^{t_1} e_S^{\mathrm{T}}(t)\tilde{Z}\, e_S(t)\, dt, \tag{7.3-27}$$

where e can be equal either to Δy or σ, then the frequency domain condition can be found as follows: Let $y = Cx$, where C is independent of α. Then $\Delta y = C\,\Delta x$ (and $\sigma = C\lambda$). Substituting this relation into Eq. (7.3-27) yields

$$\int_0^{t_1} \Delta x_R^{\mathrm{T}}\, C^{\mathrm{T}}\tilde{Z}\, C\, \Delta x_R\, dt \leq \int_0^{t_1} \Delta x_S^{\mathrm{T}}\, C^{\mathrm{T}}\tilde{Z}\, C\, \Delta x_S\, dt, \tag{7.3-28}$$

from which the sufficient frequency domain condition can be seen to be

$$\tilde{S}_p^{\mathrm{T}}(-j\omega)C^{\mathrm{T}}\tilde{Z}\, C\, \tilde{S}_p(j\omega) \leq C^{\mathrm{T}}\tilde{Z}\, C. \tag{7.3-29}$$

The weighting matrix takes here the special form $C^{\mathrm{T}}\tilde{Z}\, C$.

The importance of condition (7.3-22) in the practical application is restricted for two reasons: First, it is usually very difficult to find the general solution to Eq. (7.3-22). It will be shown in Section 7.4 that optimal linear state feedback meets the above condition. Second, the requirement that condition (7.3-22) should be held for all ω practically excludes the consideration of most output feedback systems: According to Section 7.2, the open-loop transfer operator F_0 of an output feedback system usually possesses a pole excess of at least 2. Therefore by the absolute-value-integral theorem, condition (7.3-22) cannot be satisfied for all ω.

In practice, however, the bandwidth of the input signal to a system is usually limited and it is possible to mitigate the requirement of condition (7.3-22) by limiting the range of frequency, in which Eq. (7.3-22) should hold, to the frequency range of interest, i.e., the range in which all important harmonics of the input signal lie. Thus Theorem 7.3-1 can be modified as follows:

Theorem 7.3-2 In order that a closed control loop be less parameter-sensitive than its nominally equivalent open-loop in the sense of the I_M criterion [Eq. (7.3-1)], the condition

$$S_p^{\mathrm{T}}(-j\omega)Z\, S_p(j\omega) \leq Z \tag{7.3-30}$$

must be satisfied for all frequencies of interest, i.e., the frequency range in which all the harmonics of the vector e lie.

In the practical application, the frequency range of interest must be more precisely defined. Note that by this criterion we do not have to rule out output-feedback systems from our sensitivity considerations. On the other hand, for systems with state feedback, inequality (7.3-30) may be satisfied for all ω. Thus condition (7.3-30) represents a generalization of condition (7.2-54) derived in Section 7.2.

7.4 SENSITIVITY COMPARISON OF OPTIMAL CONTROL SYSTEMS

Optimal control systems are commonly designed as an open-loop configuration and then realized by a closed loop. Therefore the question of the sensitivity comparison between the closed-loop and the nominally equivalent open-loop control is of primary interest.

The crucial point in the sensitivity comparison of optimal systems is the choice of a suitable measure of sensitivity. Since optimal systems are designed due to a certain performance index, it would be quite logical to employ the performance index sensitivity as a sensitivity measure. However just shortly after the introduction of this sensitivity measure in 1963 [36], it gave rise to an extensive discussion on its usefulness. Pagurek and Witsenhausen [75, 109] were the first to show that for optimal systems with free final values the performance-index sensitivity of the closed-loop structure is just as large as that of the open-loop one and therefore useless for comparison. This phenomenon is hereafter called the Pagurek–Witsenhausen Paradox. Besides this, in all cases in which the performance index does not contain the output or the states of the control system, e.g., in time-optimal or minimal fuel control problems, performance-index sensitivity fails to serve as a measure for the sensitivity comparison because in this case the variation δI of the performance index of the open-loop system is always equal to zero and the sensitivity depends only on the final value.

Because of these difficulties it is often the trajectory sensitivity function which is used instead as a measure of sensitivity comparison of optimal systems. The change of the performance index at finite parameter deviations may serve as an even more efficient measure for sensitivity comparison. This change is in general not the same for the open- and closed-loop systems. It indicates how "wide" or how "narrow" the optimum in both cases will be. This is precisely what is aimed at in performing a sensitivity comparison. Unfortunately, by considering finite deviations of parameters, the most important advantage of sensitivity theory, namely, linearity of the equations, will be lost and the mathematical treatment becomes very elaborate.

A great number of publications coping with the sensitivity problem in optimal control systems have appeared in recent years [20, 36, 38, 49, 50, 57, 59, 62, 75, 76, 84, 98, 105, 109, 110]. In this section, sensitivity comparison of optimal control systems will be briefly discussed by means of performance-index sensitivity and trajectory sensitivity.

7.4.1 Sensitivity Comparison by Means of Performance-Index Sensitivity

Consider a general system characterized by the state equation

$$\dot{\boldsymbol{x}}(t) = \boldsymbol{f}[\boldsymbol{x}(t), \boldsymbol{u}(t), \boldsymbol{\alpha}, t], \qquad \boldsymbol{x}(t_0) = \boldsymbol{x}^0, \tag{7.4-1}$$

where the control function $\boldsymbol{u}(t)$ is chosen to be $\boldsymbol{u}^*(t)$ in such a way that, at nominal $\boldsymbol{\alpha}$ value $\boldsymbol{\alpha}_0$ the performance index

$$J = G[\boldsymbol{x}(t_e), t_e] + \int_{t0}^{t_e} L[\boldsymbol{x}(t), \boldsymbol{u}(t), t]\, \mathrm{d}t \tag{7.4-2}$$

assumes its minimum. Let $\boldsymbol{\alpha}$ be a time-variant parameter vector which changes by infinitesimal amounts and which does not appear in the expression for J explicitly. Let the optimal control function be in one case generated independently by an open-loop structure and in the other case by an appropriate feedback configuration according to the control law

$$\boldsymbol{u}^*(t) = -\boldsymbol{k}[\boldsymbol{x}(t), \boldsymbol{\alpha}_0, t]. \tag{7.4-3}$$

In the latter case, the optimal control function depends on $\boldsymbol{x}(t)$ and thus also on the parameter deviations $\Delta\boldsymbol{\alpha}(t) = \varepsilon\, \delta\boldsymbol{\alpha}(t)$. Comparing the sensitivity between the open- and closed-loop structures by means of the performance index which is minimized, the following theorem applies:

Theorem 7.4-1 In optimal systems with free final value, the first variation δJ of the performance index caused by the first variation $\delta\boldsymbol{\alpha}$ of the parameters is the same for both the closed-loop system and its nominally equivalent open-loop counterpart, i.e.,

$$\delta J_R = \delta J_S. \tag{7.4-4}$$

For time-invariant parameter deviations, this can be expressed as

$$\left.\frac{\partial J_R}{\partial \alpha_j}\right|_{\boldsymbol{\alpha}_0} = \left.\frac{\partial J_S}{\partial \alpha_j}\right|_{\boldsymbol{\alpha}_0}, \qquad j = 1, 2, \ldots, r. \tag{7.4-5}$$

Therefore the norms derived from δJ or $(\partial J/\partial \alpha_j)_{\boldsymbol{\alpha}_0}$ will be equal as well. That is,

$$\|\delta J_R\| = \|\delta J_S\|, \qquad \left\| \left.\frac{\partial J_R}{\partial \alpha_j}\right|_{\boldsymbol{\alpha}_0} \right\| = \left\| \left.\frac{\partial J_S}{\partial \alpha_j}\right|_{\boldsymbol{\alpha}_0} \right\|. \tag{7.4-6}$$

We shall first interpret this theorem intuitively and then prove it.

According to Chapter 5, the sensitivity equation in the general case of time-variant parameter deviations $\Delta\boldsymbol{\alpha}(t) = \varepsilon\, \delta\boldsymbol{\alpha}(t)$ is given by

$$\delta\dot{\boldsymbol{x}} = \left.\frac{\partial \boldsymbol{f}}{\partial \boldsymbol{x}}\right|_{\boldsymbol{\alpha}0} \delta\boldsymbol{x} + \left.\frac{\partial \boldsymbol{f}}{\partial \boldsymbol{\alpha}}\right|_{\boldsymbol{\alpha}0} \delta\boldsymbol{\alpha} + \left.\frac{\partial \boldsymbol{f}}{\partial \boldsymbol{u}}\right|_{\boldsymbol{\alpha}0} \delta\boldsymbol{u} \tag{7.4-7}$$

with $\delta \boldsymbol{x}(t_0) = \delta \boldsymbol{x}^0 = \mathbf{0}$. Furthermore, the first variation δJ of the performance index with respect to the deviations $\delta \boldsymbol{x}$ and $\delta \boldsymbol{u}$ can be seen to be

$$\delta J = \left.\frac{\partial \boldsymbol{G}}{\partial \boldsymbol{x}}\right|_{\boldsymbol{\alpha}_0} \delta \boldsymbol{x}(\mathbf{t}_e) + \int_{t_0}^{te} \left[\left.\frac{\partial L}{\partial \boldsymbol{x}}\right|_{\boldsymbol{\alpha}_0} \delta \boldsymbol{x} + \left.\frac{\partial L}{\partial \boldsymbol{u}}\right|_{\boldsymbol{\alpha}_0} \delta \boldsymbol{u} \right] dt, \qquad (7.4\text{-}8)$$

where the gradients $\partial G/\partial \boldsymbol{x}$, $\partial L/\partial \boldsymbol{x}$, and $\partial L/\partial \boldsymbol{u}$ are defined as row vectors.

Now, the difference between the open- and closed-loop control lies in the fact that $\delta \boldsymbol{u}$ is equal to zero for the open-loop case because $\boldsymbol{u}$ is independent of $\boldsymbol{\alpha}$ whereas for the closed-loop control $\delta \boldsymbol{u}$ can be calculated from Eq. (7.4-3) to be

$$\delta \boldsymbol{u} = -\frac{\partial \boldsymbol{k}}{\partial \boldsymbol{x}} \delta \boldsymbol{x}. \qquad (7.4\text{-}9)$$

Thus at first glance, it seems that δJ_R would not be equal to δJ_S. However by the definition of optimization, J is made to be minimal with respect to $\boldsymbol{u}$ irrespective of how $\boldsymbol{u}$ is generated. Thus the term $(\partial L/\partial \boldsymbol{u})$ in Eq. (7.4-8) is equal to zero and consequently $\delta J_R = \delta J_S$.

Note that Theorem 7.4-1 also holds if controller parameter deviations are allowed, but it would no longer be valid if J is not minimized. This is for example the case if we use for the comparison purpose a performance index other than the one which was minimized in the design of the optimal system.

Proof For the rigid proof of Theorem 7.4-1, let us consider the Hamiltonian function (see., eg., Section 8.3)

$$H = L + \boldsymbol{p}\boldsymbol{f}, \qquad (7.4\text{-}10)$$

where the auxiliary function

$$\dot{\boldsymbol{p}} = -\frac{\partial H}{\partial \boldsymbol{x}} = -\frac{\partial L}{\partial \boldsymbol{x}} - \boldsymbol{p}\frac{\partial \boldsymbol{f}}{\partial \boldsymbol{x}} \qquad (7.4\text{-}11)$$

is defined as a row vector. Multiplying Eq. (7.4-11) from the right by $\delta \boldsymbol{x}$ and Eq. (7.4-7) from the left by $\boldsymbol{p}$ and then adding the two equations yields for $\boldsymbol{\alpha} = \boldsymbol{\alpha}_0$:

$$\boldsymbol{p}\,\delta \dot{\boldsymbol{x}} + \dot{\boldsymbol{p}}\,\delta \boldsymbol{x} = -\frac{\partial L}{\partial \boldsymbol{x}} \delta \boldsymbol{x} + \boldsymbol{p}\frac{\partial \boldsymbol{f}}{\partial \boldsymbol{\alpha}} \delta \boldsymbol{\alpha} + \boldsymbol{p}\frac{\partial \boldsymbol{f}}{\partial \boldsymbol{u}} \delta \boldsymbol{u}. \qquad (7.4\text{-}12)$$

Subtracting the expression $(\partial L/\partial \boldsymbol{u})\,\delta \boldsymbol{u}$ from the first term on the right-hand side of Eq. (7.4-12) and then adding it back to the last term, and noting that

$$\boldsymbol{p}\frac{\partial \boldsymbol{f}}{\partial \boldsymbol{\alpha}} \delta \boldsymbol{\alpha} = \frac{\partial H}{\partial \boldsymbol{\alpha}} \delta \boldsymbol{\alpha} \qquad (7.4\text{–}13)$$

(because L does not depend on $\boldsymbol{\alpha}$ explicitly), we obtain

$$\frac{d}{dt}(\boldsymbol{p}\,\delta \boldsymbol{x}) = -\left[\frac{\partial L}{\partial \boldsymbol{x}} \delta \boldsymbol{x} + \frac{\partial L}{\partial \boldsymbol{u}} \delta \boldsymbol{u}\right] + \frac{\partial H}{\partial \boldsymbol{\alpha}} \delta \boldsymbol{\alpha} + \boldsymbol{p}\frac{\partial \boldsymbol{f}}{\partial \boldsymbol{u}} \delta \boldsymbol{u} + \frac{\partial L}{\partial \boldsymbol{u}} \delta \boldsymbol{u}. \quad (7.4\text{–}14)$$

The expression in brackets on the right-hand side is the same as the integrand of δJ in Eq. (7.4-8). The sum of the last two terms on the right-hand side is (at $\boldsymbol{\alpha} = \boldsymbol{\alpha}_0$) equal to

$$\boldsymbol{p} \frac{\partial \boldsymbol{f}}{\partial \boldsymbol{u}} \delta \boldsymbol{u} + \frac{\partial L}{\partial \boldsymbol{u}} \delta \boldsymbol{u} = \frac{\partial H}{\partial \boldsymbol{u}} \partial \boldsymbol{u} = 0 \tag{7.4-15}$$

because $(\partial H/\partial \boldsymbol{u})_{\boldsymbol{\alpha}_0} = \mathbf{0}$ due to optimization. Substituting the expression in brackets of Eq. (7.4-14) into Eq. (7.4-8) yields, with $\delta \boldsymbol{x}(t_0) = 0$,

$$\delta J = \left[\frac{\partial G}{\partial \boldsymbol{x}} \bigg|_{\boldsymbol{\alpha}_0} - \boldsymbol{p} \right] \delta \boldsymbol{x} \big|_{t_e} + \int_{t_0}^{t_e} \frac{\partial H}{\partial \boldsymbol{\alpha}} \delta \boldsymbol{\alpha} \, dt. \tag{7.4-16}$$

This equation is valid for both the open- and closed-loop control. Thus, we obtain

$$\delta J_S - \delta J_R = \left[\frac{\partial G}{\partial \boldsymbol{x}} \bigg|_{\boldsymbol{\alpha}_0} - \boldsymbol{p} \right] \left[\delta \boldsymbol{x}_S - \delta \boldsymbol{x}_R \right] \bigg|_{t_e}. \tag{7.4-17}$$

If the set of initial states consists of only one point M_0 and the set of end states of the whole state space (free final value), then, by the transversality condition $\boldsymbol{p}(t_e) = (\partial G/\partial \boldsymbol{x})_{\boldsymbol{\alpha}_0}$, Eq. (7.4-17) becomes

$$\delta J_S - \delta J_R = 0. \quad \text{Q.E.D.} \tag{7.4-18}$$

Moreover, we note that in this case

$$\delta J_R = \delta J_S = \int_{t_0}^{t_e} \frac{\partial H}{\partial \boldsymbol{\alpha}} \delta \boldsymbol{\alpha} \, dt. \tag{7.4-19}$$

Theorem 7.4-1 is valid under the following assumptions:

(1) if J is the performance index which was optimized,
(2) if the end value $\boldsymbol{x}(t_e)$ is free,
(3) if only infinitesimal parameter deviations are considered (for finite deviations $J_R \neq J_S$),
(4) if the integral J exists. If for example $t_e = \infty$ and the open-loop control is unstable whereas the closed-loop control is stable, then δJ_S will be equal to infinity and δJ_R will remain finite so that $\delta J_R \neq \delta J_S$.

Thus we see that it is of little use to employ the performance-index sensitivity for comparing sensitivities of open- and closed-loop optimal systems. As a remedy, one may use the trajectory sensitivity for the comparison purpose in order to get a feeling of how sensitive the open- and closed-loop systems are, though this is not the natural comparative measure for optimal systems. Alternatively, we may use a performance index different from the one which was optimized in the design of the system.

7.4.2 Sensitivity Comparison by Means of the Trajectory Sensitivity Function

In Section 7.3 it was shown that the parameter-induced deviations $\Delta \boldsymbol{x}(t)$ as well as the trajectory sensitivity functions $\boldsymbol{\lambda}(t, \boldsymbol{\alpha}_0)$ of a closed-loop and its nominally equivalent open-loop system are related to each other in the frequency domain by

$$\Delta \boldsymbol{X}_R(s) = \tilde{\boldsymbol{S}}_p(s)\, \Delta \boldsymbol{X}_S(s), \tag{7.4-20}$$

where

$$\tilde{\boldsymbol{S}}_p(s) = [\boldsymbol{I} + \tilde{\boldsymbol{F}}_0(s)]^{-1} = [\boldsymbol{I} + \boldsymbol{\phi}_S(s)\boldsymbol{B}\boldsymbol{K}]^{-1} \tag{7.4-21}$$

is the comparison sensitivity matrix and $\tilde{\boldsymbol{F}}_0$ the nominal transfer matrix of the open-loop system.

Furthermore it was found that a sufficient condition such that the closed-loop system is less sensitive than the open-loop system in the sense of the criterion

$$\int_{t_0}^{t_1} \Delta \boldsymbol{x}_R^{\mathrm{T}}(t) \boldsymbol{Z}\, \Delta \boldsymbol{x}_R(t)\, dt \le \int_{t_0}^{t_1} \Delta \boldsymbol{x}_S^{\mathrm{T}}(t)\, \boldsymbol{Z}\, \Delta \boldsymbol{x}_S(t)\, dt \qquad \text{for all} \quad t_0 < t_1 \le t_e, \tag{7.4-22}$$

where $\boldsymbol{Z}$ is a symmetrical positive semidefinite matrix, is given by

$$\tilde{\boldsymbol{S}}_p^{\mathrm{T}}(-j\omega) \boldsymbol{Z}\, \tilde{\boldsymbol{S}}_p(j\omega) \le \boldsymbol{Z} \qquad \text{for all } \omega. \tag{7.4-23}$$

In view of Eq. (7.4-21), this can also be written as

$$[\boldsymbol{I} + \boldsymbol{\phi}_S(-j\omega)\boldsymbol{B}\,\boldsymbol{K}]^{\mathrm{T}} \boldsymbol{Z} [\boldsymbol{I} + \boldsymbol{\phi}_S(j\omega)\boldsymbol{B}\,\boldsymbol{K}] \ge \boldsymbol{Z} \qquad \text{for all } \omega.$$

These formulas are all derived without any limitation with respect to optimality and they are therefore also valid for optimal systems.

The question is now whether and under what conditions the requirement (7.4-23) can be satisfied for an optimal system. To anticipate the result, it is not possible to show that a relation such as Eq. (7.4-23) holds for the general case of an arbitrarily chosen $\boldsymbol{Z}$. But under certain limitations, which can occur in different forms, sufficient conditions can be derived such that an optimal closed-loop control is less sensitive than the corresponding open-loop one. Kalman [49] has shown for single variable systems and Anderson [20] for multivariable systems that for $\boldsymbol{Z} = \boldsymbol{I}$ and for a system optimized according to the performance index

$$J = \tfrac{1}{2} \int_0^{\infty} (\boldsymbol{x}^{\mathrm{T}} \boldsymbol{Q}\, \boldsymbol{x} + \boldsymbol{u}^{\mathrm{T}} \boldsymbol{R}\, \boldsymbol{u})\, dt, \tag{7.4-24}$$

where $\boldsymbol{R} = \boldsymbol{M}^{\mathrm{T}}\boldsymbol{M}$ is a positive definite symmetrical matrix with $\boldsymbol{M}$ nonsingular, $\boldsymbol{S}_p(j\omega)$ has to be modified to $[\boldsymbol{I} + \boldsymbol{M}\boldsymbol{K}\boldsymbol{\phi}_S(j\omega)\boldsymbol{B}\boldsymbol{M}^{-1}]$ in order that condition (7.4–23) can be satisfied. This means that for the above type of optimization the inequality

$$[\boldsymbol{I} + \boldsymbol{MK}\boldsymbol{\phi}_S(-j\omega)\boldsymbol{BM}^{-1}]^{\mathrm{T}}[\boldsymbol{I} + \boldsymbol{MK}\boldsymbol{\phi}_S(j\omega)\boldsymbol{BM}^{-1}] \geq \boldsymbol{I}$$

holds for all ω.

Based on this result, Kreindler [59] has derived a condition on $\boldsymbol{Z}$ while keeping $\boldsymbol{S}_p$ unchanged so that condition (7.4-23) is satisfied. We restate this result here without giving its proof.

Theorem 7.4-2 Consider a closed-loop control system characterized by the plant equation (7.4-1) and the control law (7.4-3). Let this system be optimized according to the performance index (7.4-2) and let the Jacobian matrices $\partial \boldsymbol{f}/\partial \boldsymbol{x}$ and $\partial \boldsymbol{f}/\partial \boldsymbol{u}$ be completely controllable and the rows of $\partial k/\partial \boldsymbol{x}$ be linear independent. Then this system will be less sensitive than its nominally equivalent open-loop system in the sense of criterion (7.4-22) if $\boldsymbol{Z}$ is of the form

$$\mathbf{Z} = \left(\frac{\partial \boldsymbol{k}}{\partial \boldsymbol{x}}\right)^{\mathrm{T}} \boldsymbol{M} \frac{\partial \boldsymbol{k}}{\partial \boldsymbol{x}}, \tag{7.4-25}$$

where $\boldsymbol{M} = \boldsymbol{M}(t)$ represents a given differentiable symmetrical positive semidefinite matrix.

Therefore,

$$\int_{t_0}^{t_1} \Delta \boldsymbol{x}_R{}^{\mathrm{T}} \left(\frac{\partial \boldsymbol{k}}{\partial \boldsymbol{x}}\right)^{\mathrm{T}} \boldsymbol{M} \frac{\partial \boldsymbol{k}}{\partial \boldsymbol{x}} \Delta \boldsymbol{x}_R \, dt \leq \int_{t_0}^{t_1} \Delta \boldsymbol{x}_S{}^{\mathrm{T}} \left(\frac{\partial \boldsymbol{k}}{\partial \boldsymbol{x}}\right)^{\mathrm{T}} \boldsymbol{M} \frac{\partial \boldsymbol{k}}{\partial \boldsymbol{x}} \Delta \boldsymbol{x}_S \, dt \tag{7.4-26}$$

for all t_1 in the interval $t_0 < t_1 \leq t_e$. This means that a reduction of the trajectory sensitivity by introducing feedback can only be shown if the sensitivity measure I_M is chosen such that the terms $\Delta \boldsymbol{x}^T \boldsymbol{Z} \Delta \boldsymbol{x}$ are weighted in a certain manner determined by the performance index used for optimization.

For example if $\boldsymbol{Z}$ is taken to be the unity matrix $\boldsymbol{I}$, a sensitivity reduction by the introduction of feedback can no longer be proven for the general case.

A more precise characterization of $\boldsymbol{M}$ for some optimization problems of fairly general nature can be found in Cruz [5]. We shall now show how the matrix $\boldsymbol{M}$ looks in the special case of optimal *linear state feedback*.

Consider a system whose plant differential equation is given by

$$\dot{\boldsymbol{x}} = \boldsymbol{Ax} + \boldsymbol{Bu}, \qquad \boldsymbol{x}(t_0) = \boldsymbol{x}^0, \tag{7.4-27}$$

and which is optimized according to the performance index

$$J = \tfrac{1}{2}\, \boldsymbol{x}^{\mathrm{T}}(t_e)\boldsymbol{G}\, \boldsymbol{x}(t_e) + \tfrac{1}{2} \int_{t_0}^{t_e} (\boldsymbol{x}^{\mathrm{T}}\boldsymbol{Q}\, \boldsymbol{x} + \boldsymbol{u}^{\mathrm{T}}\boldsymbol{R}\, \boldsymbol{u})\, dt. \tag{7.4-28}$$

Let the matrices $\boldsymbol{A} = \boldsymbol{A}(t, \boldsymbol{\alpha})$, $\boldsymbol{B} = \boldsymbol{B}(t, \boldsymbol{\alpha})$, $\boldsymbol{Q}(t)$, and $\boldsymbol{R}(t)$ be continuously differentiable with respect to t and $\boldsymbol{\alpha}$, respectively; let $\boldsymbol{R}(t)$ be a positive definite and $\boldsymbol{G}$ and $\boldsymbol{Q}(t)$ be a positive semidefinite symmetrical matrix; let t_e be a fixed final time and let $\boldsymbol{x}(t_e)$ be free and $\boldsymbol{u}$ be piecewise continuous and not bounded. It is well-known that under these conditions the optimal control law is given by

$$\boldsymbol{u}(t) = -\boldsymbol{K}(t)\boldsymbol{x}(t), \tag{7.4-29}$$

where

$$\boldsymbol{K} = \boldsymbol{R}^{-1}\boldsymbol{B}^{\mathrm{T}}\boldsymbol{P} \tag{7.4-30}$$

and $\boldsymbol{P}$ satisfies the Riccati matrix equation

$$-\dot{\boldsymbol{P}} = \boldsymbol{P}\boldsymbol{A} + \boldsymbol{A}^{\mathrm{T}}\boldsymbol{P} - \boldsymbol{P}\boldsymbol{B}\boldsymbol{R}^{-1}\boldsymbol{B}^{\mathrm{T}}\boldsymbol{P} + \boldsymbol{Q} \tag{7.4-31}$$

with $\boldsymbol{P}(t_e) = \boldsymbol{G}$.

In this case, $\boldsymbol{M} = \boldsymbol{R}$, i.e., $\boldsymbol{Z}$ takes the form

$$\boldsymbol{Z} = \boldsymbol{K}^{\mathrm{T}}\boldsymbol{R}\,\boldsymbol{K}, \tag{7.4-32}$$

and Theorem 7.4-2 becomes

Theorem 7.4-3 An optimal linear state feedback, defined by Eqs. (7.4-27)–(7.4-31), is less parameter-sensitive than the nominally equivalent open-loop control in the sense of the criterion

$$\int_{t_0}^{t_1} \Delta\boldsymbol{x}_R^{\mathrm{T}}\,\boldsymbol{K}^{\mathrm{T}}\boldsymbol{R}\boldsymbol{K}\,\Delta\boldsymbol{x}_R\,dt < \int_{t_0}^{t_1} \Delta\boldsymbol{x}_S^{\mathrm{T}}\,\boldsymbol{K}^{\mathrm{T}}\boldsymbol{R}\boldsymbol{K}\,\Delta\boldsymbol{x}_S\,dt \tag{7.4-33}$$

for all t_1 in the interval $t_0 < t_1 \leq t_e$ provided that $\boldsymbol{K}\,\Delta\boldsymbol{x} \not\equiv 0$ for all t in the interval $t_0 \leq t \leq t_1$. Under these conditions, the following inequality holds:

$$[\boldsymbol{I} + \phi_S(-j\omega)\boldsymbol{B}\boldsymbol{K}]^{\mathrm{T}}\boldsymbol{K}^{\mathrm{T}}\boldsymbol{R}\,\boldsymbol{K}[\boldsymbol{I} + \phi_S(j\omega)\boldsymbol{B}\,\boldsymbol{K}] \geq \boldsymbol{K}^{\mathrm{T}}\boldsymbol{R}\boldsymbol{K} \tag{7.4-34}$$

for all ω.

By this theorem we see that the weighting matrix $\boldsymbol{Z}$ in the L_2-norm employed for the comparison of the state sensitivity is fixed by the performance index of the optimization problem. In this way, it is no longer possible to deduce a sensitivity reduction, for example, for the case $\boldsymbol{Z} = \boldsymbol{I}$.

Equation (7.4-33) can also be interpreted as follows: there is a reduction of the sensitivity for the feedback signal $\boldsymbol{u} = -\boldsymbol{K}\boldsymbol{x}$ according to $\boldsymbol{I}_M$ with $\boldsymbol{Z} = \boldsymbol{R}$ but there is no reduction like this for the state $\boldsymbol{x}$ itself. It is also impossible to prove the existence of a sensitivity reduction for the output variable $\boldsymbol{y} = \boldsymbol{C}\boldsymbol{x}$ in terms of $\Delta\boldsymbol{y}^{\mathrm{T}}\,\tilde{\boldsymbol{Z}}\,\Delta\boldsymbol{y}\,dt$, where $\tilde{\boldsymbol{Z}}$ is arbitrary. This is due to the fact that in this case

$$\boldsymbol{Z} = \boldsymbol{C}^{\mathrm{T}}\tilde{\boldsymbol{Z}}\,\boldsymbol{C} \neq \boldsymbol{K}^{\mathrm{T}}\boldsymbol{R}\,\boldsymbol{K}.$$

The above says that in terms of the L_2-norm of the output variable, the closed-loop optimal control can be more sensitive than the open-loop one. This is a serious limitation of the importance of Theorem 7.4-3.

Example With the aid of the results obtained above, it is possible to explain why $|S_P{}^G(j\omega)| \leq 1$ in Example 7.2-3 for $K = 2$ and for all ω. The state feedback is optimal for $K = 2$ with respect to the performance index

$$J = \int_0^\infty (x_1{}^2 + \tfrac{1}{4}u^2)\, dt. \tag{7.4-35}$$

This implies that $\boldsymbol{R} = \frac{1}{4}$ and $\boldsymbol{K} = [2, 2]$. In addition, $\boldsymbol{S}_p$ is here identical with the Bode sensitivity function $S_P{}^G$. Therefore by Theorem 7.4–3,

$$[S_P{}^G(-j\omega)]\boldsymbol{K}^{\mathrm{T}}\boldsymbol{R}\,\boldsymbol{K}\, S_P{}^G(j\omega) \leq \boldsymbol{K}^{\mathrm{T}}\boldsymbol{R}\,\boldsymbol{K} \tag{7.4-36}$$

for all ω. Hence owing to $\boldsymbol{K}^{\mathrm{T}}\boldsymbol{R}\,\boldsymbol{K} \neq \boldsymbol{0}$,

$$|S_P{}^G(j\omega)|^2 \leq 1 \qquad \text{for all } \omega \tag{7.4-37}$$

and thus $|S_P{}^G(j\omega)| \leq 1$.

It becomes apparent by this example that there exist a relation of the type (7.4-20) and an inequality of the type (7.4-23) for every component of $\Delta\boldsymbol{x}$. This is true for any linear optimal state-feedback system with a scalar input as long as it is the phase vector which is being fed back. One can show this easily by considering the case of the general multivariable system and substituting $\boldsymbol{K}^{\mathrm{T}}\boldsymbol{R}\,\boldsymbol{K}$. This result can be summarized as follows:

Theorem 7.4-4 Consider a linear time-invariant optimal state-feedback system with a scalar input variable u. Let the plant be represented in the Frobenius canonical form, i.e., by $\dot{\boldsymbol{x}} = \boldsymbol{A}\boldsymbol{x} + \boldsymbol{b}u$, where

$$\boldsymbol{A} = \begin{bmatrix} 0 & 1 & & 0 \\ & & \ddots & \\ & & & 1 \\ -a_1 & -a_2 & \cdots & -a_n \end{bmatrix}, \qquad \boldsymbol{b} = \begin{bmatrix} 0 \\ \vdots \\ 0 \\ 1 \end{bmatrix} \tag{7.4-38}$$

and the state $\boldsymbol{x}$ be fed back. Let the performance index to be minimized be

$$J = \tfrac{1}{2}\int_0^\infty (\boldsymbol{x}^{\mathrm{T}}\boldsymbol{Q}\,\boldsymbol{x} + u^2)\, dt. \tag{7.4-39}$$

Then the closed-loop system will be less sensitive than its nominally equivalent open-loop system in the sense

$$\int_0^{t_1} (\Delta x_R{}^i)^2\, dt \leq \int_0^{t_1} (\Delta x_S{}^i)^2\, dt \tag{7.4-40}$$

for all $t_1 > 0$ and $i = 1, \ldots, n$. Equation (7.4-40) of course also holds for $\lambda_R{}^i$ and $\lambda_S{}^i$ instead of $\Delta x_R{}^i$ and $\Delta x_S{}^i$, respectively.

That means that $\boldsymbol{Z}$ can be arbitrarily chosen in this case. This can easily be shown by multiplying Eq. (7.4-40) by arbitrary nonnegative constants for each i and then adding up all expressions [99].

The above implies that for every component Δx^i of the deviations of the phase vector $\Delta \boldsymbol{x}$ there exists a corresponding relation of the form

$$\Delta X_R{}^i(s) = \frac{1}{1 + \boldsymbol{K}\,\boldsymbol{\phi}(s)\boldsymbol{b}}\,\Delta X_S{}^i(s), \tag{7.4-41}$$

with

$$|1 + \boldsymbol{K}\,\boldsymbol{\phi}(s)\boldsymbol{b}| \geq 1, \tag{7.4-42}$$

and, in view of $\boldsymbol{\phi}(s) = (s\boldsymbol{I} - \boldsymbol{A})^{-1}$,

$$|1 + \boldsymbol{K}(j\omega\,\boldsymbol{I} - \boldsymbol{A})^{-1}\boldsymbol{b}| \geq 1 \tag{7.4-43}$$

for all ω, where the sign of equality does not hold for all ω. Since every completely controllable state equation can be brought into the Frobenius canonical form, a set of statevariables can always be constructed for which Theorem 7.4-4 applies.

In the practical application, however, the situation is not as simple as it appears. In most cases, the phase variables are not available for the feedback purpose and thus the requirement in Theorem 7.4-4 regarding the state being fed back is not satisfied.

Moreover, linearity is a very important assumption for the sensitivity reduction. As soon as there is a relatively slight limitation to the control variable, e.g., a weak limitation of the form $\int_0^\infty u^2\,dt \leq c$, the sensitivity improvement gained by the feedback will be considerably reduced and become meaningless.

To summarize, we have so far been dealing with systems with *one* degree of freedom. For systems with several degrees of freedom, a series of new aspects has to be considered. Moreover, we have only compared closed-loop with open-loop systems. It is also possible and sometimes important to compare closed-loop systems with other equivalent closed-loop systems. Finally it should be mentioned that in the case of optimal feedback design it might be reasonable to define nominal equivalence in a more relaxed manner, i.e., in terms of a performance index rather than by full trajectory equality. Such a relaxed definition was given in section 3.3.3 and is discussed in more detail by Kreisselmeier and Grübel. [61].

PROBLEMS

7.1 Consider the control system shown in Fig. 7.P-1. Assume that K_2 is very large ($KK_2 \gg 1$). Calculate the Bode sensitivity functions $S_{K_1}^G$, $S_{K_2}^G$, and S_K^G, and the corresponding parameter-induced changes of the output $\Delta y/y_0$.

7.2 Consider a control system described by the input–output differential equation $\dot{y} + (a + b)y = bu$.

(a) Determine the Bode sensitivity function with respect to b. Express the result in terms of the transfer function $G_a = Y(s, b_0)/U(s)$.

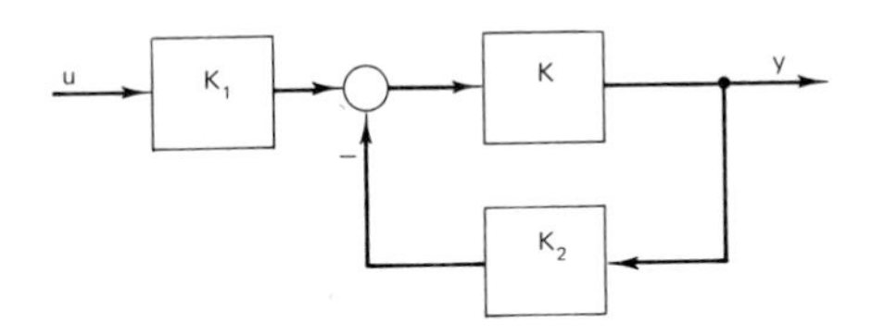

FIG. 7.P–1 Control system of Problem 7.1.

(b) Determine the Bode sensitivity function with respect to b for the case that the system is given by the differential equation $\dot{y} + (a + b)y = u$. Can the result, in terms of the transfer function $G_b = Y(s, b_0/U(s))$, be brought into the same form as in case (a)?

7.3 Given a unity feedback system with the input–output differential equation $a\dot{y} + (b + 1)y = bu$. Determine the Bode sensitivity function with respect to a and b. For what values of a and b is this feedback system, over the whole frequency range, less sensitive than the nominally equivalent open-loop system?

7.4 Given a unity feedback system with the complex loop gain $k(j\omega) = 2 + 4j\omega$.

(a) Suppose the magnitude $|k(j\omega)|$ changes by 10% at $\omega = 1$. Determine the corresponding change in the magnitude $|G|$ and the phase φ of the overall transfer function G of the control system.

(b) Find the values of the frequencies for which the closed-loop system is less sensitive than the nominally equivalend open-loop system.

(c) Why is there no contradiction to the absolute-value-integral theorem of Bode (Theorem 7.2–2)?

7.5 Show the validity of Eq. (7.3–37).

7.6 Consider two realizations of a transistor amplifier as shown in Fig. 7. P-2. Fig. 7.P–2a shows a single-stage open-loop amplifier with the voltage gain $K = -10$. Fig. 7.P-2b shows three equal stages of the same type as in (a), each having the voltage gain $K = -10$, the whole system being fed back by a gain factor R.

(a) Determine R such that that both configurations are nominally equivalent, i.e., configuration (b) has an overall gain of $K_{tot} = -10$.

(b) Compare the sensitivity of both configurations in terms of the comparison sensitivity.

(c) Suppose the gain of the stages decrease by 10% each year. In how many years has the three-stage amplifier lost as much gain as the one stage amplifier after one year? (Give the exact solution.)

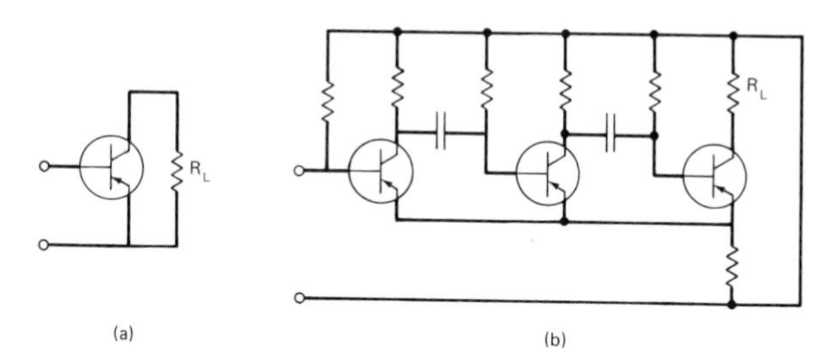

FIG. 7.P–2. Two equivalent realizations of an amplifier with a voltage gain of -10. (a) One-stage (open-loop) configuration. (b) Three-stage configuration with feedback (Problem 7.6).

7.7 Given a two-variable control system with the block diagram shown in Fig. 7.P-3. The matrices **A** and **H** are

$$\mathbf{A} = \begin{bmatrix} s+1 & s+4 \\ 4 & s^2+s+1 \end{bmatrix}, \qquad \mathbf{H} = \begin{bmatrix} -1 & 0 \\ 0 & 1 \end{bmatrix}.$$

Give the frequency range within which the trajectory sensitivity of this control system is smaller than that of the nominally equivalent open-loop system due to the L_2 norm.

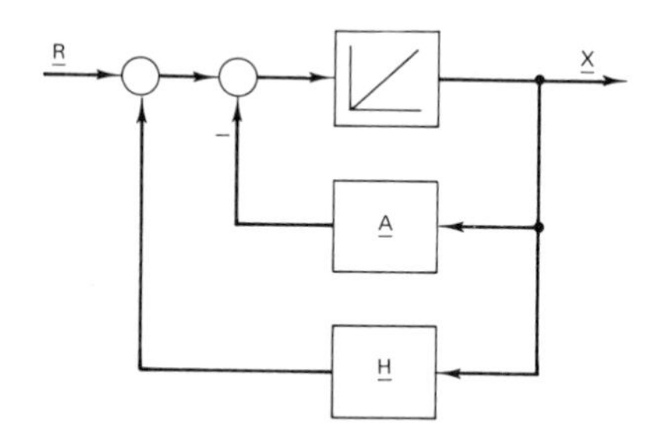

FIG. 7.P–3. Control system of Problem 7.7.

7.8 The rollers in an aluminum factory that are driven by armature controlled dc motors have the transfer function $P(s) = 0.5/(1 + s)$. Fig. 7.P-4 shows four control systems with the same overall transfer function $G(s) = (1 + 0.5s)^{-1}$ for standardized ingots. In practice, the size of the ingots vary from piece to piece, so that the parameters of $P(s)$ vary from piece to piece.

(a) Determine the comparison sensitivity functions S_p of the configurations (b), (c), and (d).

(b) Compare $|S_p|$ of (b), (c), and (d).

(c) Relate the results to the dc gains of the compensators $R(s)$.

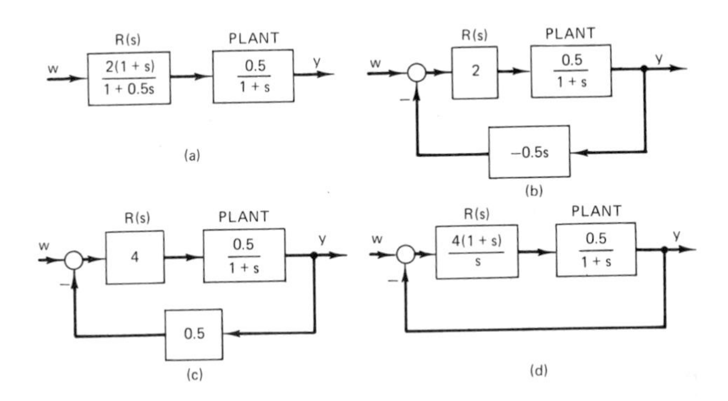

FIG. 7.P-4. Four control system configurations for the same plant having the same overall transfer function $G(s) = (1 + 0.5s)^{-1}$ (Problem 7.8).

7.9 Given the state feedback shown in Fig. 7.P-5a. The nominal parameters of the plant are $A = 10$, $K_1 = 1$, $K_2 = 5$, $K_3 = 2$, $s_2 = -5$, $s_3 = -1$. The feedback parameters are $k_1 = 1$, $k_2 = -3$, $k_3 = 5$. Consider small plant parameter variations.

(a) Find the nominally equivalent configurations of Fig. 7.P-5b and c, as well as the nominally equivalent open-loop configuration.

(b) How can the comparison sensitivity function of the state feedback, Fig. 7.P-5a, be calculated in terms of the Bode sensitivity function?

(c) Calculate all comparison sensitivity functions and compare the sensitivities.

(d) Determine for each configuration the upper frequency limit up to

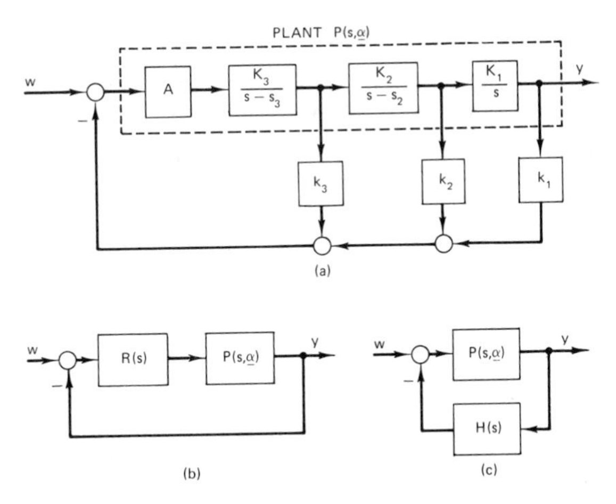

FIG. 7.P-5. The three nominally equivalent configurations of Problem 7.9.

which the corresponding configuration is less sensitive than the open-loop configuration.

7.10 Consider the second-order system shown in Fig. 7.p-6, whose nominal parameters are $K_0 = 1$ and $T_0 = 1$.

(a) Find the optimal state feedback matrix $\boldsymbol{K} = [K_1 K_2]$ minimizing the performance index $J = \frac{1}{2}\int_0^\infty (\boldsymbol{x}^T\boldsymbol{x} + u^2)\, dt$.

(b) Show that the L_2 norm of the modified state $\boldsymbol{Kx}$ of this system is smaller than the L_2 norm of the nominally equivalent open-loop configuration.

(c) Give the frequency range for which the parameter-induced output error of this control system is smaller than that of the nomainally equivalent open-loop system.

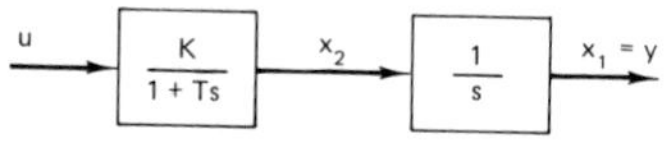

FIG. 7.P–6. Given plant for Problem 7.10 ($K_0 = 1$, $T_0 = 1$).

7.11 Given the second-order system with the state equations

$$\dot{\boldsymbol{x}} = \boldsymbol{Ax} + \boldsymbol{b}u = \begin{bmatrix} -a & 1 \\ 0 & -1 \end{bmatrix} \boldsymbol{x} + \begin{bmatrix} 0 \\ 1 \end{bmatrix} u$$

$$y = \boldsymbol{cx} = [1 \quad 0]\, \boldsymbol{x}.$$

Suppose there is a state feedback with $\boldsymbol{K} = [K_1\ K_2] = [+0.1896\ +0.5425]$. Let the nominal parameter value of the plant be $a_0 = 1$, and consider small parameter variations of a.

(a) Show that the given state feedback is less sensitive than the nominally equivalent open-loop configuration in the sense that $\tilde{\boldsymbol{S}}_p^T(-j\omega)\, \boldsymbol{K}^T\boldsymbol{K}\, \boldsymbol{S}_p(j\omega) \leq \boldsymbol{K}^T\boldsymbol{K}$ for all ω where $\tilde{\boldsymbol{S}}_p$ is the comparison sensitivity matrix.

(b) Interprete the result in terms of optimization theory.

(c) Accomplish a similar sensitivity comparison with regard to the output y. Find the frequency range at which the parameter-induced output error is smaller than that of the nominally equivalent open-loop configuration.

Chapter 8

Sensitivity Analysis of Optimal Systems

8.1 INTRODUCTION

Sensitivity considerations are among the fundamental aspects of the analysis and synthesis of optimal systems.

Optimal systems are characterized by being designed in such a way that a certain performance index or cost functional is extremized with respect to the control function or some control parameters. The usual starting point of this process is the establishment of a mathematical model of the given process whose nominal parameter values will never coincide exactly with the values of the process. Discrepancies between the nominal values of the mathematical model and the actual physical values are also inevitable in the physical realization of the optimal control law. However, the latter are far less important because of the high standard of modern technology.

Since the control function is designed to be optimal with particular regard to the nominal parameters, it appears quite logical that optimal systems are particularly sensitive to parameter variations from their nominal values. Optimal systems are often compared with a suit specially tailored for a person and, therefore, better fitted to the figure than a ready-to-wear suit; but, for the same reasons, it is also more sensitive to any changes in the person's figure.

The most important aspects of sensitivity considerations in connection with optimal systems are the following,

(1) sensitivity analysis in order to judge whether the solution of an optimization problem is of practical use in view of the given parameter tolerances, comparison with nonoptimal solutions,

(2) comparison of open-loop and closed-loop optimal systems,

(3) design of insensitive systems by taking into account the parameter variations.

In this chapter, we shall mainly deal with the methodic basis of the first aspect, i.e., the sensitivity analysis of optimal systems. In this case, because the system behavior is characterized by its performance index, the natural measure for the sensitivity will be the performance-index sensitivity defined in Section 3.4 or its generalization for finite parameter deviations.

The performance-index sensitivity was first introduced in 1963 by Dorato [36] and was further investigated by many authors mainly in connection with the comparison of optimal closed-loop and open-loop systems. In these investigations, infinitesimal plant parameter deviations were considered. Later, vestigations were extended to the design of optimal systems with finite or large parameter deviations.

In this chapter, we shall describe the most important methods for the determination of the performance-index sensitivity of a closed-loop system with respect to infinitesimal plant parameter variations. In addition, a measure for the sensitivity of the degree of optimality with respect to parameter deviations will be given. First introduced by Rohrer and Sobral [86], this definition can also be used for finite deviations and is suitable for the design of insensitive systems.

Before devoting ourselves to the different methods of analysis, we shall first note a basic property of the performance-index sensitivity that has important consequences in practical applications.

Changes of the performance index of an optimal system can be caused either by parameter changes of the mathematical model of the given process (the plant) or by changes in the control law. For a feedback control system, changes in the control law are equivalent to changes of the controller parameters. Since an optimization is achieved by minimizing the performance index J with respect to the control variable, it is evident that the performance-index sensitivity $J_{\alpha R}$ with respect to changes in the controller parameters α_R vanish as long as the minimum is a relative one (i.e., unbounded control function u). On the other hand, the performance-index sensitivity $J_{\alpha s}$ with respect to changes in the plant parameters α_s is not necessarily equal to zero. It can take on any real value. The same is true for simultaneous changes of the plant and controller parameters as may be encountered in ideal optimal controllers that provide optimal control for any set of actual parameters.

This circumstance is illustrated in Fig. 8.1-1. The values of J are plotted versus the deviations of the plant parameter α_s and the controller parameter α_R. The derivative at α_{R0}, α_{s0} in the $\Delta\alpha_R$ direction is zero (horizontal tangent a). In the $\Delta\alpha_s$ direction the derivative is not equal to zero (tangent b with slope ϑ). Thus

$$J_{\alpha R} \triangleq \left.\frac{\partial J}{\partial \alpha_R}\right|_{\alpha R0} \equiv 0, \qquad J_{\alpha s} \triangleq \left.\frac{\partial J}{\partial \alpha_s}\right|_{\alpha s_0} \not\equiv 0.$$

Therefore for the investigation of the sensitivity with respect to the con-

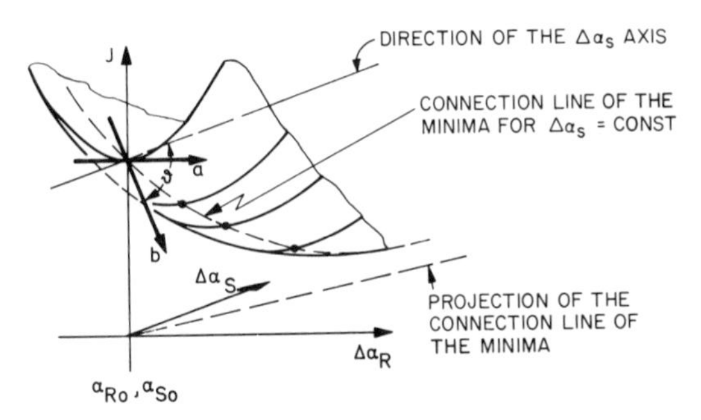

FIG. 8.1-1. Illustration of the fact that the performance-index sensitivity with respect to controller parameter changes is zero.

troller parameters, we have to apply other methods, e.g., sensitivity functions of higher orders [23] or for finite deviations.

8.2 ELEMENTARY METHODS FOR THE DETERMINATION OF PERFORMANCE-INDEX SENSITIVITY WITH RESPECT TO PLANT PARAMETER VARIATIONS

In this section, it will be shown how the performance-index sensitivity $J_\alpha = (\partial J/\partial\alpha)_{\alpha_0}$ of continuous optimal systems with respect to plant parameter variations can be determined by elementary methods.

Let the control system, for which an optimal open-loop input or an optimal closed-loop control is to be designed, be given by the following general vector differential equation

$$\dot{\boldsymbol{x}}(t) = \boldsymbol{f}[\boldsymbol{x}(t), \boldsymbol{\alpha}, t, \boldsymbol{u}(t)], \qquad \boldsymbol{x}(t_0) = \boldsymbol{x}^0, \tag{8.2-1}$$

where $\boldsymbol{x}$ is an n-dimensional state vector, $\boldsymbol{u}$ a p-dimensional control vector, and $\boldsymbol{\alpha}$ an r-dimensional parameter vector of the plant with the nominal value $\boldsymbol{\alpha}_0$.

Now the optimization problem consists in finding a control vector $\boldsymbol{u}(t)$ which brings the system from the initial state $\boldsymbol{x}(t_0)$ to the final state $\boldsymbol{x}(t_f)$ and, in doing so, delivers a minimum value of a performance index of the form

$$J = G[\boldsymbol{x}(t_f), t_f] + \int_{t_0}^{t_f} L(\boldsymbol{x}, t, \boldsymbol{\alpha}, \boldsymbol{x})\, dt, \tag{8.2-2}$$

where G and L are nonnegative functions. Note that the plant parameter vector $\boldsymbol{\alpha}$ appears explicitly in the criterion. Usually it is equal to the nominal value $\boldsymbol{\alpha}_0$, but this is not claimed here. Let the variables $\boldsymbol{u}(t)$, $\boldsymbol{x}(t_0)$, and $\boldsymbol{x}(t_f)$ in the optimization time interval $t_0 \le t \le t_f$ be confined by the following relations

$$\boldsymbol{u}(t) \in \Omega, \qquad \boldsymbol{x}(t_0) \in \Omega_1, \qquad \boldsymbol{x}(t_f) \in \Omega_2,$$

where Ω is a subspace of the p-dimensional Euclidean space and Ω_1 and Ω_2 are subspaces of the n-dimensional space. We shall concern ourselves with the case where Ω_1 consists of a single point and Ω_2 the whole space (free final value).

The set Ω represents an amplitude restraint of the control vector. For example, if the field current I of a dc motor is limited to I_{max}, then Ω will be the interval $-|I_{max}| \leq I \leq |I_{max}|$. By assuming the above form for the performance index, a great number of optimization problems of practical importance will be taken into account, e.g.,

minimal energy control:

$$G = 0, \qquad L = \tfrac{1}{2}\boldsymbol{u}^{\mathrm{T}}\boldsymbol{R}\boldsymbol{u},$$

the linear regulator problem:

$$G = \tfrac{1}{2}\boldsymbol{x}^{\mathrm{T}}(t_f)\boldsymbol{S}\boldsymbol{x}(t_f), \qquad L = \tfrac{1}{2}(\boldsymbol{u}^{\mathrm{T}}\boldsymbol{R}\,\boldsymbol{u} + \boldsymbol{x}^{\mathrm{T}}\boldsymbol{Q}\,\boldsymbol{x}),$$

where the weighting matrices $\boldsymbol{S}$ and $\boldsymbol{Q}$ are symmetrical and positive semidefinite while $\boldsymbol{R}$ is symmetrical and positive definite. $\boldsymbol{Q}$, $\boldsymbol{R}$, and $\boldsymbol{S}$ can be functions of time t and the parameter vector $\boldsymbol{\alpha}$.

The time optimal control ($G = 0$, $L = 1$) has to be excluded from consideration since in this case the sensitivity definition loses its meaning.

Now suppose that the optimal control function denoted by $\boldsymbol{u}^*(t)$ has been found for nominal parameter values by one of the usual methods, e.g., the maximum principle or the dynamic programming. In the case of an open-loop control this function is of the form

$$\boldsymbol{u}_0^*(t) = \boldsymbol{\theta}(\boldsymbol{\alpha}_0, t, \boldsymbol{x}^0, t_0) \tag{8.2-3}$$

and, in the case of a closed-loop control, of the form

$$\boldsymbol{u}_0^*(t) = \boldsymbol{K}[\boldsymbol{x}(t), \boldsymbol{\alpha}_0, t]. \tag{8.2-4}$$

If now the plant parameter value $\boldsymbol{\alpha}$ differs from the nominal value $\boldsymbol{\alpha}_0$ by $\Delta\boldsymbol{\alpha}$ whereas the control law remains the same as given by Eqs. (8.2-3) and (8.2-4), respectively, the vector differential equation of the open-loop control system will be given by

$$\dot{\boldsymbol{x}} = \boldsymbol{f}[\boldsymbol{x}, \boldsymbol{\alpha}, t, \boldsymbol{\theta}(\boldsymbol{\alpha}_0, t, \boldsymbol{x}^0, t_0)], \qquad \boldsymbol{x}(t_0) = \boldsymbol{x}^0, \tag{8.2-5}$$

and that of the closed-loop control system by

$$\dot{\boldsymbol{x}} = \boldsymbol{f}[\boldsymbol{x}, \boldsymbol{\alpha}, t, \boldsymbol{k}(\boldsymbol{x}, \boldsymbol{\alpha}_0, t)], \qquad \boldsymbol{x}(t_0) = \boldsymbol{x}^0. \tag{8.2-6}$$

The trajectories as well as the control function can be calculated by solving these state equations. By substituing the result into Eq. (8.2-2), the actual value of the performance index can be obtained. We denote the actual performance index by $J(\boldsymbol{x}^0, t_0, \boldsymbol{\alpha}, \boldsymbol{\alpha}_0)$ and the nominal one by $J(\boldsymbol{x}^0, t_0, \boldsymbol{\alpha}_0, \boldsymbol{\alpha}_0)$, the latter corresponding to $\boldsymbol{\alpha} = \boldsymbol{\alpha}_0$. In these terms, the deviation of the performance index from its nominal value due to plant parameter variations is given by

$$\Delta J = J(\boldsymbol{x}^0, t_0, \boldsymbol{\alpha}, \boldsymbol{\alpha}_0) - J(\boldsymbol{x}^0, t_0\, \boldsymbol{\alpha}_0, \boldsymbol{\alpha}_0). \tag{8.2-7}$$

Note that in this case the sign of ΔJ need not be positive because, for plant parameter changes, J can become even smaller than at nominal parameter values. Note, however, that for parameter changes in the control function or the control law, the value of J will by the definition of optimality always be increased.

For infinitesimal parameter changes $\Delta\boldsymbol{\alpha} = d\boldsymbol{\alpha}$, the performance-index deviation ΔJ can be written as

$$\Delta J = \sum_{j=1}^{r} \frac{\partial J}{\partial \alpha_j}\bigg|_{\boldsymbol{\alpha}_0} \Delta\alpha_j = \frac{\partial J}{\partial \boldsymbol{\alpha}}\bigg|_{\boldsymbol{\alpha}_0} \Delta\boldsymbol{\alpha} = J_\alpha\, \Delta\boldsymbol{\alpha}, \tag{8.2-8}$$

where

$$J_\alpha \triangleq [J_\alpha^1 \quad \cdots \quad J_\alpha^r] = \left[\frac{\partial J}{\partial \alpha_1} \quad \cdots \quad \frac{\partial J}{\partial \alpha_r}\right]_{\boldsymbol{\alpha}_0} \tag{8.2-9}$$

represents the vector of the performance-index sensitivities defined in Section 3.4. Thus J_α can be calculated by determining the actual value of the performance index $J(\boldsymbol{x}^0, t_0, \boldsymbol{\alpha}, \boldsymbol{\alpha}_0)$ according to Eq. (8.2-2) and then taking the partial derivative with respect to $\boldsymbol{\alpha}$ at $\boldsymbol{\alpha} = \boldsymbol{\alpha}_0$ (see Example 8.2-1).

Dorato [36] has proposed another method for the determination of J_α. Expressing G and L in Eq. (8.2-2) in terms of $\boldsymbol{x}$ and taking the partial derivative with respect to $\boldsymbol{\alpha}$ at $\boldsymbol{\alpha} = \boldsymbol{\alpha}_0$ yields

$$\bar{J}_\alpha = \frac{\partial J}{\partial \boldsymbol{\alpha}}\bigg|_{\boldsymbol{\alpha}_0} = \frac{\partial G}{\partial \boldsymbol{x}(t_f)}\bigg|_{\boldsymbol{\alpha}_0} \frac{\partial \boldsymbol{x}(t_f)}{\partial \boldsymbol{\alpha}}\bigg|_{\boldsymbol{\alpha}_0} + \int_{t_0}^{t} \left(\frac{\partial L}{\partial \boldsymbol{x}}\frac{\partial \boldsymbol{x}}{\partial \boldsymbol{\alpha}} + \frac{\partial L}{\partial \boldsymbol{\alpha}}\right)_{\boldsymbol{\alpha}_0} dt. \tag{8.2-10}$$

Here, $\partial G/\partial \boldsymbol{x}$, $\partial L/\partial \boldsymbol{x}$, and $\partial L/\partial \boldsymbol{\alpha}$ are defined as row vectors with elements $\partial G/\partial x_i$, $\partial L/\partial x_i$, and $\partial L/\partial \alpha_i$, respectively, taken at $\boldsymbol{\alpha} = \boldsymbol{\alpha}_0$. The term $(\partial \boldsymbol{x}/\partial \boldsymbol{\alpha})_{\boldsymbol{\alpha}_0}$ is the trajectory sensitivity matrix $\boldsymbol{\lambda}(t, \boldsymbol{\alpha}_0)$ defined in Section 3.2.3 and $(\partial \boldsymbol{x}(t_f)/\partial \boldsymbol{\alpha})_{\boldsymbol{\alpha}_0}$ its value at $t = t_f$. Thus Eq. (8.2-10) can be written as

$$\bar{J}_\alpha = \frac{\partial G}{\partial \boldsymbol{x}(t_f)}\bigg|_{\boldsymbol{\alpha}_0} \boldsymbol{\lambda}(t_f, \boldsymbol{\alpha}_0) + \int_{t_0}^{t_f} \left(\frac{\partial L}{\partial \boldsymbol{x}}\bigg|_{\boldsymbol{\alpha}_0} \boldsymbol{\lambda}(t, \boldsymbol{\alpha}_0) + \frac{\partial L}{\partial \boldsymbol{\alpha}}\bigg|_{\boldsymbol{\alpha}_0}\right) dt. \tag{8.2-11}$$

$\boldsymbol{\lambda}$ can be determined from the trajectory sensitivity equation

$$\dot{\boldsymbol{\lambda}} = \frac{\partial \boldsymbol{f}}{\partial \boldsymbol{x}}\bigg|_{\boldsymbol{\alpha}_0} \boldsymbol{\lambda} + \frac{\partial \boldsymbol{f}}{\partial \boldsymbol{\alpha}}\bigg|_{\boldsymbol{\alpha}_0}, \qquad \boldsymbol{\lambda}(t_0, \boldsymbol{\alpha}_0) = \boldsymbol{0}, \tag{8.2-12}$$

by one of the methods described in Chapter 5. The solution is given by (see Chapter 5)

$$\boldsymbol{\lambda}(t, \boldsymbol{\alpha}_0) = \int_{t_0}^{t} \phi(t, \tau) \frac{\partial \boldsymbol{f}[\boldsymbol{x}^*(\tau), \boldsymbol{\alpha}, \tau, \boldsymbol{u}^*(\tau)]}{\partial \boldsymbol{\alpha}}\bigg|_{\boldsymbol{\alpha}_0} d\tau. \tag{8.2-13}$$

Here $\boldsymbol{x}^*$ is the trajectory of the actual (open- or closed-loop) system determined from Eq. (8.2-5) or (8.2-6). $\boldsymbol{u}^*$ represents the actual control variable determined from Eq. (8.2-3) or (8.2-4). In the latter case it is calculated from the optimal control law at $\boldsymbol{\alpha} = \boldsymbol{\alpha}_0$, but using the actual trajectory $\boldsymbol{x}^*$. $\boldsymbol{\phi}(t, \tau)$ is the transition matrix obeying $\boldsymbol{\phi}(\tau, \tau) = \boldsymbol{I}$ and

$$\frac{\partial \boldsymbol{\phi}(t, \tau)}{\partial t} = \frac{\partial \boldsymbol{f}[\boldsymbol{x}^*, \boldsymbol{\alpha}, t, \boldsymbol{u}^*]}{\partial \boldsymbol{x}^*} \boldsymbol{\phi}(t, \tau)\bigg|_{\boldsymbol{\alpha}_0}. \tag{8.2-14}$$

In general, digital computers have to be employed for the evaluation of these equations. The principles of this method will be illustrated by a simple example which can be treated analytically.

Example 8.2-1 Let us consider the system that has been treated on several earlier occasions and that is given by the following (scalar) differential equation:

$$\dot{x}(t) = -a_0 x(t) + u(t), \qquad x(0) = x^0, \tag{8.2-15}$$

where u is the input variable, x the output variable, and the coefficient a_0 the nominal plant parameter. The system is to be transferred to the origin of the state space by u so that the performance index

$$J = \int_0^\infty [x^2(t) + u^2(t)]\, dt \tag{8.2-16}$$

will be minimized.

This problem can be solved by differential calculus. The optimal control function at the nominal parameter is given by

$$u_0^*(t) = x^0(a_0 - \sqrt{1 + a_0^2}) \exp(-\sqrt{1 + a_0^2}\, t) \tag{8.2-17}$$

and the corresponding optimal trajectory by

$$x_0^*(t) = x^0 \exp(-\sqrt{1 + a_0^2}\, t). \tag{8.2-18}$$

In the case of the *closed loop* control, we obtain, for the generation of the optimal control variable, the (optimal) control law:

$$u_g^*(t) = k_0 x(t), \qquad k_0 = a_0 - \sqrt{1 + a_0^2}. \tag{8.2-19}$$

Now let us assume that the plant parameter takes on an actual value $a = a_0 + \Delta a$, so that the state equation becomes

$$\dot{x}(t) = -a x(t) + u(t), \qquad x(0) = x^0, \tag{8.2-20}$$

whereas the control variable and the control law, respectively, remain unchanged. In the case of the *open loop* control, the control function is given by Eq.(8.2-17). The corresponding trajectory can be calculated as the response of the system described by Eq.(8.2-20) to the above control function. The result is

$$x_0^*(t) = x^0 \exp(-at) - x^0 \frac{a_0 - \sqrt{1 + a_0^2}}{a - \sqrt{1 + a_0^2}} [\exp(-at) - \exp(-\sqrt{1 + a_0^2}\, t)]. \tag{8.2-21}$$

In the case of the *closed loop* control, the trajectory is determined from the actual plant equation (8.2-20) and the control law given by Eq.(8.2-19). It is found to be

$$x_g^*(t) = x^0 \exp(a_0 - a - \sqrt{1 + a_0^2}\, t) \tag{8.2-22}$$

from which, with Eq.(8.2–19), the control function follows

$$u_g^*(t) = (a_0 - \sqrt{1 + a_0^2})\, x_g^*(t). \tag{8.2-23}$$

Substituting $x_g^*(t)$ and $u_g^*(t)$ into Eq.(8.2-16) of the performance index of the closed-loop system, we have

$$J = \frac{1 + a_0^2 - a_0\sqrt{1 + a_0^2}}{a - a_0 + \sqrt{1 + a_0^2}}\, x^2(0). \tag{8.2-24}$$

Hence the performance index sensitivity is given by

$$J_a = \left.\frac{\partial J}{\partial a}\right|_{a_0} = \left(\frac{a_0}{\sqrt{1 + a_0^2}} - 1\right) x^2(0). \tag{8.2-25}$$

This result differs from that of Sage [15] from which this example is taken. In his Eq. (12), Section 12.2-5, instead of calculating the performance-index sensitivity with respect to the *plant* parameter $\tilde{a}$ and then letting $\tilde{a} \to a$, it is calculated with respect to the *control* parameter a. It is clear that the expression for the sensitivity must then go to zero for $\tilde{a} \to a$. This follows from the optimization of the control system (see remarks at the end of Section 8.1).

Instead of the above procedure, namely, evaluating first the performance index and then taking the partial derivative, one can first take the partial derivative under the integral sign at $a = a_0$ and then evaluate the integral. In this case we obtain

$$J_a = \frac{\partial}{\partial a}\left(\int_0^\infty (1 + k_0^2) x_g^{*2}(t)\, dt\right)_{a_0} = 2(1 + k_0^2) \int_0^\infty x_g^*|_{a_0}\, \lambda\, dt, \tag{8.2-26}$$

with $\lambda = (\partial x_g^*/\partial a)_{a_0}$. The sensitivity function λ can be calculated from a sensitivity equation. By this procedure, one obtains

$$x_g^*|_{a_0} = x^0 \exp(-\sqrt{1 + a_0^2}\, t) \tag{8.2-27}$$

and

$$\lambda = \left.\frac{\partial x_g^*}{\partial a}\right|_{a_0} = -x^0 t \exp(-\sqrt{1 + a_0^2}\, t). \tag{8.2-28}$$

Substituting these expressions into Eq.(8.2–26) and then performing the integration, we arrive, of course, at the same result as in Eq.(8.2–25).

As can be seen from Eq.(8.2-25), J_a is negative for all $a_0 > 0$ and its absolute value decreases with increasing a_0. For quantitative investigations, it may be of advantage to normalize the performance index upon its nominal value.

8.3 DETERMINATION OF THE PERFORMANCE-INDEX SENSITIVITY BY MEANS OF A HAMILTON–JACOBI EQUATION

The performance-index sensitivity can also be determined by means of a Hamilton-Jacobi equation which is needed anyway to solve the optimization problem. This was first shown by Pagurek [75].

In order to derive the Hamilton–Jacobi equation, let us consider a system with the state equation

$$\dot{\boldsymbol{x}} = \boldsymbol{f}(\boldsymbol{x}, \boldsymbol{\alpha}, t, \boldsymbol{u}), \qquad \boldsymbol{x}(t_0) = \boldsymbol{x}^0. \tag{8.3-1}$$

This system is to be optimized according to a certain performance index of the form of Eq.(8.2-2). The Hamilton–Jacobi equation for the determination of the performance-index sensitivities $J_\alpha{}^j$, $j = 1, 2, \ldots, r$, will now be derived for the cases of closed-loop and open-loop control. It is therefore assumed that the optimal control function

$$\boldsymbol{u}_0{}^*(t) = \boldsymbol{\theta}(\boldsymbol{\alpha}_0, t, \boldsymbol{x}^0, t_0) \tag{8.3-2}$$

(in the open-loop case) or the optimal control law

$$\boldsymbol{u}_0{}^*(t) = \boldsymbol{k}[\boldsymbol{x}(t), \boldsymbol{\alpha}_0, t] \tag{8.3-3}$$

(in the closed-loop case) are known for nominal parameter values $\boldsymbol{\alpha} = \boldsymbol{\alpha}_0$.

8.3.1 Closed-Loop Control

Now suppose that the actual plant parameter values $\boldsymbol{\alpha}$ differ from the nominal ones $\boldsymbol{\alpha}_0$ whereas the control law remains unaltered. Then the dynamic behavior of the actual control system will be characterized by the state equation

$$\dot{\boldsymbol{x}}(t) = \boldsymbol{f}[\boldsymbol{x}(t), \boldsymbol{\alpha}, t, \boldsymbol{k}(\boldsymbol{x}, \boldsymbol{\alpha}_0, t)], \qquad \boldsymbol{x}(t_0) = \boldsymbol{x}^0. \tag{8.3-4}$$

Let the solution of this equation be

$$\boldsymbol{x}(t) = \boldsymbol{\xi}(t, \boldsymbol{\alpha}, \boldsymbol{\alpha}_0, \boldsymbol{x}^0, t_0), \qquad t_0 \le t \le t_f. \tag{8.3-5}$$

The actual trajectory depends on $\boldsymbol{\alpha}_0$ as well as $\boldsymbol{\alpha}$. Substituting this equation into Eq.(8.3-3), we obtain the following dependence of $\boldsymbol{u}^*(t)$ upon $\boldsymbol{\alpha}$:

$$\boldsymbol{u}^*(t) = \boldsymbol{k}(\boldsymbol{\xi}, \boldsymbol{\alpha}_0, t) = \boldsymbol{\psi}(t, \boldsymbol{\alpha}, \boldsymbol{\alpha}_0, \boldsymbol{x}^0, t_0). \tag{8.3-6}$$

Now let $(\boldsymbol{x},t)$ be an arbitrary point on the trajectory between t_0 and t_f. The section of the trajectory in the interval $t \leq \tau \leq t_f$ will be given by

$$\boldsymbol{x}(\tau) = \boldsymbol{\xi}(\tau, \boldsymbol{\alpha}, \boldsymbol{\alpha}_0, \boldsymbol{x}, t), \qquad t_0 \leq \tau \leq t_f. \tag{8.3-7}$$

If we evaluate the performance index of Eq.(8.2.2) along the trajectory $\boldsymbol{x}(\tau)$ from t onward, then without introducing new symbols and with the abbreviations

$$G[\boldsymbol{\xi}(t_f, \boldsymbol{\alpha}, \boldsymbol{\alpha}_0, \boldsymbol{x}, t), t_f] \triangleq G[\boldsymbol{x}(t_f), t_f],$$

$$L[\boldsymbol{\xi}(\tau, \boldsymbol{\alpha}, \boldsymbol{\alpha}_0, \boldsymbol{x}, t), \boldsymbol{\alpha}, \tau, \boldsymbol{k}(\boldsymbol{\xi}, \boldsymbol{\alpha}_0, \tau)] \triangleq L[\boldsymbol{x}, \tau, \boldsymbol{\alpha}, \boldsymbol{\alpha}_0],$$

we obtain

$$J(\boldsymbol{x}, t, \boldsymbol{\alpha}, \boldsymbol{\alpha}_0) = G[\boldsymbol{x}(t_f), t_f] + \int_t^{t_f} L(\boldsymbol{x}, \tau, \boldsymbol{\alpha}, \boldsymbol{\alpha}_0), d\tau. \tag{8.3-8}$$

As can be seen J is now a time function. If we assume that the solution of Eq. (8.3-6) is unique, then

$$\boldsymbol{\xi}(\tau, \boldsymbol{\alpha}, \boldsymbol{\alpha}_0, \boldsymbol{x}, t) = \boldsymbol{\xi}(\tau, \boldsymbol{\alpha}, \boldsymbol{\alpha}_0, \boldsymbol{x}^0, t_0).$$

From Eq.(8.3-8), the time derivative of $J(\boldsymbol{x},t,\boldsymbol{\alpha},\boldsymbol{\alpha}_0)$ along the trajectory is obtained as

$$\dot{J}(\boldsymbol{x}, t, \boldsymbol{\alpha}, \boldsymbol{\alpha}_0) = -L(\boldsymbol{x}, t, \boldsymbol{\alpha}, \boldsymbol{\alpha}_0). \tag{8.3-9}$$

On the other hand, since $\mathscr{F}$ is a function of both t and $\boldsymbol{x}(t)$, the derivative can also be written as

$$\dot{J} = \frac{\partial J}{\partial \boldsymbol{x}} \dot{\boldsymbol{x}} + \frac{\partial J}{\partial t}. \tag{8.3-10}$$

Substituting for $\dot{J}$ the right-hand side of Eq.(8.3-9) and for $\dot{\boldsymbol{x}}$ the right-hand side of Eq.(8.3-4), where $\boldsymbol{f}(\boldsymbol{x}, \boldsymbol{\alpha}, t, \boldsymbol{k}(\boldsymbol{x}, \boldsymbol{\alpha}_0, t)]$ is abbreviated to $\boldsymbol{f}(\boldsymbol{x}, t, \boldsymbol{\alpha}, \boldsymbol{\alpha}_0)$, then we obtain the Hamilton–Jacobi equation

$$\frac{\partial J(\boldsymbol{x}, t, \boldsymbol{\alpha}, \boldsymbol{\alpha}_0)}{\partial t} = -H\left(\boldsymbol{x}, \frac{\partial J}{\partial \boldsymbol{x}}, \boldsymbol{\alpha}, \boldsymbol{\alpha}_0, t\right) \tag{8.3-11}$$

with Hamiltonion function

$$H\left(\boldsymbol{x}, \frac{\partial J}{\partial \boldsymbol{x}}, \boldsymbol{\alpha}, \boldsymbol{\alpha}_0, t\right) = \frac{\partial J(\boldsymbol{x}, t, \boldsymbol{\alpha}, \boldsymbol{\alpha}_0)}{\partial \boldsymbol{x}} \boldsymbol{f}(\boldsymbol{x}, t, \boldsymbol{\alpha}, \boldsymbol{\alpha}_0) + L(\boldsymbol{x}, \boldsymbol{t}, \boldsymbol{\alpha}, \boldsymbol{\alpha}_0) \tag{8.3-12}$$

and the supplementary condition [by Eq. (8.3-8)]:

$$J[\boldsymbol{x}(t_f), t_f, \boldsymbol{\alpha}, \boldsymbol{\alpha}_0] \equiv G[\boldsymbol{x}(t_f), t_f]. \tag{8.3-13}$$

In order to arrive at a similar formula for the *performance index sensitivity* $J_\alpha{}^j$, let us take the partial derivatives of Eqs.(8.3-11)–(8.3-13) with respect to

$\boldsymbol{\alpha}$ and then set $\boldsymbol{\alpha} = \boldsymbol{\alpha}_0$. Assuming that the order of differentiation with respect t and $\boldsymbol{\alpha}$ can be interchanged and denoting

$$\left.\frac{\partial J(\boldsymbol{x}, t, \boldsymbol{\alpha}, \boldsymbol{\alpha}_0,)}{\partial \alpha_j}\right|_{\boldsymbol{\alpha}_0} \triangleq J_\alpha{}^j(\boldsymbol{x}, t, \boldsymbol{\alpha}_0), \tag{8.3-14}$$

we obtain a Hamilton–Jacobi equation for the partial derivatives $J_\alpha{}^j(\boldsymbol{x}, t, \boldsymbol{\alpha}_0)$. This is given by

$$\frac{\partial J_\alpha{}^j(\boldsymbol{x}, t, \boldsymbol{\alpha}_0)}{\partial t} = -H_\alpha\left(\boldsymbol{x}, \frac{\partial J_\alpha{}^j}{\partial \boldsymbol{x}}, t, \boldsymbol{\alpha}_0\right) \tag{8.3-15}$$

with Hamiltonian function

$$\begin{aligned} H_\alpha\left(\boldsymbol{x}, \frac{\partial J_\alpha{}^j}{\partial \boldsymbol{x}}, t, \boldsymbol{\alpha}_0\right) &= \frac{\partial J_\alpha{}^j(\boldsymbol{x}, t, \boldsymbol{\alpha}_0)}{\partial \boldsymbol{x}} \boldsymbol{f}(\boldsymbol{x}, t, \boldsymbol{\alpha}_0) \\ &+ \left.\frac{\partial J(\boldsymbol{x}, t, \boldsymbol{\alpha}_0)}{\partial \boldsymbol{x}} \frac{\partial \boldsymbol{f}(\boldsymbol{x}, t, \boldsymbol{\alpha}, \boldsymbol{\alpha}_0)}{\partial \alpha_j}\right|_{\boldsymbol{\alpha}_0} \\ &+ \left.\frac{\partial L(\boldsymbol{x}, t, \boldsymbol{\alpha}, \boldsymbol{\alpha}_0)}{\partial \alpha_j}\right|_{\boldsymbol{\alpha}_0} \end{aligned} \tag{8.3-16}$$

and the supplementary condition

$$J_\alpha{}^j[\boldsymbol{x}(t_f), t_f, \boldsymbol{\alpha}_0] \equiv 0, \qquad j = 1, 2, \ldots, r. \tag{8.3-17}$$

By solving Eq.(8.3-11), $J(\boldsymbol{x}, t, \boldsymbol{\alpha}_0)$ as well as $\partial J(\boldsymbol{x}, t, \boldsymbol{\alpha}_0)/\partial \boldsymbol{x}$ can be determined. Substituting $\partial J(\boldsymbol{x}, t, \boldsymbol{\alpha}_0)/\partial \boldsymbol{x}$ into Eq.(8.3-15) and solving the latter at $t = t_0$, $\boldsymbol{x} = \boldsymbol{x}_0$, one obtains the desired performance-index sensitivities $J_\alpha{}^j(\boldsymbol{x}^0, t_0, \boldsymbol{\alpha}_0)$.

8.3.2 Open-Loop Control

The Hamilton–Jacobi equation for the open-loop case can be derived in a manner similar to that above. In this case the optimal control function is given by Eq.(8.3-2). Substituting this function into the state equation (8.3-1), we obtain the state equation of the actual open-loop control system,

$$\dot{\boldsymbol{x}}(t) = \boldsymbol{f}[\boldsymbol{x}(t), \boldsymbol{\alpha}, t, \boldsymbol{\theta}(\boldsymbol{\alpha}_0, t, \boldsymbol{x}^0, t_0)], \qquad \boldsymbol{x}(t_0) = \boldsymbol{x}^0. \tag{8.3-18}$$

To avoid using new symbols, we write for the sake of simplicity

$$\boldsymbol{f}[\boldsymbol{x}(t), \boldsymbol{\alpha}, t, \boldsymbol{\theta}(\boldsymbol{\alpha}_0, t, \boldsymbol{x}^0, t_0)] \triangleq \boldsymbol{f}(\boldsymbol{x}, t, \boldsymbol{\alpha}, \boldsymbol{\alpha}_0, \boldsymbol{x}^0, t_0).$$

If we denote by t an arbitrary instant between t_0 and t_f, the resulting trajectory of Eq.(8.3-18) in the interval $t \leq \tau \leq t_f$ is given by

$$\boldsymbol{x}(\tau) = \boldsymbol{\xi}(\boldsymbol{x}, \tau, t, \boldsymbol{\alpha}, \boldsymbol{\alpha}_0, \boldsymbol{x}^0, t_0). \tag{8.3-19}$$

The performance index evaluated along this trajectory $\boldsymbol{x}(t)$ from the point t onward is again a time function given by

$$J(\boldsymbol{x}, t, \boldsymbol{\alpha}, \boldsymbol{\alpha}_0, \boldsymbol{x}^0, t_0) = G[\boldsymbol{x}(t_f), t_f] + \int_{t_0}^{t_f} L(\boldsymbol{x}, \tau, \boldsymbol{\alpha}, \boldsymbol{\alpha}_0)\, d\tau. \tag{8.3-20}$$

If $\boldsymbol{\theta}(\boldsymbol{\alpha}_0, t, \boldsymbol{x}^0, t_0)$ is, at least piecewise, a continuous function of t, and if $\boldsymbol{x}^0$ and t_0 are not the parameters of interest, we can proceed in the same manner as in the case of the closed-loop control.

Thus the Hamilton–Jacobi equation for J is obtained as

$$\frac{\partial J(\boldsymbol{x}, t, \boldsymbol{\alpha}, \boldsymbol{\alpha}_0, \boldsymbol{x}^0, t_0)}{\partial t} = -H\left(\boldsymbol{x}, \frac{\partial J}{\partial \boldsymbol{x}}, \boldsymbol{\alpha}, \boldsymbol{\alpha}_0, t\right) \tag{8.3-21}$$

with the Hamilton function

$$H = L(\boldsymbol{x}, t, \boldsymbol{\alpha}, \boldsymbol{\alpha}_0) + \frac{\partial J(\boldsymbol{x}, t, \boldsymbol{\alpha}, \boldsymbol{\alpha}_0, \boldsymbol{x}^0, t_0)}{\partial \boldsymbol{x}} \boldsymbol{f}(\boldsymbol{x}, t, \boldsymbol{\alpha}, \boldsymbol{\alpha}_0, \boldsymbol{x}^0, t_0) \tag{8.3-22}$$

and the supplementary condition

$$J[\boldsymbol{x}(t_f), t_f, \boldsymbol{\alpha}, \boldsymbol{\alpha}_0, \boldsymbol{x}^0, t_0] = G[\boldsymbol{x}(t_f), t_f)]. \tag{8.3-23}$$

Likewise, we obtain the Hamilton–Jacobi equations for $J_\alpha{}^j = [\partial J(\boldsymbol{x}, t, \boldsymbol{\alpha}, \boldsymbol{\alpha}_0, \boldsymbol{x}^0, t_0)/\partial \alpha_j]_{\boldsymbol{\alpha}_0}$ by differentiating Eqs. (8.3-21)–(8.3-23) with respect to α_j as

$$\frac{\partial J_\alpha{}^j(\boldsymbol{x}, t, \boldsymbol{\alpha}_0, \boldsymbol{x}^0, t_0)}{\partial t} = -H_\alpha\left(\boldsymbol{x}, \frac{\partial J}{\partial \boldsymbol{x}}, \boldsymbol{\alpha}, \boldsymbol{\alpha}_0, t\right) \tag{8.3-24}$$

with Hamiltonian function

$$H_\alpha = \frac{\partial J(\boldsymbol{x}, t, \boldsymbol{\alpha}_0, \boldsymbol{x}^0, t_0)}{\partial \mathbf{x}} \frac{\partial \boldsymbol{f}(\boldsymbol{x}, t, \boldsymbol{\alpha}, \boldsymbol{\alpha}_0, \boldsymbol{x}^0, t_0)}{\partial \alpha_j}\bigg|_{\boldsymbol{\alpha}_0} + \frac{\partial L(\boldsymbol{x}, t, \boldsymbol{\alpha}, \boldsymbol{\alpha}_0)}{\partial \alpha_j}\bigg|_{\boldsymbol{\alpha}_0}$$
$$+ \frac{\partial J_\alpha{}^j(\boldsymbol{x}, t, \boldsymbol{\alpha}_0, \boldsymbol{x}^0, t_0)}{\partial \boldsymbol{x}} \boldsymbol{f}(\boldsymbol{x}, t, \boldsymbol{\alpha}_0, \boldsymbol{x}^0, t_0) \tag{8.3-25}$$

and the supplementary condition

$$J_\alpha{}^j[\boldsymbol{x}(t_f), t_f, \boldsymbol{\alpha}_0, \boldsymbol{x}^0, t_0] \equiv 0. \tag{8.3-26}$$

By solving Eqs.(8.3-21) and (8.3-24) with $t = t_0$ and $\boldsymbol{x} = \boldsymbol{x}^0$, all the performance-index sensitivities $J_\alpha{}^j$, $j = 1, 2, \ldots, r$, can be found.

If we examine the results more closely, we shall discover a very close analogy with the methods of determining the trajectory sensitivity functions insofar as the performance-index sensitivities can also be obtained from sensitivity models that are similar to the original models. Original models and sensitivity models in the present case are the Hamilton–Jacobi differential equations.

One disadvantage of this formulation of the performance-index sensitivity problem is that, with a few exceptions, it is very difficult to solve the Hamilton–Jacobi differential equations. One exception to this is the optimal linear state-feedback problem. In the next two subsections we shall derive its solution for the linear regulator problem and the follow-up problem.

8.3.3 PERFORMANCE-INDEX SENSITIVITY OF THE LINEAR REGULATOR PROBLEM

Let us consider a linear plant characterized by the state differential equation

$$\dot{\boldsymbol{x}} = \boldsymbol{A}(\boldsymbol{\alpha})\boldsymbol{x} + \boldsymbol{B}(\boldsymbol{\alpha})\boldsymbol{u}, \qquad \boldsymbol{x}(t_0) = \boldsymbol{x}^0,$$
$$\boldsymbol{y} = \boldsymbol{C}(\boldsymbol{\alpha})\mathrm{x}, \tag{8.3-27}$$

where $\boldsymbol{x}$ is an $n \times 1$ state vector, $\boldsymbol{u}$ a $p \times 1$ control vector, $\boldsymbol{y}$ a $q \times 1$ output vector, $\boldsymbol{\alpha}$ an $r \times 1$ parameter vector, $\boldsymbol{A}$ is an $n \times n$ matrix, $\boldsymbol{B}$ an $n \times p$ matrix, and $\boldsymbol{C}$ a $q \times n$ matrix. With the exception of $\boldsymbol{\alpha}$, all the quantities are functions of time. For the sake of simplicity of notation, we shall omit t in Eq.(8.3–27). Further, let the performance index be of the form

$$J = \tfrac{1}{2}\, \boldsymbol{x}^{\mathrm{T}}(t_f)\boldsymbol{S}\, \boldsymbol{x}(t_f) + \tfrac{1}{2} \int_{t_0}^{t_f} (\boldsymbol{y}^{\mathrm{T}}\boldsymbol{Q}\, \boldsymbol{y} + \boldsymbol{u}^{\mathrm{T}}\boldsymbol{R}\, \boldsymbol{u})\, dt, \tag{8.3-28}$$

where $\boldsymbol{Q}$ and $\boldsymbol{R}$ are symmetrical positive definite metrices of order $p \times p$ and $q \times q$, respectively, and $\boldsymbol{S}$ is a symmetrical positive semidefinite constant matrix.

It is well known from optimization theory that the optimal control vector for the above problem at nominal parameter values $\boldsymbol{\alpha} = \boldsymbol{\alpha}_0$, i.e, for $\boldsymbol{A}(\boldsymbol{\alpha}_0) = \boldsymbol{A}_0$ etc. in Eq.(8.3-27), is given by

$$\boldsymbol{u}_0{}^* = -\boldsymbol{R}_0{}^{-1}\boldsymbol{B}_0{}^{\mathrm{T}}\boldsymbol{M}_0\boldsymbol{x}, \tag{8.3-29}$$

where the symmetrical positive definite matrix $\boldsymbol{M}_0$ is the solution of the Riccati matrix equation

$$\dot{\boldsymbol{M}}_0 + \boldsymbol{A}_0{}^{\mathrm{T}}\boldsymbol{M}_0 + \boldsymbol{M}_0\boldsymbol{A}_0 - \boldsymbol{M}_0\boldsymbol{B}_0\boldsymbol{R}_0{}^{-1}\boldsymbol{B}_0{}^{\mathrm{T}}\boldsymbol{M}_0 + \boldsymbol{C}_0{}^{\mathrm{T}}\boldsymbol{Q}_0\boldsymbol{C}_0 = \boldsymbol{0}, \tag{8.3-30}$$
$$\boldsymbol{M}_0(\mathrm{t}_f) = \boldsymbol{S}.$$

The subscript 0 denotes that the solution is to be determined at nominal parameter values $\boldsymbol{\alpha}_0$. The performance index is then given by

$$J = \tfrac{1}{2}\, \boldsymbol{x}^{\mathrm{T}}(t_0)\boldsymbol{M}_0(t_0)\, \boldsymbol{x}(t_0). \tag{8.3-31}$$

Now let us assume that the plant parameter vector $\boldsymbol{\alpha}$ deviates from its nominal value $\boldsymbol{\alpha}_0$ by $\Delta\boldsymbol{\alpha}$, i.e., $\boldsymbol{A}_0$ becomes $\boldsymbol{A}$, etc., and the weighting matrices $\boldsymbol{Q}_0$, $\boldsymbol{R}_0$ become $\boldsymbol{Q}$, $\boldsymbol{R}$. Substituting the optimal control vector of Eq.(8.3-29) into the actual state equation of the plant yields the following state equation characterizing the actual overall system:

$$\dot{\boldsymbol{x}} = (\boldsymbol{A} - \boldsymbol{B}\boldsymbol{R}_0{}^{-1}\boldsymbol{G}_0{}^{\mathrm{T}}\boldsymbol{M}_0)\boldsymbol{x} \triangleq \boldsymbol{F}\boldsymbol{x}. \tag{8.3-32}$$

With the notation of Section 8.3.2, we see that

$$\boldsymbol{F}\boldsymbol{x} = \boldsymbol{f}(\boldsymbol{x}, t, \boldsymbol{\alpha}, \boldsymbol{\alpha}_0). \tag{8.3-33}$$

Further, the function L of the performance index is, in this case, given by

$$L(\boldsymbol{x}, t, \boldsymbol{\alpha}, \boldsymbol{\alpha}_0) = \tfrac{1}{2}\, \boldsymbol{x}^{\mathrm{T}}\tilde{\boldsymbol{Q}}\, \boldsymbol{x} \tag{8.3-34}$$

with

$$\tilde{\boldsymbol{Q}} = \boldsymbol{M}_0 \boldsymbol{B}_0 \boldsymbol{R}_0^{-1} \boldsymbol{R} \boldsymbol{R}_0^{-1} \boldsymbol{B}_0^{\mathrm{T}} \boldsymbol{M}_0 + \boldsymbol{C}^{\mathrm{T}} \boldsymbol{Q}\, \boldsymbol{C}. \tag{8.3-35}$$

By assuming that the solution of the performance index is of the form

$$J(\boldsymbol{x}, t, \boldsymbol{\alpha}, \boldsymbol{\alpha}_0) = \tfrac{1}{2}\, \boldsymbol{x}^{\mathrm{T}} \boldsymbol{M} \boldsymbol{x} \tag{8.3-36}$$

and substituting this into the Hamilton–Jacobi equation (8.3-11), we find that $\boldsymbol{M}$ must satisfy the following special form of the Riccati equation:

$$\dot{\boldsymbol{M}} + \boldsymbol{F}^{\mathrm{T}} \boldsymbol{M} + \boldsymbol{M} \boldsymbol{F} + \tilde{\boldsymbol{Q}} = \boldsymbol{0}, \qquad \boldsymbol{M}(t_f) = \boldsymbol{S}. \tag{8.3-37}$$

After substituting the expressions obtained above into Eq.(8.3-16), the performance-index sensitivities can be determined from the Hamilton–Jacobi equation for $J_\alpha{}^j$, Eq.(8.3-16), in a manner similar to that above. From Eq. (8.3-32),

$$\boldsymbol{f}(\boldsymbol{x},t,\boldsymbol{\alpha},\boldsymbol{\alpha}_0) = \boldsymbol{F}_0 \boldsymbol{x} = (\boldsymbol{A}_0 - \boldsymbol{B}_0 \boldsymbol{R}_0^{-1} \boldsymbol{B}_0^{\mathrm{T}} \boldsymbol{M}_0)\, \boldsymbol{x}. \tag{8.3-38}$$

Furthermore, with Eqs.(8.3-34)–(8.3-37), the Hamilton–Jocobi equation of the present problem can be expressed as

$$\frac{\partial J_\alpha{}^j(\boldsymbol{x}, t, \boldsymbol{\alpha}_0)}{\partial t} = -\frac{\partial J_\alpha{}^j(\boldsymbol{x}, t, \boldsymbol{\alpha}_0)}{\partial \boldsymbol{x}} \boldsymbol{F}_0 \boldsymbol{x}$$

$$- \frac{1}{2} \boldsymbol{x}^{\mathrm{T}} \left[\frac{\partial \boldsymbol{F}^{\mathrm{T}}}{\partial \alpha_j} \boldsymbol{M}_0 + \boldsymbol{M}_0 \frac{\partial \boldsymbol{F}}{\partial \alpha_j} + \frac{\partial \tilde{\boldsymbol{Q}}}{\partial \alpha_j} \right]_{\boldsymbol{\alpha}0} \boldsymbol{x}. \tag{8.3-39}$$

By assuming the solution to be of the form

$$J_\alpha{}^j(\boldsymbol{x}, t, \boldsymbol{\alpha}_0) = \tfrac{1}{2}\, \boldsymbol{x}^{\mathrm{T}} \boldsymbol{P}_j \boldsymbol{x} \tag{8.3-40}$$

and substituting this into Eq.(8.3-39), we obtain the following equation for determing the symmetrical matrix $\boldsymbol{P}_j$:

$$\dot{\boldsymbol{P}}_0 + \boldsymbol{F}_0^{\mathrm{T}} \boldsymbol{P}_j + \boldsymbol{P}_j \boldsymbol{F}_0 + \boldsymbol{Q}^* = \boldsymbol{0}, \qquad \boldsymbol{P}_j(t_f) = \boldsymbol{0}, \tag{8.3-41}$$

with

$$\boldsymbol{Q}^* = \left[\frac{\partial \boldsymbol{F}^{\mathrm{T}}}{\partial \alpha_j} \boldsymbol{M}_0 + \boldsymbol{M}_0 \frac{\partial \boldsymbol{F}}{\partial \alpha_j} + \frac{\partial \tilde{\boldsymbol{Q}}}{\partial \alpha_j} \right]_{\boldsymbol{\alpha}0}. \tag{8.3-42}$$

If there are r parameters of interest, we have r equations of the above form for the determination of all $\boldsymbol{P}_j$. From the r $\boldsymbol{P}_j$'s and Eq. (8.3-40), the expressions for $J_\alpha{}^j(\boldsymbol{x}, t, \boldsymbol{\alpha}_0)$ can be determined, and by setting $t = t_0$ and $\boldsymbol{x} = \boldsymbol{x}^0$, the performance-index sensitivites $J_\alpha{}^j(\boldsymbol{x}^0, t_0, \boldsymbol{\alpha}_0)$, $j = 1, 2, \ldots, r$, are obtained.

This approach leads, for very simple systems, to complicated equations, the solutions of which are very cumbersome. The solution procedure will now be illustrated by the simple example of Section 8.2.

Example 8.3-1 Consider the nominal plant equation (8.2-15) and the performance index (8.2-16) to be minimized. Let us calculate the performance-index sensitivity of the optimized control system with respect to variations of the coefficient a_0 by means of the Hamilton–Jacobi equation.

Solution In the general matrix Riccati equation (8.3-40) we have to set $\boldsymbol{A}_0 = -a_0$, $\boldsymbol{B}_0 = 1$, $\boldsymbol{C}_0 = 1$, $t_0 = 0$, $t_f = \infty$, $\boldsymbol{Q}_0 = 1$, $\boldsymbol{R}_0 = 1$, and $\boldsymbol{S}_0 = \infty$ (because the origin is the final state to be arrived at [15]). Thus, the matrix Riccati equation becomes

$$\dot{M}_0 - 2a_0 M_0 - M_0^2 + 1 = 0, \qquad M_0(t_f) = S_0 = \infty. \tag{8.3-43}$$

Because of the infinite end value, it is necessary to solve the *inverse* Riccati equation. By observing that $\boldsymbol{M}_0\boldsymbol{M}_0^{-1} = \boldsymbol{I}$, differentiating this to become $\boldsymbol{M}_0\dot{\boldsymbol{M}}_0^{-1} + \dot{\boldsymbol{M}}_0\boldsymbol{M}_0^{-1} = \boldsymbol{0}$, solving for $\dot{\boldsymbol{M}}_0^{-1} = -\mathbf{M}_0^{-1}\dot{\mathbf{M}}_0\mathbf{M}_0^{-1}$, and substituting $\boldsymbol{M}_0$ of Eq.(8.3–43) into this expression, the inverse Riccati equation reads

$$\dot{M}^{-1} = -2a_0 M_0^{-1} - 1 + M_0^{-2}, \qquad M_0^{-1}(t_f) = S_0^{-1} = 0, \tag{8.3-44}$$

the solution of which is $M_0 = -a_0 + \sqrt{1 + a_0^2}$. The optimal control function is then given by

$$u_0^* = -(-a_0 + \sqrt{1 + a_0^2})\, x, \tag{8.3-45}$$

and the optimal value of the performance index becomes $J = M_0 x^2(0)$.

Now Eq.(8.3-37) will be solved. With $\boldsymbol{F} = -a - M$ and $\tilde{\boldsymbol{Q}} = M_0^2 + 1$ [according to Eq.(8.3-35)], the Riccati equation reads

$$\dot{M} - 2(a + M_0)M + M_0^2 + 1 = 0, \qquad M(t_f) = S_0 = \infty. \tag{8.3-46}$$

Again, in view of the infinite final value, the equation has to be brought to the inverse form:

$$\dot{M}^{-1} = -2(a + M_0)M^{-1} + M_0^2 M^{-2} + M^{-2}, \qquad M^{-1}(t_f) = 0. \tag{8.3-47}$$

The solution is given by

$$M = \frac{1 + a_0^2 - a_0\sqrt{1 + a_0^2}}{a - a_0 + \sqrt{1 + a_0^2}}, \tag{8.3-48}$$

Thus for the parameter a not at nominal value, $J = Mx^2(0)$. We see that $M \to M_0$ for $a \to a_0$.

Finally we have to solve E1.(8.3-41). With the above data it becomes

$$\dot{P} - 2(a_0 + M_0)P - 2M_0 = 0, \qquad P(t_f) = 0. \tag{8.3-49}$$

The solution of this equation is

$$P = \frac{a_0 - \sqrt{1 + a_0^2}}{\sqrt{1 + a_0^2}}. \tag{8.3-50}$$

Since the form of the solution of the Hamilton–Jacobi equation was assumed to be $J_a(x, t, a_0) = Px^2$, the resulting solution is given by

$$J_a(x_0, t_0, a_0) = Px^2(0) = \left(-1 + \frac{a_0}{\sqrt{1 + a_0^2}}\right) x^2(0), \tag{8.3-51}$$

which is in agreement with the result of Example 8.2-1.

8.3.4 Performance-Index Sensitivity of the Follow-Up Problem

Let the plant be given by

$$\dot{\boldsymbol{x}} = \boldsymbol{A}(\boldsymbol{\alpha})\boldsymbol{x} + \boldsymbol{B}(\boldsymbol{\alpha})\boldsymbol{u} + \boldsymbol{e}, \qquad \boldsymbol{x}(t_0) = \mathbf{x}_0,$$
$$\boldsymbol{y} = \boldsymbol{C}(\boldsymbol{\alpha})\boldsymbol{x} + \boldsymbol{D}(\boldsymbol{\alpha})\boldsymbol{u}, \tag{8.3-52}$$

where $\boldsymbol{x}$ is $n \times 1$, $\boldsymbol{u}$ is $p \times 1$, the disturbance vector $\boldsymbol{e}$ is $n \times 1$, the reference input vector $\boldsymbol{r}$ is $q \times 1$, and the matrices $\boldsymbol{A}$, $\boldsymbol{B}$, $\boldsymbol{C}$, $\boldsymbol{D}$ are of dimensions $(n \times n)$, $(n \times p)$, $(q \times n)$, and $(n \times p)$, respectively. Let us assume that a feedback control is to be designed such that the performance index

$$J = \tfrac{1}{2}\int_{t_0}^{t_f} [(\boldsymbol{r} - \boldsymbol{y})^{\mathrm{T}}\boldsymbol{Q}(\boldsymbol{r} - \boldsymbol{y}) + \boldsymbol{u}^{\mathrm{T}}\boldsymbol{R}\,\boldsymbol{u}]\,dt \tag{8.3-53}$$

is minimized, where the $q \times q$ matrix $\boldsymbol{Q}$ and the $p \times p$ matrix $\boldsymbol{R}$ are symmetrical and positive definite.

This optimization problem can be solved by a Hamilton–Jacobi equation of the form [113]

$$\frac{\partial J}{\partial t} + \min_{\boldsymbol{u}\in\Omega} H = 0, \qquad J[\boldsymbol{x}(t_f),t_f] = 0, \tag{8.3-54}$$

where the Hamiltonian function is given by

$$H = \tfrac{1}{2}(\boldsymbol{r} - \boldsymbol{y})^{\mathrm{T}}\boldsymbol{Q}(\boldsymbol{r} - \boldsymbol{y}) + \tfrac{1}{2}\boldsymbol{u}^{\mathrm{T}}\boldsymbol{R}\,\boldsymbol{u} + \boldsymbol{p}^{\mathrm{T}}(\boldsymbol{A}\boldsymbol{x} + \boldsymbol{B}\boldsymbol{u} + \boldsymbol{e}) \tag{8.3-55}$$

and the Lagrange multiplier is

$$\boldsymbol{p}^{\mathrm{T}} = \frac{\partial J}{\partial \boldsymbol{x}}. \tag{8.3-56}$$

Now we assume that the solution of Eq. (8.3-54) can be written in the form

$$J = \tfrac{1}{2}\,\boldsymbol{x}^{\mathrm{T}}\boldsymbol{M}_0\boldsymbol{x} - \boldsymbol{M}_{10}{}^{\mathrm{T}}\boldsymbol{x} + \boldsymbol{M}_{20},$$

where the subscript 0 denotes nominal values. Substituting this expression into Eq.(8.3-54) yields three equations for the determination of $\boldsymbol{M}_0$, $\boldsymbol{M}_{10}$, and $\boldsymbol{M}_{20}$. With the abbreviation

$$\boldsymbol{V} = (\boldsymbol{D}^{\mathrm{T}}\boldsymbol{Q}\,\boldsymbol{D} + \boldsymbol{R})^{-1} \tag{8.3-57}$$

and $\boldsymbol{V} = \boldsymbol{V}^{\mathrm{T}}$, these equations read

$$\begin{aligned}\dot{\boldsymbol{M}}_0 + \boldsymbol{A}_0{}^{\mathrm{T}}\boldsymbol{M}_0 + \boldsymbol{M}_0\boldsymbol{A}_0 - \boldsymbol{C}_0{}^{\mathrm{T}}\boldsymbol{Q}_0\boldsymbol{D}_0\boldsymbol{V}_0\boldsymbol{B}_0{}^{\mathrm{T}}\boldsymbol{M}_0 - \boldsymbol{M}_0\boldsymbol{B}_0\boldsymbol{V}_0\boldsymbol{D}_0{}^{\mathrm{T}}\boldsymbol{Q}_0\boldsymbol{C}_0\\ -\boldsymbol{C}_0{}^{\mathrm{T}}\boldsymbol{Q}_0\boldsymbol{D}_0\boldsymbol{V}_0\boldsymbol{D}_0{}^{\mathrm{T}}\boldsymbol{Q}_0\boldsymbol{C}_0 + \boldsymbol{C}_0{}^{\mathrm{T}}\boldsymbol{Q}_0\boldsymbol{C}_0 - \boldsymbol{M}_0\boldsymbol{B}_0\boldsymbol{V}_0\boldsymbol{B}_0{}^{\mathrm{T}}\boldsymbol{M}_0 = \boldsymbol{0},\end{aligned} \tag{8.3-58}$$

with $\boldsymbol{M}_0(t_f) = \boldsymbol{0}$;

$$\begin{aligned}\dot{\boldsymbol{M}}_{10}{}^{\mathrm{T}} = \boldsymbol{M}_{10}{}^{\mathrm{T}}(\boldsymbol{B}_0\boldsymbol{V}_0\boldsymbol{B}_0{}^{\mathrm{T}}\boldsymbol{M}_0 + \boldsymbol{B}_0\boldsymbol{V}_0\boldsymbol{D}_0{}^{\mathrm{T}}\boldsymbol{Q}_0\boldsymbol{C}_0 - \boldsymbol{A}_0) + \boldsymbol{e}^{\mathrm{T}}\boldsymbol{M}_0\\ + \boldsymbol{r}^{\mathrm{T}}(\boldsymbol{Q}_0\boldsymbol{C}_0 - \boldsymbol{Q}_0\boldsymbol{D}_0\boldsymbol{V}_0\boldsymbol{B}_0{}^{\mathrm{T}}\boldsymbol{M}_0 - \boldsymbol{Q}_0\boldsymbol{D}_0\boldsymbol{V}_0\boldsymbol{D}_0{}^{\mathrm{T}}\boldsymbol{Q}_0\boldsymbol{C}_0),\end{aligned} \tag{8.3-59}$$

with $\boldsymbol{M}_{10}(t_f) = \mathbf{0}$,

$$\dot{\boldsymbol{M}}_{20} = \tfrac{1}{2}\boldsymbol{M}_{10}{}^{\mathrm{T}}\boldsymbol{B}_0\boldsymbol{V}_0\boldsymbol{B}_0{}^{\mathrm{T}}\boldsymbol{M}_{10} + (\boldsymbol{r}^{\mathrm{T}}\boldsymbol{Q}_0\boldsymbol{D}_0\boldsymbol{V}_0\boldsymbol{B}_0{}^{\mathrm{T}} + \boldsymbol{e}^{\mathrm{T}})\boldsymbol{M}_{10} + \tfrac{1}{2}\boldsymbol{r}^{\mathrm{T}}(\boldsymbol{Q}_0\boldsymbol{D}_0\boldsymbol{V}_0\boldsymbol{D}_0{}^{\mathrm{T}} - \boldsymbol{I})\boldsymbol{Q}_0\boldsymbol{r}, \tag{8.3-60}$$

with $\boldsymbol{M}_{20}(t_f) = \mathbf{0}$.
The optimal control vector is given by

$$\boldsymbol{u}_0{}^* = -\boldsymbol{K}_0\boldsymbol{x} + \boldsymbol{V}_0\boldsymbol{D}_0{}^{\mathrm{T}}\boldsymbol{Q}_0\boldsymbol{r} + \boldsymbol{V}_0\boldsymbol{B}_0{}^{\mathrm{T}}\boldsymbol{M}_{10}, \tag{8.3-61}$$

where

$$\boldsymbol{K}_0 = \boldsymbol{V}_0(\boldsymbol{B}_0{}^{\mathrm{T}}\boldsymbol{M}_0 + \boldsymbol{D}_0{}^{\mathrm{T}}\boldsymbol{Q}_0\boldsymbol{C}_0), \tag{8.3-62}$$

and the optimal performance index is

$$J = \tfrac{1}{2}\boldsymbol{x}^{\mathrm{T}}(t_0)\boldsymbol{M}_0(t_0)\,\boldsymbol{x}(t_0) - \boldsymbol{M}_{10}{}^{\mathrm{T}}(t_0)\boldsymbol{x}(t_0) + \boldsymbol{M}_{20}(t_0). \tag{8.3-63}$$

For $t_f \to \infty$ and the given process being time-invariant with constant input signals, the matrices $\boldsymbol{M}_0$ and $\boldsymbol{M}_{10}$ are time-invariant as well, i.e., $\dot{\boldsymbol{M}}_{10} = \mathbf{0}$. Then $\boldsymbol{M}_{10}$ found from Eq.(8.3-59) can be expressed by constant weighting of $\boldsymbol{r}$ and $\boldsymbol{e}$ so that the control vector becomes

$$\boldsymbol{u}_0{}^* = -\boldsymbol{K}_0\boldsymbol{x} + \boldsymbol{W}_0\boldsymbol{r} - \boldsymbol{E}_0\boldsymbol{e}. \tag{8.3-64}$$

The weighting matrices $\boldsymbol{W}_0$ and $\boldsymbol{E}_0$ are given by the solution of Eq.(8.3-59) and will not be carried out in more detail here.

As a result, we arrive in that case at a stationary control law as illustrated in Fig. 8.3-1.

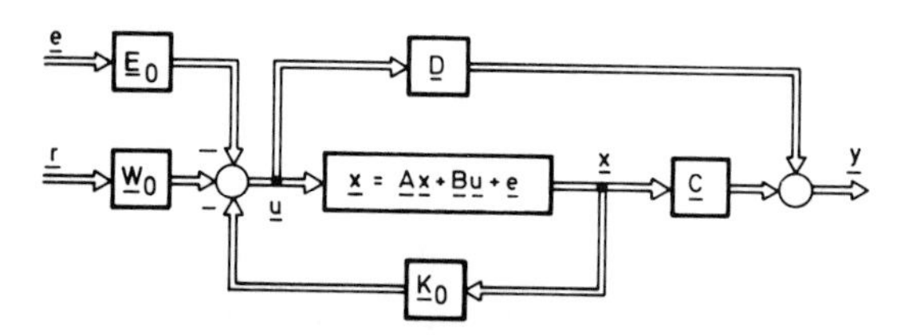

FIG. 8.3-1. Structure of the optimal control system.

Let us now assume that the parameter vector $\boldsymbol{\alpha}$ deviates from its nominal value $\boldsymbol{\alpha}_0$ by $\Delta\boldsymbol{\alpha}$ so that the matrices $\boldsymbol{A}_0$, $\boldsymbol{B}_0$, $\boldsymbol{C}_0$, etc. go over into $\boldsymbol{A}$, $\boldsymbol{B}$, $\boldsymbol{C}$, etc. However, the controller may remain unaltered. Then the state equations of the control system as it is depicted in Fig. 8.3-1 become

$$\dot{\boldsymbol{x}} = \boldsymbol{F}\boldsymbol{x} + \boldsymbol{B}\boldsymbol{N}_0 + \boldsymbol{e}, \tag{8.3-65}$$

where the following abbreviations are used

$$\boldsymbol{F} = \boldsymbol{A} - \boldsymbol{B}\boldsymbol{N}_0 \tag{8.3-66}$$

$$\boldsymbol{N}_0 = \boldsymbol{V}_0(\boldsymbol{D}_0{}^{\mathrm{T}}\boldsymbol{Q}_0\boldsymbol{r} + \boldsymbol{B}_0{}^{\mathrm{T}}\boldsymbol{M}_{10}) \tag{8.3-67}$$

and $\boldsymbol{K}_0$ and $\boldsymbol{V}_0$ are as defined in Eqs. (8.3-62) and (8.3-57), respectively. In these terms the control vector becomes

$$\boldsymbol{u}_0{}^* = -\boldsymbol{K}_0\boldsymbol{x} + \boldsymbol{N}_0, \tag{8.3-68}$$

and the integrand L of the performance index takes the form

$$\begin{aligned} L = {} & \tfrac{1}{2}\,\boldsymbol{r}^{\mathrm{T}}\boldsymbol{Q}\boldsymbol{r} - \boldsymbol{r}^{\mathrm{T}}\boldsymbol{Q}\boldsymbol{C}\boldsymbol{x} + \tfrac{1}{2}\,\boldsymbol{x}^{\mathrm{T}}\boldsymbol{C}^{\mathrm{T}}\boldsymbol{Q}\boldsymbol{C}\boldsymbol{x} + \boldsymbol{r}^{\mathrm{T}}\boldsymbol{Q}\boldsymbol{D}\boldsymbol{K}_0\boldsymbol{x} + \boldsymbol{r}^{\mathrm{T}}\boldsymbol{Q}\boldsymbol{D}\boldsymbol{N}_0 \\ & - \boldsymbol{x}^{\mathrm{T}}\boldsymbol{C}^{\mathrm{T}}\boldsymbol{Q}\boldsymbol{D}\boldsymbol{K}_0\boldsymbol{x} + \boldsymbol{N}_0{}^{\mathrm{T}}\boldsymbol{D}^{\mathrm{T}}\boldsymbol{Q}\boldsymbol{C}\boldsymbol{x} + \tfrac{1}{2}\,\boldsymbol{x}^{\mathrm{T}}\boldsymbol{K}_0{}^{\mathrm{T}}\boldsymbol{V}^{-1}\boldsymbol{K}_0\boldsymbol{x} - \boldsymbol{N}_0{}^{\mathrm{T}}\boldsymbol{V}^{-1}\boldsymbol{K}_0\boldsymbol{x} \\ & + \tfrac{1}{2}\,\boldsymbol{N}_0{}^{\mathrm{T}}\boldsymbol{V}^{-1}\boldsymbol{N}_0. \end{aligned} \tag{8.3-69}$$

Now we assume that the solution of Eq.(8.3-11) can be written in the form

$$J(\boldsymbol{x}, t, \boldsymbol{\alpha}, \boldsymbol{\alpha}_0) = \tfrac{1}{2}\,\boldsymbol{x}^{\mathrm{T}}\boldsymbol{M}\boldsymbol{x} - \boldsymbol{M}_1{}^{\mathrm{T}}\boldsymbol{x} + \boldsymbol{M}_2, \tag{8.3-70}$$

where $\boldsymbol{M}$ is symmetrical. Forming

$$\frac{\partial J}{\partial t} = \tfrac{1}{2}\,\boldsymbol{x}^{\mathrm{T}}\dot{\boldsymbol{M}}\boldsymbol{x} - \dot{\boldsymbol{M}}_1{}^{\mathrm{T}}\boldsymbol{x} + \dot{\boldsymbol{M}}_2 \tag{8.3-71}$$

and

$$\frac{\partial J}{\partial \boldsymbol{x}} = \boldsymbol{x}^{\mathrm{T}}\boldsymbol{M} - \boldsymbol{M}_1{}^{\mathrm{T}} \tag{8.3-72}$$

and substituting Eqs.(8.3-70)–(8.3-72) into Eq.(8.3-11), we obtain the following three equations for the determination of the actual performance index:

$$\text{(1)}\quad \dot{\boldsymbol{M}} + \boldsymbol{M}\boldsymbol{F} + \boldsymbol{F}^{\mathrm{T}}\boldsymbol{M} + \tilde{\mathbf{Q}} = \mathbf{0}, \qquad \boldsymbol{M}(\mathrm{t}_f) = \mathbf{0}, \tag{8.3-73}$$

with

$$\tilde{\mathbf{Q}} = \boldsymbol{C}^{\mathrm{T}}\boldsymbol{Q}\boldsymbol{C} - \boldsymbol{C}\boldsymbol{Q}\boldsymbol{D}\boldsymbol{K}_0 - \boldsymbol{K}_0{}^{\mathrm{T}}\boldsymbol{D}^{\mathrm{T}}\boldsymbol{Q}\boldsymbol{C} + \boldsymbol{K}_0{}^{\mathrm{T}}\boldsymbol{V}^{-1}\boldsymbol{K}_0; \tag{8.3-74}$$

$$\begin{aligned} \text{(2)}\quad \dot{\boldsymbol{M}}_1{}^{\mathrm{T}} = {} & -\boldsymbol{M}_1{}^{\mathrm{T}}\boldsymbol{F} + (\boldsymbol{N}_0{}^{\mathrm{T}}\boldsymbol{B} + \boldsymbol{e}^{\mathrm{T}})\boldsymbol{M} - \boldsymbol{r}^{\mathrm{T}}(\boldsymbol{Q}\boldsymbol{C} - \boldsymbol{Q}\boldsymbol{D}\boldsymbol{K}_0) \\ & + \boldsymbol{N}_0{}^{\mathrm{T}}\boldsymbol{D}^{\mathrm{T}}\boldsymbol{Q}\boldsymbol{C} - \boldsymbol{N}_0{}^{\mathrm{T}}\boldsymbol{V}^{-1}\boldsymbol{K}_0, \qquad \boldsymbol{M}_1(\mathrm{t}_f) = \mathbf{0}; \end{aligned} \tag{8.3-75}$$

$$\begin{aligned} \text{(3)}\quad \dot{\boldsymbol{M}}_2 = {} & \boldsymbol{M}_1{}^{\mathrm{T}}(\boldsymbol{B}\boldsymbol{N}_0 + \boldsymbol{e}) - \tfrac{1}{2}\,\boldsymbol{r}^{\mathrm{T}}\boldsymbol{Q}\boldsymbol{r} + \boldsymbol{r}^{\mathrm{T}}\boldsymbol{Q}\boldsymbol{D}\boldsymbol{N}_0 \\ & - \tfrac{1}{2}\,\boldsymbol{N}_0{}^{\mathrm{T}}\boldsymbol{V}^{-1}\boldsymbol{N}_0, \qquad \boldsymbol{M}_2(\mathrm{t}_f) = \mathbf{0}. \end{aligned} \tag{8.3-76}$$

In a similar manner the equations for the determination of the performance-index sensitivities can be found. The Hamilton–Jacobi equation (8.3-15) reads, in this case,

$$\begin{aligned} \frac{\partial J_\alpha{}^j}{\partial t} = {} & -\frac{\partial J_\alpha{}^j}{\partial \boldsymbol{x}}(\boldsymbol{F}_0\boldsymbol{x} + \boldsymbol{B}_0\boldsymbol{N}_0 + \boldsymbol{e}) \\ & -(\boldsymbol{x}^{\mathrm{T}}\boldsymbol{M}_0 - \boldsymbol{M}_{10}{}^{\mathrm{T}})\left(\frac{\partial \boldsymbol{F}}{\partial \alpha_j}\boldsymbol{x} + \frac{\partial \boldsymbol{B}}{\partial \alpha_j}\boldsymbol{N}_0\right)\bigg|_{\boldsymbol{\alpha}_0} + \frac{\partial L}{\partial \alpha_j}\bigg|_{\boldsymbol{\alpha}_0}. \end{aligned} \tag{8.3-77}$$

To solve the above equation, we assume that the solution can be written in the form

$$J_\alpha^j = \tfrac{1}{2}\, \boldsymbol{x}^{\mathrm{T}}\boldsymbol{P}_j\boldsymbol{x} - \boldsymbol{P}_{j1}{}^{\mathrm{T}} + \boldsymbol{P}_{j2}, \qquad \boldsymbol{P}_j \text{ symmetrical.} \tag{8.3-78}$$

Substituting this into Eq.(8.3-77) yields the following three equations:

$$\text{(1)}\quad \dot{\boldsymbol{P}}_j + \boldsymbol{P}_j\boldsymbol{F}_0 + \boldsymbol{F}_0{}^{\mathrm{T}}\boldsymbol{P}_j + \boldsymbol{Q}^* = \boldsymbol{0}, \qquad \boldsymbol{P}_j(\mathrm{t}_f) = \boldsymbol{0}, \tag{8.3-79}$$

with

$$\boldsymbol{Q}^* = \left[\frac{\partial \boldsymbol{F}^{\mathrm{T}}}{\partial \alpha_j}\boldsymbol{M}_0 + \boldsymbol{M}_0\frac{\partial \boldsymbol{F}}{\partial \alpha_j} + \frac{\partial \tilde{\boldsymbol{Q}}}{\partial \alpha_j}\right]_{\boldsymbol{\alpha}_0}; \tag{8.3-80}$$

$$\begin{aligned}
\text{(2)}\quad \dot{\boldsymbol{P}}_{j1}{}^{\mathrm{T}} = {} & -\boldsymbol{P}_{j1}{}^{\mathrm{T}}\boldsymbol{F}_0 + (\boldsymbol{N}_0{}^{\mathrm{T}}\boldsymbol{B}_0{}^{\mathrm{T}} + \boldsymbol{e}^{\mathrm{T}})\boldsymbol{P}_j + \boldsymbol{N}_0{}^{\mathrm{T}}\left.\frac{\partial \boldsymbol{B}^{\mathrm{T}}}{\partial \alpha_j}\right|_{\boldsymbol{\alpha}_0}\boldsymbol{M}_0 - \boldsymbol{M}_{10}{}^{\mathrm{T}}\left.\frac{\partial \boldsymbol{F}}{\partial \alpha_j}\right|_{\boldsymbol{\alpha}_0} \\
& -\boldsymbol{r}^{\mathrm{T}}\left(\frac{\partial \boldsymbol{Q}}{\partial \alpha_j}\boldsymbol{C}_0 + \boldsymbol{Q}_0\frac{\partial \boldsymbol{C}}{\partial \alpha_j}\right)_{\boldsymbol{\alpha}_0} + \boldsymbol{r}^{\mathrm{T}}\left(\frac{\partial \boldsymbol{Q}}{\partial \alpha_j}\boldsymbol{D}_0 + \boldsymbol{Q}_0\frac{\partial \boldsymbol{D}}{\partial \alpha_j}\right)_{\boldsymbol{\alpha}_0}\boldsymbol{K}_0 \\
& +\boldsymbol{N}_0{}^{\mathrm{T}}\left(\frac{\partial \boldsymbol{D}^{\mathrm{T}}}{\partial \alpha_j}\boldsymbol{Q}_0\boldsymbol{C}_0 + \boldsymbol{D}_0{}^{\mathrm{T}}\frac{\partial \boldsymbol{Q}}{\partial \alpha_j}\boldsymbol{C}_0 + \boldsymbol{D}_0{}^{\mathrm{T}}\boldsymbol{Q}_0\frac{\partial \boldsymbol{C}}{\partial \alpha_j}\right)_{\boldsymbol{\alpha}_0} \\
& -\boldsymbol{N}_0{}^{\mathrm{T}}\left.\frac{\partial \boldsymbol{V}^{-1}}{\partial \alpha_j}\right|_{\boldsymbol{\alpha}_0}\boldsymbol{K}_0
\end{aligned}$$

$$\boldsymbol{P}_{j1}(\mathrm{t}_f) = \boldsymbol{0}; \tag{8.3-81}$$

$$\begin{aligned}
\text{(3)}\quad & \dot{\boldsymbol{P}}_{j2} - \boldsymbol{P}_{j1}{}^{\mathrm{T}}(\boldsymbol{B}_0\boldsymbol{N}_0 + \boldsymbol{e}) - \boldsymbol{M}_{10}{}^{\mathrm{T}}\left.\frac{\partial \boldsymbol{B}}{\partial \alpha_j}\right|_{\boldsymbol{\alpha}_0}\boldsymbol{N}_0 + \tfrac{1}{2}\,\boldsymbol{r}^{\mathrm{T}}\left.\frac{\partial \boldsymbol{Q}}{\partial \alpha_j}\right|_{\boldsymbol{\alpha}_0}\boldsymbol{r} \\
& - \boldsymbol{r}^{\mathrm{T}}\left(\frac{\partial \boldsymbol{Q}}{\partial \alpha_j}\boldsymbol{D}_0 + \boldsymbol{Q}_0\frac{\partial \boldsymbol{D}}{\partial \alpha_j}\right)_{\boldsymbol{\alpha}_0}\boldsymbol{N}_0 + \tfrac{1}{2}\,\boldsymbol{N}_0{}^{\mathrm{T}}\left.\frac{\partial \boldsymbol{V}^{-1}}{\partial \alpha_j}\right|_{\boldsymbol{\alpha}_0}\boldsymbol{N}_0 = \boldsymbol{0},
\end{aligned}$$

$$\boldsymbol{P}_{j2}(\mathrm{t}) = \boldsymbol{0}. \tag{8.3-82}$$

In general, $\boldsymbol{M}$ and $\boldsymbol{M}_1$ as well as $\boldsymbol{P}_j$ and $\boldsymbol{P}_{j1}$ are time-variant. If, however, $t_f \to \infty$, and the given process is time-invariant, $\boldsymbol{F}$ and $\boldsymbol{F}_0$ have only eigenvalues with negative real parts, and $\boldsymbol{r}$ and $\boldsymbol{e}$ are constant, then $\boldsymbol{M}$, $\boldsymbol{M}_1$, $\boldsymbol{M}_2$ and $\boldsymbol{P}_j$, $\boldsymbol{P}_{j1}$, $\boldsymbol{P}_{j2}$ are time-invariant. Instead of differential equations, we then have algebraic equations so that for simple systems analytic solutions can be found. In any case the digital computer will be mandatory to solve the above equations. A computer program for the calculation of the optimal feedback coefficients and the performance-index sensitivities according to the above equations is given by Thieme [113].

8.4 QUANTITATIVE VALUATION OF OPTIMAL SYSTEMS BY MEANS OF PERFORMANCE-INDEX SENSITIVITY

As mentioned before, the performance-index sensitivity represents the most logical measure of sensitivity of optimal systems. It can as well be used as a sensitivity measure for nonoptimal systems and, therefore, is apt to serve as a comparison criterion between both.

For practical applications it is reasonable to normalize both the performance-index variation ∂J and the parameter variation $\partial \alpha_j$ on their nominal values $J(\boldsymbol{x}^0, t_0, \boldsymbol{\alpha}_0)$ and α_{j0}, repsectively, i.e., to use the relative sensitivity measure

$$\bar{J}_\alpha^j \triangleq \left.\frac{\partial \ln J}{\partial \ln \alpha_j}\right|_{\boldsymbol{\alpha}_0} = J_\alpha^{\,j}(\boldsymbol{x}^0, t_0, \boldsymbol{\alpha}_0) \frac{\alpha_{j0}}{J(\boldsymbol{x}^0, \mathrm{t}_0, \boldsymbol{\alpha}_0)} \tag{8.4-1}$$

provided that $J(\boldsymbol{x}^0, t, \boldsymbol{\alpha}_0) \neq 0$.

For an overall valuation of the performance-index sensitivity of an optimal system with respect to various parameters $\alpha_1, \ldots, \alpha_r$, a norm of the $J_\alpha^{\,j}$'s or $\bar{J}_\alpha^{\,j}$'s can be employed, such as the Euclidean norm

$$\|\bar{J}_\alpha\| = [(\bar{J}_\alpha^{\,1})^2 + \cdots + (\bar{J}_\alpha^{\,r})^2]^{1/2}. \tag{8.4-2}$$

The performance-index sensitivity has been used in a min–max procedure for the design of insensitive optimal control systems [37, 50, 51]. Here the basic idea is to choose $\boldsymbol{u}$ such that the maximal change of the performance index over the interval of possible parameter variations is minimized.

The use of the performance-index sensitivity in optimal systems has three salient disadvantages:

(1) If the performance-index sensitivity is defined in terms of the performance index that is optimized, then for a wide class of optimal systems the value of J_α is the same for both the open- and closed-loop controls so that a comparison between them in these terms is of no use.

(2) As already mentioned at the beginning of this chapter, the values of $J_\alpha^{\,j}$ are trivially zero for parameter changes of the controller if the performance index has a *relative* minimum at $\boldsymbol{\alpha}_0$ and thus has a horizontal tangent at this point. This is due to the fact that the system has been optimized with respect to the control function, i.e., to the parameters of the controller.

(3) Owing to the differential character of $J_\alpha^{\,j}$, one can only obtain an idea as to the alternating tendency of the performance index. Information about how wide or narrow the minimum would be in the presence of finite parameter variations cannot be deduced from such a sensitivity measure.

In order to avoid the above drawbacks, it is customary to use the *trajectory sensitivity* as a means of sensitivity comparison or for the design of insensitive optimal systems. Dorato [36] suggested using for the sensitivity characterization a performance index other than the one according to which the system is optimized. It is, however, difficult to justify such alternatives. Once it has turned out to be reasonable to optimize a system according to a certain performance index, it is reasonable also to measure the sensitivity against this same performance index.

A more promising definition of sensitivity in optimal systems is the consideration of performance index changes due to *finite* parameter variations. This allows a realistic description of the performance index behavior around the nominal value and generally discovers considerable differences between open and closed loop controls. The price however is a much more involved mathematical treatment.

In the following section, a sensitivity measure will be described that expresses the degree of optimality in dependence upon the parameter variations. This sensitivity measure is particularly suited for optimal systems and applies to finite parameter variations also.

8.5 A RELATIVE SENSITIVITY MEASURE OF OPTIMALITY (OPTIMALITY LOSS)

8.5.1 DEFINITIONS

For the characterization of the parameter sensitivity of an optimal system at infinitesimal or finite parameter deviations, Rohrer and Sobral [86] have proposed a relative sensitivity measure that expresses the loss of optimality at actual plant parameters, or, in other words, the relative increase of the performance index compared with its reachable minimum at actual parameters.

To deduce this sensitivity measure, let us consider a system with the state equation

$$\dot{\boldsymbol{x}} = \boldsymbol{f}(\boldsymbol{x}, t, \boldsymbol{\alpha}_0, \boldsymbol{u}), \qquad \boldsymbol{x}(t_0) = \boldsymbol{x}^0. \tag{8.5-1}$$

[The meaning of the symbols is the same as in Eq.(8.2-1).] Consider the control function to be chosen in such a way that a certain positive performance index

$$J = J(\boldsymbol{\alpha}_0, \boldsymbol{u}) > 0 \tag{8.5-2}$$

is minimized. Let $\boldsymbol{\alpha}$ be the actual value of the parameter vector and $\boldsymbol{u} = \boldsymbol{u}(t, \boldsymbol{\alpha}, \boldsymbol{\alpha}_0)$ the corresponding control vector, which is optimal with respect to the nominal parameter value $\boldsymbol{\alpha} = \boldsymbol{\alpha}_0$. Thus the actual value of the performance index is $J(\boldsymbol{\alpha}, \boldsymbol{u})$.

Now let the *optimal* control vector with respect to the *actual* parameter value be denoted by $\boldsymbol{u}^* = \boldsymbol{u}(t, \boldsymbol{\alpha}, \boldsymbol{\alpha}_0)$ and the corresponding value of the performance index by

$$J(\boldsymbol{\alpha}, \boldsymbol{u}^*) = \min_{\boldsymbol{u}} \{J(\boldsymbol{\alpha}, \boldsymbol{u})\} > 0. \tag{8.5-3}$$

Then a relative measure of optimality can be defined as

$$\bar{S}(\boldsymbol{\alpha}, \boldsymbol{u}) = \frac{J(\boldsymbol{\alpha}, \boldsymbol{u}) - J(\boldsymbol{\alpha}, \boldsymbol{u}^*)}{J(\boldsymbol{\alpha}, \boldsymbol{u}^*)}, \tag{8.5-4}$$

i.e., as the difference between the actual value of the performance index and its smallest possible value (at the actual parameter vector $\boldsymbol{\alpha}$) divided by the latter for normalization. In other words, $\bar{S}(\boldsymbol{\alpha}, \boldsymbol{u})$ indicates the relative deviation from the reachable minimum at actual parameter values and will therefore be termed *optimality loss*. Thus, the degree of achieved optimality is characterized by a number between 0 and ∞. If $\bar{S}(\boldsymbol{\alpha}, \boldsymbol{u})$ is close to zero over a wide range of parameter variations, the system is evidently insensitive to those variations.

Comparing this sensitivity definition with the performance index sensitivity as defined by Eq.(8.4-1), we see that the only difference is in the reference of the performance index. Whereas performance index sensitivity is referred to the optimal value of J at *nominal* parameters, sensitivity loss is referred to the optimal value of J at *actual* parameters. Thus the latter does not consider the part of the performance increase that is inevitable anyway. The situation is illustrated in Fig. 8.5-1.

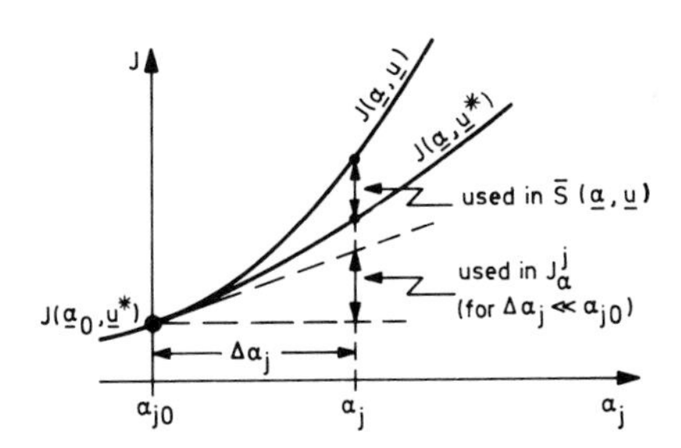

FIG. 8.5-1. The difference between performance-index sensitivity J^j_α and optimality loss $\bar{S}(\boldsymbol{\alpha}, \boldsymbol{u})$.

For small parameter variations $\|\delta\boldsymbol{\alpha}\| = \|\boldsymbol{\alpha} - \boldsymbol{\alpha}_0\| \ll \boldsymbol{\alpha}_0$ and hence small parameter-induced deviations $\|\delta\boldsymbol{u}\| = \|\boldsymbol{u} - \boldsymbol{u}^*\| \ll \boldsymbol{u}^*$, the numerator of $\bar{S}(\boldsymbol{\alpha}, \boldsymbol{u})$ can be expanded into a series:

$$J(\boldsymbol{\alpha}, \boldsymbol{u}) - J(\boldsymbol{\alpha}, \boldsymbol{u}^*) = \delta J(\boldsymbol{\alpha}, \boldsymbol{u}^*, \delta\boldsymbol{u}) + \delta^2 J(\boldsymbol{\alpha}, \boldsymbol{u}^*, \delta\boldsymbol{u}) + \cdots. \tag{8.5-5}$$

In the case such that $\boldsymbol{u}^*$ lies in the interior of the permissible set (i.e., for unbounded $\mathbf{u}$), the first variation in Eq.(8.5-5) must vanish by the definition of optimality and the second variation must be positive. The optimality loss is then approximately given by

$$\bar{S}(\boldsymbol{\alpha}, \boldsymbol{u}) \approx \frac{\delta^2 J(\boldsymbol{\alpha}, \boldsymbol{u}^*, \delta\boldsymbol{u})}{J(\boldsymbol{\alpha}, \boldsymbol{u}^*)}. \tag{8.5-6}$$

If the optimal control function lies on the boundary of Ω (i.e., in the case of bounded $\boldsymbol{u}$) the first variation does not vanish and the approximate formula of the optimality loss is given by

$$\bar{S}(\boldsymbol{\alpha}, \boldsymbol{u}) \approx \frac{\delta J(\boldsymbol{\alpha}, \boldsymbol{u}^*, \delta \boldsymbol{u})}{J(\boldsymbol{\alpha}, \boldsymbol{u}^*)}. \tag{8.5-7}$$

If the above sensitivity definition is to be used as a basis for the design of insensitive optimal control systems with respect to a set of allowable plant parameters $\boldsymbol{\alpha} \in B_\alpha$, it is suitable to define the sensitivity measure in the following way

$$\bar{S}^M(\boldsymbol{u}) \triangleq \max_{\boldsymbol{\alpha} \in B_\alpha} \{\bar{S}(\boldsymbol{\alpha}, \boldsymbol{u}). \tag{8.5-8}$$

It expresses the maximum value (worst case) of the optimality loss within the permissible parameter range. The design problem then consists in finding a control function $\boldsymbol{u} = \boldsymbol{u}^*$ such that $\boldsymbol{u}^*$ satisfies the minimax condition

$$\bar{S}_m(\boldsymbol{u}^*) = \min_{\boldsymbol{u} \in \Omega} \{\bar{S}^M(\boldsymbol{u})\}. \tag{8.5-9}$$

Instead of minimizing $\bar{S}^M$, we can also require that the expectation of $\bar{S}^M$ be minimized.

The application of this sensitivity measure will now be illustrated by the following example [86]:

Example 8.5-1 Given a second order system described by the state equation

$$\begin{aligned} \dot{x}_1 &= x_2, & x_1(0) &= 1, \\ \dot{x}_2 &= -\alpha x_1 + u, & x_2(0) &= 0, \end{aligned} \tag{8.5-10}$$

where the parameter α is assumed to lie in the interval $0 < \alpha \leq 2$.

(a) Determine the optimal control $\boldsymbol{u}_0{}^*$ for a nominal parameter α_0 such that the performance index

$$J(\alpha_0, u^*) = \int_0^\infty (x_1{}^2 + x_2{}^2 + u^2)\, dt \tag{8.5-11}$$

is minimized.

(b) Give the optimality loss for $\alpha \neq \alpha_0$.

(c) Find the optimal value of u according to the criterion $S_m(u^*)$.

Solution (*a*) It is easily seen from optimization theory that the optimal trajectory for α_0 satisfies the equations

$$\begin{aligned} \dot{x}_{10} &= x_{20}, \\ \dot{x}_{20} &= -\sqrt{\alpha_0{}^2 + 3}\, x_{20} - x_{10}. \end{aligned} \tag{8.5-12}$$

The corresponding optimal control law is given by

$$u_0{}^* = k_{10} x_{10} + k_{20} x_{20}, \tag{8.5-13}$$

where, by comparison of Eqs.(8.5–10) and (8.5–12) for $\alpha = \alpha_0$,

$$k_{10} = -1,$$
$$k_{20} = \alpha_0 - \sqrt{\alpha_0^2 + 3}. \tag{8.5-14}$$

(b) Now let $\alpha_0 \to \alpha$, keeping $k_{10} = -1$ and k_{20} constant. The actual control u will then be

$$u = -x_1 + k_{20}x_2, \tag{8.5–15}$$

which is under the given circumstances no more optimal. Substituting the solution of Eq.(8.5-12) for actual parameters α and u of Eq.(8.5-15) into Eq.(8.5-11) yields the actual performace index

$$J(\alpha, k_{20}) = \frac{(\alpha - k_{20})^2 + \alpha^2 + 3}{2(\alpha - k_{20})} x_1^2(0) + 2x_1(0)x_2(0) + \frac{k_{20}^2 + 3}{2(\alpha - k_{20})} x_2^2(0) \tag{8.5-16}$$

or, with the given values of $x_1(0) = 1$, $x_2(0) = 0$,

$$J(\alpha, k_{20}) = \frac{k_{20}^2 + 3}{2(\alpha - k_{20})} + \alpha. \tag{8.5-17}$$

The optimal value of J at α is obtained, according to (a), by replacing k_{20} in Eq.(8.5-17) by

$$k_2 = \alpha - \sqrt{\alpha^2 + 3}. \tag{8.5-18}$$

This gives

$$J(\alpha,k_2) = \sqrt{\alpha^2 + 3}. \tag{8.5-19}$$

From Eqs.(8.5-17) and (8.5-19) we obtain the optimality loss

$$\bar{S}(\alpha, k_{20}) = \frac{J(\alpha, k_{20}) - J(\alpha, k_2)}{J(\alpha, k_2)} = \frac{(\alpha - k_{20} - \sqrt{\alpha^2 + 3})^2}{2(\alpha - k_{20})\sqrt{\alpha^2 + 3}} \tag{8.5-20}$$

or by substituting k_{20}

$$\bar{S}(\alpha, \alpha_0) = \frac{(\alpha - \alpha_0 + \sqrt{\alpha_0^2 + 3} - \sqrt{\alpha^2 + 3})^2}{2(\alpha - \alpha_0 + \sqrt{\alpha_0^2 + 3})\sqrt{\alpha^2 + 3}} \tag{8.5-21}$$

(c) The value k_{20} in Eq.(8.5–20) has to be chosen such that

$$\bar{S}^M(\alpha,k_{20}) \triangleq \max_{\alpha} \{\bar{S}(\alpha,k_{20})\} \tag{8.5-22}$$

becomes minimal. The plotting of $\bar{S}(\alpha, k_{20})$ versus α for various values of k_{20} reveals [86] that the optimal value is $k_{20} = -1.3$. Thus the control system has to be optimized for a nominal parameter value $\alpha_0 \approx 0.5$. In this case $\bar{S}^M$ is about 0.04 and the performance index J varies from 1.77 to 2.72 as α varies from 0 to 2.

8.5.2 The Calculation of Optimality Loss

We shall now show for the linear regulator problem how the optimality loss can be calculated. Let the given process be characterized by the state equation

$$\dot{\boldsymbol{x}} = \boldsymbol{A}\boldsymbol{x} + \boldsymbol{B}\boldsymbol{u}, \qquad \boldsymbol{x}(t_0) = \boldsymbol{x}^0,$$
$$\boldsymbol{y} = \boldsymbol{C}\boldsymbol{x} \tag{8.5-23}$$

and let the performance index be

$$J = \tfrac{1}{2}\,\boldsymbol{x}^{\mathrm{T}}(t_f)\boldsymbol{S}\,\boldsymbol{x}(t_f) + \tfrac{1}{2}\int_{t_0}^{t_f} (\boldsymbol{x}^{\mathrm{T}}\boldsymbol{Q}\,\boldsymbol{x} + \boldsymbol{u}^{\mathrm{T}}\boldsymbol{R}\,\boldsymbol{u})\,dt. \tag{8.5-24}$$

The actual performance index is then

$$J(\boldsymbol{\alpha}, \boldsymbol{u}) = \tfrac{1}{2}\,\boldsymbol{x}^{\mathrm{T}}(t_0)\boldsymbol{M}(\boldsymbol{\alpha}, t_0)\boldsymbol{x}(t_0), \tag{8.5-25}$$

where $\boldsymbol{M}$ is found from Eq.(8.3-37). The *optimal* actual performance index is given by

$$J(\boldsymbol{\alpha}, \boldsymbol{u}^*) = \tfrac{1}{2}\,\boldsymbol{x}^{\mathrm{T}}(t_0)\boldsymbol{M}^*(\boldsymbol{\alpha}, t_0)\boldsymbol{x}(t_0); \tag{8.5-26}$$

$\boldsymbol{M}^*$ follows from the matrix Riccati equation (8.3-30) after having replaced all the nominal parameters by actual ones so that

$$\dot{\boldsymbol{M}}^* + \boldsymbol{A}^{\mathrm{T}}\boldsymbol{M}^* + \boldsymbol{M}^*\boldsymbol{A} - \boldsymbol{M}^*\boldsymbol{B}^{\mathrm{T}}\boldsymbol{R}^{-1}\boldsymbol{B}^{\mathrm{T}}\boldsymbol{M}^* + \boldsymbol{C}^{\mathrm{T}}\boldsymbol{Q}\boldsymbol{C} = \boldsymbol{0}. \tag{8.5-27}$$

Thus the optimality loss becomes

$$\bar{S}(\boldsymbol{\alpha},\boldsymbol{u}) = \left.\frac{\boldsymbol{x}^{\mathrm{T}}(\boldsymbol{M} - \boldsymbol{M}^*)\boldsymbol{x}}{\boldsymbol{x}^{\mathrm{T}}\boldsymbol{M}^*\boldsymbol{x}}\right|_{t_0}. \tag{8.5-28}$$

This is a formula to calculate $\bar{S}(\boldsymbol{\alpha}, \boldsymbol{u})$ by means of the Riccati equations derived for the calculation of the performance index sensitivity $J_\alpha^{\ j}$.

Now the question is whether $\bar{S}(\boldsymbol{\alpha},\boldsymbol{u})$ can be determined by an extrapolation of the performance index from nominal parameter values, making use of the performance index sensitivity and its derivatives. This is identical with a series expansion of $\bar{S}(\boldsymbol{\alpha}, \boldsymbol{\mu})$ around $\boldsymbol{\alpha}_0$.

As far as the first derivative at $\boldsymbol{\alpha}_0$ is concerned, it is clear that it is the same for $J(\boldsymbol{\alpha}, \boldsymbol{u})$ and $J(\boldsymbol{\alpha}, \boldsymbol{u}^*)$. This can formally be shown, e.g., by taking the first derivatives of the corresponding Riccati equations [113]. Since the derivative of $J(\boldsymbol{\alpha}, \boldsymbol{u})$ at $\boldsymbol{\alpha}_0$ was found to be [see Eq.(8.3-40)]

$$J_\alpha^{\ j} = \tfrac{1}{2}\,\boldsymbol{x}^{\mathrm{T}}(t_0)\boldsymbol{P}_j(\boldsymbol{\alpha}_0, t_0)\,\boldsymbol{x}(t_0), \tag{8.5-29}$$

the derivative of $\boldsymbol{M}(\boldsymbol{\alpha},t_0)$ at $\boldsymbol{\alpha}_0$ must be

$$\left.\frac{\partial \boldsymbol{M}(\boldsymbol{\alpha}, t_0)}{\partial \alpha_j}\right|_{\boldsymbol{\alpha}_0} = \boldsymbol{P}_j(\boldsymbol{\alpha}, t_0). \tag{8.5-30}$$

Hence

$$\left.\frac{\partial \boldsymbol{M}^*(\boldsymbol{\alpha}, t_0)}{\partial \alpha_j}\right|_{\boldsymbol{\alpha}_0} = \left.\frac{\partial \boldsymbol{M}(\boldsymbol{\alpha}, t_0)}{\partial \alpha_j}\right|_{\boldsymbol{\alpha}_0} = \boldsymbol{P}_j(\boldsymbol{\alpha}_0, t_0). \tag{8.5-31}$$

This means that the optimality loss disappears at $\boldsymbol{\alpha}_0$ in a first order approximation so that the second-order approximation is needed for the extrapolation.

The second-order approximations of $\boldsymbol{M}(\boldsymbol{\alpha}, t_0)$ and $\boldsymbol{M}^*(\boldsymbol{\alpha}, t_0)$ are

$$\mathbf{M}(\boldsymbol{\alpha}, t_0) = \mathbf{M}_0(\boldsymbol{\alpha}_0, t_0) + \boldsymbol{P}_j(\boldsymbol{\alpha}_0, t_0)\, \Delta\alpha_j + \frac{1}{2} \frac{\partial^2 \boldsymbol{M}(\boldsymbol{\alpha}, t_0)}{\partial \alpha_j^2}\bigg|_{\boldsymbol{\alpha}_0} \Delta\alpha_j^2,$$

$$\boldsymbol{M}^*(\boldsymbol{\alpha}, t_0) = \boldsymbol{M}_0(\boldsymbol{\alpha}_0, t_0) + \boldsymbol{P}_j(\boldsymbol{\alpha}_0, t_0)\, \Delta\alpha_j + \frac{1}{2} \frac{\partial^2 \boldsymbol{M}^*(\boldsymbol{\alpha}_0, t_0)}{\partial \alpha_j^2}\bigg|_{\boldsymbol{\alpha}_0} \Delta\alpha_j^2.$$

Therefore the difference is

$$\boldsymbol{M} - \boldsymbol{M}^* = \boldsymbol{Z}\, \Delta\alpha_j^2 \tag{8.5-32}$$

with

$$\boldsymbol{Z} = \frac{1}{2}\left(\frac{\partial^2 \boldsymbol{M}}{\partial \alpha_j}\bigg|_{\boldsymbol{\alpha}_0} - \frac{\partial^2 \boldsymbol{M}^*}{\partial \alpha_j^2}\bigg|_{\boldsymbol{\alpha}_0}\right). \tag{8.5-33}$$

The vector $\boldsymbol{Z}$ can be found from the linear differential equation

$$\dot{\boldsymbol{Z}} + \boldsymbol{F}_0^{\mathrm{T}}\boldsymbol{Z} + \boldsymbol{Z}\boldsymbol{F}_0 + \boldsymbol{P}_j\, \boldsymbol{B}_0 \boldsymbol{R}_0^{-1} \frac{\partial \boldsymbol{B}^{\mathrm{T}}}{\partial \alpha_j}\bigg|_{\boldsymbol{\alpha}_0} \boldsymbol{M}_0 + \boldsymbol{M}_0 \frac{\partial \boldsymbol{B}}{\partial \alpha_j}\bigg|_{\boldsymbol{\alpha}_0} \boldsymbol{R}_0^{-1}\boldsymbol{B}_0^{\mathrm{T}}\boldsymbol{P}_j$$

$$+\, \boldsymbol{P}_j\boldsymbol{B}_0\boldsymbol{R}_0^{-1}\boldsymbol{B}_0^{\mathrm{T}}\boldsymbol{P}_j + \boldsymbol{M}_0 \frac{\partial \boldsymbol{B}}{\partial \alpha_j}\bigg|_{\boldsymbol{\alpha}_0} \boldsymbol{R}_0^{-1} \frac{\partial \boldsymbol{B}^{\mathrm{T}}}{\partial \alpha_j}\bigg|_{\boldsymbol{\alpha}_0} \boldsymbol{M}_0 = \mathbf{0},$$

$$\boldsymbol{Z}(t_f) = \mathbf{0} \tag{8.5-34}$$

which can be deduced by taking the second partial derivatives of Eqs.(8.3-41) and (8.5-27) for actual $\boldsymbol{\alpha}$ and then letting $\boldsymbol{\alpha} \to \boldsymbol{\alpha}_0$.

With the aid of the above results, the nominator of the optimality loss (i.e., the "absolute" optimality loss) is obtained in a second-order approximation as

$$S(\boldsymbol{\alpha}, \boldsymbol{u}) \approx \tfrac{1}{2}\, \boldsymbol{x}^{\mathrm{T}}(t_0)\boldsymbol{Z}(\boldsymbol{\alpha}_0, t_0)\boldsymbol{x}(t_0)\, \Delta\alpha_j^2. \tag{8.5-35}$$

The calculation of $S(\boldsymbol{\alpha}, \boldsymbol{u})$ according to Eqs.(8.5-34) and (8.5-35) is easier than that of $\bar{S}(\boldsymbol{\alpha}, \boldsymbol{u})$ based on Eq.(8.5-28) since only one linear equation has to be solved. $S(\boldsymbol{\alpha}, \boldsymbol{u})$ can thus be found from the performance-index sensitivity at nominal parameters. Whereas the performance-index sensitivity considers a single point only, $S(\boldsymbol{\alpha}, \boldsymbol{u})$ considers a finite interval. Another advantage of Eq.(8.5-34) compared to Eq.(8.5-28) concerns the accuracy of numerical calculations. If, for example $\Delta\alpha_j = 0.1$, then the numbers for calculating $\boldsymbol{Z}$ are 10^2 times larger than for calculating $\boldsymbol{M} - \boldsymbol{M}^*$, thus yielding a much higher accuracy.

The advantage of simplicity of calculation gets lost when $S(\boldsymbol{\alpha}, \boldsymbol{u})$ is normalized on $J(\boldsymbol{\alpha}, \boldsymbol{u}^*)$ since then $\boldsymbol{M}^*$ has to be calculated in addition so that again a linear and a Riccati equation have to be solved as in the case of Eq.(8.5-28).

To avoid this, the absolute optimality loss $\boldsymbol{S}(\boldsymbol{\alpha}, \boldsymbol{u})$ can be normalized on the optimal value of the performance index $J(\boldsymbol{\alpha}, \boldsymbol{u}^*)$ at nominal parameters rather than upon the reachable value $J(\boldsymbol{\alpha}, \mathbf{u}^*)$ at actual values. $J(\boldsymbol{\alpha}_0, \boldsymbol{u}^*)$ is always known from the solution of the optimization problem and therefore requires no further calculations.

Based on these considerations, the definition of optimality loss is proposed as follows:

$$\bar{S}_0(\boldsymbol{\alpha}, \boldsymbol{u}) \triangleq \frac{J(\boldsymbol{\alpha}, \boldsymbol{u}) - J(\boldsymbol{\alpha}, \boldsymbol{u}^*)}{J(\boldsymbol{\alpha}_0, \mathbf{u}^*)} = \frac{S(\boldsymbol{\alpha}, \boldsymbol{u})}{J(\boldsymbol{\alpha}_0, \boldsymbol{u}^*)}. \tag{8.5-36}$$

Similar to S, S_0 takes on values between 0 and ∞ and can thus be used for design purposes. In terms of the above results, $\bar{S}_0$ becomes

$$\bar{S}_0(\boldsymbol{\alpha}, \boldsymbol{u}) = \frac{\boldsymbol{x}^{\mathrm{T}}(t_0)\boldsymbol{Z}(\boldsymbol{\alpha}_0, t_0)\boldsymbol{x}(t_0)}{\boldsymbol{x}^{\mathrm{T}}(t_0)\boldsymbol{M}_0(\boldsymbol{\alpha}_0, t_0)\boldsymbol{x}(t_0)}. \tag{8.5-37}$$

Minimal sensitivity of a system means a minimal value of $\bar{S}_0$.

A certain disadvantage of the criterion of optimality loss lies in the fact that it does not directly indicate the absolute or relative increase of the performance index which is felt to be the most relevant quantity in judging an optimal system: If an optimal system is designed with respect to a certain performance index, then it is logical that its sensitivity should also be optimized with respect to this performance index.

A more detailed treatment of sensitivity in connection with the design of optimal systems and large parameter variations is outside the scope of this book and the reader is referred to more particular literature [27,67].

PROBLEMS

Comment: Problems 8.1–8.6 are to be solved analytically, the rest numerically.

8.1 Consider a system characterized by the transfer function

$$F(s) = \frac{X(s)}{U(s)} = \frac{K}{1 + Ts}.$$

(a) Determine the optimal control function $u_0^*(t)$ and the corresponding feedback control law $u_C^*(t) = -k\, x(t)$ such that the system is brought from an arbitrary fixed initial state $x(0)$ to the origin $x(0) = 0$, thereby minimizing the performance index

$$J = \int_0^\infty [x^2(t) + ru^2(t)]\, dt.$$

(b) Give the optimal value of the performance index $J(\boldsymbol{\alpha}_0, \boldsymbol{u}^*)$.

(c) Determine the absolute and relative performance-index sensitivity with respect to K and T for $K_0 = 1$, $T_0 = 1$, $\Delta K = 0.1$, $\Delta T = 0$, and $r = 1$.

(d) Determine the optimality loss $\bar{S}(\boldsymbol{\alpha}, \boldsymbol{u})$ and $\bar{S}_0(\boldsymbol{\alpha}, \boldsymbol{u})$ for the same values.

(e) Determine the approximate optimality loss $\bar{S}_0(\boldsymbol{\alpha}, \boldsymbol{u})$ according to Eq. (8.5-37).

8.2 Referring to problem 8.1,

(a) determine the performance-index sensitivity J_K and J_T for the open-loop control and compare with the results of Problem 8.1,

(b) determine the optimality loss $\bar{S}$ $(\boldsymbol{\alpha}, \boldsymbol{u})$ for open-loop control and compare with the results of Problem 8.1.

8.3 Given the state feedback system shown in Fig. 8.P-1. The nominal parameter values are $K_0 = 1$, $T_0 = 1$.

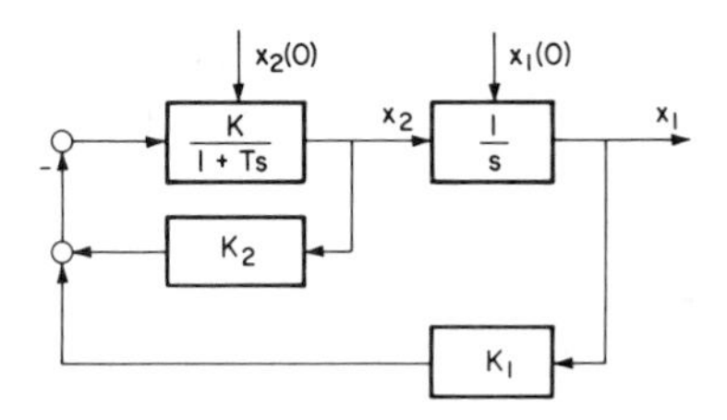

FIG. 8.P-1. State feedback system of Problem 8.3.

(a) Determine the feedback gains K_1, K_2 such that the state of the system is brought from the initial state $x_1(0)$, $x_2(0)$ to the origin with a minimum of the performance index

$$J = \tfrac{1}{2} \int_0^\infty (\boldsymbol{x}^{\mathrm{T}} \boldsymbol{Q} \boldsymbol{x} + R u^2)\, dt,$$

where

$$\boldsymbol{Q} = \begin{bmatrix} q_{11} & 0 \\ 0 & q_{22} \end{bmatrix} = \begin{bmatrix} 1 & 0 \\ 0 & 1 \end{bmatrix}, \qquad R = 1.$$

(b) Determine the absolute and relative performance-index sensitivities $\bar{J}_K$ and $\bar{J}_T$ with respect to K and T, respectively.

(c) Assume that $x_2(0) = n x_1(0)$, where $0 \leq n \leq \infty$, and give the corresponding intervals for $\bar{J}_K$ and $\bar{J}_T$ as n goes from 0 to ∞.

(d) Calculate the optimality loss for the same data as in (c) and $\Delta T = 0.1$, $\Delta K = 0$.

8.4 Given the state feedback system shown in Fig. 8.P-2 with $\dot{\boldsymbol{x}} = \boldsymbol{Ax} + \boldsymbol{bu}$, $\boldsymbol{y} = \boldsymbol{cx}$, and

$$\boldsymbol{A} = \begin{bmatrix} 0 & 1 \\ 0 & -1/T \end{bmatrix}, \qquad \boldsymbol{b} = \begin{bmatrix} 0 \\ 1 \end{bmatrix}, \qquad \boldsymbol{c} = [k/T \quad 0].$$

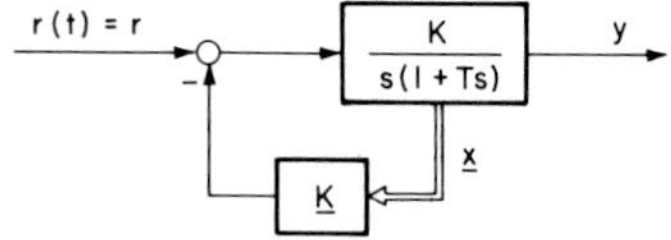

FIG. 8.P-2. State feedback of the follow-up system of Problem 8.4.

Suppose the initial state is zero but the system is to follow a constant reference input $r(t) = r_1$ such that the performance index

$$J = \tfrac{1}{2}\int_0^\infty [(y - r_1)^2 + Ru_0^2]\, dt$$

is minimized.

(a) Determine the optimal control function u^* and the corresponding feedback gain vector $\boldsymbol{K}$.

(b) Give the optimal performance index for moninal parameters K_0, T_0.

(c) Give the performance index sensitivities J_K and J_T.

8.5 Given the system shown in Fig. 8.P-3 with the nominal parameters $K_0 = 1$ and $T_0 = 1$. The state $\boldsymbol{x} = [x_1\ x_2]^{\mathrm{T}}$ is to be fed back by gain factors

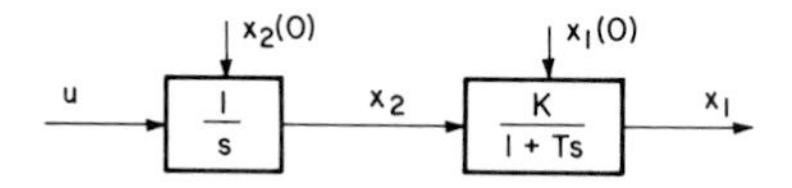

FIG. 8.P-3. Given system for Problem 8.5.

K_1 and K_2 such that the system is brought from an arbitrary fixed initial state $\boldsymbol{x}(0)$ to the origin 0 thereby minimizing the performance index

$$J = \tfrac{1}{2}\int_0^\infty (\boldsymbol{x}^{\mathrm{T}}\boldsymbol{Q}\boldsymbol{x} + Ru^2)\, dt$$

with $\boldsymbol{Q} = \boldsymbol{I}$ (unity matrix) and $R = 1$.

(a) Determine the feedback gain factors K_1 and K_2.

(b) Determine the optimal performance index J_0.

(c) Find the absolute and relative performance-index sensitivities J_T, J_K and $\bar{J}_T$, $\bar{J}_K$, respectively.

(d) Assume that $x_2(0) = nx_1(0)$, where $0 \leq n \leq \infty$ and give the dependence of $\bar{J}_T$ and $\bar{J}_K$ upon n.

(e) Determine the value of n for which $\bar{J}_T$ becomes 0.

(f) Determine the optimality loss $\bar{S}(\boldsymbol{\alpha}, u)$ for $\Delta K = 0$, $\Delta T = 0.1$ and give the interval in which $\bar{S}(\boldsymbol{\alpha}, u)$ varies as n goes from 0 to ∞.

(g) Calculate the approximations of $\bar{S}(\boldsymbol{\alpha}, u)$ and $\bar{S}_0(\boldsymbol{\alpha}, u)$ according to the Eq.(8.5-37).

8.6 Given the state feedback system shown in Fig. 8.P-4 and the performance index $J = \frac{1}{2}\int_0^\infty(\mathbf{x}^T\boldsymbol{Q}\mathbf{x} + Ru^2)\,dt$. The initial state $\boldsymbol{x}(0)$ is arbitrary but fixed and the final state is free. Let the nominal parameter be a_0.

(a) Find the expressions for the optimal gain factors K_1, K_2 and the matrix $\boldsymbol{M}_0$. Find the expression for the performance-index sensitivity J_a.

(c) Assume that $x_2(0) = n\,x_1(0)$ and calculate K_1, K_2 and the values of the relative performance-index sensitivites $\bar{J}_a$ for $n = 0$ and $n = \infty$ at $a_0 = 1$ in each of the following cases:

(1) $\boldsymbol{Q} = \boldsymbol{I}$, $R = 1$, (2) $\boldsymbol{Q} = \boldsymbol{I}$, $R = 2$,

(3) $\boldsymbol{Q} = \begin{bmatrix} 0.5 & -0.5 \\ -0.5 & 1 \end{bmatrix}$, $R = 2$, (4) $\boldsymbol{Q} = \begin{bmatrix} 1 & 0 \\ 0 & 0 \end{bmatrix}$, $R = 1$.

8.7 For the system shown in Fig. 8.P-5 the feedback gains K_1, K_2, K_3 are to be chosen such that the performance index

$$J = \tfrac{1}{2}\int_0^\infty (y^2 + u^2)\,dt$$

is minimized. The initial state is fixed, the final state is free, and u is unbounded. Compute the optimal values of K_1, K_2, K_3 and J_0, $\bar{J}_{T_1}$, $\bar{J}_{T_2}$, for $T_{10} = 5$, $T_{20} = 2$ and either of the initial states:

(a) $x_1(0) = 1$, $x_2(0) = 0$, $x_3(0) = 0$,

(b) $x_1(0) = 0$, $x_2(0) = 1$, $x_3(0) = 0$.

8.8 Consider the state feedback system shown in Fig. 8.P–6. The nominal values of the parameters are $V_{10} = 2.5$, $V_{20} = 3.2$, $V_{30} = 3$; $T_{10} = 5$, $T_{20} = 2$,

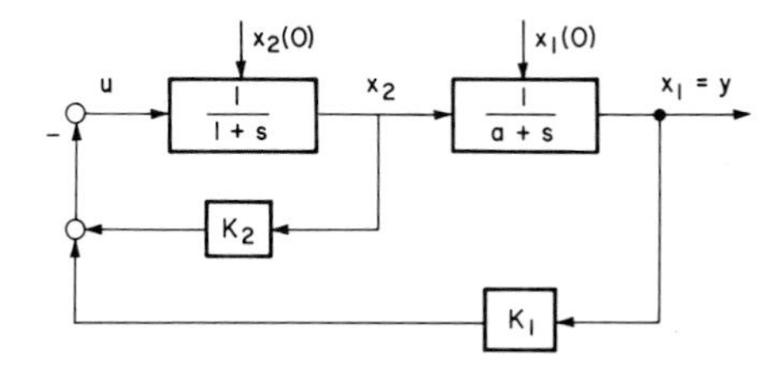

FIG. 8.P-4. State feedback system of Problem 8.6.

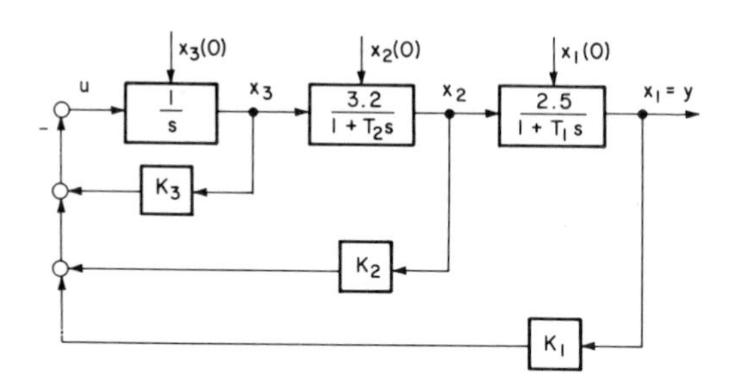

FIG. 8.P-5. Feedback system considered in Problem 8.7.

$T_{30} = 0.1$. The system is to be optimized according to the performance index

$$J = \tfrac{1}{2}\int_0^\infty (x_1^2 + u^2)\,dt$$

with unbounded u and free final value. Compute the optimal gain factors K_1, K_2, K_3, the optimal performance index J_0, and the performance-index sensitivities $\bar{J}_{T_1}, \bar{J}_{T_2}, \bar{J}_{T_3}, \bar{J}_{V_1}, \bar{J}_{V_2}, \bar{J}_{V_3}$ for the following initial states:

(a) $x_1 = 1,\ x_2 = 0,\ x_3 = 0$;
(b) $x_1 = 0,\ x_2 = 1,\ x_3 = 0$;
(c) $x_1 = 0,\ x_2 = 0,\ x_3 = 1$.

8.9 Consider the state-feedback system shown in Fig. 8.P-7 which may be optimal according to the performance index $J = \int_0^\infty (x_1^2 + u^2)\,dt$. The nominal parameter values are $V_{10} = 2.5$, $V_{20} = 3.2$, $V_{30} = 6$, $V_{40} = 3$, $V_{50} = 3$, $T_{10} = 5$, $T_{20} = 2$, $T_{30} = 0.075$, $T_{40} = 0.04$, $T_{50} = 0.1$.

(a) Compute the feedback gains K_1, K_2, K_3, K_4, K_5, the optimal performance index J_0, the performance-index sensitivities $\bar{J}_{V_1}, \bar{J}_{V_2}, \bar{J}_{V_3}, \bar{J}_{V_4}, \bar{J}_{V_5}, \bar{J}_{T_1}, \bar{J}_{T_2}, \bar{J}_{T_3}, \bar{J}_{T_4}, \bar{J}_{T_5}$ for $x_1(0) = 1$, $x_2(0) = x_3(0) = x_4(0) = x_5(0) = 0$.

(b) Compute the optimality loss $\bar{S}$ and $\bar{S}_0$ for the following cases: $\Delta T_1/T_{10} = 20\%$, $\Delta T_1/T_{10} = 50\%$, $\Delta T_4/T_{40} = 50\%$.

8.10 Consider the same system as in Problem 8.9, however with $\boldsymbol{x}(0) = \mathbf{0}$.

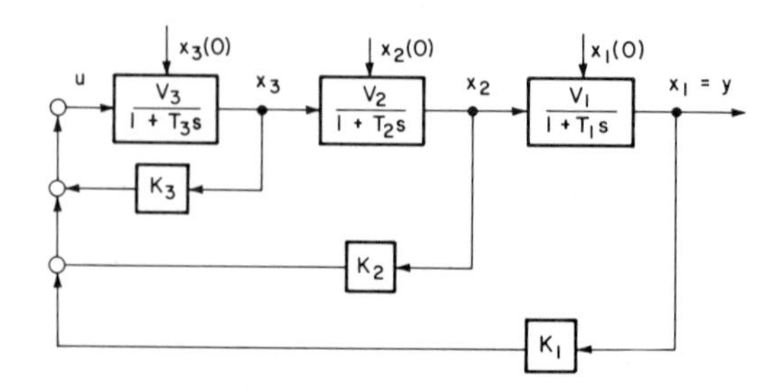

FIG. 8.P-6. Feedback system of Problem 8.8.

The output is to be brought to the constant reference input $r_1 = 1$ such that the performance index

$$J = \frac{1}{2} \int_0^{t_f} [(r_1 - y)^2 + Ru^2]\, dt$$

is minimized. Compute the relative performance-index sensitivities $\bar{J}_{V_1} \cdots \bar{J}_{V_5}$ and $\bar{J}_{T_1} \cdots \bar{J}_{T_5}$ for $t_f = 3$.

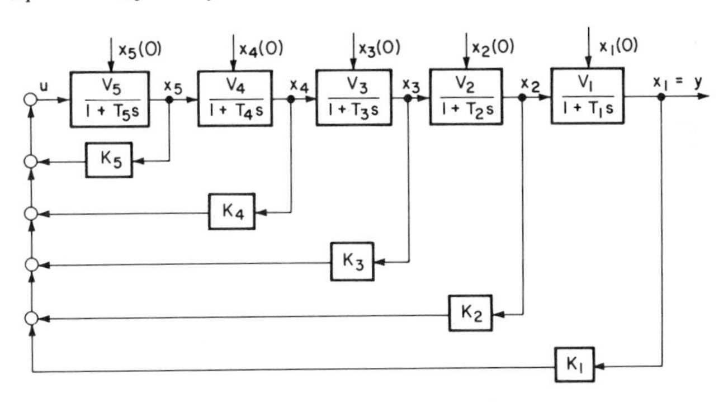

FIG. 8.P-7. State feedback system of Problem 8.9.

Chapter 9

Applications

9.1 INTRODUCTION

This chapter is intended to illustrate the utility of sensitivity theory by means of some examples of application in the field of control engineering. Here the sensitivity model is of particular importance. It can be applied in different forms, namely, either as a physical system or as a fictive model. As a physically realized system it is utilized, for example, in parameter identification schemes using self-adaptive models, in automatic optimization schemes, in adaptive and learning control systems, as well as in optimal feedback control systems where the sensitivity functions are part of the performance index. As a fictive model it is generally used for the synthesis of parameter-insensitive feedback control systems in the time domain.

Four examples will be considered in this chapter. The first two are concerned with automatic optimization and parameter identification. In the first example, the sensitivity model is employed for the on-line generation of the sensitivity functions, while in the second example it serves as a basis for the design of the best input signal for the identification of the parameter. In the next two examples it is applied to the design of parameter-insensitive feedback control systems.

9.2 AUTOMATIC OPTIMIZATION AND PARAMETER IDENTIFICATION USING A SENSITIVITY MODEL

It is a common practice in parameter identification, automatic optimization, and self-adaptation to use a self-adaptive model. In order to adjust the parameters of the model, the partial derivatives with respect to the model parameters, i.e., the sensitivity functions, are needed. These can be generated simultaneously by means of a sensitivity model, e.g., by application of the method of sensitivity points or the method of Wilkie and Perkins that is outlined in Chapters 4 and 5. This section describes a probate scheme for the application of the sensitivity model to parameter identification.

The basic idea is shown in Fig. 9.2-1. Let the system to be identified be given by the general state equation

$$\dot{\boldsymbol{x}} = \boldsymbol{f}(\boldsymbol{x}, \boldsymbol{u}, \boldsymbol{\alpha}), \tag{9.2-1}$$

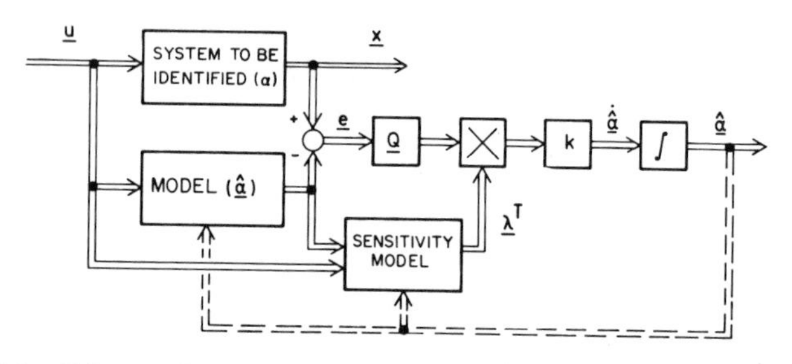

FIG. 9.2-1. Scheme of parameter identification using a self-adaptive model including a sensitivity model.

where $\boldsymbol{\alpha}$ represents the vector of the unknown parameters that are to be identified. It is assumed that $\boldsymbol{\alpha}$ is time-invariant or only slowly varying with time. The state vector $\boldsymbol{x}$ will be considered as the output vector of the system.

As an adaptive model a system with the same state equation is used. Its parameter and state vector will, in general, more or less differ from those of the original system model. Denoting the parameter vector and the state vector of the model by $\hat{\boldsymbol{\alpha}}$ and $\boldsymbol{x} = \boldsymbol{x}(\hat{\boldsymbol{\alpha}})$, respectively, the state equation of the model reads

$$\dot{\hat{\boldsymbol{x}}} = \boldsymbol{f}(\hat{\boldsymbol{x}}, \boldsymbol{u}, \hat{\boldsymbol{\alpha}}). \tag{9.2-2}$$

From both systems, the state error $\boldsymbol{e} = \boldsymbol{x} - \hat{\boldsymbol{x}}$ is calculated. It is assumed that $\boldsymbol{e} = \boldsymbol{e}(t, \hat{\boldsymbol{\alpha}})$ is a unique function of $\hat{\boldsymbol{\alpha}}$ and that $\boldsymbol{e} = \boldsymbol{0}$ as $\hat{\boldsymbol{\alpha}} = \boldsymbol{\alpha}$. Therefore, the goal is to change $\hat{\boldsymbol{\alpha}}$ until $\boldsymbol{e}$ is either zero or if there is noise minimal. A method of how to achieve this can be based on the Liapunov function

$$V(\boldsymbol{e}) = \frac{1}{2}\, \boldsymbol{e}^{\mathrm{T}} \boldsymbol{Q} \boldsymbol{e}, \tag{9.2-3}$$

where $\boldsymbol{Q}$ is a symmetric positive definite matrix that weights the individual components $e_i = x_i - \hat{x}_i$ of the state error in a suitable manner and V is a positive definite scalar function. It is clear that if the time derivative $\dot{V}$ can be made negative definite, then $\boldsymbol{e}(t, \hat{\boldsymbol{\alpha}})$ approaches the origin (or the minimum of $\boldsymbol{e}$) asymptotically. Determining V from Eq. (9.2-3), we obtain

$$\dot{V} = \frac{\partial V}{\partial \boldsymbol{e}}\, \dot{\boldsymbol{e}} = \boldsymbol{e}^{\mathrm{T}} \boldsymbol{Q}\, \dot{\boldsymbol{e}} \tag{9.2-4}$$

and because

$$\dot{\boldsymbol{e}} = \frac{\partial \boldsymbol{e}}{\partial \hat{\boldsymbol{\alpha}}} \dot{\hat{\boldsymbol{\alpha}}}, \tag{9.2-5}$$

we have

$$\dot{V} = \boldsymbol{e}^{\mathrm{T}} \boldsymbol{Q} \frac{\partial \boldsymbol{e}}{\partial \hat{\boldsymbol{\alpha}}} \dot{\hat{\boldsymbol{\alpha}}}. \tag{9.2-6}$$

If now $\hat{\boldsymbol{\alpha}}$ is changed according to the gradient of V, that is,

$$\dot{\hat{\boldsymbol{\alpha}}} = -k\left(\frac{\partial V}{\partial \hat{\boldsymbol{\alpha}}}\right)^{\mathrm{T}} = -k\left(\frac{\partial \boldsymbol{e}}{\partial \hat{\boldsymbol{\alpha}}}\right)^{\mathrm{T}} \boldsymbol{Q}^{\mathrm{T}} \boldsymbol{e}, \tag{9.2-7}$$

where the gradient is again defined as a row vector and k is a positive constant, then one obtains by substituting Eq. (9.2-7) into Eq. (9.2-6),

$$\dot{V} = -k \boldsymbol{e}^{\mathrm{T}} \boldsymbol{Q} \frac{\partial \boldsymbol{e}}{\partial \hat{\boldsymbol{\alpha}}} \left(\frac{\partial \boldsymbol{e}}{\partial \hat{\boldsymbol{\alpha}}}\right)^{\mathrm{T}} \boldsymbol{Q}^{\mathrm{T}} \boldsymbol{e}. \tag{9.2-8}$$

This is a quadratic form which is certainly not negative definite but negative semidefinite. Experience has shown that such a choice of $\dot{\hat{\boldsymbol{\alpha}}}$ is satisfactory in many practical situations, although the general conditions for the convergence of this method are not yet fully explored.

This means that $\dot{\hat{\boldsymbol{\alpha}}}$ has to be produced according to Eq. (9.2-7). For the Jacobi matrix $\partial \boldsymbol{e}/\partial \hat{\boldsymbol{\alpha}}$ in Eq. (9.2-7) we obtain, since $\boldsymbol{x}$ is not a function of $\hat{\boldsymbol{\alpha}}$,

$$\frac{\partial \boldsymbol{e}}{\partial \hat{\boldsymbol{\alpha}}} = \frac{\partial(\boldsymbol{x} - \hat{\boldsymbol{x}})}{\partial \hat{\boldsymbol{\alpha}}} = -\frac{\partial \hat{\boldsymbol{x}}}{\partial \hat{\boldsymbol{\alpha}}}. \tag{9.2-9}$$

If the gain factor k is small so that $\hat{\boldsymbol{\alpha}}$ can be considered quasi constant, then $\partial \hat{\boldsymbol{x}}/\partial \hat{\boldsymbol{\alpha}}$ represents the sensitivity matrix $\boldsymbol{\lambda}$ of the model to be adapted, where the value $\hat{\boldsymbol{\alpha}}$ of the parameter reached at any time represents the nominal value. The sensitivity matrix, which is the solution of the state sensitivity equation

$$\dot{\boldsymbol{\lambda}} = \frac{\partial \boldsymbol{f}}{\partial \hat{\boldsymbol{x}}} \boldsymbol{\lambda} + \frac{\partial \boldsymbol{f}}{\partial \hat{\boldsymbol{\alpha}}}, \qquad \boldsymbol{\lambda}(0) = \boldsymbol{0}, \tag{9.2-10}$$

can be generated by a sensitivity model. This is illustrated in Fig. 9.2-1.

The adaptation procedure starts with arbitrarily chosen initial conditions $\hat{\boldsymbol{\alpha}}(0) = \hat{\boldsymbol{\alpha}}^0$. With the aid of the sensitivity model, the expression

$$\dot{\hat{\boldsymbol{\alpha}}} = k \boldsymbol{\lambda}^{\mathrm{T}} \boldsymbol{Q}^{\mathrm{T}} \boldsymbol{e} = k \boldsymbol{\lambda}^{\mathrm{T}} \boldsymbol{Q} \boldsymbol{\lambda} \tag{9.2-11}$$

is formed. The gain factor k has to be chosen carefully as in all gradient methods. If k is chosen too large, the procedure diverges; if it is chosen too samll, the identification takes too much time. Commonly, k as well as $\boldsymbol{Q}$ are chosen empirically.

As a concrete example, let us consider the third-order system already studied in previous chapters. Its differential equation is given by

$$\dddot{y} + \alpha_3\ddot{y} + \alpha_2\dot{y} + \alpha_1 y = u, \tag{9.2-12}$$

with the initial conditions $y(0) = 2$, $\dot{y}(0) = \ddot{y}(0) = 0$. Here α_1, α_2, and α_3 are the parameters to be identified.

The corresponding model differential equation is

$$\dddot{\hat{y}} + \hat{\alpha}_3\ddot{\hat{y}} + \hat{\alpha}_2\dot{\hat{y}} + \hat{\alpha}_1\hat{y} = u, \tag{9.2-13}$$

with $\hat{y}(0) = \dot{\hat{y}}(0) = \ddot{\hat{y}}(0) = 0$. Let the state vector be defined as $\boldsymbol{x} \triangleq [y\ \dot{y}\ \ddot{y}]^{\mathrm{T}}$. Then the state error becomes

$$\boldsymbol{e} = [e_1\ \ e_2\ \ e_3]^{\mathrm{T}} = [y - \hat{y}\ \ \dot{y} - \dot{\hat{y}}\ \ \ddot{y} - \ddot{\hat{y}}]^{\mathrm{T}}. \tag{9.2-14}$$

The rule for the generation of the vector $\dot{\hat{\boldsymbol{\alpha}}} = [\dot{\hat{\alpha}}_1\ \dot{\hat{\alpha}}_2\ \dot{\hat{\alpha}}_3]^{\mathrm{T}}$ is

$$\dot{\hat{\boldsymbol{\alpha}}} = k\boldsymbol{\lambda}^{\mathrm{T}}\boldsymbol{Q}\boldsymbol{e}, \tag{9.2-15}$$

where $\boldsymbol{\lambda} = (\lambda_{ij}) = (\partial\hat{\boldsymbol{x}}/\partial\hat{\boldsymbol{\alpha}})_{\boldsymbol{\alpha}_0}$ is a 3×3 sensitivity matrix.

According to the results of Section 5.6, the trajectory sensitivity functions λ_{ij} of $\boldsymbol{\lambda}$ can be generated simultaneously from a structure as shown in Figs. 5.6-3 and 5.6-4. When by such a procedure the sensitivity matrix is known, the expression (9.2-15) can be formed.

The complete identification scheme resulting from this procedure is illustrated in Fig. 9.2-2. From the four signals available in the sensitivity model

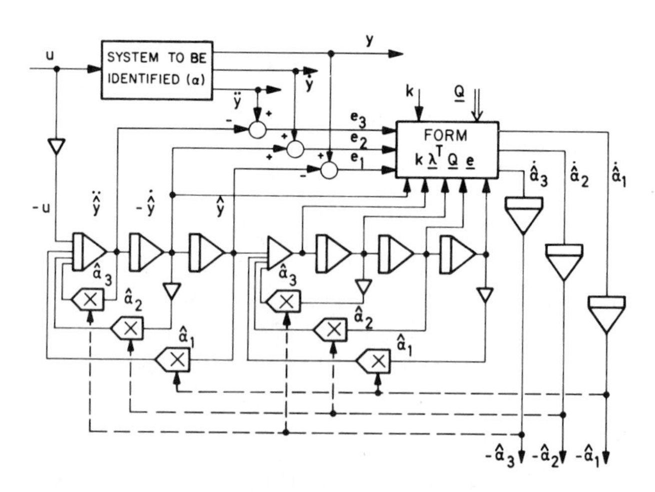

FIG. 9.2-2. Circuit for the identification of the parameters $\alpha_1, \alpha_2, \alpha_3$ of the system of Eq. (9.2-12).

and the state variable $-\hat{y}$, all components of the sensitivity matrix $\boldsymbol{\lambda}$ can be determined due to the theorems of Wilkie and Perkins (see Section 5.6). If, in particular, $\boldsymbol{Q}$ is chosen as the unity matrix, the parameters can be found from the formula

$$\dot{\hat{\alpha}}_j = k(\lambda_{1j}e_1 + \lambda_{2j}e_2 + \lambda_{3j}e_3) = k\boldsymbol{\lambda}_j^{\mathrm{T}}\boldsymbol{e}, \tag{9.2-16}$$

i.e., the jth parameter can be found from the jth sensitivity vector $\boldsymbol{\lambda}_j$ and all errors e_i $(i = 1, 2, 3)$.

For the particular case in which $\alpha_1 = 2$, $\alpha_2 = 4$, $\alpha_3 = 3$, and

$$u(t) = 1 + \sin(1.5t), \tag{9.2–17}$$

Brogan [2] proposes $k = 10$ as a proper empirical value for k. Choosing $k = 10$ and the initial values of all parameters as $\alpha_1 = \alpha_2 = \alpha_3 = 5$, one obtains an adaptation procedure as plotted in Fig. 9.2-3 [2]. This shows that the procedure of adaptation will be finished after a period of about 15 summed time constants α_2/α_1.

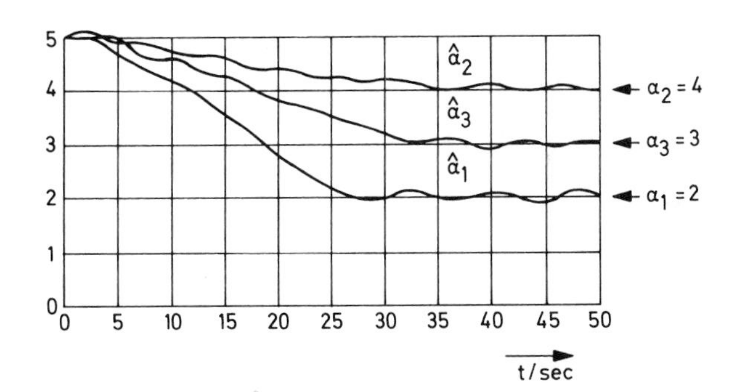

FIG. 9.2-3. Time history of the identification of $\alpha_1, \alpha_2, \alpha_3$ of the considered third-order example.

As a final comment on this identification scheme, let us note that the identification time and the expenditure of realization are considerable: A model as well as a sensitivity model are needed. However, in the presence of uncorrelated noise at the output of the unknown system, the parameters can be identified without any bias. This is easy to see if one pursues the path of noise during the circuit up to the point where the parameter values appear. Since the noise is not squared along this way, the estimate will remain unbiased.

It is evident that the above scheme can also be applied if instead of the state error the output error $e = e_1 = y - \hat{y}$ is used. This is why this method is commonly termed *output error method.* Alternatively the identification problem can be solved in a more elegant way by means of the so-called *equation error method.* Both definitions of the error are illustrated in Fig. 9.2-4 for a system with the transfer function

$$G(s) = \frac{\beta_2 s + \beta_1}{s^2 + \alpha_2 s + \alpha_1}, \tag{9.2-18}$$

where α_1, α_2, β_1, β_2 are the parameters to be identified. The advantage of the equation error method is that the equation error e is linear in the parameters. Therefore the identification is faster than in the case of the output error method. Furthermore, as was shown in Section 4.6, the sensitivity functions in this case can be taken directly from the model without the need of an additional sensitivity model.

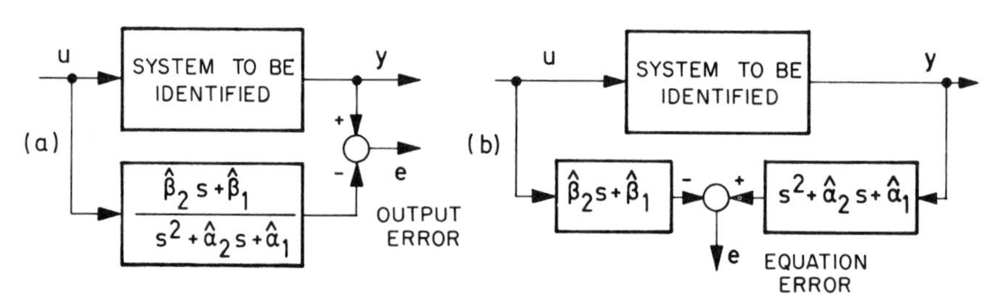

FIG. 9.2-4. Two possible methods of error generation: (a) output error method, (b) equation error method.

The practical realization of the equation error method requires the application of low pass filters since pure differentiations as shown in Fig. 9.2-4b are not realizable. The complete method including the low pass filters will be illustrated by the following example.

In this example the filter can be of the type of a second order low pass with the transfer function

$$F(s) = \frac{1}{s^2 + a_2 s + a_1}. \tag{9.2-19}$$

With the aid of this filter the model transfer functions in Fig. 9.2–4b take the forms

$$\frac{\hat{\beta}_2 s + \hat{\beta}_1}{s^2 + a_2 s + a_1} \quad \text{and} \quad \frac{s^2 + \hat{\alpha}_2 s + \hat{\alpha}_1}{s^2 + a_2 s + a_1}, \tag{9.2-20}$$

respectively. The resulting circuit for the identification of the parameter α_1 is shown in Fig. 9.2-5. We must proceed in a similar way for every parameter.

Note that the sensitivity functions $\partial e/\partial \hat{\alpha}_i$ and $\partial e/\partial \hat{\beta}_i$ that are needed for the identification of all parameters, according to the relations

$$\dot{\hat{\alpha}}_i = -ke\frac{\partial e}{\partial \hat{\alpha}_i}, \qquad \dot{\hat{\beta}}_i = -ke\frac{\partial e}{\partial \hat{\beta}_i}, \tag{9.2-21}$$

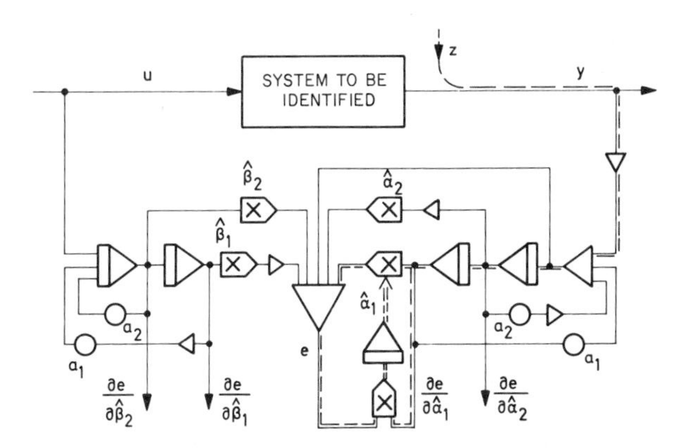

FIG. 9.2-5. Identification scheme based on the equation error using the sensitivity functions (carried out for α_1 only).

can be taken here directly from the adaptive model without the use of an additional sensitivity model. This is an important advantage of this method.

The above shows that the equation error method is superior to the output error method as far as the time of identification and the realization expenditure are concerned. However, a serious shortcoming of this method is that in the case of uncorrelated output noise of the plant the parameter identification will be biased since the noise signal is multiplied by itself (i.e., squared) and then integrated. This is illustrated by the dashed lines in Fig. 9.2-5. This is why nowadays the slower but biasfree output error method illustrated in Fig. 9.2–1 is getting more and more attractive again.

Severe theoretical problems concerning the convergence of this method emerge from the fact that the gradient has to be taken continuously, i.e., simultaneously with the setting of the parameters. Therefore, in order to simplify the theory, an intermittent operation is often preferred: The parameters are kept constant while the gradient is formed. Then the parameters are set without measuring the gradient and so forth [7].

The identification method described serves as the basis for many adaptive, self-optimizing, and learning feedback control systems.

9.3 OPTIMIZATION OF THE INPUT SIGNAL FOR IDENTIFICATION

As another example of the application of sensitivity theory, it will be shown in this section how an optimal input signal for the parameter identification can be found with the aid of a sensitivity model.

Figure 9.3-1 shows an identification scheme using a comparative model. The problem of optimizing the input signal can here be stated as follows:

To optimize the accuracy for the measurement of the parameters of the actual system with respect to the measurement errors that are commonly present, the input signal $u(t)$ has to be chosen such that the output error $\Delta \mathbf{z}$ becomes maximal even for small parameter variations $\Delta\alpha_j$. For $\Delta \mathbf{z}$ we have

$$\Delta \mathbf{z} = \mathbf{z}(t, \boldsymbol{\alpha}_0 + \Delta\boldsymbol{\alpha}) - \mathbf{z}(t, \boldsymbol{\alpha}_0) = \sum_{j=1}^{r} \boldsymbol{\sigma}_j \Delta\alpha_j = \sum_{j=1}^{r} \frac{\partial \mathbf{g}}{\partial \mathbf{x}}\bigg|_{\boldsymbol{\alpha}_0} \boldsymbol{\lambda}_j \Delta\alpha_j, \quad (9.3\text{–}1)$$

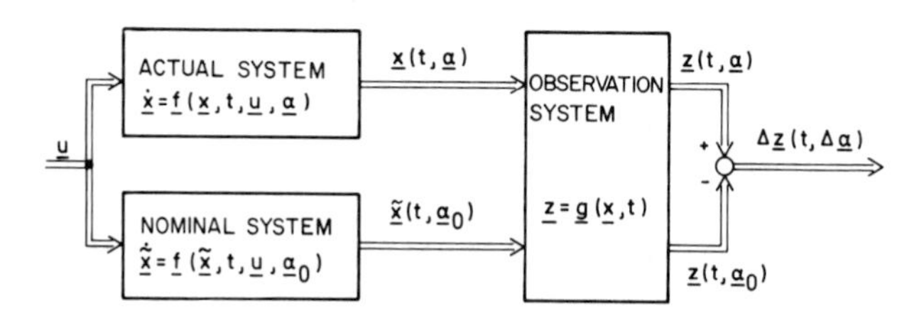

FIG. 9.3-1. Scheme of identification.

where $\boldsymbol{\sigma}_j = (\partial \mathbf{z}/\partial\alpha_j)_{\boldsymbol{\alpha}_0}$ denotes the output sensitivity vector and $\boldsymbol{\lambda}_j = (\partial \mathbf{x}/\partial\alpha_j)_{\boldsymbol{\alpha}_0}$ the trajectory sensitivity vector.

Therefore the problem can be stated as follows: Choose the input $\mathbf{u}(t)$ such that the output sensitivity vector $\boldsymbol{\sigma}_j$ is maximized. Or in mathermatical terms: Given the combined system

$$\begin{aligned} \dot{\tilde{\mathbf{x}}} &= \mathbf{f}(\tilde{\mathbf{x}}, t, \mathbf{u}, \boldsymbol{\alpha}_0), \qquad \tilde{\mathbf{x}}(0) = \tilde{\mathbf{x}}^0, \\ \dot{\boldsymbol{\lambda}}_j &= \frac{\partial \mathbf{f}}{\partial \mathbf{x}} \boldsymbol{\lambda}_j + \frac{\partial \mathbf{f}}{\partial \alpha_j}, \qquad \boldsymbol{\lambda}_j(0) = \mathbf{0}, \quad j = 1, 2, \ldots, r, \qquad (9.3\text{-}2) \\ \boldsymbol{\sigma}_j &= \frac{\partial \mathbf{g}}{\partial \mathbf{x}} \boldsymbol{\lambda}_j, \qquad j = 1, 2, \ldots, r. \end{aligned}$$

Determine the input vector $\mathbf{u}$ such that at the instant T the performance index

$$J = J[\boldsymbol{\sigma}_1(\mathbf{x}, T, \boldsymbol{\lambda}_1), \ldots, \boldsymbol{\sigma}_r(\mathbf{x}, T, \boldsymbol{\lambda}_r)] \qquad (9.3\text{-}3)$$

is maximized under a certain restraint, such as

$$J_N = \int_0^T \mathbf{u}^{\mathrm{T}} \mathbf{R} \mathbf{u}\, dt \leq E, \qquad (9.3\text{-}4)$$

where E is the allowable energy for the identification. If instead of a single instant the identification is to be optimized for the whole time interval $0 \leq t \leq T$, then $\boldsymbol{\sigma}_j$ in Eq.(9.3-3) has to be replaced by

$$\boldsymbol{v}_j = \int_0^T \boldsymbol{\sigma}_j\, dt. \qquad (9.3\text{-}5)$$

A structural interpretation of Eq. (9.3-2) is given in Fig. 9.3-2. As we see, a sensitivity model is needed to produce the output sensitivity vector $\boldsymbol{\sigma}_j$. By the above formulation, the problem is reduced to a classical optimization problem that will not be treated further.

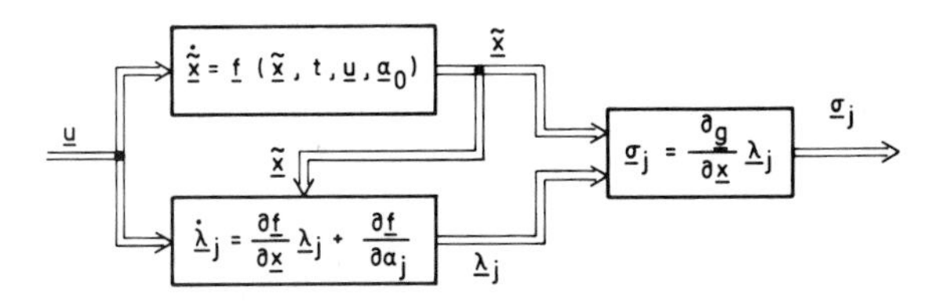

FIG. 9.3-2. Combined system to determine $\boldsymbol{\sigma}_j$.

This method yields good results in the case of a single parameter identification. If multiple parameters are to be identified simultaneously, it has to be required in addition that the sensitivity vectors $\boldsymbol{\sigma}_j$ $(j = 1, 2, \ldots, r)$ differ from each other as much as possible, that is, they should be orthogonal.

Example 9.3-1 The execution of the above procedure shall be illustrated by means of an example. Given a bilinear single-variable system with the state equation

$$\dot{x} = ax + bxu, \qquad x(0) = 1, \quad b \neq 0. \tag{9.3-6}$$

Let the observation matrix be the identity, i.e.,

$$z = x \tag{9.3-7}$$

and the parameter of interest be a. The input u is to be optimized with respect to a single measurement at $t = T$.

The combined system here has the form

$$\dot{x} = ax + bux, \qquad x(0) = 1, \tag{9.3-8}$$

$$\dot{\lambda}_a = x + a\lambda_a + b\lambda_a u, \qquad \lambda_a(0) = 0. \tag{9.3-9}$$

Instead of maximizing $J = |\lambda_a|$ with the boundary condition

$$J_N = \int_0^T u^2\, dt \leq E, \tag{9.3-10}$$

it is also possible to solve the inverse problem, namely, to minimize the energy

$$J_N = \int_0^T u^2\, dt = \min \tag{9.3-11}$$

and impose for λ_a the boundary condition

$$\lambda_a = \lambda_e. \tag{9.3-12}$$

For all bilinear systems this problem has the simple solution

$$u = \text{const.} \tag{9.3-13}$$

Especially for $E = 1$, one obtains $u = \pm 1$. The relation between the sensitivity value λ_e and the corresponding value of the control function u that is optimal in the sense of minimal input energy is plotted in Fig. 9.3-3a along with the energy. From this diagram, knowing the input energy, one can determine the optimal value of u that minimizes λ_e. For example, for $E = 1$, the optimal value providing maximal sensitivity is $u = 1$, while $u = -1$ provides minimal sensitivity.

Figure 9.3-3b shows the corresponding plots of $x(t)$ for nominal parameters, i.e., $\Delta a = 0$, as well as for $\Delta \alpha = 0.2$. Since the signals x and $\tilde{x}$ are subtracted at the output, the amount of Δx is an immediate indication of what has been gained by optimization.

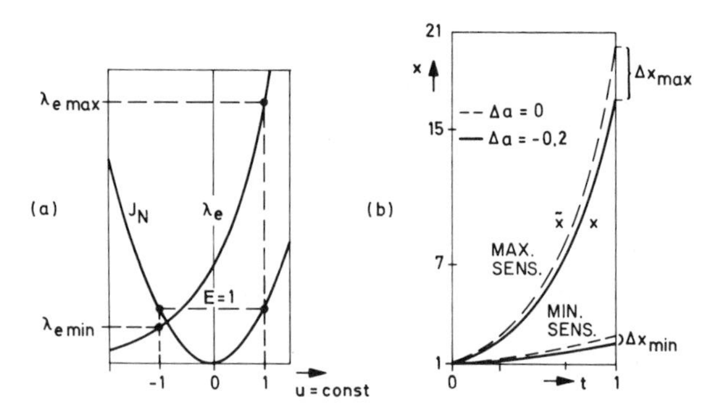

FIG. 9.3-3. Results of Example 9.3-1. (a) Plots of minimal input energy and boundary sensitivity, (b) plots of output $x(t)$ for maximal and minimal sensitivity for nominal parameter ($\Delta\, a = 0$) and $\Delta\, a = -0.2$.

9.4 DESIGN OF OPTIMAL FEEDBACK CONTROLS USING THE SENSITIVITY FUNCTIONS IN THE PERFORMANCE INDEX

In order to decrease the parameter sensitivity of an optimal feedback control system, one can use the sensitivity functions in the optimization procedure already, i.e., as part of the performance index in addition to the control and state variables. This can be done in several ways. For example, the state vector $\boldsymbol{x}$ in the performance index can simply be extended by the semirelative sensitivity vectors $\tilde{\boldsymbol{\lambda}}_j$ with respect to the parameters of interest α_j.

In the case of a quadratic performance index, for example, this gives

$$J = \int_{t_0}^{t_f} (\tilde{\boldsymbol{x}}^{\mathrm{T}} \tilde{\boldsymbol{Q}} \tilde{\boldsymbol{x}} + \boldsymbol{u}^{\mathrm{T}} \boldsymbol{R} \boldsymbol{u})\, dt, \tag{9.4-1}$$

where $\tilde{\boldsymbol{x}} = [x_1 \cdots x_n\ \lambda_{11} \cdots \lambda_{nr}]^{\mathrm{T}}$, $\tilde{\lambda}_{ij} = (\partial x_i / \partial \ln \alpha_j)_{\boldsymbol{\alpha}_0}$ and $\boldsymbol{Q}$ is a weighting matrix.

If $\mathbf{Q}$ is chosen such that there are no products of sensitivity functions with state functions and no products between sensitivity functions with different numbers of j, Eq.(9.4-1) may be rewritten in the form

$$J = \int_{t_0}^{t_f} \left[\boldsymbol{x}^{\mathrm{T}} \boldsymbol{Q} \boldsymbol{x} + \boldsymbol{u}^{\mathrm{T}} \boldsymbol{R} \boldsymbol{u} + \sum_{j=1}^{r} (\boldsymbol{\lambda}_j{}^{\mathrm{T}} \boldsymbol{S} \boldsymbol{\lambda}_j) \right] dt. \tag{9.4-2}$$

This shows that the L_2-norms of the sensitivity vectors $\tilde{\boldsymbol{\lambda}}_j$ $(j = 1, 2, \ldots, r)$ are taken into account in the performance index.

It is evident that by such a modification of the performance index a compromise is obtained between optimality in the sense of the criterion

$$J = \int_{t_0}^{t_f} (\boldsymbol{x}^{\mathrm{T}} \boldsymbol{Q} \boldsymbol{x} + \boldsymbol{u}^{\mathrm{T}} \boldsymbol{R} \boldsymbol{u})\, dt \tag{9.4-3}$$

and insensitivity in the sense of the L_2-norm I_M(Definition 3.2-11). This compromise can be influenced in a wide range by a proper choice of the weighting matrices $\boldsymbol{Q}$, $\boldsymbol{R}$, $\boldsymbol{S}$.

Because of the similar character of the expression of Eqs.(9.4-1) and (9.4-3), it is evident that the solution of the extended problem can again be found from a Riccati equation. However, now $\tilde{\boldsymbol{\lambda}}_j$ has to be fed back in addition to the state since the resulting control law will be of the form

$$\boldsymbol{u} = \boldsymbol{K}(\tilde{\boldsymbol{x}}) = \boldsymbol{K}(\boldsymbol{x}, \tilde{\boldsymbol{\lambda}}_j). \tag{9.4-4}$$

In the case of a linear time-invariant plant with a single input, the solution becomes, particulary,

$$\boldsymbol{u} = -\boldsymbol{K}_x{}^{\mathrm{T}} \boldsymbol{x} - \boldsymbol{K}_\lambda{}^{\mathrm{T}} \tilde{\boldsymbol{\lambda}}_j, \tag{9.4-5}$$

where $\boldsymbol{K}_x{}^{\mathrm{T}} = [K_{x1} \cdots K_{xn}]$ and $\boldsymbol{K}_\lambda{}^{\mathrm{T}} = [K_{\lambda 1} \cdots K_{\lambda n}]$ designate constant $1 \times n$ coefficient matrices that can be obtained from equations of the Riccati type.

The structure of the feedback control system resulting from this procedure is shown in Fig. 9.4-1. As a basic difference from the result of conventional optimization, a sensitivity model is used in the feedback path. Thus instead of proportional state feedback, dynamic elements have to be provided in the feedback path.

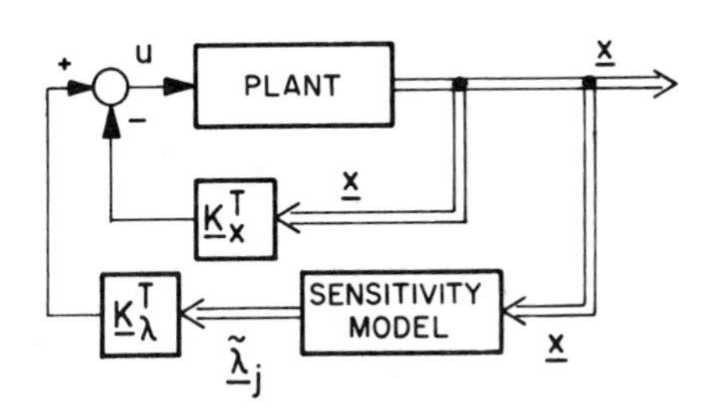

FIG. 9.4-1. Optimal control of a linearfirst-order plant using the sensitivity vector in the performance index.

There is a great variety of other ways to incorporate the sensitivity functions into the performance index J[6].

The analytic or numerical solution of the extended optimization problem based on Eq. (9.4-2) creates considerable mathematical difficulties for plants higher than second order. Therefore, Rilling and Roy [85] developed a method including hybrid computation that works even in the case of more complicated systems where analytic methods fail. This method shall be briefly discussed.

Suppose a plant is given by the following state equation:

$$\dot{\boldsymbol{x}} = \boldsymbol{A}\boldsymbol{x} + \boldsymbol{B}\boldsymbol{u}, \qquad \boldsymbol{x}(0) = \boldsymbol{x}^0, \tag{9.4-6}$$

where $\boldsymbol{A}$ and $\boldsymbol{B}$ are functions of a parameter α_j. Let this plant be fed back according to the control law

$$\boldsymbol{u} = -\boldsymbol{K}^{\mathrm{T}}\boldsymbol{x}, \tag{9.4-7}$$

where $\boldsymbol{K} = [K_1 \ . \ . \ . \ K_n]^{\mathrm{T}}$ is a matrix of gain factors. In contrast to the method described above, only the state $\mathbf{x}$ is fed back and not the sensitivity vector $\tilde{\boldsymbol{\lambda}}_j$. However, the elements $K_1, \ . \ . \ ., K_n$ are determined such that the sensitivity vector $\tilde{\boldsymbol{\lambda}}_j$ is taken into account. They are found by minimizing the performance index

$$J = \int_0^{t_f} (\boldsymbol{x}^{\mathrm{T}}\boldsymbol{Q}\boldsymbol{x} + \boldsymbol{u}^{\mathrm{T}}\boldsymbol{R}\boldsymbol{u})\, dt + \int_0^{t_f} \tilde{\boldsymbol{\lambda}}_j^{\mathrm{T}}\, \boldsymbol{S}\tilde{\boldsymbol{\lambda}}_j\, dt. \tag{9.4-8}$$

Substituting Eq. (9.4-7) into (9.4-6), one obtains for the feedback system a homogeneous state equation of the form

$$\dot{\boldsymbol{x}} = (\boldsymbol{A} - \boldsymbol{B}\boldsymbol{K}^{\mathrm{T}})\boldsymbol{x}, \qquad \boldsymbol{x}(0) = \boldsymbol{x}^0 \tag{9.4-9}$$

and a corresponding trajectory sensitivity equation of the form

$$\dot{\boldsymbol{\lambda}}_j = (\boldsymbol{A} - \boldsymbol{B}\boldsymbol{K}^{\mathrm{T}})\boldsymbol{\lambda}_j + \left(\frac{\partial \boldsymbol{A}}{\partial \alpha_j} - \frac{\partial \boldsymbol{B}}{\partial \alpha_j}\boldsymbol{K}^{\mathrm{T}}\right)\boldsymbol{x}, \tag{9.4-10}$$

where all expressions have to be taken for nominal parameter values. From Eq. (9.4-10) the semirelative sensitivity vector can be determined.

Equations (9.4-9) and (9.4-10) are now solved simultaneously by an analog computer. The analog computer diagram containing a nominal original model and a sensitivity model of the feedback system is shown in Fig. 9.4-2. From the resulting values of $\boldsymbol{x}$ and $\tilde{\boldsymbol{\lambda}}_j$ the value of J can be computed by direct integration.

Since all the variables $\boldsymbol{x}$ and $\tilde{\boldsymbol{\lambda}}_j$ of the combined system depend upon $\boldsymbol{K}$, one can now minimize J by varying $\boldsymbol{K}$. This is done according to the flowchart shown in Fig. 9.4-3, by which the logic device of the hybrid computer is programmed.

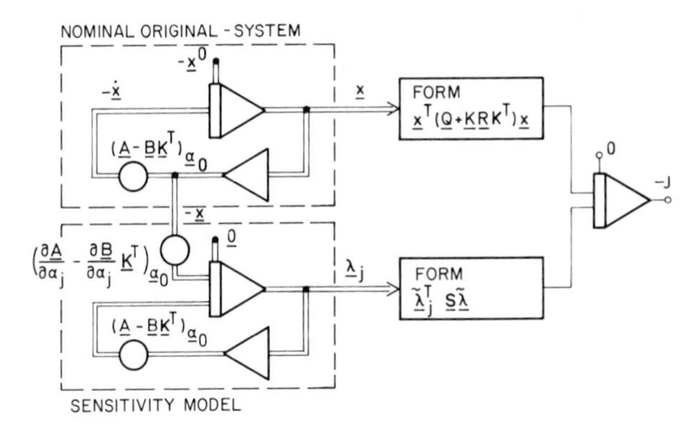

FIG. 9.4-2. Analog computer diagram for the determination of J using a sensitivity model.

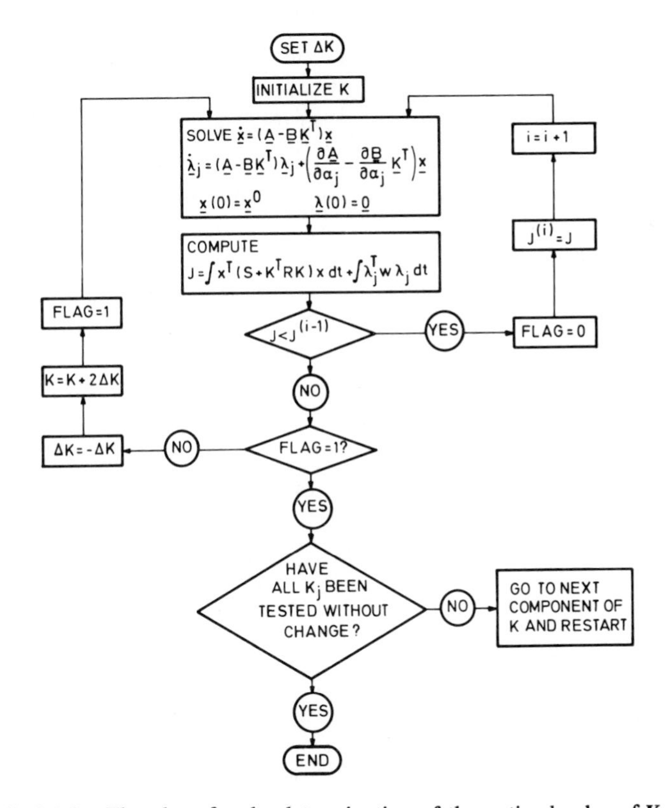

FIG. 9.4-3. Flowchart for the determination of the optimal value of **K**.

The procedure is as follows: Starting with an initial value of $\mathbf{K}$, the performance index J is minimized by first changing one element K_1 of $\boldsymbol{K}$ by a constant step size ΔK_1 in the positive or negative direction. This procedure is now done in series for all elements of $\boldsymbol{K}$ until no decrease of the performance index can be observed. It is thereby assumed that the procedure converges to the true value. It is possible now to decrease the step size ΔK_ν in order to enlarge the accuracy with which $\boldsymbol{K} = \boldsymbol{K}_{opt}$ is finally found.

Rilling and Roy [83] applied this procedure to the design of a Saturn V–Apollo pitch channel control system. The pitch displacement ϕ of the booster was to be controlled by feeding back ϕ itself as well as its derivative $\dot{\phi}$ with the gains K_1 and K_2, respectively. The flexibility of the booster body had to be taken into account; however, only the first bending mode was considered. The aim of the investigation was to select the gain factors K_1 and K_2 such that in the case of a deviation of the bending mode frequency from its nominal value ω_0 by as much as 20% a stable control was still achieved.

In order to give a rough impression of the power of the method, the results are summarized in Fig. 9.4-4 without entering into details. The plots show the time histories of the pitch displacement after an initial pitch displacement of 5° for the following cases:

(a) for the optimal system without considering the sensitivity vector $\tilde{\boldsymbol{\lambda}}_j$ and for nominal bending mode frequency $\omega = \omega_0$,

(b) same as (a), however for $\omega = 0.8\ \omega_0$,

(c) for the optimal system at $\omega = \omega_0$, considering the sensitivity vector $\tilde{\boldsymbol{\lambda}}_j = [\partial\phi/\partial \ln \omega \quad \partial\dot{\phi}/\partial \ln \omega]_{\omega_0}{}^{\mathrm{T}}$ according to Eq.(9.4-2),

(d) same as (c), however at $\omega = 0.8\ \omega_0$.

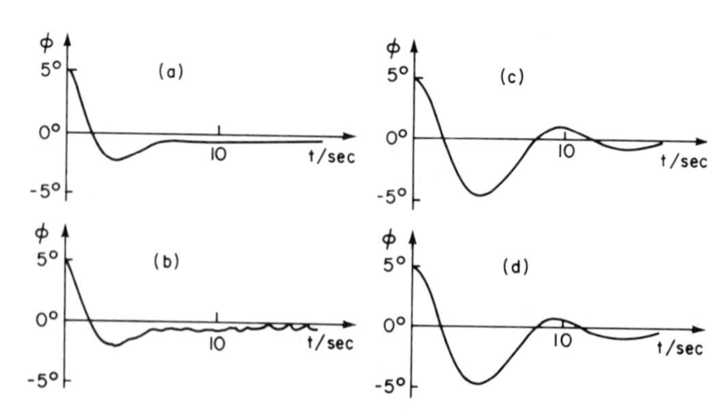

FIG. 9.4-4. Time histories of the pitch angle displacement; (a) for the optimized system without considering the sensitivity $\omega = \omega_0$, (b) same as (a) but $\omega = 0.8\ \omega_0$, (c) considering the sensitivity $\omega = \omega_0$, (d) same as (c) but $\omega = 0.8\omega_0$.

It is seen that in the case of nominal parameter values $\omega = \omega_0$, the optimal system (a) is slightly better than the desensitized one (b). However, as ω changes by 20%, it turns out to be already unstable [superimposed increasing oscillation in (b)], whereas the desensitized systems (c) and (d) do not change their behavior considerably.

9.5 SENSITIVITY REDUCTION IN SAMPLED-DATA SYSTEMS

9.5.1 INTRODUCTION

A great number of publications describing the various design procedures of sampled-data systems have appeared during the last two decades. Many of these methods are based on the dead-beat criterion and all assume the knowledge of the exact mathematical model of the plant. According to this, under the assumption of an exact mathematical model of the plant, the controlled variable reaches its equilibrium value at predetermined finite time. As a shortcoming of the dead-beat design, the system response is highly sensitive to deviations of the plant parameters from their nominal values. However, those deviations have to be considered in a realistic design since there is always a discrepancy between the parameters of the mathematical model and its physical counterpart.

There are several methods to cope with this difficulty. Either the methods of adaptive control or those of parameter insensitive design can be applied in order to reduce the sensitivity of the deadbeat response to parameter deviations. The adaptive control approach requires cumbersome mathematical developments and highly sophisticated instrumentation. A considerable simplication with comparable results can be achieved by application of parameter insensitive design. For a sampled-data plant with a single parameter deviation only, Schmidt [94] has solved this problem using z-domain. On the other hand Grübel and Kreisselmeier [39, 40] and Kreisselmeier and Grübel [61] have treated the more general problem of several plant parameter variations for a continuously acting control system, considering the worst case of the parameter deviations.

In this section a design procedure is presented for parameter insensitive state feedback for linear single-input single-output sampled-data plants of nth order with up to n parameter deviations. This procedure was developed by Münzner and co-workers [71, 72]. Making use of a two degree of freedom configuration for the control system, both the output error of the nominal control system and its parameter sensitivity functions are minimized simultaneously according to the dead-beat criterion. This guarantees that the output error is also minimized according to the dead-beat criterion even in case of several small parameter deviations of the plant.

The design procedure consists of three steps: First, the nominal plant is optimized according to deadbeat by a prefilter. Then a complementary control sequence is calculated, which minimizes the output sensitivity functions according to dead-beat, taking into account the *unavoidable sensitivity*. From this result the parameters of the feedback controller are determined.

It is assumed that the plant is completely controllable and observable and that the state variables are measurable. At first glance it might appear that in this case the parameters could be calculated directly from the measured state variables so that the method described could be avoided. However, this would require an adaptive procedure for the adjustment of the controller which is felt to be too complicated to be of practical value. In contrast to this, the proposed discrete-time design procedure is no more complex than that of the conventional dead-beat design and is applicable to plants with transfer functions of arbitrary order. This method is also applicable for plants in which m $(m < n)$ state variables are available for control. In this case only m parameter deviations can be taken into account.

For ease of presentation and for reduction of computational complexity, the design procedure will be described for plants with no zeros and with real poles only. However, this is no restriction of the proposed method which is equally valid for general linear plants.

9.5.2 Statement of the Problem

Consider a linear nth order single-input single-output sampled-data plant that for mathematical convenience may contain n series components with the transfer functions $S_j(s, \alpha_j) = k_j/(1 + T_j s)$ defining n state variables $x_1, \ldots, x_n$ as shown in Fig. 9.5-1. As a whole there can be considered n parameter deviations. Therefore one coefficient of each component, either the time-constant T_j or the gain k_j, can be considered as a parameter, denoted in abbreviated form by α_j. Since the gain factors affect the behavior of the closed loop more strongly than the time constants, they should normally be considered as the parameters.

First consider nominal parameter values denoted here by $\tilde{\alpha}_j$. Let us assume that the corresponding nominal control sequence $\tilde{\boldsymbol{u}}$ is computed by $R_0(z)$ from the reference input w such that the nominal output variable $\tilde{x}_1$ reaches the equilibrium value in finite time and then remains constant according to the dead-beat criterion. The design of $R_0(z)$ is a well-known procedure which will not be treated here in more detail. $R_0(z)$ is assumed to be known.

In the case of parameter deviations of the plant $\Delta\alpha_j = \alpha_j - \tilde{\alpha}_j$, the output of the actual system x_1 will differ from that of the nominal system $\tilde{x}_1$ thereby violating the requirement of the dead-beat criterion. The difference $\Delta x_1 = x_1 - \tilde{x}_1$ is due to the parameter deviations that are assumed to be unknown.

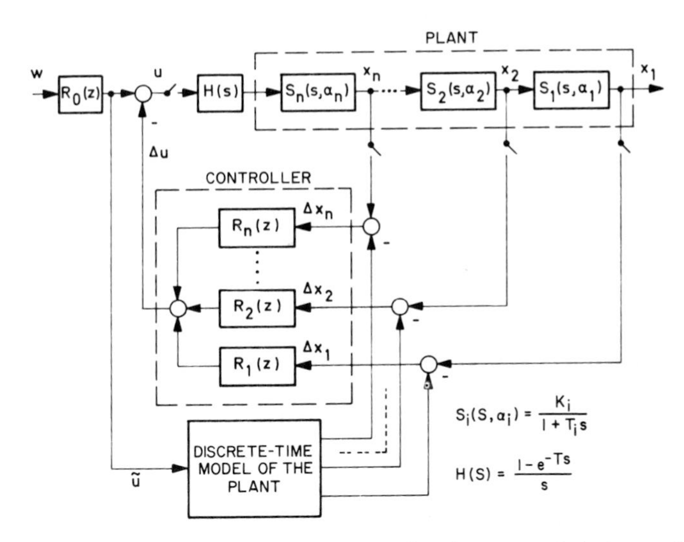

FIG. 9.5-1. Basic control loop configuration for the proposed design method.

According to Fig. 9.5-1, to compensate for x_1 a correcting control sequence $\Delta u(kT)$, $k = 0, 1, \ldots$, is applied to the plant input. $\Delta u(kT)$ is generated from the state errors $\Delta x_j = x_j - \tilde{x}_j$ by dynamic elements $R_1(z), \ldots, R_n(z)$, for the sake of abbreviation called the controller. The Δx_j are produced using a discrete-time model of the plant which can readily be implemented on a digital computer.

With the assumption of such a configuration, the sensitivity reduction problem can be stated as follows: Determine $\Delta u(kT)$ such that Δx_1 is minimized in some manner. Here we will treat the case such that it is minimized in the sense of dead-beat. Since the considered control system of Fig. 9.5-1 is a two-degree-of-freedom system, this requirement can be fulfilled independently of the dead-beat design of the nominal system. Once Δu is known, the elements $R_1(z), \ldots, R_n(z)$ can be calculated from $\Delta x_1, \ldots, \Delta x_n$ and Δu.

In the case of small parameter deviations, i.e., $|\Delta\alpha_j| \ll |\tilde{\alpha}_j|$, the state error vector $\Delta \mathbf{x} = [\Delta x_1 \quad \cdots \quad \Delta x_n]^{\mathrm{T}}$ can be approximately represented by

$$\Delta \mathbf{x} = \sum_{j=1}^{n} \boldsymbol{\lambda}_j \, \Delta\alpha_j \tag{9.5-1}$$

where $\boldsymbol{\lambda}_j = (\partial \mathbf{x}/\partial \alpha_j)_{\tilde{\boldsymbol{\alpha}}}$ is the trajectory sensitivity vector of the plant. It can be seen from Eq.(9.5-1) that the error in the output Δx_1 due to small parameter deviations is minimized according to the dead-beat criterion whenever the sensitivity functions $\lambda_{1,j}(j = 1, \ldots, n)$ are minimized according to this criterion. Thus the design can be accomplished using the sensitivity functions (instead of Δx_j and Δu) which can be determined from the sensitivity equations.

9.5.3 The Sensitivity Equations of the Plant

Using physical state variables x_j as defined in Fig. 9.5-1, the plant can be represented by the following vector state equation:

$$\dot{\boldsymbol{x}}(t,\boldsymbol{\alpha}) = \boldsymbol{A}(\boldsymbol{\alpha})\boldsymbol{x}(t, \boldsymbol{\alpha}) + \boldsymbol{b}(\boldsymbol{\alpha})u(t, \boldsymbol{\alpha}), \tag{9.5-2}$$

where $\boldsymbol{A}(\boldsymbol{\alpha})$ is an $n \times n$ system matrix (the $\boldsymbol{A}(\boldsymbol{\alpha})$ matrix is triangular since the plant transfer function is assumed to contain no zeros), $\boldsymbol{b}(\boldsymbol{\alpha})$ is an $n \times 1$ input matrix corresponding to the parameter vector $\boldsymbol{\alpha}$, $\boldsymbol{x}(t, \boldsymbol{\alpha})$ is a state vector (set of physical variables $x_1 \cdots x_n$ as shown in Fig. 9.5-1), $u(t, \boldsymbol{\alpha})$ is the control variable; $x_1(t)$ the output variable (first component of $\boldsymbol{x}(t)$) and $\boldsymbol{\alpha}$ the parameter vector with n components $\alpha_1, \ldots, \alpha_n$.

Due to parameter deviations, the matrices $\boldsymbol{A}$ and $\boldsymbol{b}$ and, consequently, the variables x_j and u are not only time functions but also functions of the parameter vector $\boldsymbol{\alpha}$. The dependence of u upon $\boldsymbol{\alpha}$ must be considered because of the existence of the feedback. Let $\tilde{\boldsymbol{\alpha}}$ represent the *nominal* parameter vector. Denoting all variables and matrices depending upon the *nominal* parameter vector by a tilde, e.g., $\mathbf{x}(t, \tilde{\boldsymbol{\alpha}}) = \tilde{\boldsymbol{x}}(t)$, the nominal state equation reads

$$\dot{\tilde{\boldsymbol{x}}}(t) = \tilde{\boldsymbol{A}}\tilde{\boldsymbol{x}}(t) + \tilde{\boldsymbol{b}}\tilde{u}(t). \tag{9.5-3}$$

Since the control variable of the discrete-time system $u(t)$ is constant during the sampling period T, the state transition equation of the nominal system at the sampling instants is given by

$$\tilde{\boldsymbol{x}}[(k+1)T] = \tilde{\boldsymbol{\phi}}(T)\tilde{\boldsymbol{x}}(kT) + \tilde{\boldsymbol{h}}(T)\tilde{u}(kT), \tag{9.5-4}$$

where

$$\tilde{\boldsymbol{\phi}}(T) = \exp(\tilde{\boldsymbol{A}}T),$$
$$\tilde{\boldsymbol{h}}(T) = \int_0^T \tilde{\boldsymbol{\phi}}(T-\tau)\tilde{\boldsymbol{b}}\, d\tau. \tag{9.5-5}$$

In abbreviated form Eq.(9.5-4) can be written as

$$\tilde{\boldsymbol{x}}_{k+1} = \tilde{\boldsymbol{\phi}}\tilde{\boldsymbol{x}}_k + \tilde{\boldsymbol{h}}\tilde{u}_k. \tag{9.5-6}$$

Note that the $\tilde{\boldsymbol{\phi}}$ matrix in Eq.(9.5-6) is also triangular since $\tilde{\boldsymbol{A}}$ is triangular. If $\boldsymbol{x}$, u, $\boldsymbol{h}$, and $\boldsymbol{\phi}$ are continuously differentiable with respect to the parameters $\alpha_j (j = 1, \ldots, n)$, the sensitivity equation of the sampled-data system with respect to the parameter α_j can be found by partial differentiation of the state equation with respect to α_j, and then setting $\boldsymbol{\alpha} = \tilde{\boldsymbol{\alpha}}$. This yields

$$\left.\frac{\partial \boldsymbol{x}_{k+1}}{\partial \alpha_j}\right|_{\tilde{\boldsymbol{\alpha}}} = \tilde{\boldsymbol{\phi}} \left.\frac{\partial \mathbf{x}_k}{\partial \alpha_j}\right|_{\tilde{\boldsymbol{\alpha}}} + \tilde{\boldsymbol{h}} \left.\frac{\partial u_k}{\partial \alpha_j}\right|_{\tilde{\boldsymbol{\alpha}}} + \left.\frac{\partial \boldsymbol{\phi}}{\partial \alpha_j}\right|_{\tilde{\boldsymbol{\alpha}}} \boldsymbol{x}_k + \left.\frac{\partial \boldsymbol{h}}{\partial \alpha_j}\right|_{\tilde{\boldsymbol{\alpha}}} \tilde{u}_k. \tag{9.5-7}$$

Note that since u_k is a function of x_k due to the feedback (Fig. 9.5-1), the term $(\partial u_k/\partial \alpha_j)_{\tilde{\boldsymbol{\alpha}}}$ in Eq.(9.5-7) generally does not vanish.

There are n such sensitivity equations to describe the sensitivity of the system with respect to all parameters α_j. Using the abbreviations

$$\boldsymbol{\lambda}_{k,j} = \left.\frac{\partial \boldsymbol{x}_k}{\partial \alpha_j}\right|_{\bar{\alpha}}, \qquad v_{k,j} = \left.\frac{\partial u_k}{\partial \alpha_j}\right|_{\bar{\alpha}},$$

$$\boldsymbol{\mu}_j = \left.\frac{\partial \boldsymbol{\phi}}{\partial \alpha_j}\right|_{\bar{\alpha}}, \qquad \boldsymbol{\nu}_j = \left.\frac{\partial \boldsymbol{h}}{\partial \alpha_j}\right|_{\bar{\alpha}}, \tag{9.5-8}$$

the sensitivity equation (9.5-7) can be written as

$$\boldsymbol{\lambda}_{k+i,j} = \tilde{\boldsymbol{\phi}}\, \boldsymbol{\lambda}_{k,j} + \tilde{\boldsymbol{h}}\, v_{k,j} + \boldsymbol{f}_{k,j}, \qquad j = 1, 2, \ldots, n, \tag{9.5-9}$$

where $\boldsymbol{\lambda}_{0,j} = \boldsymbol{0}$ and

$$\boldsymbol{f}_{k,j} = \boldsymbol{\mu}_j \tilde{\boldsymbol{x}}_k + \nu_j \tilde{u}_k. \tag{9.5-10}$$

Equation (9.5-9) represents the set of n sensitivity equations of the sampled-data plant. A graphical representation of Eqs. (9.5-6) and (9.5-9) is given in Fig. 9.5-2. A comparison of Eq. (9.5-9) with the state equation of the original system, Eq. (9.5-6), reveals that the $\boldsymbol{\phi}$ and $\boldsymbol{h}$ matrices, i.e., the *structures* of the two systems in Fig. 9.5-2 are the same. However the sensitivity model [Eq.(9.5-9)] contains an additional input $\boldsymbol{f}_{k,j}$ due to the parameter deviations. This term which will be called parameter-induced disturbance vector depends only upon nominal values of the original system as can be seen from Fig. 9.5-2. It can be calculated in advance from Eq.(9.5-10) and can therefore be considered to be known for the further developments. It should be noted that if the parameters α_j are the time-constants T_j, then the sensitivity equations (9.5-9) are linearly independent.

The n sensitivity equations (9.5-9) can be written as a single matrix transition equation

$$\boldsymbol{\lambda}_{k+1} = \tilde{\boldsymbol{\phi}}\, \boldsymbol{\lambda}_k + \tilde{\boldsymbol{h}}\, \boldsymbol{V}_k^{\mathrm{T}} + \boldsymbol{F}_k, \qquad \boldsymbol{\lambda}_0 = \boldsymbol{0}, \tag{9.5-11}$$

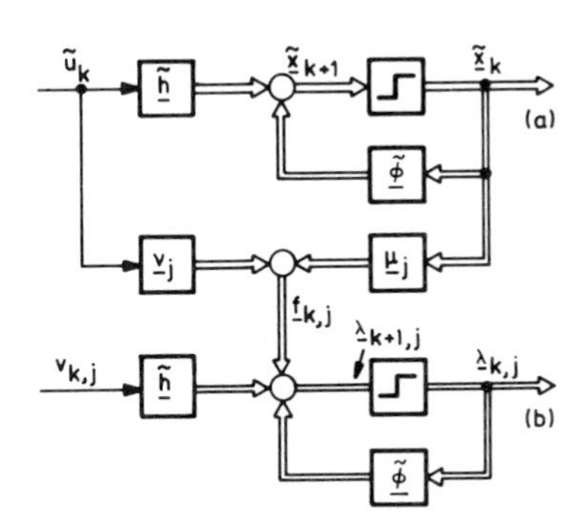

FIG. 9.5-2. Structural representation of the sensitivity equation (9.5-9); (a) nominal system, (b) sensitivity model; Problem: choose $v_{k,j}$ such that $\lambda_{k,j}$ is dead-beat.

where

$$\boldsymbol{\lambda}_k = \begin{bmatrix} \lambda_{1,k,1} & \lambda_{1,k,2} & \cdots & \lambda_{1,k,n} \\ 0 & \lambda_{2,k,2} & & \\ \vdots & \vdots & \ddots & \vdots \\ 0 & 0 & \cdots & \lambda_{n,k,n} \end{bmatrix} \tag{9.5-12}$$

is the sensitivity matrix and

$$\boldsymbol{F}_k = \begin{bmatrix} f_{1,k,1} & f_{1,k,2} & \cdots & f_{1,k,n} \\ 0 & f_{2,k,2} & \cdots & \\ \vdots & \vdots & \ddots & \vdots \\ 0 & 0 & \cdots & f_{n,k,n} \end{bmatrix} \tag{9.5-13}$$

is the parameter-induced disturbance matrix known from Eq. (9.5-10) and

$$\boldsymbol{V}_k = [v_{k,1} \quad \cdots \quad v_{k,n}]^{\mathrm{T}} \tag{9.5-14}$$

is the sensitivity control vector. Note that the matrices (9.5-12) and (9.5-13) are triangular because of the assumption that each component of the plant transfer function $S_j(s, \alpha_j)$ depends only upon one parameter α_j.

9.5.4 Calculation of the Optimal Control Sequence

Now the sensitivity control vector $\boldsymbol{V}_k$ has to be determined such that the sensitivity function $\lambda_{1,k,j}$ of Eq.(9.5-9), i.e., the output variable of the sensitivity model in Fig. 9.5-2, is minimized according to the dead-beat criterion. This means that the following requirements must be fulfilled:

(1) The output sensitivity function $\lambda_{1,k,j}$ must go to zero in finite time, say $k_0 T$, and must keep this value thereafter.

(2) Requirement (1) must be satisfied not only at the sampling instants but also between them.

From (1) it follows that

$$\lambda_{1,k,j} = 0 \qquad \text{for all} \quad k \geq k_0, \qquad j = 1, 2, \ldots, n, \tag{9.5-15}$$

$$\lambda_{1,j}(z) = \sum_{k=0}^{k_0-1} \lambda_{1,k,j} z^{-k}.$$

A necessary and sufficient condition to fulfill requirement (2) is, for the special type of the plant,

$$\boldsymbol{\lambda}_{k+1,j} = \boldsymbol{\lambda}_{k,j} \qquad \text{for all} \quad k \geq k_0, \tag{9.5-16}$$

where k_0 is an integer yet to be determined.

The repeated application of the recursion in Eq. (9.5-9) from $k = 0$ up to

$k = k_0$, according to condition (9.5-15), yields

$$\boldsymbol{\lambda}_{k_0, j} = \tilde{\boldsymbol{\phi}}^{k_0}\boldsymbol{\lambda}_{0, j} + \sum_{k=0}^{k_0-1} \tilde{\boldsymbol{\phi}}^{k_0-k-1}\boldsymbol{f}_{k, j} + \sum_{k=0}^{k_0-1} \tilde{\boldsymbol{\phi}}^{k_0-k-1}\, \tilde{\boldsymbol{h}} v_{k, j}. \qquad (9.5\text{-}17)$$

Recalling that $\boldsymbol{\lambda}_{0, j} = \mathbf{0}$ and $v_{0, j} = 0$ and solving Eq.(9.5-17) for $v_{k,j}$ yields the general solution

$$\boldsymbol{V}_j = [\tilde{\boldsymbol{\phi}}^{k_0-2}\tilde{\boldsymbol{h}}, . . ., \tilde{\boldsymbol{h}}]^{-1}\, [\boldsymbol{\lambda}_{k_0, j} - \sum_{k=0}^{k_0-1} \tilde{\boldsymbol{\phi}}^{k_0-k-1}\boldsymbol{f}_{k, j}], \qquad (9.5\text{-}18)$$

where $\boldsymbol{V}_j = [v_{1, j} \cdot \cdot \cdot v_{k-1, j}]^{\mathrm{T}}$ is the vector of the components of the sensitivity control vector $\boldsymbol{V}_k$ at different sampling instants for constant j ($j = 1, . . ., n$). Thus for each α_j, Eq.(9.5-18) represents a set of $k_0 - 1$ equations. A solution to this set of equations can exist only if $\boldsymbol{V}_j$ is of the same dimension as the vectors of the right-hand side, which is n. As a consequence, $k_0 - 1 = n$, i.e., the sensitivity sequence can be zero at the earliest after $k_0 = n + 1$ steps.

The matrix $(\tilde{\boldsymbol{\phi}}^{k_0-2}\tilde{\boldsymbol{h}}, . . ., \tilde{\boldsymbol{h}})$ is the controllability matrix which is invertible by the presumption of the controllability of the plant. Furthermore it is necessary that the sensitivity vectors $\boldsymbol{\lambda}_{k_0, j}$ in Eq.(9.5-18) be known. These can be determined by introducing the conditions (9.5-16) into Eq.(9.5-9), which yields

$$\boldsymbol{\lambda}_{k_0, j} = \tilde{\boldsymbol{\phi}}\, \boldsymbol{\lambda}_{k_0, j} + \tilde{\boldsymbol{h}}\, v_{k_0, j} + \boldsymbol{f}_{k_0, j}. \qquad (9.5\text{-}19)$$

Since the nominal system is optimized according to the dead-beat criterion, as well, the following equations are valid:

$$\boldsymbol{f}_{k_0, j} = \boldsymbol{f}_{k_0-1, j} = \boldsymbol{f}_{\infty, j}, \qquad (9.5\text{-}20)$$

$$v_{k_0+1, j} = v_{k_0, j} = v_{\infty, j}, \qquad (9.5\text{-}21)$$

$$\lambda_{k_0+1, j} = \lambda_{k_0, j} = \lambda_{\infty, j}. \qquad (9.5\text{-}22)$$

Thus from Eq.(9.5-19) it follows

$$[(\boldsymbol{I} - \tilde{\boldsymbol{\phi}}) \quad - \tilde{\boldsymbol{h}}]\, [\boldsymbol{\lambda}^{\mathrm{T}}{}_{\infty, j} \quad v_{\infty, j}]^{\mathrm{T}} = \boldsymbol{f}_{\infty, j}. \qquad (9.5\text{-}23)$$

Because of the requirement $\lambda_{1, \infty, j} = 0$, there results from Eq.(9.5-23) a solvable system of equations for the remaining components of $\boldsymbol{\lambda}_{\infty, j}$ and $v_{\infty, j}$ by crossing out the first column of the matrix $[\boldsymbol{I} - \tilde{\boldsymbol{\phi}} \quad -\tilde{\boldsymbol{h}}]$ and the first row of the vector $[\boldsymbol{\lambda}^{\mathrm{T}}{}_{\infty, j} \quad v_{\infty, j}]^{\mathrm{T}}$. Thus Eq.(9.5-18) can be evaluated.

From this derivation the following observations can be made:

(1) The sensitivity control sequence $v_{k, j}$ can be calculated using the same algorithm as for the control sequence $\tilde{\boldsymbol{u}}_k$ of the original system.

(2) The sensitivity sequence of the output function $\lambda_{1, k, j}$ can be brought to zero at the earliest after $n + 1$ sampling periods.

(3) As shown by Münzner [71], if a time constant T_j is a parameter, the condition (9.5-16) especially reads

$$\boldsymbol{\lambda}_{n+1,\,j} = \mathbf{0}. \tag{9.5-24}$$

9.5.5 The Limits of Sensitivity Reduction

To achieve stable control, no poles of the control system to be designed may lie in the right s-plane, i.e., no poles of $R_i(z)$ may lie outside the unit circle in the z-plane. However, these poles are the zeros of the sensitivity matrix $\boldsymbol{\lambda}_k$. Therefore these zeros have to be singled out in advance. This causes an error in the result, which means that there are certain limits in the achievable insensitivity that are of the same nature as unavoidable errors in the Wiener approach to the design of control systems [8, 28].

In order to discuss the limits of achievable insensitivity, consider the z-transforms of the difference sequences of vectors or matrices for $k = 0, \ldots, \infty$:

$$\begin{aligned} \boldsymbol{V}(z) &= \sum_{k=0}^{\infty} \boldsymbol{V}_k z^{-k}, \\ \boldsymbol{\lambda}(z) &= \sum_{k=0}^{\infty} \boldsymbol{\lambda}_k z^{-k}, \\ \boldsymbol{F}(z) &= \sum_{k=0}^{\infty} \boldsymbol{F}_k z^{-k}. \end{aligned} \tag{9.5-25}$$

Let the controller elements be represented by the controller vector in the z-plane,

$$\boldsymbol{R}(z) = [R_1(z) \quad \cdots \quad R_n(z)]^{\mathrm{T}}. \tag{9.5-26}$$

From Fig. 9.5–1 it follows that for the feedback path $\Delta u(z) = \mathbf{R}^{\mathrm{T}}(z)\,\Delta \boldsymbol{x}(z)$. With $\Delta u(z) = \boldsymbol{V}^{\mathrm{T}}(z)\,\Delta\boldsymbol{\alpha}$ and $\Delta\boldsymbol{x}(z) = \boldsymbol{\lambda}(z)\,\Delta\boldsymbol{\alpha}$, the following relation is obtained:

$$\boldsymbol{V}^{\mathrm{T}}(z) = \boldsymbol{R}^{\mathrm{T}}(z)\boldsymbol{\lambda}(z). \tag{9.5-27}$$

Introducing Eq.(9.5-27) into Eq.(9.5-11), the z-transformed sensitivity function of the closed-loop system becomes

$$\boldsymbol{\lambda}(z) = [\boldsymbol{I} - z^{-1}(\tilde{\boldsymbol{\phi}} + \tilde{\boldsymbol{h}}\,\boldsymbol{R}^{\mathrm{T}}(z))]^{-1} z^{-1}\boldsymbol{F}(z) = \boldsymbol{G}(z)\boldsymbol{F}(z), \tag{9.5-28}$$

where $\boldsymbol{I}$ is the identity matrix and $\boldsymbol{G}(z)$ is the z-transfer matrix of the closed-loop system which can be written as

$$\boldsymbol{G}(z) = \boldsymbol{\lambda}(z)\,\boldsymbol{F}^{-1}(z). \tag{9.5-29}$$

From this it follows that the closed-loop system is stable only if the zeros of $\det \boldsymbol{F}(z) = 0$ lie inside the unit circle in the z-plane. Since in the present case $\boldsymbol{F}(z)$ is triangular, $\det \boldsymbol{F}(z)$ is a product only of the elements of the main diagonal of $\boldsymbol{F}(z)$. Therefore the system will be stable if all zeros of the main diagonal of $\boldsymbol{F}(z)$ lie inside the unit circle.

It can happen that for certain values of plant parameters some zeros of $\boldsymbol{F}(z)$ come to lie outside the unit circle. For these sets the design conditions [Eqs. (9.5-15) and (9.5-16)] cannot be fulfilled, which means that dead-beat cannot be reached after $n + 1$ sampling periods. In order to find an optimal solution in this case, the matrix $\boldsymbol{F}(z)$ is factorized by two matrices:

$$\boldsymbol{F}(z) = \boldsymbol{F}^*(z)\, \boldsymbol{F}^+(z), \tag{9.5-30}$$

where

(1) $\boldsymbol{F}^*(z)$ is a matrix with eigenvalues only inside the unit circle, and
(2) $\boldsymbol{F}^+(z)$ is a matrix with eigenvalues only outside the unit circle.

A method of factorizing $\boldsymbol{F}$ in the above sense is given by Münzner; see also Frank [8] and Bongiorno [28].

The design is now based on $\boldsymbol{F}^*(z)$, i.e., $\boldsymbol{F}_k$ is replaced by $\boldsymbol{F}_k{}^*$ in all equations of the previous sections. From this procedure there results a fictitious sensitivity matrix that will be denoted by $\boldsymbol{\lambda}^*(z)$. However, since in the real system the parameter-induced disturbance matrix will still be $\boldsymbol{F}(z)$, the corresponding *true* sensitivity matrix $\boldsymbol{\lambda}(z)$ will generally differ from the fictitious sensitivity matrix $\boldsymbol{\lambda}^*(z)$ used as a design basis only. Replacing $\boldsymbol{F}(z)$ by $\boldsymbol{F}^*(z)$ in Eq.(9.5-28), we obtain

$$\boldsymbol{\lambda}^*(z) = \boldsymbol{G}(z)\boldsymbol{F}^*(z). \tag{9.5-31}$$

Postmultiplying Eq.(9.5-31) by $\boldsymbol{F}^+(z)$ and reusing Eqs. (9.5-28) and (9.5-30), the true sensitivity matrix is found to be

$$\boldsymbol{\lambda}(z) = \boldsymbol{\lambda}^*(z)\boldsymbol{F}^+(z). \tag{9.5-32}$$

Thus the true sensitivity matrix contains the eigenvalues of $\boldsymbol{F}(z)$ outside the unit circle. Each such eigenvalue causes an additional step for reaching dead-beat. The difference

$$\boldsymbol{\lambda}^u(z) = \boldsymbol{\lambda}(z) - \boldsymbol{\lambda}^*(z) \tag{9.5-33}$$

is called *unavoidable* sensitivity because under the given conditions no better sensitivity can be obtained. It can be shown that for second-order systems all eigenvalues of $\boldsymbol{F}(z)$ come to lie inside the unit circle in the z-plane regardless of the sampling period; therefore, $\boldsymbol{F}^+ = \boldsymbol{I}$ and $\boldsymbol{\lambda}^u = \boldsymbol{0}$. For higher-order systems $\boldsymbol{\lambda}^u(z)$ can be regarded as a system variable depending upon the structure of the system and the sampling period T of the controller. By increasing the sampling period T all eigenvalues of $\boldsymbol{F}(z)$ can be brought to lie inside the unit circle in the z-plane thereby improving the sensitivity of the control system with respect to the number of sampling periods. The optimal sensitivity control vector determined in the way described above is denoted by $\boldsymbol{V}^*(z)$.

9.5.6 Determination of the Controller Coefficients

The coefficients of the controller vector $\boldsymbol{R}(z)$ can now be computed from Eq.(9.5-27) replacing $\boldsymbol{V}(z)$ by $\boldsymbol{V}^*(z)$ and $\boldsymbol{\lambda}(z)$ by $\boldsymbol{\lambda}^*(z)$. Transposing Eq. (9.5-27) yields

$$\boldsymbol{V}^*(z) = \boldsymbol{\lambda}^{*\mathrm{T}}(z)\boldsymbol{R}(z). \tag{9.5-34}$$

The components of the controller vector $\boldsymbol{R}(z)$ are now chosen to be of the general form

$$R_i(z) = \frac{r_{i1} + r_{i2}z^{-1} + \cdots + r_{i,w+1}z^{-w}}{1 - r_{01}z^{-1} - \cdots - r_{0w}z^{-w}}, \tag{9.5-35}$$

where the denominator polynomial is the same for all components. The order is chosen as $w = n^2 - 2n + z_{kj} + 1$, where n is the order of the plant. $z_{kj} = 1$ if the gains k_j are considered to be the parameters and $z_{kj} = 0$ if the time constants T_j are considered to be the parameters.

Introducing $\boldsymbol{R}(z)$ with elements of Eq. (9.5-35) into Eq.(9.3-34) and taking the inverse z-transform, we obtain

$$\boldsymbol{V}_k^* = \sum_{\nu=1}^{k} r_{0,\nu}\boldsymbol{V}_{k-\nu}^* + \sum_{i=1}^{n}\sum_{\nu=1}^{w+1} r_{i,\nu}\boldsymbol{\lambda}_{i,k-\nu+1}^{*T}, \qquad k = 0,\ldots,\infty, \tag{9.5-36}$$

where $\boldsymbol{\lambda}_{i,q}^*$ is the ith row of $\boldsymbol{\lambda}_k^*$. From this equation a system of equations can be derived from which the coefficients $r_{i,\nu}$ ($i = 0,\ldots, n$, $\nu = 1,\ldots, w + 1$) can be calculated.

Defining the controller coefficient vector $\boldsymbol{r}$ as

$$\boldsymbol{r} = [r_{0,1} \quad \cdots \quad r_{0,w} \quad r_{1,1} \quad \cdots \quad r_{1,w+1} \quad \cdots \quad r_{n,1} \quad \cdots \quad r_{n,w+1}]^{\mathrm{T}} \tag{9.5-37}$$

and the matrix

$$\boldsymbol{M}_k = [\boldsymbol{V}_{k-1}^* \quad \cdots \quad \boldsymbol{V}_{k-w}^* \quad \boldsymbol{\lambda}_{1,k}^{*\mathrm{T}} \quad \cdots \quad \boldsymbol{\lambda}_{1,k-2}^{*\mathrm{T}} \quad \cdots \quad \boldsymbol{\lambda}_{n,k}^{*\mathrm{T}} \quad \cdots \quad \boldsymbol{\lambda}_{n,k-w}^{*\mathrm{T}}], \tag{9.5-38}$$

Eq.(9.5-36) can be written as

$$\boldsymbol{V}_k^* = \boldsymbol{M}_k\boldsymbol{r}. \tag{9.5-39}$$

Since Eq.(9.5-39) must be satisfied for all sampling instants,

$$\begin{bmatrix} \boldsymbol{V}_1^* \\ \vdots \\ \boldsymbol{V}_k^* \end{bmatrix} = \begin{bmatrix} \boldsymbol{M}_1 \\ \vdots \\ \boldsymbol{M}_k \end{bmatrix} \boldsymbol{r}. \tag{9.5-40}$$

If k is chosen such that $\boldsymbol{M} = (\boldsymbol{M}_1 \cdots \boldsymbol{M}_k)^{\mathrm{T}}$ is square, then Eq. (9.5-40) can be solved for the vector $\boldsymbol{r}$ either in a straight forward way by a matrix inversion or by a recursive algorithm [71].

9.5.7 Example

Consider a third-order plant with the transfer function

$$S(s, T_1, T_2, T_3) = [\prod_{j=1}^{3} (1 + T_j s)]^{-1}, \tag{9.5-41}$$

where the time constants T_1, T_2, T_3 are regarded as the parameters. Let the nominal values be $\tilde{T}_1 = 3.6$ sec, $\tilde{T}_2 = 2.5$ sec, $\tilde{T}_3 = 1.2$ sec, and the sampling period 1 sec. For this plant the semirelative output sensitivity functions with respect to the parameters T_1, T_2, T_3, defined as

$$\tilde{\lambda}_{1,k,j} = \left.\frac{\partial x_{1,k}}{\partial \ln T_j}\right|_{\tilde{T}} = \boldsymbol{\lambda}_{1,k,j}\tilde{T}_j, \qquad j = 1, 2, 3, \tag{9.5-42}$$

are determined for the open-loop dead-beat system and the insensitive control system designed according to the above method. The results are illustrated in Fig. 9.5-3 and Fig. 9.5-4.

The results show that for T_1 one zero of $\boldsymbol{\lambda}(z)$ lies outside the unit circle in the z-domain for the assumed values of the parameters and sampling period. Therefore according to the theory, the sensitivity function in Fig. 9.5-4 does not reach zero prior to the fifth sampling period. The corresponding functions for T_2 and T_3, for which all eigenvalues lie inside the unit circle in the z-domain, already reach zero at the fourth sampling period. The curve labeled $\tilde{\lambda}^*_{1,k,1}$ shows the avoidable semirelative sensitivity sequence when T_1 is the parameter. The difference $\tilde{\lambda}_{1,k,1} - \tilde{\lambda}^*_{1,k,1}$ at the sampling instants reveals the unavoidable sensitivity in this case. It is zero for the other two cases.

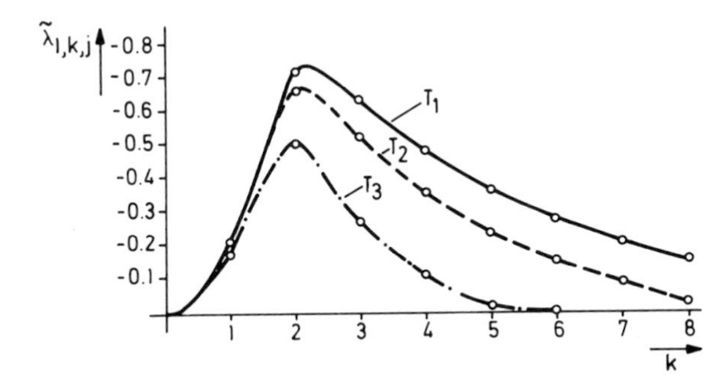

FIG. 9.5-3. Semirelative sensitivity functions with respect to T_1, T_2, and T_3 for the open-loop system (without feedback) for the discussed third-order example.

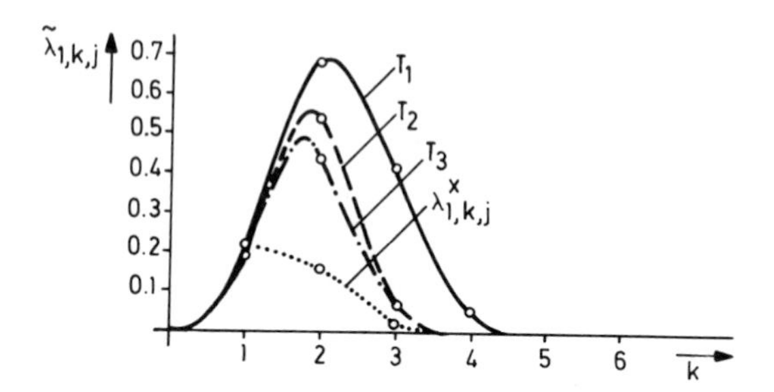

FIG. 9.5-4. Same functions as in Fig. 9.5-3, however, for the parameter insensitive design; in addition: avoidable semirelative sensitivity sequence $\lambda^*_{1,k,1}$.

A comparison of the sensitivity curves of the insensitive control loop, Fig. 9.5–4, and the open-loop case, Fig. 9.5-3, shows that a sensitivity reduction is not possible prior to the third sampling instant if one of the zeros lies outside the unit circle. In the other two cases, the sensitivity is reduced starting even as soon as the second sampling instant.

The step response of the parameter-insensitive control loop is plotted in Fig. 9.5-5 for the nominal set of parameters ($\tilde{T}_1 = 3.6$ sec, $\tilde{T}_2 = 2.5$ sec, $\tilde{T}_3 = 1.2$ sec) and for the two worst cases where all three parameters deviate in the same direction by $+20\%$ or by -20%. All curves of a mixed type of parameter deviation lie between the curves I and II. For a comparison of these results with a conventional dead-beat control, the corresponding curves of the latter are plotted in Fig. 9.5-6. It can be seen that because of the unavoidable sensitivity the system error during the first two sampling periods is approximately the same as that of a system with conventional dead-beat

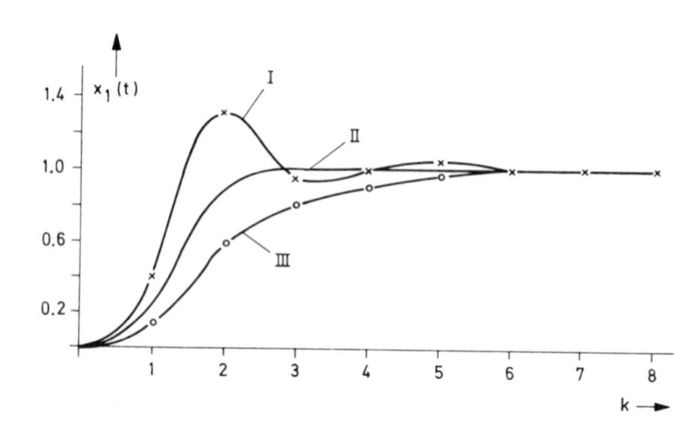

FIG. 9.5-5. Step response of the insensitive control system for the discussed example for the two worst cases: (I) $\Delta T_1 = -20\%$, $\Delta T_2 = -20\%$, $\Delta T_3 = -20\%$, (II) $\Delta T_1 = +20\%$, $\Delta T_2 = 20\%$, $\Delta T_3 = 20\%$, and (III) the case of nominal parameters.

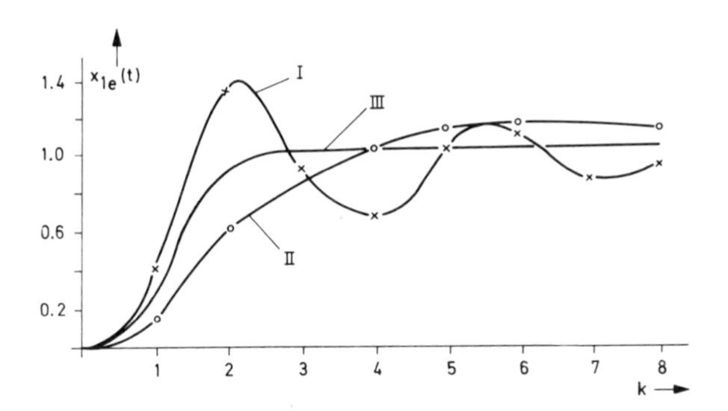

FIG. 9.5-6. Same functions as in Fig. 9.5-5, however, for conventional dead-beat design.

control. However, for the subsequent period, a considerable improvement of the parameter-insensitive design is apparent.

The price which has to be paid for the sensitivity reduction is an increase in complexity of the digital controller. In conventional dead-beat control the overall transfer function of the control system is

$$G(z) = \frac{S(z)R(z)}{1 + S(z)R(z)}. \tag{9.5-43}$$

For a third-order plant, the order of the overall transfer function is therefore 6 and the order of the controller to be simulated on the computer is 3. In case of the described insensitive control, the structure of the control loop is that shown in Fig. 9.5–1 with $n = 3$ in the case of a third-order plant. The controller components are of the form

$$R_i(z) = \frac{r_{i1} + r_{i2}z^{-1} + r_{i3}z^{-2} + r_{i4}z^{-3} + r_{i5}z^{-4}}{1 + r_{01}z^{-1} + r_{02}z^{-2} + r_{03}z^{-3} + r_{04}z^{-4}}, \qquad i = 1, 2, 3. \tag{9.5-44}$$

The denominator is the same for R_1, R_2, and R_3. In realization of the controller, $R_0(z)$, $R_1(z)$, $R_2(z)$, $R_3(z)$, and the discrete nominal plant model have to be programmed on the digital computer or a special purpose digital controller constructed to carry out these functions. Therefore if all state variables are fed back, the order of the structure to be programmed on the computer is 18. The order could be reduced by choosing different denominators for the R_i's. However this would mean a loss of the clearness of the program.

9.5.8 Conclusions

In this section, a design procedure for parameter insensitive state feedback for linear plants has been formulated. This procedure assumes that the single-input single-output plant is controllable and observable and that there are as many state variables available as there are parameters to be considered. The maximum number of parameters to be considered is therefore n for an nth order plant. In principle, this method can be extended to include multivariable systems.

The design presented herein is based on the dead-beat criterion although other criteria are applicable and have been used [71]. The dead-beat criterion is used twice, first for optimization of the nominal system and second for the optimization of the sensitivity. As a result, the system error can be brought to zero at the earliest time, which is after $(n + 1)$ sampling periods for an nth order plant. In the case of unavoidable sensitivity, which can exist for plants of higher than second order, the number of sampling periods for reaching zero system error is larger.

It has been shown that for a given plant there exists an upper limit for sensitivity reduction that is directly dependent upon the unavoidable sensitivity. The latter is predetermined by the sampling period and by the parameter values of the plant.

The improvements of the method over the conventional dead-beat design can be interpreted as an increase of relative stability in the case of parameter deviations.

A study of over 20 examples has shown [71] that the proposed method, although its mathematical justification is based on infinitesimally small parameter deviations, is applicable for parameter deviations up to 30%. A change of parameters by 50% proved the systems designed by the described method to be still stable, whereas the conventional dead-beat design yielded already instable systems without exception.

In this procedure, the sensitivity model plays the same role for the reduction of the sensitivity to parameter variations as the nominal original system model for the reduction of the output errors due to disturbances.

9.6 FINAL REMARKS

These four examples show that the sensitivity model has to be considered as a necessary complement of the original system in all those cases where the sensitivity is of any concern. This is true for many engineering systems. In those cases where the sensitivity functions are needed continuously, the sensitivity model takes on a physical form. For design purposes, however, where the sensitivity is taken into account without changing the structure of the system, the sensitivity model is of a pure fictitious nature.

As the final example shows particulary, all manipulations undertaken in the signal domain with the original system model have a parallel in the sensitivity domain with the sensitivity model.

Methods for the approximate calculation of sensitivity functions on a hybrid computer are given by Schönfeld [95].

References

Books, Course Materials

[1] BODE, H. W., "Network Analysis and Feedback Amplifier Design." Van Nostrand —Reinhold, Princeton, New Jersey, 1945.

[2] BROGAN, W. L., "Modern Control Theory." Quantum Publ. Inc., 1974.

[3] CALAHAN, D. A., "Computer-Aided Network Design." McGraw-Hill, New York, 1972.

[4] CHANG, S. S. L., "Synthesis of Optimum Control Systems." McGraw-Hill, New York, 1961.

[5] CRUZ, J. B., "Feedback Systems." McGraw-Hill, New York, 1972.

[6] CRUZ, J. B., "System Sensitivity Analysis." Dowden Hutchinson, & Ross, Stroudsburg, Pennsylvania, 1973.

[7] EYKHOFF, P., "System Identification." Wiley, New York, 1974.

[8] FRANK, P. M., "Entwurf von Regelkreisen mit vorgeschriebenem Verhalten," Braun Verlag, Karlsruhe, 1974.

[9] GÉHER, K., "Theory of Network Tolerances." Akadémiai Kiadó, Budapest, 1971.

[10] GRÜBEL, G., Zur Beurteilung der Parameterempfindlichkeit linearer Systeme, *in* "Einführung in die Regelungstheorie," Chap. VII. Carl Cranz Gesellschaft, 1969.

[11] HOROWITZ, I. M., "Synthesis of Feedback Systems." Academic Press, New York, 1963.

[12] KOKOTOVIC, P. V., *et al.,* Singular Perturbations: Order Reduction in Control System Design, *JACC, 1972. Am. Soc.* of *Mech. Eng.,* New York, 1972.

[13] PERKINS, W. R., and CRUZ, J. B., "Engineering of Dynamic Systems." Wiley, New York, 1969.

[14] RADANOVIC, L., Sensitivity Methods in Control Theory, *Proc. Int. Symp. Sensitivity, Dubrovnik, Yugoslavia, 1964.* Pergamon Press, Oxford, 1966.

[15] ŠAGE, A. P., "Optimum Systems." Prentice-Hall, Englewood cliffs, New Jersey, 1968.

[16] ŠILIAK, D. D., "Nonlinear Systems." Wiley New York, 1969.

[17] System Sensitivity and Adaptivity, Preprints *IFAC Symp. and, Dubrovnik, Yugoslavia, August 1968.*

[18] TOMOVIC, R., "Sensitivity Analysis of Dynamic Systems." McGraw-Hill, New York, 1963.

[19] TOMOVIC, R., and VUCOBRATOVIC, M., "General Sensitivity Theory." Amer. Elsevier, New York, 1972.

Articles, Dissertations, Reports

[20] ANDERSON, B. D. O., The Inverse Problem of Optimal Control, *Ann. Allerton Conf., 4th, Urbana, Illinois, October 1966.*

[21] BARNETT, S., Insensitivity of Control Systems, *Int. J. Contr.* **10,** 665–675 (1969).

[22] BECKEY, G., and TOMOVIC, R., Sensitivity of Discrete Systems to Variation of Sampling Interval, *IEEE Trans. Autom. Contr.* **11,** 284–287 (1966).

[23] BÉLANGER, P. R., Some Aspects of Control Tolerances and First-Order Sensitivity in Optimal Control Systems, *IEEE Trans. Autom. Contr.* **11,** 77–83 (1966).

[24] BIKHOVSKI, M. L., "Dynamic Accuracy of Electrical and Mechanical Circuits." Akademia Nauk USSR, Moscow, 1958.

[25] BIKHOVSKI, M. L., Sensitivity and Dynamic Accuracy of Control Systems, *Eng. Cybern. (USSR),* 121–134 (1964).

[26] BINGULAC, S. P., Simultaneous Generation of the Second-Order Sensitivity Functions, *IEEE Trans. Autom. Contr.,* **11,** 563–566 (1966).

[27] BISWAS, R. N , and KUH, E. S., A Multiparameter Sensitivity Measure for Linear Systems, *IEEE Trans.* CT, 718–719 (1971).

[28] BONGIORNO J. J. Jr., Minimum Sensitivity Designs, *IEEE Int. Conv. Rec.* **15,** 129–135 (1967).

[29] CROSSLEY, T. R., and PORTER, B., Eigenvalue and Eigenvector Sensitivities in Linear Systems Theory, *Int. J. Contr.* **10,** 163–170 (1969).

[30] CRUZ, J. B., and PERKINS, W. R., A New Approach to the Sensitivity Problem in Multivariable Feedback System Design, *IEEE Trans. Autom. Contr.* **9,** 216–223 (1964).

[31] CRUZ, J. B., and PERKINS, W. R., Sensitivity Comparison of Open-Loop and Closed-Loop Systems, *Proc. Ann. Allerton Conf., 3rd, October 1965.*

[32] CRUZ, J. B., and PERKINS, W. R., Criteria for System Sensitivity to Parameter Variations, *IFAC Cong., 3rd, London, 1966,* Paper 18C.

[33] DE BACKER, W., Jump Conditions for Sensitivity Coefficients, *Proc. Int. Symp. Sensitivity, Dubrovnik, Yugoslavia, 1964,* pp. 168–175. Pergamon Press, Oxford, 1966.

[34] DENERY, D. G., Simplification in the Computation of the Sensitivity Functions, *IEEE Trans. Autom. Contr.* **16,** 348–350 (1971).

[35] DIRECTOR, S. W., and ROHRER, R. A., A Generalized Adjoint Network and Network Sensitivities, *IEEE Trans.* CT **16,** 330–336 (1969).

[36] DORATO, P., On Sensitivity in Optimal Control Systems, *IEEE Trans. Autom. Contr.* **8,** 256–257 (1963).

[37] DORATO, P., and KESTENBAUM, A., Application of Game Theory to the Sensitivity Design of Optimal Systems, *IEEE Trans. Autom. Contr.* 85–87 (1967).

[38] DUNN, C. J., Further Results on the Sensitivity of Optimally Controlled Systems, *Trans. IEEE Autom. Contr.* **12,** 324–326 (1967).

[39] GRÜBEL, G., and KREISSELMEIER, G., Effective Parameter Sensitivity Reduction through Minimization of a Sensitivity Measure, *JACC St. Louis, August 1971,* pp. 79–86, Paper 2-El.

[40] GRÜBEL, G., and KREISSELMEIER, G., A Generalized Comparison Sensitivity Concept for Sensitivity Reduction in Control System Design, *Proc. JACC, Austin, Texas, 15th, June 18–21, 1974,* pp. 328–332.

[41] GUARDABASSI, G., LOCATELLI, A., and RINALDI, S., On the Optimality of the Wilkie–Perkins Low-Order Sensitivity Model, *IEEE Trans. Autom. Contr.* 382–384 (1970).

[42] GUARDABASSI, G., LOCATELLI, A., and RINALDI, S., Structural Uncontrollability of Sensitivity Coefficients, Preprints *IFAC Symp., 2nd, Dubrovnik, Yugoslavia, August 1968,* pp. A55–A69.

[43] GUARDABASSI, G., LOCATELLI, A., and RINALDI, S., Structural Uncontrollability of Sensitivity System, *Automatica* **5**, 297–301 (1969).

[44] HOLTZMAN, J. M., and HORING, S., The Sensitivity of Terminal Conditions of Optimal Control Systems to Parameter Variations, *IEEE Trans. Autom. Contr.* **10**, 420–426 (1965).

[45] HOROWITZ, I. M., and SHAKED, U., Superiosity of Transfer Function over State Variable Methods in Linear Time-Invariant Feedback System Design, *IEEE Trans. Autom. Contr.* 84–97 (1975).

[46] HOROWITZ, I. M., Plant Adaptive Systems versus Ordinary Feedback Systems; *IRE Trans. Autom. Control* 48–56 (1962).

[47] HUANG, R. Y., The Sensitivity of the Poles of Linear, Closed-Loop Systems, *Trans. Am. Inst. Electr. Eng. Part 2* **77**, 182–187 (1958).

[48] JACOBI, C. G. J., *J. Reine Angew. Math.* **30**, 51–95 (1846).

[49] KALMAN, R E., When Is a Linear Control System Optimal?, *Trans. ASME, J. Basic Eng.* **86**, 51–60 (1964).

[50] KIOVO, A J., Performance Sensitivity of Dynamical Systems, *Proc. Inst. Electr. Eng.* **117**, 825–830 (1970).

[51] KIOVO, A J., Performance Sensitivity of Sampling Systems—A Unified Approach, *J. Franklin Inst.* **287**, 209–221 (1969).

[52] KLINSMANN, L., Empfindlichkeitsfunktionen linearer Systeme mit Totzeit, msr 15, **10**, 365–369 (1972).

[53] KOKOTOVIC, P. V., and SANNUTI, P., Singular Perturbation Method for Reducing the Model Order in Optimal Control Design, *IEEE Trans. Autom. Contr.* **13**, 377–384 (1968).

[54] KOKOTOVIC, P., and RUTMAN, R. S., On the Determination of Sensitivity Functions with Respect to the Change of System Order, *Proc. Int. Symp. Sensitivity, Dubrovnik, Yugoslavia, 1964,* pp. 131–137. Pergamon Press, Oxford, 1966.

[55] KOKOTOVIC, P. V., Method of Sensitivity Points in the Investigation and Optimization of Linear Control Systems, *Autom. Remote Control (USSR)* 1512–1518 (1964).

[56] KOKOTOVIC, P. V., and RUTMAN, R. S., Sensitivity of Automatic Control Systems (Survey), *Autom. Remote Control (USSR)* **26**, 727–749 (1965).

[57] KOKOTOVIC, P., HELLER, J., and SANNUTI, P., Sensitivity Comparison of Optimal Controls, *Int. J. Contr.* 111–117 (1969).

[58] KREINDLER, E., On the Definition and Application of the Sensitivity Function, *J. Franklin Inst.* **285**, 26–36 (1968).

[59] KREINDLER, E., Closed-Loop Sensitivity Reduction of Linear Optimal Control Systems, *IEEE Trans. Autom. Contr.* **13**, 254–262 (1968).

[60] KREINDLER, E., Formulation of the Minimum Trajectory Sensitivity Problem, *IEEE Trans. Autom. Contr.* **14**, 206–207 (1969).

[61] KREISSELMEIER, G., and GRÜBEL, G., The Design of Optimally Parameter Insensitive Control Systems, *IFAC Cong., Paris, France, 1972.*

[62] KREISSELMEIER, G., Empfindlichkeitseigenschaften beim Riccati–Entwurf, Lehrgang OR 3 Lineare Optimale Regelung, Carl Cranz Gesellschaft, 1975, pp. 1–13.

[63] LEE, C.–K., and CHEN, C.–T., Sensitivity Comparisons of Various Analogue Computer Simulations, *Int. J. Contr.* **10**, 227–233 (1969).

[64] MALEY, C. E., The Effect of Parameters on the Roots of an Equation System, *Comput. J.* **4**, 62–63 (1963).

[65] MANTEY, P. E., Eigenvalue Sensitivity and State Variable Selection, *IEEE Trans. Autom. Contr.* **13**, (1968).

[66] MAZER, W. M., Specification of the Linear Feedback Sensitivity Function, *IRE Trans. Autom. Control* 85–93 (1960).

[67] MC CLAMROCH, N. H., CLARK, L. G., and AGGARWAL, J. K., Sensitivity of Linear Control System to Large Parameter Variations, *IFAC Symp. System Sensitivity Adaptivity, Dubrovnik, Yugoslama, 1968,* pp. B8–B17.

[68] MILLER, K. S., and MURRAY, F. J., The Mathematical Basis for the Error Analysis of Differential Analyzers, *J. Math. Phys. (Cambridge, Mass.),* (1953).

[69] MILLER, K. S., and MURRAY, F. J., Error Analysis for Differential Analyzers, *WADC TR* 54–252 (1954).

[70] MORGAN, B. S., Jr., Sensitivity Analysis and Synthesis of Multivariable Systems, *IEEE Trans. Autom. Contr.* **11**, 506–512 (1966).

[71] MÜNZNER, W., Ein Beitrag zur Synthese parameterunempfindlicher Abtastsysteme mit endlicher Einstellzeit, Dissertation. Univ. of Karlsruhe, 1973.

[72] MÜNZNER, W., FRANK, P. M., and NOGES, E., Synthesis of Parameter-Insensitive Sampled-Data Systems According to Dead-Beat Criterion Assuming Several Parameter Uncertainties, *IFAC Symp. Discrete Syst. Riga, 1974.*

[73] NEUMANN, D. P., and SOOD, A. K., Sensitivity Models of Canonical Form Delay-Differential Systems, *IEEE Trans. Autom. Contr.* **16,** 365–366 (1971).

[74] NUGUYEN THUONG NGO, Sensitivity of Automatic Control Systems (Survey), *Autom. Remote Control (USSR)* **32,** 735–762 (1971).

[75] PAGUREK, B., Sensitivity of the Performance of Optimal Linear Control Systems to Parameter Variations (Short Paper), *IEEE Trans. Autom. Contr.* **10,** 178–180 (1965).

[76] PALLOD, R., and WOMACK, B. F., Allowable Range of Plant Parameter Variation to the Terminal Margin in Closed Loop Optimal Control Systems. Univ. of Texas, Austin, Rep. TM5 AFOSR-69-0348 TR AD-685722, 1969.

[77] PERKINS, W. R., and CRUZ, J. B., Sensitivity Operators for Linear Time-Varying Systems, *Proc. Int. Symp. Sensitivity, Dubrovnik, Yugoslavia, 1964,* pp. 67–77. Pergamon Press, Oxford, 1966.

[78] PERKINS, W. R., and CRUZ, J. B., The Parameter Variation Problem in State Feedback Control Systems, *Trans. ASME* **87,** 120–124 (1965).

[79] PERKINS, W. R., Sensitivity Models for Discrete-Time Linear Systems, *2nd, Proc. IFAC Symp.* on *Multivariable Technical Control Systems, Düsseldorf, 1971*, pp. 1–9, Paper 1.5.3.

[80] PERKINS, W. R., CRUZ, J. B., and KOKOTOVIC, P. V., The Sensitivity Techniques for Adaptive Control, *Proc. Int. Conf. Cybern. Soc., Washington, D. C., October 1972,* pp. 396–400.

[81] PERLIS, H. J., On the Residue of a Sensitivity Function, *IEEE Trans. Autom. Contr.* **10,** 496–497 (1965).

[82] REDDY, D. C., Eigenfunction Sensitivity and the Parameter Variation Problem, *Int. J. Contr.* **9,** 561–568 (1969).

[83] RILLING, J. H., and ROY, J. R., Analog Sensitivity Design of Saturn V Launch Vehicle, *IEEE Trans. Autom. Contr.* **15,** 437–442 (1970).

[84] RISSANEN, J., Performance Deterioration of Optimum Systems, *IEEE Trans. Autom. Contr.* **12,** 530–532 (1967).

[85] ROBERTS, J. D., Special Problems in the Synthesis of Sensitivity Networks, *Proc. Int. Symp. Sensitivity, Dubrovnik, Yugoslavia, 1964,* pp. 176–194. Pergamon Press, Oxford, 1966.

[86] ROHRER, R. A., and SOBRAL, M., Sensitivity Considerations in Optimal System Design, *IEEE Trans. Autom. Contr.* 43–48 (1965).

[87] ROSENBROCK, H. H., Sensitivity of an Eigenvalue to Changes in the Matrix, *Electr. Let. Inst. Electr. Eng.* **1**, 278 (1965).

[88] ROSKA, T., Summed-Sensitivity Invariants and Their Generation, *Electron. Lett. Inst. Electr. Eng.* **4**, 281–282 (1968).

[89] ROZENVASSER, E. N., General Sensitivity Equations of Discontinuous Systems, *Autom. Remote Control (USSR)* **28**, 400–404 (1967).

[90] ROZENVASSER, E. N., and YUSUPOV, R. M., Sensitivity Equations of Pulse Control Systems, *Autom. Remote Control* **30**, 526–536 (1969).

[91] RUTMAN, R. S., The Method of Three Points in Sensitivity Theory, *Eng. Cybern.* (USSR) 131–141 (1968).

[92] SARMA, V. V. S., and SINGH, S. N., Comparison of Open- and Closed-Loop Sensitivities for Systems with Stochastic Inputs, *IEEE Trans. Autom. Contr.* **15**, 253–254 (1970).

[93] SCHMIDT, G., Parameterempfindlichkeit linearer Regelsysteme Theoretische Grundlagen und spezielle Untersuchungen, Dissertation. Univ. of Darmstadt, 1966.

[94] SCHMIDT, G., Parameterempfindlichkeit von Regelkreisen, msr 7, **3**, 101–106 (1964).

[95] SCHÖNFELD, R., Die näherungsweise Berechnung von Empfindlichkeitsfunktionen dynamischer Systeme auf einem hybriden Analogrechner, *Wiss. Z. Tech. Hochsch. Ilmenau* **5**, 25–32 (1970).

[96] SIBUL, L. H., Sensitivity Analysis of Linear Control Systems with Random Plant Parameters, *IEEE Trans. Autom. Contr.* **15**, 459–462 (1970).

[97] SINGER, R. A., Selecting State Variables to Minimize Eigenvalue Sensitivity of Multivariable Systems, *Automatica,* **5**, 85–93 (1969).

[98] SOBRAL, M., Jr., Sensitivity in Optimal Control Systems, *Proc. IEEE* **56**, 1644–1652 (1968).

[99] SUNDARARAJAN, N., and CRUZ, J. B., Trajectory Insensitivity of Optimal Feedback Systems, *IEEE Trans. Autom. Contr.* **15**, 663–665 (1970).

[100] TOMOVIC, R., and BEKEY, G. A., Adaptive Sampling Based on Amplitude Sensitivity, *IEEE Trans. Autom. Contr.* **11**, 282–284 (1966).

[101] TSYPKIN, Y. Z., and RUTMAN, R. S., Sensitivity Equations for Discontinuous Systems, *Proc. Int. Symp. Sensitivity, Dubrovnik, Yugoslavia, 1964,* pp. 195–196. Pergamon Press, Oxford, 1966.

[102] UUSPÄÄ, P. T., Modelling and Parameter Sensitivity of Pulse Frequency Modulated Control Systems, Ph. D. Dissertation. University of Washington, Seattle, 1973.

[103] VAN NESS, J. E., BOYLE, J. M., and IMAD, F. P., Sensitivity of Large, Multiple-Loop Control Systems, *IEEE Trans. Autom. Contr.* **10**, 308–315 (1965).

[104] VASILEVA, A. B., On the differentiation of the Solution of Differential Equations Containing a Small Parameter, *Dokl. Akad. Nauk SSSR* **11**, (1948).

[105] VENGEROV, A. A., ROSHANSKII, V. L., and ULANOV, G. M., Estimation of Sensitivity of Integral Performance Criteria of Systems of Variable Structure, *Autom. Remote Control,* **32**, 245–250 (1971).

[106] VUSKOVIC, M., and CIRCIC, V., Structural Rules for the Determination of Sensitivity Functions of Nonlinear Nonstationary Systems, *Proc. Int. Symp. Sensitivity, Dubrovnik, Yugoslavaia, 1964,* pp. 154–165. Pergamon Press, Oxford, 1966.

[107] WILKIE, D. F., and PERKINS, W. R., Generation of Sensitivity Functions for Linear Systems Using Low-Order Models, *IEEE Trans Autom. Contr.* **14**, 123–129 (1969).

[108] WILKIE, D. F., and PERKINS, W. R., Essential Parameters in Sensitivity Analysis, *Automation (Cleveland)* **5**, 191–197 (1969).

[109] WITSENHAUSEN, H. S., On the Sensitivity of Optimal Control System *IEEE Trans. Autom. Contr.* **10**, (1965).

[110] YOULA, D. C., and DORATO, P., On the Comparison of the Sensitivities of Open-Loop and Closed-Loop Optimal Control Systems, *IEEE Trans. Autom. Contr.* **13,** 186–188 (1968).

Master's and Diploma Theses

[111] LITTY, E. D., Sensitivity Functions of a Type I Pulse Frequency Modulated Control System. Dept. Electr. Eng., University of Washington, Seattle, 1975.

[112] DILLMANN, R., Berechnung der Trajektorienempfindlichkeitsfunktionen eines Lageregelkreises mit PFM-Regler vom Type II Institut für Regelungs- und Steuerungssysteme, Univ. of Karlsruhe, Karlsruhe, 1976.

[113] THIEME, G., Untersuchung der Parameterempfindlichkeit optimaler Systeme anhand von Beispielen. Institut für Regelungs- und Steuerungssysteme, Univ. of Karlsruhe, Karlsruhe, 1976.

Solutions to Problems

CHAPTER 2

2.1 Rewriting the equations in terms of partial derivatives, we have

$$\frac{\partial(\zeta_1\zeta_2)}{\partial\alpha} = \zeta_1\frac{\partial\zeta_2}{\partial\alpha} + \zeta_2\frac{\partial\zeta_1}{\partial\alpha}, \qquad \text{from} \quad (2.4\text{-}2)$$

$$\frac{\partial(\zeta_1/\zeta_2)}{\partial\alpha} = \frac{\zeta_2\dfrac{\partial\zeta_1}{\partial\alpha} - \zeta_1\dfrac{\partial\zeta_2}{\partial\alpha}}{\zeta_2^2} = \frac{1}{\zeta_2}\frac{\partial\zeta_1}{\partial\alpha} - \frac{\zeta_1}{\zeta_2^2}\frac{\partial\zeta_2}{\partial\alpha}, \qquad \text{from} \quad (2.4\text{-}3)$$

$$\frac{\partial\zeta[\beta(\alpha)]}{\partial\alpha} = \frac{\partial\zeta}{\partial\beta}\frac{\partial\beta}{\partial\alpha}. \qquad \text{from} \quad (2.4\text{-}4)$$

These relationships hold by virtue of the rules of differentiation. In an analogous manner it can be shown that

$$\frac{\partial(\zeta_1\zeta_2)}{\partial\alpha}\frac{\alpha}{\zeta_1\zeta_2} = \frac{\alpha}{\zeta_2}\frac{\partial\zeta_2}{\partial\alpha} + \frac{\alpha}{\zeta_1}\frac{\partial\zeta_1}{\partial\alpha},$$

$$\frac{\partial(\zeta_1/\zeta_2)}{\partial\alpha}\frac{\alpha}{\zeta_1/\zeta_2} = \frac{\alpha}{\zeta_1}\frac{\partial\zeta_1}{\partial\alpha} - \frac{\alpha}{\zeta_2}\frac{\partial\zeta_2}{\partial\alpha},$$

$$\frac{\partial\zeta[\beta(\alpha)]}{\partial\alpha}\frac{\alpha}{\zeta} = \left(\frac{\beta}{\zeta}\frac{\partial\zeta}{\partial\beta}\right)\left(\frac{\alpha}{\beta}\frac{\partial\beta}{\partial\alpha}\right).$$

2.2 By the application of the rules of differentiation,

$$\tilde{S}_{\alpha}^{\zeta_1\zeta_2} = \zeta_{10}\tilde{S}_{\alpha}^{\zeta_2} + \zeta_{20}\tilde{S}_{\alpha}^{\zeta_1}, \qquad \tilde{S}_{\alpha}^{\zeta_1/\zeta_2} = \frac{1}{\zeta_{20}}\tilde{S}_{\alpha}^{\zeta_1} - \frac{\zeta_{10}}{\zeta_{20}^2}\tilde{S}_{\alpha}^{\zeta_2},$$

$$\tilde{S}_{\alpha}^{\zeta} = \frac{1}{\beta_0}\tilde{S}_{\beta}^{\zeta}\tilde{S}_{\alpha}^{\beta}, \qquad \underset{\sim}{S}_{\alpha}^{\zeta_1\zeta_2} = \underset{\sim}{S}_{\alpha}^{\zeta_1} + \underset{\sim}{S}_{\alpha}^{\zeta_2},$$

$$\underset{\sim}{S}_{\alpha}^{\zeta_1/\zeta_2} = \underset{\sim}{S}_{\alpha}^{\zeta_1} - \underset{\sim}{S}_{\alpha}^{\zeta_2}, \qquad \underset{\sim}{S}_{\alpha}^{\zeta} = \beta_0\underset{\sim}{S}_{\beta}^{\zeta}\underset{\sim}{S}_{\alpha}^{\beta}.$$

2.3 Denoting the derivative with respect to t by a dot, we have

(a) $\dot{f} = (2x + z)\dot{x} + 4y\dot{y} + x\dot{z},$

(b) $\dfrac{df}{dx} = 2x + z + 4y\dfrac{dy}{dx} + x\dfrac{dz}{dx},$

(c) $\dfrac{\partial f}{\partial x} = 2x + z + x\dfrac{\partial z}{\partial x}.$

2.4 (a) $S_K{}^y = S_{\alpha 1}^y + R_0 C S_{\alpha 1}^y, \qquad S_R{}^y = K_0 C S_{\alpha 2}^y,$

$\Delta y = S_{\alpha_1}^y \,\Delta K + S_{\alpha 1}^y (R_0 C\,\Delta K + K_0 C\,\Delta R),$

(b) $S_K{}^y = S_{\alpha 1}^y + R_0 C_0 S_{\alpha 2}^y, \qquad S_R{}^y = K_0 C_0 S_{\alpha 2}^y, \qquad S_C{}^y = K_0 R_0 S_{\alpha 2}^y,$

$\Delta y = S_{\alpha_1}^y \,\Delta K + S_{\alpha 2}^y (R_0 C_0\,\Delta K + K_0 C_0\,\Delta R + K_0 R_0\,\Delta C).$

2.5 (a) $S_\nu{}^T = \pi\alpha\sqrt{l_0/g},$

(b) $\Delta t = 4.32$ sec.

2.6 By application of the chain rule [according to Eq. (2.4-20)],

$$\frac{\partial \dot{x}}{\partial \alpha} = \frac{\partial \boldsymbol{f}}{\partial \boldsymbol{x}}\frac{\partial \boldsymbol{x}}{\partial \alpha} + \frac{\partial \boldsymbol{f}}{\partial \alpha}.$$

2.7 By application of the chain rule and Definition 2.4-1 we obtain

$$\frac{\partial f}{\partial y^{(n)}}\frac{\partial y^{(n)}}{\partial \boldsymbol{\alpha}} + \cdots + \frac{\partial f}{\partial \dot{y}}\frac{\partial \dot{y}}{\partial \boldsymbol{\alpha}} + \frac{\partial f}{\partial y}\frac{\partial y}{\partial \boldsymbol{\alpha}} + \frac{\partial f}{\partial \boldsymbol{\alpha}} = \mathbf{0}.$$

2.8 (a) Since $\partial I/\partial\alpha = c/\cos^2\alpha$, we obtain

$$\frac{\Delta I}{I_0} = \frac{c\alpha_0}{\cos^2\alpha_0\, c\tan\alpha_0}\frac{\Delta\alpha}{\alpha_0} = \frac{2\alpha_0}{\sin 2\alpha_0}\frac{\Delta\alpha}{\alpha_0}, \qquad 0 \le \alpha_0 \le \frac{\pi}{2},$$

or

$$\frac{\Delta\alpha}{\alpha_0} = \frac{\sin 2\alpha_0}{2\alpha_0}\frac{\Delta I}{I_0} = \bar{S}_I{}^\alpha \frac{\Delta I}{I_0}, \qquad 0 \le \alpha_0 \le \frac{\pi}{2}.$$

From this we see that the sensitivity $\bar{S}_I{}^\alpha$ reaches its maximum when $\alpha_0 = 0$.

(b) In a similar way to that above we obtain

$$\frac{\Delta I}{I_0} = \frac{2}{\sin 2\alpha_0}\Delta\alpha = \underset{\sim}{S}_I{}^\alpha\,\Delta\alpha, \qquad 0 \le \alpha_0 \le \frac{\pi}{2}.$$

From this we see that the maximum accuracy for a given error of deflection is obtained when $\alpha_0 = 4/\pi$ or 45°.

2.9 (a) $\dfrac{\Delta u}{u_0} = \dfrac{1}{2}\dfrac{\Delta p}{p_0}, \qquad$ (b) $\underset{\sim}{S}_u{}^p = \dfrac{2}{u_0}, \; u_0 = 10$ V.

CHAPTER 3

3.1 According to Eq. (3.3-18), we have

$$S_\alpha{}^G = S_\alpha{}^N - S_\alpha{}^D,$$

where $N = G_1 + \alpha G_2$ and $D = G_3 + \alpha G_4$. Thus

$$S_\alpha^{\ G} = \frac{\alpha G_2}{G_1 + \alpha G_2}\bigg|_{\alpha_0} - \frac{\alpha G_4}{G_3 + \alpha G_4}\bigg|_{\alpha_0} = \frac{\alpha(G_2 G_3 - G_1 G_4)}{(G_1 + \alpha G_2)(G_3 + \alpha G_4)}\bigg|_{\alpha_0}.$$

3.2 Differentiation of the output $y = 1 - \cos \omega t$ with respect to ω yields $\sigma = t \sin \omega_0 t$. The envelope is t which diverges with time.

3.3 The solution of the differential equation is

$$y(t, \omega) = 1 - \frac{1}{\sqrt{1 - \delta^2}} e^{-\delta \omega t} \sin(\omega \sqrt{1 - \delta^2}\, t + \cos^{-1} \delta).$$

Taking the partial derivative with respect to ω and setting $\omega = \omega_0$, we obtain

$$\sigma(t, \omega_0) = \frac{te^{-\delta \omega_0 t}}{\sqrt{1 - \delta^2}} [\delta \sin(\omega_0 \sqrt{1 - \delta^2}\, t + \cos^{-1} \delta) - \cos(\omega_0 \sqrt{1 - \delta^2}\, t + \cos^{-1} \delta)].$$

As we see, $\sigma(t, \omega_0)$ diverges for $\delta \leq 0$, i.e., if the system is either unstable or executes sustained oscillations.

3.4 Equation (3.3-7) has to be evaluated. With G denoting the transfer function of the closed loop, we have $S_S^{\ G} = 1/(1 + RS)$ and $S_T^S = -T_0 s/(1 + T_0 s)$. Hence

$$\frac{\Delta y}{y_0} = -\frac{T_0 s}{1 + T_0 s + KR(s)} \frac{\Delta T}{T_0}.$$

3.5 For a_2 as the parameter of interest, one has $G_1 = a_1(s + a_3)$, $G_2 = 0$, $G_3 = s^3 + a_4 s^2 + a_1 s + a_1 a_3$, $G_4 = s(s + a_4)$. Hence

$$S_{a2}^G = -\left[1 + \frac{s^3 + a_{40} s^2 + a_{10} s + a_{10} a_{20}}{a_{20} s(s + a_{40})}\right]^{-1}.$$

For a_3 as the parameter of interest, one has $G_1 = a_1 s$, $G_2 = a_1$, $G_3 = s^3 + (a_2 + a_4)s^2 + (a_1 + a_2 a_4)s$, $G_4 = a_1$. Consequently,

$$S_{a3}^G = \frac{a_{30}}{s + a_{30}} \left[1 + \frac{a_{10}(s + a_{30})}{s^3 + (a_{20} + a_{40})s^2 + a_{20} a_{40} s}\right]^{-1}.$$

3.6 The transfer function of the network is given by $G(s, C) = (1 + CRs)/(2 + CRs)$. Comparison with Eq. (3.3-19) reveals that $G_1 = 1$, $G_2 = Rs$, $G_3 = 2$, $G_4 = Rs$, $\alpha = C$. Hence according to Eq. (3.3-20),

$$S_C^G = \frac{C_0 Rs}{(2 + C_0 Rs)(1 + C_0 Rs)}.$$

3.7 (i) The Bode sensitivity functions are

(a) $$S_{G2}^G = \frac{1}{1 + KG_1 G_{20} G_3},$$

$$\text{(b)}\quad S_{G_2}^{G} = \frac{1 + G_1 G_4}{1 + G_1 G_4 + G_1 G_{20} G_3},$$

$$\text{(c)}\quad S_{G_2}^{G} = \frac{1 - G_1 G_3 G_4}{1 - G_1 G_3 G_4 + G_1 G_{20} G_4}.$$

(ii) The expressions for comparison sensitivity functions are the same as above except G_{20} is replaced by the actual plant transfer function G_2.

(iii) (a) $K = \infty$, (b) $G_1 G_4 = -1$, (c) $G_1 G_3 G_4 = 1$.

3.8 Determination of h:

$$\left(\frac{Y}{U}\right)_{\min} = \frac{K/2}{1 + h/2} K_2, \qquad \left(\frac{Y}{U}\right)_{\max} = \frac{2K}{1 + 2h} K_2,$$

$$\frac{(Y/U)_{\max} - (Y/U)_{\min}}{(Y/U)_{\min}} \overset{!}{=} 0.2.$$

Evaluation for h yields $h = 7$.

Determination of K_2 from the requirement of equivalence for $K_1 = K$:

$$K \overset{!}{=} K_2 \frac{K}{1 + h}.$$

From this, with $h = 7$, we obtain $K_2 = 8$.

3.9

$$S_{h11}^{K} = -\frac{h_{11}(h_{22} + Y_S)}{\Delta}, \qquad S_{h21}^{K} = \frac{(h_{11} + Z_S)(h_{22} + Y_L)}{\Delta},$$

$$S_{h12}^{K} = \frac{h_{12} h_{21}}{\Delta}, \qquad S_{h22}^{K} = -\frac{h_{22}(h_{11} + Z_S)}{\Delta},$$

where $\Delta = (h_{11} + Z_S)(h_{22} + Y_L) - h_{12} h_{21}$, $Y_S = 1/Z_S$, $Y_L = 1/Z_L$.

3.10 (a) The actual solution is $u_2(t, R) = 1/RC \exp[-(t/RC)]$. Differentiating with respect to R and setting $R = R_0$ yields

$$\sigma(t, R_0) = \frac{1}{R_0^2 C} e^{-t/(R_0 C)} \left(-1 + \frac{t}{R_0 C}\right),$$

(see Fig. S-1a).

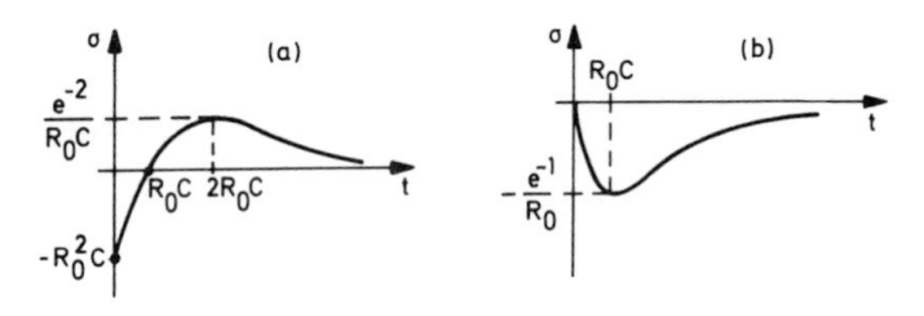

FIG. S-1. (a) Graph of σ in case (a), (b) graph of σ in case (b).

(b) The actual solution is $u_2(t, R) = -\exp[-(t/RC)] + 1$. Differentiation with respect to R and setting $R = R_0$ yields

$$\sigma(t, R_0) = -\frac{t}{R_0^2 C} e^{-t/(R_0 C)},$$

(see Fig. S-1b).

Discussion: R affects only the transient of u_2 since $\sigma(t \to \infty) \to 0$. In case (a), the sensitivity is zero at $t = T_0 = R_0 C$, whereas in case (b), the sensitivity is maximum at T_0.

3.11 The output variables are

$$y_1(t, K, M) = \frac{5}{\sqrt{K/M}} \sin\sqrt{\frac{K}{M}}\, t, \qquad y_2(t, K, M) = 5 \cos\sqrt{\frac{K}{M}}\, t.$$

The output sensitivity matrix is given by

$$\boldsymbol{\sigma} = \begin{bmatrix} \dfrac{\partial y_1}{\partial K} & \dfrac{\partial y_1}{\partial M} \\[2ex] \dfrac{\partial y_2}{\partial K} & \dfrac{\partial y_2}{\partial M} \end{bmatrix}_{K=K_0, M=M_0} = \begin{bmatrix} \sigma_{11} & \sigma_{12} \\ \sigma_{21} & \sigma_{22} \end{bmatrix},$$

where

$$\sigma_{11} = \frac{5}{2K_0}\left(-\sqrt{\frac{M_0}{K_0}} \sin\sqrt{\frac{K_0}{M_0}}\, t + t \cos\sqrt{\frac{K_0}{M_0}}\, t\right),$$

$$\sigma_{12} = \frac{5}{2\sqrt{M_0}}\left(\frac{1}{\sqrt{K_0}} \sin\sqrt{\frac{K_0}{M_0}}\, t - \frac{t}{\sqrt{M_0}} \cos\sqrt{\frac{K_0}{M_0}}\, t\right),$$

$$\sigma_{21} = -\frac{5t}{2\sqrt{K_0 M_0}} \sin\sqrt{\frac{K_0}{M_0}}\, t, \qquad \sigma_{22} = \frac{5\sqrt{K_0}\, t}{2M_0\sqrt{M_0}} \sin\sqrt{\frac{K_0}{M_0}}\, t,$$

3.12 The bridge voltage is $u_{12} = \frac{1}{2}(R_1 - R)/(R_1 + R)\, u$. Therefore,

(a) $\sigma(R_{10}) = \dfrac{R}{(R_{10} + R)^2}\, u,$ (b) $\Delta u_{12}(R_1) = \dfrac{R}{(R_{10} + R)^2}\, U \Delta R_1,$

(c) $\tilde{\sigma}(R_{10}) = \dfrac{R_{10} R}{(R_{10} + R)^2}\, u,$ (d) $\bar{\sigma}(R_{10}) = \dfrac{2R_{10} R}{R_{10}^2 - R^2}$

(e) $\dfrac{\partial \sigma}{\partial R} \stackrel{!}{=} 0,$ whence $R = R_{10}$.

3.13 Determining the time functions of the state variable and taking the partial derivatives yields

$$\boldsymbol{\lambda} = \begin{bmatrix} \lambda_{11} & \lambda_{12} \\ \lambda_{21} & \lambda_{22} \end{bmatrix} = \begin{bmatrix} \dfrac{\partial x_1}{\partial K} & \dfrac{\partial x_1}{\partial T} \\[2ex] \dfrac{\partial x_2}{\partial K} & \dfrac{\partial x_2}{\partial T} \end{bmatrix}_{K_0, T_0}$$

with

$$\lambda_{11} = 1 - A\,[d \sin at + aT_0 \cos at],$$
$$\lambda_{12} = -K_0(t/T_0)\,A \sin at,$$
$$\lambda_{21} = 0,$$
$$\lambda_{22} = \frac{t}{T_0^{\,2}}\,A\,\{(aT_0 - 2d)\sin at - d \cos at\},$$

where

$$A = \frac{\exp(-td/T_0)}{\sqrt{1-d^2}} \qquad \text{and} \qquad a = \frac{\sqrt{1-d^2}}{T_0}.$$

3.14 (a) The differential equation is

$$\ddot{y}(t) = -\alpha_3 y(t), \qquad y(0) = -\alpha_2, \qquad \dot{y}(0) = \alpha_1.$$

It has the solution

$$y(t,\boldsymbol{\alpha}) = \frac{\alpha_1}{\sqrt{\alpha_3}} \sin \sqrt{\alpha_3}\,t - \alpha_2 \cos \sqrt{\alpha_3}\,t.$$

(b) Partial differentiation of $y(t, \boldsymbol{\alpha})$ with respect to α_1, α_2, α_3 and going over to nominal values $\boldsymbol{\alpha} = [\alpha_{10}\ \alpha_{20}\ \alpha_{30}]^{\mathrm{T}}$ yields

$$\boldsymbol{\sigma}_1(t,\boldsymbol{\alpha}_0) = \left[\left.\frac{\partial y}{\partial \alpha_1}\right|_{\boldsymbol{\alpha}_0}\ \left.\frac{\partial y}{\partial \alpha_2}\right|_{\boldsymbol{\alpha}_0}\ \left.\frac{\partial y}{\partial \alpha_3}\right|_{\boldsymbol{\alpha}_0}\right]^{\mathrm{T}},$$

where

$$\left.\frac{\partial y}{\partial \alpha_1}\right|_{\boldsymbol{\alpha}_0} = \frac{1}{\sqrt{\alpha_{30}}} \sin \sqrt{\alpha_{30}}\,t, \qquad \left.\frac{\partial y}{\partial \alpha_2}\right|_{\boldsymbol{\alpha}_0} = -\cos\sqrt{\alpha_{30}}\,t,$$

$$\left.\frac{\partial y}{\partial \alpha_3}\right|_{\boldsymbol{\alpha}_0} = \frac{\alpha_{20}t - \alpha_{10}/\alpha_{30}}{2\sqrt{\alpha_{30}}} \sin\sqrt{\alpha_{30}}\,t + \frac{\alpha_{10}t}{2\alpha_{30}} \cos\sqrt{\alpha_{30}}\,t$$

3.15 The roots follow from $\alpha s^4 + a_2 s^2 = 0$. They are $s_1 = +j\sqrt{a_2/\alpha}$ and $s_2 = -j\sqrt{a_2/\alpha}$. Thus the root sensitivity becomes infinite.

3.16 The sensitivity measure is $m = 3.46\ \Delta R/R_0$.

3.17 The sensitivity measure is $m = \sqrt{2.5}\ \Delta a_1/a_{10}$.

CHAPTER 4

4.1 Taking the partial derivatives with respect to a and letting a approach a_0 yields, with $\sigma \triangleq (\partial y/\partial a)_{a_0}$ and $y_0 \triangleq y(t, a_0)$,

(a) $t^2\ddot{\sigma} + t\dot{\sigma} + (t^2 - a_0^{\,2})\sigma = 2a_0 y_0$, $\sigma(0) = \dot{\sigma}(0) = 0$,

(b) $t(1-t)\ddot{\sigma} + [y_0 - (a_0 + b + 1)t]\dot{\sigma} - (\dot{y}_0 + a_0 b)\sigma = t\dot{y}_0 + by_0$, $\sigma(0) = \dot{\sigma}(0) = 0$,

(c) $\ddot{\sigma} + 3t\dot{y}_0^2\dot{\sigma} - a_0 e^{y_0}\sigma = y_0 e^{a_0 y_0}$, $\sigma(0) = \dot{\sigma}(0) = 0$,
(d) $\dot{\sigma} - (2a_0 y_0 + bt)\sigma = y_0^2$, $\sigma(0) = 0$,
(e) $2\dot{y}_0\dot{\sigma} - 2a^2{}_0 y_0\sigma = 2a_0 y_0^2$, $\sigma(0) = 0$,
(f) $\ddot{\sigma} + a_0 y_0\dot{\sigma} + (a_0\dot{y}_0 + b)\sigma = y_0\dot{y}_0$, $\sigma(0) = \dot{\sigma}(0) = 0$,
(g) $(2y_0 + cy_0^2 + ca_0)\dot{\sigma} + (cy_0 - bt)\sigma = -c\dot{y}_0$, $\sigma(0) = 0$,
(h) $(2\dot{y}_0 + bt)\dot{\sigma} - \sigma = 0$, $\sigma(0) = 1$.

4.2 Application of Eq. (4.3-36) yields
(a) $\ddot{\nu} - (1 - y_0^2)\dot{\nu} + (1 + 2y_0\dot{y}_0)\nu = 2(1 - y_0^2)\dot{\sigma} - 4y_0\dot{y}_0\sigma$
$-4y_0\sigma\dot{\sigma} - 2\dot{y}_0\sigma^2$, $\nu(0) = 0$, $\dot{\nu}(0) = 0$,
(b) $\ddot{\nu} + \nu = 2(1 - y_0^2)\dot{\sigma} - 4y_0\dot{y}_0\sigma$, $\nu(0) = 0$, $\dot{\nu}(0) = 0$

4.3 (a) The sensitivity equations with respect to a_1 and $y(0)$, where $\sigma_1 = (\partial y/\partial a_1)_{a10}$ and $\sigma = \partial y/\partial y(0)_{y0(0)}$ are as follows:

$$a_{10}a_{20}\ddot{\sigma}_1 + a_{10}a_{30}\dot{\sigma}_1 + a_{30}\sigma_1 = -a_{20}y_0 - a_{30}y_0, \qquad \sigma(0) = \dot{\sigma}(0), = 0$$

$$a_{10}a_{20}\ddot{\sigma} + a_{10}a_{30}\dot{\sigma}_1 + a_{30}\sigma_1 = 0, \qquad \sigma(0) = 1, \quad \dot{\sigma}(0) = 0.$$

(b) The signal flow graph to measure σ_1 is shown in Fig. S-2,

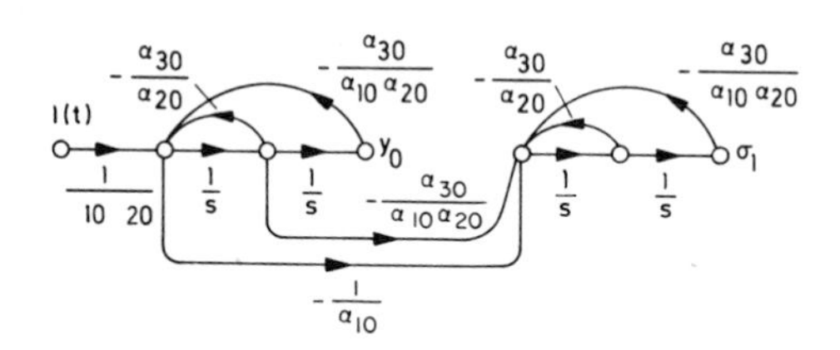

Fig. S-2. Signal flow graph to measure σ_1 according to the sensitivity equation.

(c) Σ_1 follows from (a) by Laplace transform or from (b) by block diagram manipulation:

$$\Sigma_1 = -\frac{a_{20}s + a_{30}}{(a_{10}a_{20}s^2 + a_{10}a_{30}s + a_{30})^2}.$$

4.4 The differential equation is $\ddot{y} = g$ with the initial conditions $y(0) = \beta_0$, $\dot{y}(0) = \beta_1$. The corresponding sensitivity equation becomes $\ddot{\sigma} = 0$. The initial conditions are as follows:
(a) with respect to m: $\sigma(0) = 0$, $\dot{\sigma}(0) = 0$,
(b) with respect to β_0: $\sigma(0) = 1$, $\dot{\sigma}(0) = 0$,
(c) with respect to β_1: $\sigma(0) = 0$, $\dot{\sigma}(0) = 1$.
Hence solving the sensitivity equation with the above initial conditions yields

$$\left.\frac{\partial y}{\partial m}\right|_{m_0} = 0, \qquad \left.\frac{\partial y}{\partial \beta_0}\right|_{\beta 00} = 1, \qquad \left.\frac{\partial y}{\partial \beta_1}\right|_{\beta 10} = t - t_0.$$

4.5 From the results of Problem 3.15, $s_1 = j\sqrt{a_2/\alpha}$ and $s_2 = -j\sqrt{a_2/\alpha}$, it can be seen that sustained oscillations will occur since the real parts of s_1 and s_2 are zero. Therefore the application of the Laplace transform method is not allowed.

4.6 Determining the sensitivity equation and solving it by Laplace transform yields the desired sensitivity function:

$$\sigma(t, a_{40}) = -b + \left[b - \frac{a_3 b}{a_1{}^2} + \frac{b(a_3 + a_1 a_2 + a_1{}^2)}{a_1{}^3} t\right] e^{-t/a_1}, \qquad t > 0.$$

4.7 Employing the methods of Section 3.6.4, we obtain

$$\Sigma_1 = -\frac{1}{(1 + Ts)^2}, \qquad s < \infty, \qquad \text{or} \qquad \sigma_1 = -\frac{1}{T^2} te^{-t/T}, \qquad t > 0.$$

4.8 With the aid of the sensitivity equation, the following solution is obtained:

$$\sigma_\varepsilon = \begin{cases} 0 \quad \text{for} \quad t \leq t_1, \\ \dfrac{0.0446}{\text{sec}} \left[\exp\left(\dfrac{t_1 - t}{10 \text{ sec}}\right) - \exp\left(\dfrac{t_1 - t}{5 \text{ sec}}\right)\right] - \dfrac{0.00446}{\text{sec}^2} (t - t_1) \exp\left(\dfrac{t_1 - t}{10 \text{ sec}}\right) \\ \qquad \text{for} \quad t \geq t_1 \end{cases}$$

4.9 The resulting block diagram to measure σ is shown in Fig. S-3.

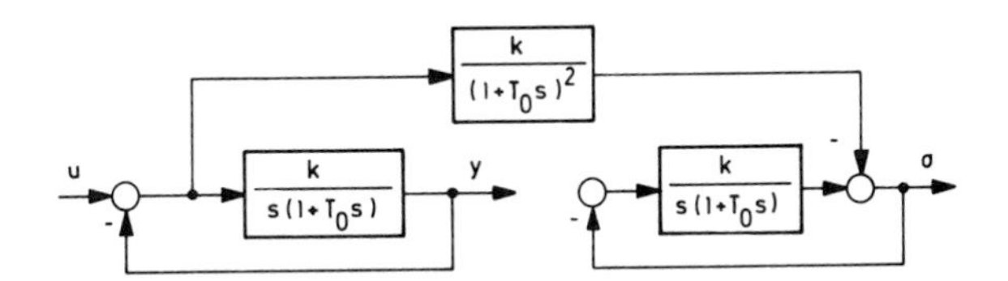

FIG. S-3. Block diagram for the measurement of σ.

4.10 The block diagram to generate σ according to the variable component method is shown in Fig. S-4.

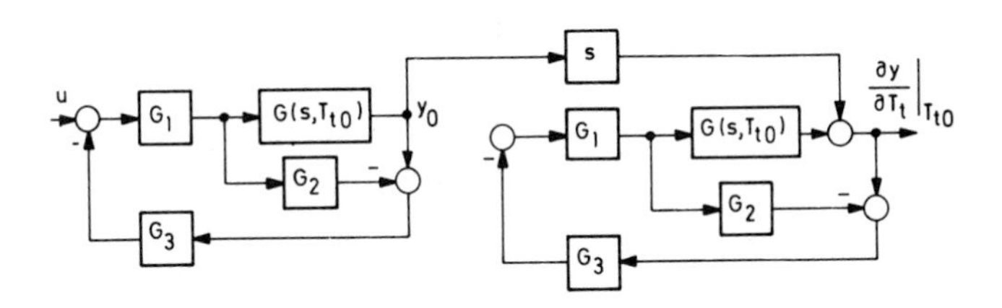

FIG. S-4. Block diagram of Problem 4.10.

4.11 The block diagram is shown in Fig. S-5. From this diagram the analytical expression for $\mathbf{\Sigma} = \mathscr{L}\{\sigma\}$ is found as

$$\mathbf{\Sigma} = -\frac{\mathrm{b}(1 + a_2 s + a_3 s^2)}{s(1 + a_1 s)^2}, \qquad \sigma(0) = 0,$$

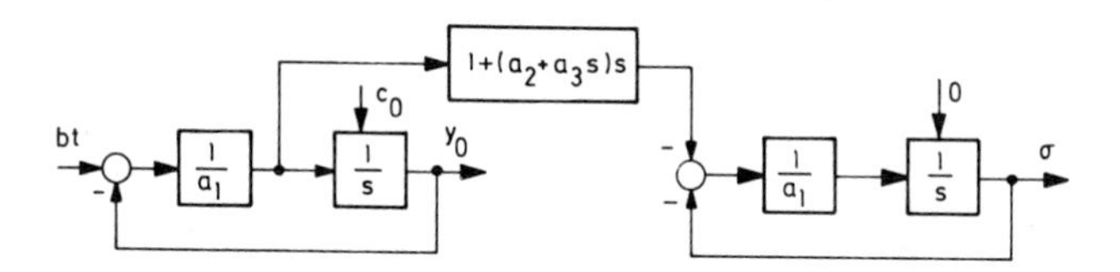

FIG. S-5. Block diagram of Problem 4.11.

which leads to the same result as that obtained in Problem 4.6.

4.12 The signal flow diagram resulting from the variable component method, after having set $a_{10} = 0$, is shown in Fig. S-6. From this diagram it follows that $\mathbf{\Sigma} = \mathscr{L}\{\sigma\} = -1/(s^2 + 4)^2$ and from this by means of the Laplace transform,

$$\sigma(t,0) = -\frac{1}{16}(\sin 2t - 2t\cos 2t).$$

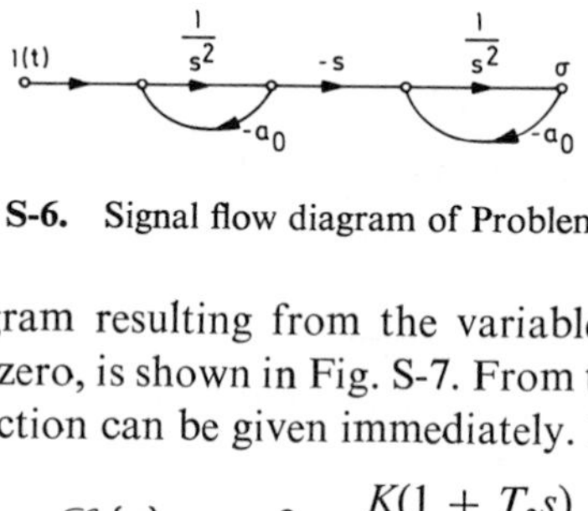

FIG. S-6. Signal flow diagram of Problem 4.12.

4.13 The block diagram resulting from the variable component method, after T_1 is set equal to zero, is shown in Fig. S-7. From this diagram the transformed sensitivity function can be given immediately. It is

$$\mathbf{\Sigma} = \mathscr{L}\{\sigma\} = -3\frac{K(1 + T_2 s)}{((1 + K) + T_2 s)^2}.$$

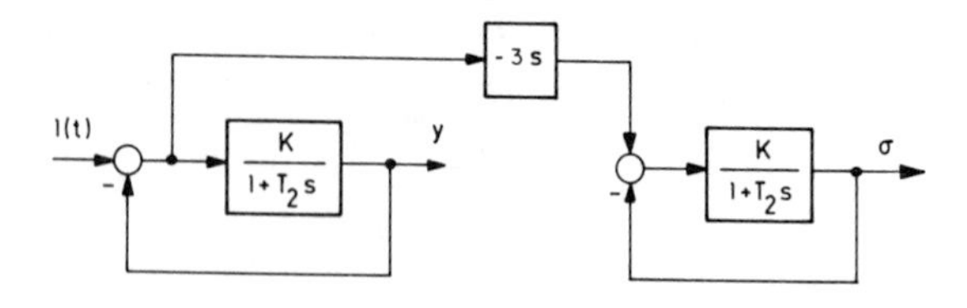

FIG. S-7. Block diagram for the generation of σ in Problem 4.13.

Applying the inverse Laplace transform yields

$$\sigma(t,0) = \left[\frac{3k^2}{T_2^{\,2}}\,t - \frac{3K}{T_2}\right] \exp\left(-\frac{1+K}{T_2}\,t\right).$$

4.14 (a) The sensitivity equation is obtained, by partial differentiation with respect to a, as $\ddot{\sigma} + (a_0 + b_0 \cos \omega_0 t)\,\sigma = -y_0$.

(b) The desired block diagram is shown in Fig. S-8.

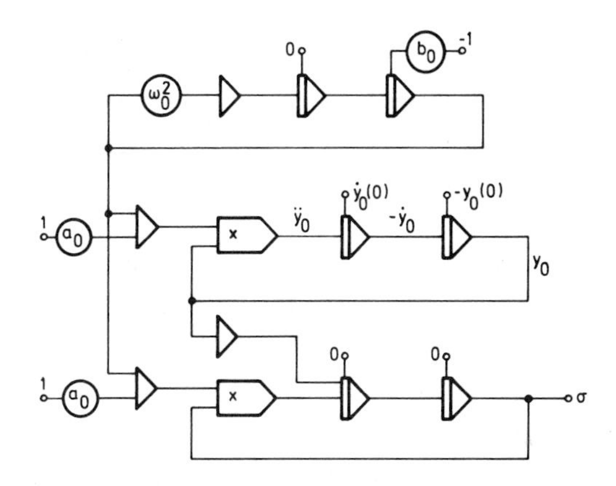

FIG. S-8. Analog computer diagram to measure σ in Problem 4.14.

4.15 (a) The block diagram to measure $\sigma_\omega = (\partial y/\partial\omega)_{\omega 0}$ is given in Fig. S-9.

(b) The resulting block diagram is shown in Fig. S-10.

4.16 The resulting block diagram is given in Fig. S-11.

4.17 The resulting block diagram is shown in Fig. S-12.

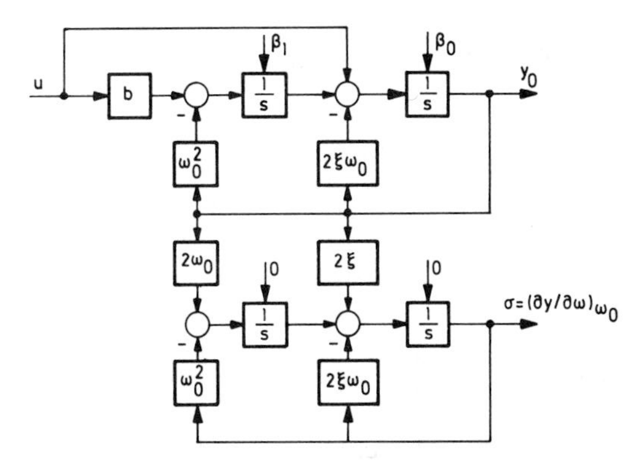

FIG. S-9. Block diagram of Problem 4.15 (a).

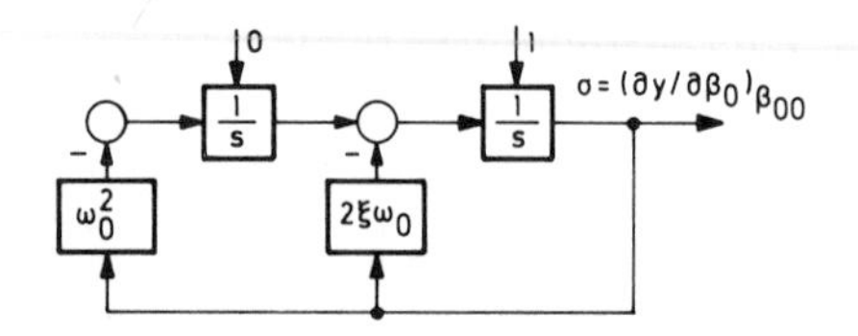

FIG. S-10. Block diagram of Problem 4.15 (b).

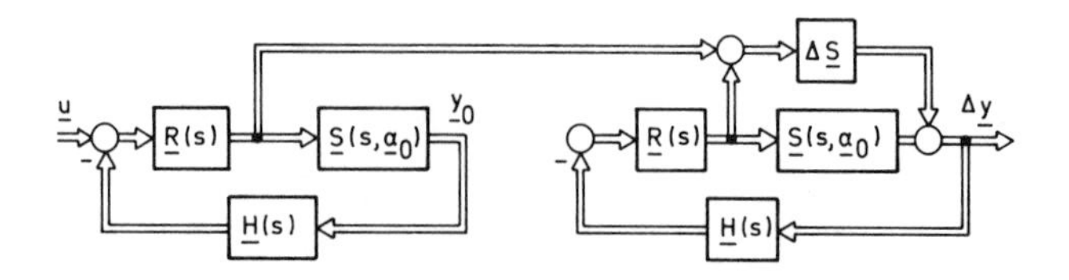

FIG. S-11. Resulting block diagram of Problem 4.16.

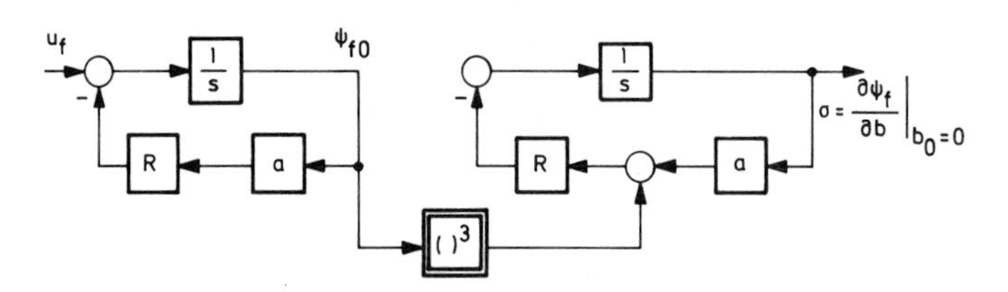

FIG. S-12. Block diagram to measure the sensitivity of ψ_f with respect to b in the dc motor of Problem 4.17.

4.18 By application of the variable component method, the block diagram of Fig. S-13 is achieved after L_a has been set equal to zero.

4.19 The solution of Problem 4.19 is shown in Fig. S-14.

4.20 (a) Taking the partial derivative with respect to a, b, c, respectively, the following sensitivity equations are obtained:

$$\ddot{\sigma}_a + \left[a_0 + 2b_0(\operatorname{sgn}\dot{\varphi}_0)\dot{\varphi}_0 + b_0\frac{\partial\operatorname{sgn}\dot{\varphi}_0}{\partial\dot{\varphi}_0}\dot{\varphi}_0{}^2\right]\dot{\sigma}_a + c_0\cos\varphi_0\sigma_a = -\dot{\varphi}_0,$$

$$\ddot{\sigma}_b + \left[a_0 + 2b_0(\operatorname{sgn}\dot{\varphi}_0)\dot{\varphi}_0 + b_0\dot{\varphi}_0{}^2\frac{\partial\operatorname{sgn}\dot{\varphi}_0}{\partial\dot{\varphi}_0}\right]\dot{\sigma}_b + c_0\cos\varphi_0\sigma_b = -(\operatorname{sgn}\dot{\varphi}_0)\dot{\varphi}_0{}^2,$$

$$\ddot{\sigma}_c + \left[a_0 + 2b_0(\operatorname{sgn}\dot{\varphi}_0)\dot{\varphi}_0 + b_0\dot{\varphi}_0{}^2\frac{\partial\operatorname{sgn}\dot{\varphi}_0}{\partial\dot{\varphi}_0}\right]\dot{\sigma}_c + c_0\cos\varphi_0\sigma_c = -\sin\varphi_0.$$

(b) The desired analog computer diagrams can be drawn immediately from the sensitivity equations of (a). The result is shown in Fig. S-15. For contact positions a,b, and c, the corresponding sensitivity functions $(\partial\varphi/\partial a)_{a_0}$, $(\partial\varphi/\partial b)_{b_0}$, and $(\partial\varphi/\partial c)_{c_0}$ are obtained, respectively, at the point marked by

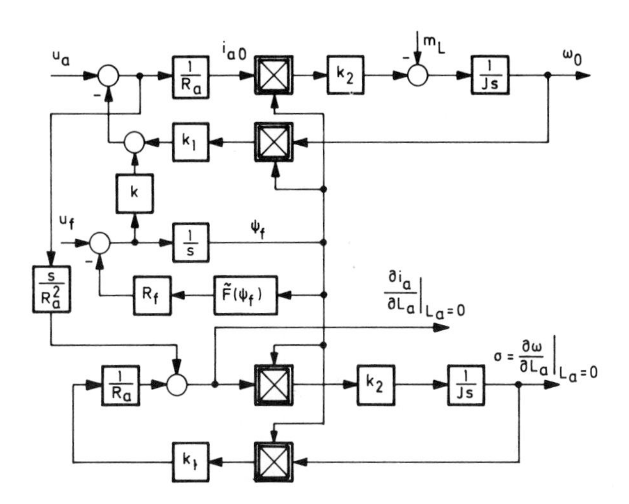

FIG. S-13. Block diagram to measure the sensitivity of the current and the angular velocity of a dc motor with respect to the neglected inductance of the armature circuit (Problem 4.18).

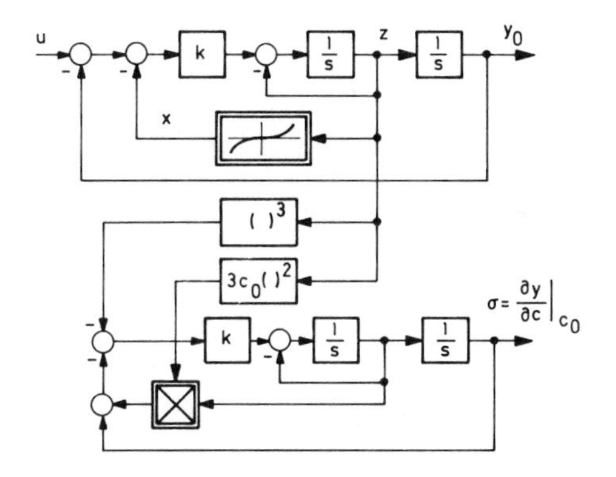

FIG. S-14. Block diagram to measure $\sigma = (\partial y/\partial c)_{c_0}$ in Problem 4.19.

σ. Note that $\delta(\dot{\varphi}_0) = \partial \operatorname{sgn} \dot{\varphi}_0/\partial\dot{\varphi}_0$ is the Dirac δ-function at $\dot{\varphi}_0$, that is, $\delta(\dot{\varphi}_0) = 0$ for $\dot{\varphi}_0 \lessgtr 0$ and $\delta(\dot{\varphi}_0) = \infty$ for $\dot{\varphi}_0 = 0$.

4.21 According to the method of sensitivity points, the analog computer diagram of Fig. S-16 is obtained. Note that all parameter settings are at nominal values.

4.22 The original block diagram first has to be brought into the form of Fig. 4.6-3, which is achieved by simple block diagram manipulations (Fig. S-17). In this diagram the sensitivity point is defined, according to Fig. 4.6-3, as indicated in Fig. S-17.

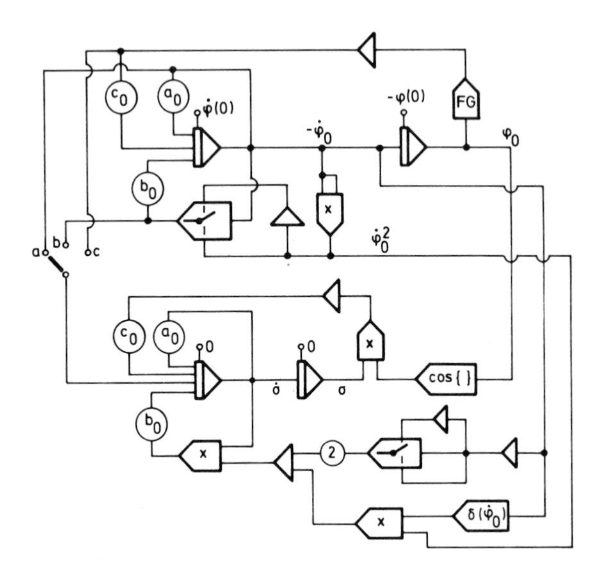

FIG. S-15. Resulting analog computer diagram of Problem 4.20.

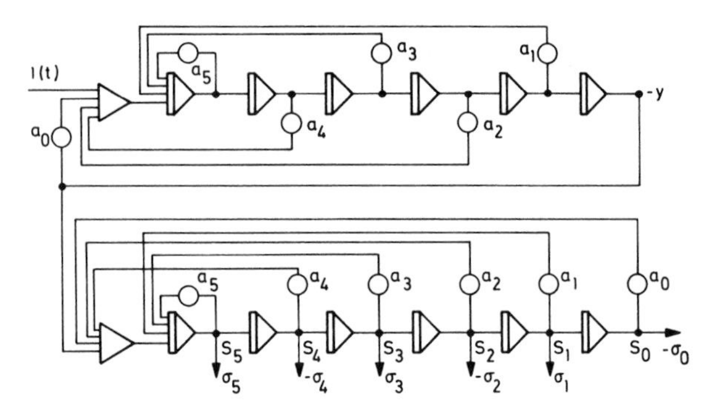

FIG. S-16. Analog computer diagram to measure simultaneously all output sensitivity functions σ_ν, $\nu = 0, 1, \ldots, 5$ (Problem 4.21).

4.23 By the application of the method of sensitivity points, we obtain the desired diagram as shown in Fig. S-18.

4.24 The desired signal flow diagram obtained by the method of sensitivity points is shown in Fig. S-19.

4.25 The diagram has first to be brought into the form of Fig. 4.6-3 in which the sensitivity point P_i with respect to K is defined. This form, as well as the resulting diagram for the calculation of the output sensitivity function $\sigma = (\partial y/\partial K)_{K_0}$, is shown in Fig. S-20.

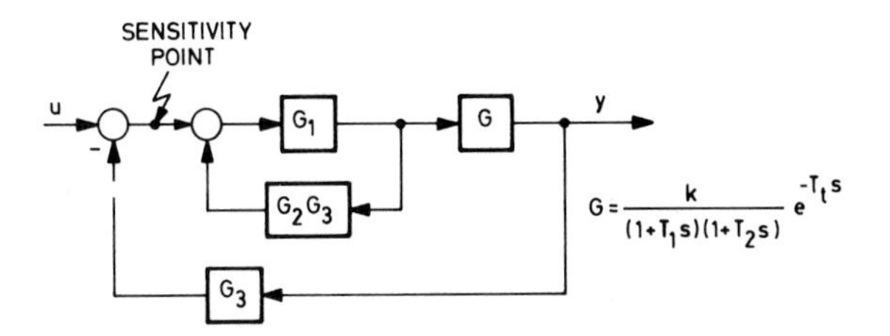

FIG. S-17. Block diagram including the sensitivity point with respect to T_t of Problem 4.22.

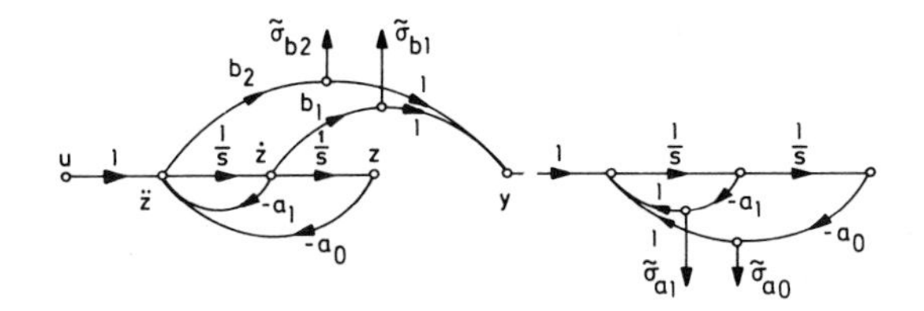

FIG. S-18. Resulting signal flow graph of Problem 4.23.

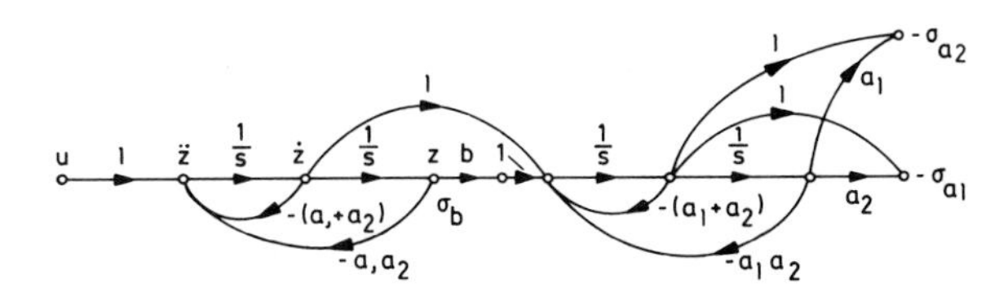

FIG. S-19 Signal flow diagram of Problem 4.24.

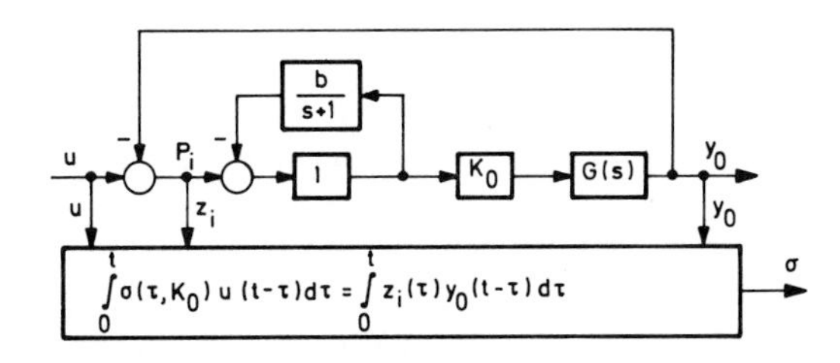

FIG. S-20. Structural diagram to calculate $\sigma = (\partial y/\partial K)_{K_0}$ of the feedback control system according to the three-point method (Problem 4.25).

4.26 The signal flow graph for the simultaneous measurement of all output sensitivity functions is shown in Fig. S-21.

4.27 For the branch of the variable component we have

$$\frac{\partial u_c}{\partial C} = \frac{1}{C}\int \frac{\partial i_c}{\partial C}\,dt - \frac{1}{C}\,u_c.$$

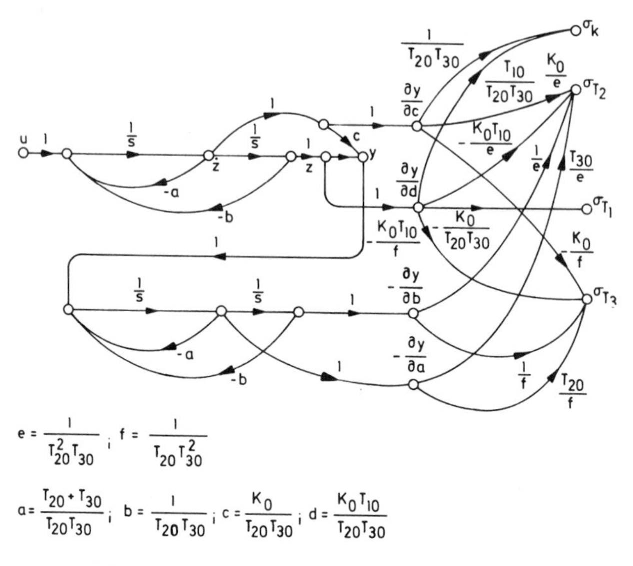

FIG. S-21. Signal flow graph of Problem 4.26.

This implies that the resulting network to measure $(\partial u_2/\partial C)_{C_0}$ is of the form shown in Fig. S-22.

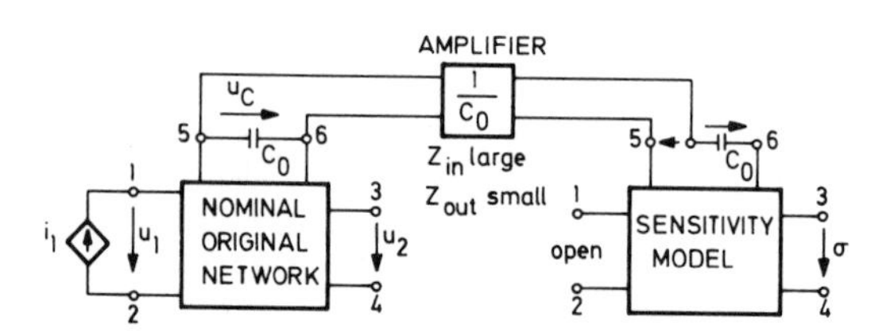

FIG. S-22. Measuring circuit of Problem 4.27.

4.28 The resulting network is shown in Fig. S-23.

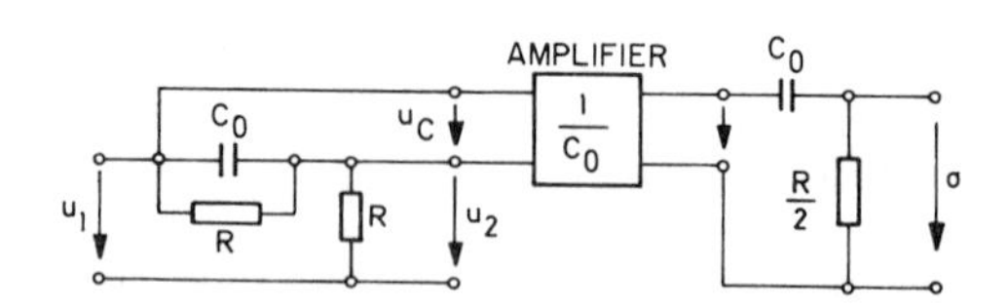

FIG. S-23. Resulting network of Problem 4.28.

4.29 The normalized networks N and $\hat{N}$ are shown in Fig. S-24. The currents i_ν and $\hat{i}_\nu$ found by network analyses of N and $\hat{N}$ are indicated. By multiply-

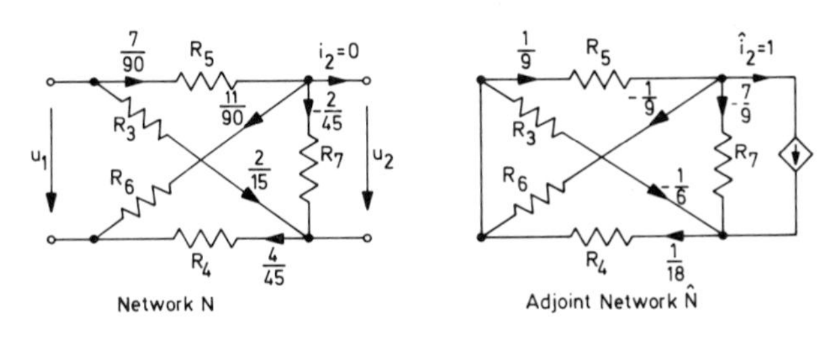

FIG. S-24. Network and adjoint network of Problem 4.29.

ing the corresponding currents i_ν and $\hat{i}_\nu$ and normalization by R_ν/u_2, the following results are obtained

$$\bar{\sigma}_3 = \frac{R_3}{u_2}\, i_3\hat{i}_3 = -\frac{50}{(-5)}\frac{2}{15}\left(-\frac{1}{6}\right) = -\frac{2}{9},$$

$$\bar{\sigma}_4 = -\frac{R_4}{u_2}\, i_4\hat{i}_4 = -\frac{150}{(-5)}\frac{4}{45}\frac{1}{18} = \frac{4}{27},$$

$$\bar{\sigma}_5 = -\frac{R_5}{u_2}\, i_5\hat{i}_5 = -\frac{100}{(-5)}\frac{7}{90}\frac{1}{9} = \frac{14}{81},$$

$$\bar{\sigma}_6 = -\frac{R_6}{u_2}\, i_6\hat{i}_6 = -\frac{100}{(-5)}\frac{1}{90}\left(-\frac{1}{9}\right) = -\frac{22}{81},$$

$$\bar{\sigma}_7 = -\frac{R_7}{u_2}\, i_7\hat{i}_7 = -\frac{25}{(-5)}\left(-\frac{2}{45}\right)\left(-\frac{7}{9}\right) = \frac{14}{81}.$$

The sum of all relative sensitivity functions is indeed

$$\sum_{\nu=3}^{7} \bar{\sigma}_\nu = -\frac{2}{9} + \frac{4}{27} + \frac{14}{81} - \frac{22}{81} + \frac{14}{81} = 0.$$

CHAPTER 5

5.1 The required trajectory sensitivity vectors are

$$\left.\frac{\partial \boldsymbol{x}}{\partial m}\right|_{m_0} = \begin{bmatrix} 0 \\ 0 \end{bmatrix}, \qquad \left.\frac{\partial \boldsymbol{x}}{\partial \beta_1}\right|_{\beta_0} = \begin{bmatrix} 1 \\ 0 \end{bmatrix}, \qquad \left.\frac{\partial \boldsymbol{x}}{\partial \beta_2}\right|_{\beta_0} = \begin{bmatrix} t - t_2 \\ 1 \end{bmatrix}$$

5.2 Using the definitions

$$\lambda_1 = \left.\frac{\partial \theta}{\partial J_F}\right|_{J_{F0}}, \qquad \eta = \left.\frac{\partial \omega}{\partial J_F}\right|_{J_{F0}}, \qquad \delta = \left.\frac{\partial \Omega}{\partial J_F}\right|_{J_{F0}},$$

the required trajectory sensitivity equations are

$$\dot{\lambda}_1 = \eta, \qquad \dot{\eta} = \frac{1}{J_{F0}}\left[D + \frac{K_1^{\,2}}{R_a}\right]\delta + \frac{1}{J_F^{\,2}}\left[D + \frac{K_1^{\,2}}{R_a}\right]\Omega_0 - \frac{K_1}{J_{F0}^2 R_a}\, v_a,$$

$$\dot{\delta} = -\left[\frac{1}{J_{F_0}} + \frac{1}{J_M}\right]\left[D + \frac{K_1^{\,2}}{R_a}\right]\delta + \frac{1}{J_{F_0}^2}\left[D + \frac{K_1^{\,2}}{R_a}\right]\Omega_0 - \frac{Q_1}{R_a J_{F_0}^2}\, v_a.$$

The corresponding block diagram is shown in Fig. S-25.

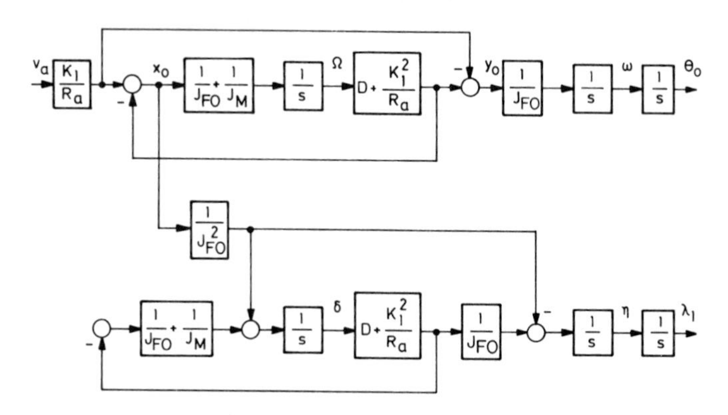

FIG. S-25. Block diagram to measure λ_1 of Problem 5.2.

5.3 Denoting $\boldsymbol{\lambda} = (\partial\boldsymbol{x}/\partial\boldsymbol{a})_{\boldsymbol{a}_0}$ and $\boldsymbol{\eta} = (\partial\boldsymbol{i}/\partial\boldsymbol{a})_{\boldsymbol{a}_0}$, we obtain

$$\dot{\boldsymbol{\lambda}} = \boldsymbol{A}_0\boldsymbol{\lambda} + \left.\frac{\partial\boldsymbol{A}}{\partial\boldsymbol{a}}\right|_{\boldsymbol{a}_0}\boldsymbol{x}_0 + \left.\frac{\partial\boldsymbol{B}}{\partial\boldsymbol{a}}\right|_{\boldsymbol{a}_0}\boldsymbol{v}, \qquad \boldsymbol{\eta} = \boldsymbol{C}_0\boldsymbol{\lambda} + \left.\frac{\partial\boldsymbol{C}}{\partial\boldsymbol{a}}\right|_{\boldsymbol{a}_0}\boldsymbol{x}_0 + \left.\frac{\partial\boldsymbol{D}}{\partial\boldsymbol{a}}\right|_{\boldsymbol{a}_0}\boldsymbol{v},$$

where, due to Eqs. (5.2-12) and (5.2-13),

$$\left.\frac{\partial\boldsymbol{A}}{\partial\boldsymbol{a}}\right|_{\boldsymbol{a}_0}\boldsymbol{x}_0 = \begin{bmatrix} \frac{1}{R_{10}^2C_{10}}x_{10} & \frac{1}{R_{10}C_{10}^2}x_{10} & 0 & 0 \\ 0 & 0 & \frac{1}{R_{20}^2C_{20}}x_{20} & \frac{1}{R_{20}C_{20}^2}x_{20} \end{bmatrix},$$

$$\left.\frac{\partial\boldsymbol{B}}{\partial\boldsymbol{a}}\right|_{\boldsymbol{a}_0}\boldsymbol{v} = \begin{bmatrix} -\frac{1}{R_{10}^2C_{20}}v_1 & -\frac{1}{R_{10}C_{20}^2}v_1 & 0 & 0 \\ 0 & 0 & -\frac{1}{R_{20}^2C_{20}}(v_1 - v_2) & -\frac{1}{R_{20}C_{20}^2}(v_1 - v_2) \end{bmatrix}$$

$$\left.\frac{\partial\boldsymbol{C}}{\partial\boldsymbol{a}}\right|_{\boldsymbol{a}_0}\boldsymbol{x}_0 = \begin{bmatrix} -\frac{1}{R_{10}^2}x_{10} & 0 & \frac{1}{R_{20}^2}x_{20} & 0 \\ 0 & 0 & -\frac{1}{R_{20}^2}x_{20} & 0 \end{bmatrix},$$

$$\left.\frac{\partial\boldsymbol{D}}{\partial\boldsymbol{a}}\right|_{\boldsymbol{a}_0}\boldsymbol{v} = \begin{bmatrix} -\frac{1}{R_{10}^2}v_1 & 0 & -\frac{1}{R_{20}^2}(v_1 - v_2) & 0 \\ 0 & 0 & \frac{1}{R_{20}^2}(v_1 - v_2) & 0 \end{bmatrix}.$$

5.4 (a) The reduced block diagram is shown in Fig. S-26.

(b) The state equations for the nonreduced system are

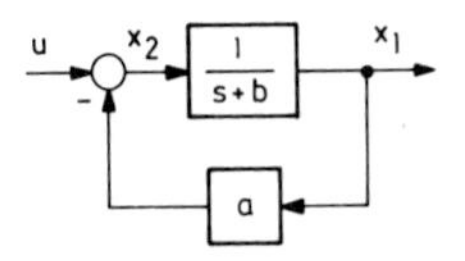

FIG. S-26. Reduced block diagram of the system of Problem 5.4.

$$\dot{x}_1 = -bx_1 + x_2, \qquad x_1(0) = 0,$$
$$\alpha\dot{x}_2 = -x_2 + x_3, \qquad x_2(0) = 0,$$
$$\alpha\dot{x}_3 = -ax_1 - x_3 + u, \qquad x_3(0) = 0.$$

The state equation for the reduced system is

$$\dot{x}_{10} = -(a + b)x_{10} + u.$$

(c) The trajectory sensitivity equation is

$$\dot{\lambda}_1 = -(a + b)\lambda_1 - \dot{x}_{20} - \dot{x}_{30}, \qquad \lambda_1(0) = -2.$$

(d) The solution of the above equation (using the variable component method) yields

$$\lambda_1 = 2(at - 1)\, e^{-(a+b)t}, \qquad t > 0,$$
$$\eta_1 = \dot{\lambda}_1 + b\lambda_1, \qquad \eta_2 = -a\lambda_1 - \dot{x}_{30}, \qquad t > 0.$$

5.5 (a) The desired trajectory sensitivity equation is

$$\begin{bmatrix} \dot{\lambda}_1 \\ \dot{\lambda}_2 \end{bmatrix} = \begin{bmatrix} 0 & 1 \\ 2\alpha_0 x_{10} x_{20} - 1 & -1 + \alpha_0 x_{10}^2 \end{bmatrix} \begin{bmatrix} \lambda_1 \\ \lambda_2 \end{bmatrix} + \begin{bmatrix} 0 \\ x_{10}^2 x_{20} \end{bmatrix}, \qquad \begin{bmatrix} \lambda_1(0) \\ \lambda_2(0) \end{bmatrix} = \mathbf{0}.$$

(b) A possible diagram to measure $\lambda_1 = (\partial x_1/\partial\alpha)_{\alpha_0}$ is shown in Fig. S-27.

(c) The desired sensitivity equation is

$$\begin{bmatrix} \dot{\lambda}_1 \\ \dot{\lambda}_2 \end{bmatrix} = \begin{bmatrix} 0 & 1 \\ 2\alpha_0 x_{10} x_{20} - 1 & -1 + \alpha_0 x_{10}^2 \end{bmatrix} \begin{bmatrix} \lambda_1 \\ \lambda_2 \end{bmatrix}, \qquad \begin{bmatrix} \lambda_1(0) \\ \lambda_2(0) \end{bmatrix} = \begin{bmatrix} 1 \\ 0 \end{bmatrix}.$$

5.6 The required sensitivity equation is

$$T\dot{\lambda} + \lambda = \frac{1}{T^2} \exp\left(-\frac{t}{U}\right), \qquad \lambda(0) = -\frac{1}{T^2}, \qquad t > 0.$$

5.7 The sensitivity equation is $\dot{\lambda} = -\lambda - e^{-t}$, $\lambda(0) = 3$. Its solution is $\lambda(t) = (3 - t)\exp(-t)$, $t > 0$.

5.8 (a) The desired sensitivity equations are

$$\dot{\boldsymbol{\lambda}}_j = (\boldsymbol{A} - \boldsymbol{bK})_{\boldsymbol{\alpha}_0} \boldsymbol{\lambda}_j + \left[\frac{\partial \boldsymbol{A}}{\partial \alpha_j} - \frac{\partial \boldsymbol{b}}{\partial \alpha_j} \boldsymbol{K}\right]_{\boldsymbol{\alpha}_0} \boldsymbol{x}_0 + \left.\frac{\partial \boldsymbol{b}}{\partial \alpha_j}\right|_{\boldsymbol{\alpha}_0} r, \qquad \boldsymbol{\lambda}(0) = \mathbf{0},$$

$$\sigma_j = (\boldsymbol{C} - d\boldsymbol{K})_{\boldsymbol{\alpha}_0} \boldsymbol{\lambda}_j + \left[\frac{\partial \boldsymbol{C}}{\partial \alpha_j} - \frac{\partial d}{\partial \alpha_j} \boldsymbol{K}\right]_{\boldsymbol{\alpha}_0} \boldsymbol{x}_0 + \left.\frac{\partial d}{\partial \alpha_j}\right|_{\boldsymbol{\alpha}_0} r.$$

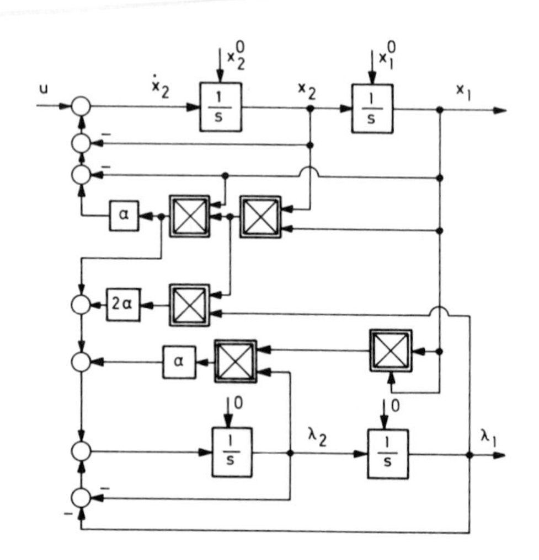

FIG. S-27. Block diagram to measure λ_1 of Problem 5.5.

(b) The graphical representation of the above equations is shown in Fig. S-28.

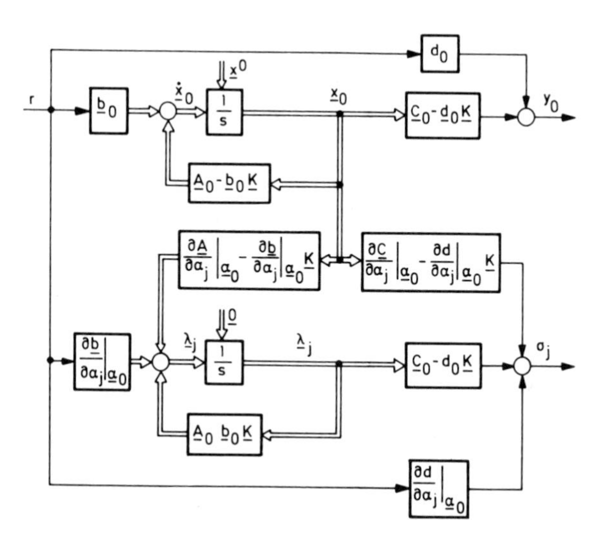

FIG. S-28. Graphical interpretation of the sensitivity equations of Problem 5.8.

5.9 According to Theorem 5.6-1, we have $\xi_{12} = \xi_{21}$, $\xi_{13} = \xi_{22} = \xi_{31}$, $\xi_{14} = \xi_{23} = \xi_{32} = \xi_{41}$, $\xi_{24} = \xi_{33} = \xi_{42}$, $\xi_{34} = \xi_{43}$. Furthermore, by virtue of Theorem 5.6-2 and Eq. (5.6-31),

$$\frac{\partial z_k}{\partial \alpha_4} = -z_{k-1} - \sum_{i=1}^{6-k} \alpha_i \frac{\partial z_{i+k-2}}{\partial \alpha_1} - \sum_{i=7-k}^{4} \alpha_i \frac{\partial z_{i-5+k}}{\partial \alpha_4}, \qquad k = 3, 4.$$

The resulting diagram is shown in Fig. S-29.

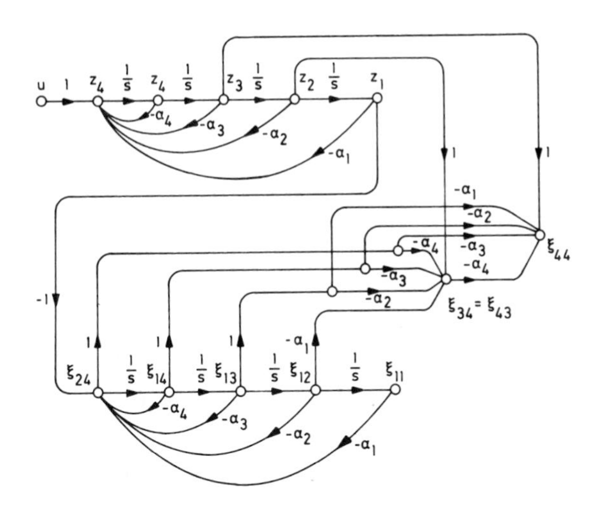

FIG. S-29. Signal flow diagram of Problem 5.9.

5.10 The nominal eigenvalues are $\lambda_{10} = -1$, $\lambda_{20} = -2$, $\lambda_{30} = -3$. Using Eq. (5.8-20), we obtain the semirelative eigenvalue sensitivities

$$\left.\frac{\partial \lambda_1}{\partial \ln \alpha_1}\right|_{\boldsymbol{\alpha}_0} = -3, \qquad \left.\frac{\partial \lambda_1}{\partial \ln \alpha_2}\right|_{\boldsymbol{\alpha}_0} = \frac{11}{2}, \qquad \left.\frac{\partial \lambda_1}{\partial \ln \alpha_3}\right|_{\boldsymbol{\alpha}_0} = -3,$$

$$\left.\frac{\partial \lambda_2}{\partial \ln \alpha_1}\right|_{\boldsymbol{\alpha}_0} = 6, \qquad \left.\frac{\partial \lambda_2}{\partial \ln \alpha_2}\right|_{\boldsymbol{\alpha}_0} = -22, \qquad \left.\frac{\partial \lambda_2}{\partial \ln \alpha_3}\right|_{\boldsymbol{\alpha}_0} = 24,$$

$$\left.\frac{\partial \lambda_3}{\partial \ln \alpha_1}\right|_{\boldsymbol{\alpha}_0} = -3, \qquad \left.\frac{\partial \lambda_3}{\partial \ln \alpha_2}\right|_{\boldsymbol{\alpha}_0} = \frac{33}{2}, \qquad \left.\frac{\partial \lambda_3}{\partial \ln \alpha_1}\right|_{\boldsymbol{\alpha}_0} = -27.$$

With these results the summed semirelative eigenvalue sensitivities become

$$\tilde{S}_A^{\lambda_1} = \frac{\alpha_{10} + |\lambda_{10}\alpha_{20}| + |\lambda_{10}^2\alpha_{30}|}{|(\lambda_{10} - \lambda_{20})(\lambda_{10} - \lambda_{30})|} = \frac{23}{2}$$

and accordingly, $\tilde{S}_A^{\lambda_2} = 52$, $\tilde{S}_A^{\lambda_3} = 93/2$. This indicates that the most sensitive eigenvalue is λ_2.

5.11 (a) The desired trajectory sensitivity equation within $t_i < t < t_i + \gamma$ and $t_i + \gamma < t < t_{i+1}$ $(i = 0, 1, \ldots)$ is

$$\dot{\boldsymbol{\lambda}}_\alpha = \begin{bmatrix} 0 & 1 & 0 \\ 0 & 0 & 0 \\ -K_{10} & -K_{20} & -c \end{bmatrix} \boldsymbol{\lambda}_\alpha + \boldsymbol{E}\boldsymbol{x}_0, \qquad \boldsymbol{\lambda}_{\alpha'}(t_0) = \mathbf{0},$$

where α stands for K_1 or K_2 and

$$\boldsymbol{E} = \begin{bmatrix} 0 & 0 & 0 \\ 0 & 0 & 0 \\ -1 & 0 & 0 \end{bmatrix} \qquad \boldsymbol{E} = \begin{bmatrix} 0 & 0 & 0 \\ 0 & 0 & 0 \\ 0 & -1 & 0 \end{bmatrix}$$

if K_1 is the parameter, if K_2 is the parameter.

(b) Expressions for dt_i/dK_1 and dt_i/dK_2:

Case (1) $$\frac{dt_i}{dK_1} = -\frac{1}{\dot{p}(t_i^-)}\frac{\partial p(t_i^-)}{\partial K_1}, \qquad \frac{dt_i}{dK_2} = -\frac{1}{\dot{p}(t_i^-)}\frac{\partial p(t_i^-)}{\partial K_2},$$

Case (2) $$\frac{dt_i}{dK_1} = \frac{dt_{i-1}}{dK_1}, \qquad \frac{dt_i}{dK_2} = \frac{dt_{i-1}}{dK_2}.$$

(c) The jump conditions at $t = t_i$ are

$$\Delta\boldsymbol{\lambda} = \left\{\begin{bmatrix} 0 & 0 & 0 \\ 0 & 0 & 0 \\ K_{10} & K_{20} & c \end{bmatrix} \boldsymbol{x}_0(t_i^+) - \begin{bmatrix} 0 \\ 0 \\ 1 \end{bmatrix} w(t_i) - \begin{bmatrix} 0 \\ K_0 \\ 0 \end{bmatrix} u(t_i^+)\right\} \frac{dt_i}{d\alpha}\bigg|_{\alpha_0}$$
$$- \begin{bmatrix} 0 & 0 & 0 \\ 0 & 0 & 0 \\ 0 & 0 & 1 \end{bmatrix} \boldsymbol{\lambda}(t_i^-),$$

where α can stand for K_1 or K_2. At $t = t_i + \gamma$,

$$\Delta\boldsymbol{\lambda} = \begin{bmatrix} 0 \\ K \\ 0 \end{bmatrix} M \operatorname{sgn}[p(t_i^-)] \frac{dt_i}{d\alpha}\bigg|_{\alpha_0},$$

where α again stands for K_1 or K_2.

5.12 (a) The desired trajectory sensitivity equations in the intervals $t_i < t < t_i + \gamma$ and $t_i + \gamma < t < t_{i+1}$ $(i = 0, 1, \ldots)$ are

$$\dot{\boldsymbol{\lambda}}_K = \begin{bmatrix} 0 & 1 & 0 & 0 \\ 0 & 0 & 1 & 0 \\ 0 & -a_{10}a_{20} & -(a_{10} + a_{20}) & 0 \\ -K_1 & -K_2 & -K_3 & -c \end{bmatrix} \boldsymbol{\lambda}_k + \begin{bmatrix} 0 \\ 0 \\ 1 \\ 0 \end{bmatrix} u(t),$$

where $\boldsymbol{\lambda}_K = (\partial \boldsymbol{x}/\partial K)_{K_0}$ and

$$u(t) = M \operatorname{sgn}[p(t_i^-)][1(t - t_i) - 1(t - t_i - \gamma)].$$

(b) The expressions for dt_i/dK are

Case (1): $$\frac{dt_i}{dK} = -\frac{1}{\dot{p}(t_i^-)}\frac{\partial p(t_i^-)}{\partial K},$$

Case (2): $$\frac{dt_i}{dK} = \frac{dt_{i-1}}{dK}$$

(c) The jump conditions are, at $t = t_i$,

$$\Delta\boldsymbol{\lambda} = \left\{ \begin{bmatrix} 0 & 0 & 0 & 0 \\ 0 & 0 & 0 & 0 \\ 0 & 0 & 0 & 0 \\ K_1 & K_2 & K_3 & c \end{bmatrix} \boldsymbol{x}_0(t_i^+) - \begin{bmatrix} 0 \\ 0 \\ 0 \\ 1 \end{bmatrix} w(t_i) - \begin{bmatrix} 0 \\ 0 \\ K_0 \\ 0 \end{bmatrix} u(t_i^+) \right\} \frac{dt_i}{dK}\bigg|_{K_0} - \begin{bmatrix} 0 & 0 & 0 & 0 \\ 0 & 0 & 0 & 0 \\ 0 & 0 & 0 & 0 \\ 0 & 0 & 0 & 1 \end{bmatrix} \boldsymbol{\lambda}(t_i^-)$$

and, at $t = t_i + \gamma$,

$$\Delta\boldsymbol{\lambda} = \begin{bmatrix} 0 \\ 0 \\ K_0 \\ 0 \end{bmatrix} M \operatorname{sgn}[p(t_i^-)] \frac{dt_i}{dK}\bigg|_{K_0}.$$

CHAPTER 6

6.1 The nominal zero is $s_{10} = -1$ and the nominal poles $s_{20} = -2$, $s_{30} = -3$. Evaluating Eq. (6.3-4) with the above numbers gives the desired semir elative root sensitivities:

$$\tilde{S}_{a_1}^{s_1} = -1, \qquad \tilde{S}_{a_1}^{s_2} = 4, \qquad \tilde{S}_{a_1}^{s_3} = -9.$$

6.2 The nominal poles are $s_{10} = -1{,}594$, $s_{20} = -1.953 + j\,1.562$, $s_{30} = -1.953 - j\,1.568$. From Eq. (6.3-4) the following semirelative root sensitivities are obtained:

$$\tilde{S}_{a_1}^{s_1} = 2.294, \qquad \tilde{S}_{a_1}^{s_2} = -1.148 + j\,3.452, \qquad \tilde{S}_{a_1}^{s_3} = -1.148 - j\,3.452.$$

Normalizing upon the corresponding real and imaginary parts of the roots yields the relative root sensitivities

$$\bar{S}_{a_1}^{s_1} = -1.439, \qquad \bar{S}_{a_1}^{s_2} = 0.588 + j\ 2.201, \qquad \bar{S}_{a_1}^{s_3} = 0.588 + j\,2.201.$$

6.3 The nominal transfer function has two simple poles at $s_{10} = -2$, $s_{20} = -4$ and a double real pole at $s_{30} = -3$. For the simple poles, we obtain

$$\tilde{S}_{a_2}^{s_1} = -106, \qquad \tilde{S}_{a_2}^{s_2} = 424,$$

and for the double real pole (from Eq. 6.3-22) with $\nu = 2$

$$\tilde{S}_{a_2}^{s_3} = 477.$$

6.4 The nominal transfer function has a simple real root $s_{10} = -4$ and a double complex root at $s_{2,3} = -2 \pm j3$. Thus the semirelative real root sensitivity becomes

$$\tilde{S}_{a3}^{s1} = 28.03$$

and the semirelative complex root sensitivities are found to be

$$\tilde{S}_{a3}^{s2} = 6.21 + j\,7.02, \qquad \tilde{S}_{a3}^{s3} = 6.21 - j\,7.02.$$

6.5 The transfer function of the closed loop is

$$G(s) = \frac{K(s + a)}{s^2 + (b + c + K)s + bc + aK}$$

and the nominal roots are

$$s_{10} = -2.5 + j\,2.3979,$$
$$s_{20} = -2.5 - j\,2.3979.$$

Thus the root sensitivities become

$$\tilde{S}_K^{s_1} = -1 + j\,1.0425 = 1.44\underline{/133.8^\circ},$$
$$\tilde{S}_K^{s_2} = -1 - j\,1.0425 = 1.44\underline{/-133.8^\circ}.$$

6.6 The transfer function of the closed loop has the roots $s_1 = -101$, $s_{2,3} = -4.5 \pm j\,8.42$. The required semirelative root sensitivities are

$$\tilde{S}_{K2}^{s_1} = -0.9795, \qquad \tilde{S}_{K2}^{s_{2,3}} = 0.4973 \pm j\;5.617 = 5.6\underline{/\pm 84.9^\circ},$$
$$\tilde{S}_a^{s_1} = -0.1077, \qquad \tilde{S}_a^{s_{2,3}} = -4.947 \mp j\,2.649 = 5.6\underline{/\mp 151.8^\circ}.$$

6.7 If the closed-loop transfer function has the pole $s_1 = -1 + j$, then K must be 5. The required semirelative root sensitivity becomes

$$\tilde{S}_a^{s_1} = -0.3125 + j\,0.3125 = 0.44\underline{/135^\circ}.$$

CHAPTER 7

7.1 The desired Bode sensitivity functions and parameter-induced changes are

$$S_{K_1}^G = 1, \qquad \Delta y/y_0 = \Delta K_1/K_{10}$$

$$S_{K_2}^G = -\frac{K_0 K_{20}}{1 + K_0 K_{20}} \approx -1, \qquad \Delta y/y_0 \approx -\Delta K_2/K_{20} \qquad (\text{for } K_{20} \to \infty)$$

$$S_K^G = \frac{1}{1 + K_0 K_{20}} \approx 0, \qquad \Delta y/y_0 \approx 0 \qquad (\text{for } K_{20} \to \infty).$$

7.2 (a) The desired Bode sensitivity functions are

$$S_b^G = \frac{s + a_0}{s + a_0 + b_0} = 1 - G_a,$$

where $G_a = b_0/(s + a_0 + b_0)$ is the overall transfer function of the control system.

(b) The Bode sensitivity functions in case (b) are

$$S_b^G = -\frac{b_0}{s + a_0 + b_0} \neq 1 - G_b,$$

where $G_b = 1/(b + a_0 + b_0)$ is the overall transfer function of the control system. In case (b), $S_b^G \neq 1 - G_b$.

7.3 The desired Bode sensitivity functions are

$$S_a^G(s) = -\frac{as}{1 + b + as}, \qquad S_b^G(s) = \frac{1 + as}{1 + b + as}.$$

The requirements $|S_a^G(j\omega)| < 1$ and $|S_b^G(j\omega)| < 1$ lead to the following conditions for a and b: a any real number, $b > 0$ (and $b < -2$).

7.4 The overall transfer function G is given by $G(j\omega) = K(j\omega)\,[1 + K(j\omega)]^{-1}$.

(a) The Bode sensitivity function of G with respect to $|K|$ becomes $S_{|K|}^G = S_K^G = [1 + K(j\omega)]^{-1} = (3 - 4j\omega)(9 + 16\omega^2)^{-1}$. From this,

$$S_{|K|}^{|G|} = \frac{3}{9 + 16\omega^2} = \frac{3}{25}, \qquad \varphi S_{|K|}^{\varphi} = -\frac{4\omega}{9 + 16\omega^2} = -\frac{4}{25},$$

or, since $\varphi = \tan^{-1} 2/11 = 0.18$,

$$S_{|K|}^{\varphi} = -0.89.$$

Thus, $\Delta|G|/|G_0| = 1.2\%$ and $\Delta\varphi/\varphi_0 = -8.9\%$.

(b) The requirement $|S_{|K|}^G(j\omega)| \leq 1$ is satisfied for all real values of ω.

(c) No contradiction because $K(s)$ has no pole excess.

7.5 Substituting $\Delta\boldsymbol{G} = \boldsymbol{G} - \boldsymbol{G}_0$, $\Delta\boldsymbol{P} = \boldsymbol{P} - \boldsymbol{P}_0$, and $\boldsymbol{G} = (\boldsymbol{I} + \boldsymbol{PR})^{-1}\boldsymbol{PR}$ into Eq. (7.2-37), and performing a number of matrix manipulations, gives $\mathbf{S}_p = \boldsymbol{I} + \boldsymbol{PR})^{-1}$. This expression is identical with Eq. (7.2-35), thus proving the validity of Eq. (7.2–37).

7.6 (a) From $K_{\text{tot}} = -10 = K^3/(1 - RK^3)$ and $K = -10$ it follows that $R = 0.099$.

(b) The comparison sensitivity function becomes

$$S_p = 3/(1 - 0.099K^3) = 0.03.$$

(c) From $-9 = K_1^3/(1 - 0.099\,K_1^3)$ we find $K_1 = 4.35$, where K_1 is the amplification that gives $K_{\text{tot}} = -9$. From $10\cdot(0.9)^t = 4.35$ we find $t = 7.9$ years as the time for reaching $K_{\text{tot}} = -9$.

7.7 With the aid of $|\boldsymbol{I} + \boldsymbol{\phi}_s(j\omega)\boldsymbol{H}| \geq \boldsymbol{I}$, where $\boldsymbol{\phi}_s = (s\boldsymbol{I} - \boldsymbol{A})^{-1}$, the desired frequency range can be determined. The result shows that the *closed-loop* configuration is less sensitive over the whole frequency range.

7.8 (a) The desired comparison sensitivity functions are $S_{p,b} = (1 + s)/(1 + 0.5s)$, $S_{p,c} = (1 + s)/(2 + s)$, $S_{p,d} = s/(2 + s)$.

(b) $|S_{p,b}| \geq 1$ for all ω; $|S_{p,c}| \leq 1$ for all ω; $|S_{p,d}| \leq |S_{p,c}| \leq 1$ for all ω.

(c) The dc gains of the compensators are 2 in case (b), 4 in the case (c), and ∞ in case (d). Hence, the dc gains have to be enlarged as sensitivity is to be reduced.

7.9 (a) In the open-loop case, the prefilter must have the transfer function

$$L(s) = s(s^2 + 6s + 5)/(s^3 + 106s^2 + 205s + 100).$$

For configurations (b) and (c) one obtains, respectively,

$$R(s) = (s^2 + 6s + 5)/(s^2 + 106s + 205) \quad \text{and} \quad H(s) = s^2 + 2s + 1.$$

(b) Calculate the Bode sensitivity functions and divide them by the Bode sensitivity functions of the open-loop configuration.

(c) In the case of state feedback (Fig. 7.P-5a), the comparison sensitivity functions with respect to the different parameters become

$$S_{p,A} = (s^3 + 6s^2 + 5s)/(s^3 + 106s^2 + 205s + 100) = S_{p,K_3} = S_{p,s3}$$
$$S_{p,K_1} = (s^3 + 106s^2 + 205s)/(s^3 + 106s^2 + 205s + 100)$$
$$S_{p,s2} = (s^3 + 106s^2 + 505s)/(s^3 + 106s^2 + 205s + 100).$$

The comparison sensitivity function for configuration (b) becomes, with respect to all parameters,

$$S_p = (s^3 + 106s^2 + 205s)/(s^3 + 106s^2 + 205s + 100) = S_{p,K_1} \quad \text{of (a)}.$$

For configuration (c) one obtains, with respect to all parameters,

$$S_p = (s^3 + 6s^2 + 5s)/(s^3 + 106s^2 + 205s + 100) = S_{p,A} \quad \text{of (a)}.$$

(d) For $S_{p,A} = S_{p,K_3} = s_{p,s_3}$ of configuration (a) and S_p of configuration (c) one obtains that $|S_{p,A}| \leq 1$ for $\omega > -0.925$. For S_{p,K_1} of configuration (a) and S_p of (b) one finds $-0.0686 \leq \omega \leq 0.0686$. For S_{p,s_2} of configuration (a) one yields $\omega \leq 0.04$ and $\omega \geq 339.33$.

7.10 (a) The desired feedback matrix is $\boldsymbol{K} = [1 \ 1]$.

(b) The comparison sensitivity matrix $\tilde{\boldsymbol{S}}_p$ is found to be

$$\tilde{\boldsymbol{S}}_p = [s\boldsymbol{I} - \boldsymbol{A} + \boldsymbol{b}\boldsymbol{K}]^{-1}[s\boldsymbol{I} - \boldsymbol{A}] =$$
$$= \frac{1}{Ts^2 + (K + 1)s + K}\begin{bmatrix} Ts^2 + (K + 1)s & -K \\ -Ks & Ts^2 + s + K \end{bmatrix}.$$

Evaluating $\tilde{\boldsymbol{S}}^{\mathrm{T}}{}_p(-j\omega)\boldsymbol{K}^{\mathrm{T}}\boldsymbol{K}\tilde{\boldsymbol{S}}_p(j\omega) \leq \boldsymbol{K}^{\mathrm{T}}\boldsymbol{K}$ for $K = T = 1$ gives $|j\omega/(1 + j\omega)| \leq 1$ for all ω. This means that the L_2 norm of $\boldsymbol{Kx}$ of the state feedback is smaller than that of the nominally equivalent open-loop control.

(c) The comparison sensitivity function with regard to y is obtained as

$S_{p,y} = S_{p,11} = (Ts^2 + (1 + K)s)/(Ts^2 + (K + 1)s + K)$. Substituting $T = 1$, $K = 1$ gives $S_{p,y} = s(s + 2)/(s + 1)^2$. The condition $|S_{p,y}(j\omega)| \leq 1$ is fulfilled only for $\omega < 0.707$. This indicates that the output of the considered feedback system is not less sensitive than the open-loop system in terms of the L_2 norm.

7.11 (a) The comparison sensitivity matrix $\tilde{\boldsymbol{S}}_p$ becomes

$$\tilde{\boldsymbol{S}}_p = [\boldsymbol{I} + (s\boldsymbol{I} - \boldsymbol{A})^{-1}\boldsymbol{bK}]^{-1} =$$

$$= \frac{1}{s^2 + (1 + a_0 + K_2)s + a_0 + K_1 + a_0K_2}$$

$$\cdot \begin{bmatrix} (s + a_0)(s + K_2 + 1) & -K_2 \\ -K_1(s + a_0) & s^2 + (1 + a_0)s + a_0 + K_1 \end{bmatrix}.$$

Setting $a_0 = 1$, the condition $\tilde{\boldsymbol{S}}^{\mathrm{T}}{}_p(-j\omega)\boldsymbol{K}^{\mathrm{T}}\boldsymbol{K}\tilde{\boldsymbol{S}}_p(j\omega) \leq \boldsymbol{K}^{\mathrm{T}}\boldsymbol{K}$ lead to the condition $|S_p(j\omega)| \leq 1$, where

$$S_p(s) = \frac{(1 + s)^2}{s^2 + (2 + K_2)s + 1 + K_1 + K_2}$$

is the comparison sensitivity function of $\boldsymbol{Kx}$. For the given values of K_1 and K_2, $|S_p(j\omega)| < 1$ for $\omega^2 > -2$, i.e., for all ω.

(b) The results verify that this system is optimal according to the performance index $J = \frac{1}{2} \int_0^\infty (\boldsymbol{x}^{\mathrm{T}}\boldsymbol{Qx} + u^2)\, dt$.

(c) The comparison sensitivity function of the output y is identical to the first component of $\tilde{\boldsymbol{S}}_p$, i.e.,

$$S_{p,y} = \tilde{S}_{p,11} = \frac{s^2 + (1 + a_0 + K_2)s + a_0 + a_0K_2}{s^2 + (1 + a_0 + K_2)s + a_0 + K_1 + a_0K_2}.$$

For the preassigned values of a_0, K_1, K_2, $|S_{p,y}(j\omega)| < 1$ for $\omega < 1.268$, i.e., not for all ω.

CHAPTER 8

8.1 (a) The optimal control function is found from the Hamilton canonical system as

$$u_0^*(t) = -\frac{x(0)\ K_0}{r + \sqrt{r^2 + K_0^2 r}}\ exp\left(-\frac{\sqrt{r^2 + K_0^2 r}}{rT_2}\, t\right).$$

The feedback control law is

$$u_c^*(t) = -kx \qquad \text{with} \quad k = \frac{K_0}{r + \sqrt{r^2 + K_0^2 r}},$$

u_c^* is not a function of the time constant T.

(b) The optimal value of J is

$$J_0 = J(\boldsymbol{\alpha}_0, u^*) = \frac{rT_0 x^2(0)}{r + \sqrt{r^2 + K_0^2 r}}.$$

(c) By any of the three methods described, the *absolute* performance-index sensitivities become

$$J_K = \left.\frac{\partial J}{\partial K}\right|_{K_0, T_0} = -\frac{rT_0 K_0 x^2(0)}{(r + \sqrt{r^2 + K_0^2 r})(r + \sqrt{r^2 + K_0^2 r} + K_0^2)} = 0.12\, x^2(0),$$

$$J_T = \left.\frac{\partial J}{\partial T}\right|_{K_0, T_0} = \frac{rx^2(0)}{r + \sqrt{r^2 + K_0^2 r}} = 0.41\, x^2(0).$$

The relative performance-index sensitivities are therefore

$$\bar{J}_K = \left.\frac{\partial \ln J}{\partial \ln K}\right|_{K_0, T_0} = -\frac{K_0^2}{r + K_0^2 + \sqrt{r^2 + K_0^2 r}} = 0.29,$$

$$\bar{J}_T = \left.\frac{\partial \ln J}{\partial \ln T}\right|_{K_0, T_0} = 1.$$

For $r > 0$, $|\bar{J}_K|$ is always smaller than 1. $\bar{J}_T$ is independent of T.

(d) The actual value of J is

$$J(\boldsymbol{\alpha}, u) = \frac{Tr(r + K_0^2 + \sqrt{r^2 + K_0^2 r})\, x^2(0)}{(r + \sqrt{r^2 + K_0^2 r})\,(r + KK_0 + \sqrt{r^2 + K_0^2 r})}.$$

For $K_0 \to K$ and $T_0 \to T$, the *optimal* actual value $J(\boldsymbol{\alpha}, u^*)$ is obtained as

$$J(\boldsymbol{\alpha}, u^*) = \frac{Trx^2(0)}{(r + \sqrt{r^2 + K^2 r})}.$$

Thus $\bar{S}(\boldsymbol{\alpha}, u)$ becomes

$$\bar{S}(\boldsymbol{\alpha}, u) = \frac{J(\boldsymbol{\alpha}, u)}{J(\boldsymbol{\alpha}, u^*)} - 1 = \frac{1}{r}\,\frac{(r + \sqrt{r^2 + K^2 r})\,\sqrt{r^2 + K_0^2 r}}{r + \sqrt{r^2 + K_0^2 r} + KK_0} - 1.$$

Setting $K_0 = 1$, $r = 1$, and $K = 1.1$ yields

$$\bar{S}(\boldsymbol{\alpha}, u) \approx 6.8 \cdot 10^{-4} \approx 0.07\%.$$

Analogously, $\bar{S}_0(\boldsymbol{\alpha}, u)$ becomes

$$\bar{S}_0(\boldsymbol{\alpha}, u) = \frac{J(\boldsymbol{\alpha}, u) - J(\boldsymbol{\alpha}, u^*)}{J(\boldsymbol{\alpha}_0, u^*)} = \frac{T}{T_0}\left[\frac{r + K_0^2 + \sqrt{r^2 + K_0^2 r}}{r + KK_0 + \sqrt{r^2 + K_0^2 r}} - \frac{r + \sqrt{r^2 + K_0^2 r}}{r + \sqrt{r^2 + K^2 r}}\right]$$

and with the given values

$$\bar{S}_0(\boldsymbol{\alpha}, u) = 6.5 \times 10^{-4} \approx 0.07\%.$$

For a 10% change of K, the performance index deviates less than 0.1% from its optimal value.

(e) The approximate evaluation gives $\boldsymbol{Z} = \boldsymbol{0}$, i.e., $\bar{S}_0(\boldsymbol{\alpha}, u) = 0$.

8.2 (a) The performance-index sensitivities of the open-loop control are found as

$$J_K = -\frac{rT_0K_0x^2(0)}{(r + \sqrt{r^2 + K_0^2r})(r + K_0^2 + \sqrt{r^2 + K_0^2r})} = 0.12x^2(0),$$

$$J_T = \frac{rx^2(0)}{r + \sqrt{r^2 + K_0^2r}} = 0.41x^2(0).$$

The performance-index sensitivities are the same as in the closed-loop case, although the end value is fixed. Reason: Since $t_f \to \infty$, the variations δx_S and δx_R are equal so that the performance-index sensitivities are also equal.

(b) The optimality loss becomes

$$\bar{S}(\boldsymbol{\alpha}, u) = \frac{r + \sqrt{r + K_0^2r}}{rTx^2(0)} J(\boldsymbol{\alpha}, u) - 1 = 0.016.$$

A comparison of this with $\bar{S}(\boldsymbol{\alpha}, u)$ of the closed loop (Problem 8.1d) shows that $\bar{S}_{\text{closed}} < \bar{S}_{\text{open}}$.

8.3 (a) The optimal gain factors are

$$K_1 = \sqrt{\frac{q_{11}}{R}} = 1, \qquad K_2 = \frac{1}{K_0}(w - 1) = 1,$$

with $w = \sqrt{1 + \dfrac{K_0^2 q_{22}}{R} + 2K_0T_0\sqrt{\dfrac{q_{11}}{R}}} = 2.$

(b) The absolute performance-index sensitivities are

$$J_K = \tfrac{1}{2}\boldsymbol{x}^{\mathrm{T}}(0)\boldsymbol{P}_K\boldsymbol{x}(0),$$

with

$$\boldsymbol{P}_K = \begin{bmatrix} -\dfrac{1}{K_0^2w}\sqrt{Rq_{11}} - \dfrac{T_0q_{11}}{K_0w} & -\dfrac{T_0}{K_0^2}\sqrt{Rq_{11}} \\ -\dfrac{T_0}{K_0^2}\sqrt{Rq_{11}} & \dfrac{T_0}{w}\left(-\dfrac{T_0}{K_0^2}\sqrt{Rq_{11}} - \dfrac{R(w-1)^2}{K_0^3}\right) \end{bmatrix}$$

$$= \begin{bmatrix} -1 & -1 \\ -1 & -1 \end{bmatrix}$$

and

$$J_T = \tfrac{1}{2}\boldsymbol{x}^{\mathrm{T}}(0)\,\boldsymbol{P}_T\boldsymbol{x}(0)$$

with

$$\boldsymbol{P}_T = \begin{bmatrix} \dfrac{q_{11}}{w} & \dfrac{1}{K_0}\sqrt{Rq_{11}} \\ \dfrac{1}{K_0}\sqrt{Rq_{11}} & \dfrac{T_0}{K_0 w}\sqrt{Rq_{11}} + \dfrac{R(w-1)}{K_0{}^2} \end{bmatrix} = \begin{bmatrix} 0.5 & 1 \\ 1 & 1.5 \end{bmatrix}.$$

The relative performance-index sensitivities are generally

$$\bar{J}_K = \frac{\boldsymbol{x}^{\mathrm{T}}(0)\boldsymbol{P}_K\boldsymbol{x}(0)}{\boldsymbol{x}^{\mathrm{T}}(0)\boldsymbol{M}_0\boldsymbol{x}(0)}, \qquad \bar{J}_T = \frac{\mathbf{x}^{\mathrm{T}}(0)\,\boldsymbol{P}_T\boldsymbol{x}(0)}{\boldsymbol{x}^{\mathrm{T}}(0)\boldsymbol{M}_0\boldsymbol{x}(0)},$$

where

$$\boldsymbol{M}_0 = \begin{bmatrix} \dfrac{w}{K_0}\sqrt{Rq_{11}} & \dfrac{T_0}{K_0}\sqrt{Rq_{11}} \\ \dfrac{T_0}{K_0}\sqrt{Rq_{11}} & \dfrac{RT_0}{K_0{}^2}(w-1) \end{bmatrix} = \begin{bmatrix} 2 & 1 \\ 1 & 1 \end{bmatrix}.$$

(c) The relative performance-index sensitivities are

$$\bar{J}_K = -1 + \frac{1}{n^2 + 2n + 2}, \qquad \bar{J}_T = 1.5 - \frac{n + 2.5}{n^2 + 2n + 2},$$

thus for $0 \leq n \leq \infty$, we have $-\frac{1}{2} \geq \bar{J}_K \geq -1$ and $\frac{1}{4} \leq \bar{J} \leq \frac{3}{2}$.

(d) The actual performance index is given by

$$J(\boldsymbol{\alpha},u) = \tfrac{1}{2}\boldsymbol{x}^{\mathrm{T}}(0)\boldsymbol{M}\boldsymbol{x}(0),$$

where the elements of $\boldsymbol{M}$ are found from the equation $\boldsymbol{F}^{\mathrm{T}}\boldsymbol{M} + \boldsymbol{M}\boldsymbol{F} + \tilde{\boldsymbol{Q}} = \mathbf{0}$ to be

$$m_{12} = m_{21} = \frac{T}{K}\sqrt{Rq_{11}} = 1.1, \qquad m_{22} = \frac{2m_{12} + (w-1)^2 R/K_0{}^2}{\dfrac{2}{T}\left(1 + \dfrac{K}{K_0}(w-1)\right)} = 0.88$$

$$m_{11} = \frac{1}{T}\left(1 + \frac{K}{K_0}(w-1)\right)m_{12} + \frac{K}{T}\sqrt{\frac{q_{11}}{R}}\,m_{22} - \frac{(w-1)}{K_0}\sqrt{rp_{11}} = 1.8.$$

Thus for $0 \leq n \leq \infty$, we have $0 \leq \bar{S}(\boldsymbol{\alpha}, u) \leq 0.47\%$ as T varies by $+10\%$.

8.4 (a) The optimal control function is

$$u^* = -\boldsymbol{K}_0\boldsymbol{x} + \frac{1}{\sqrt{R}}r_1, \qquad \boldsymbol{K}_0 = \left[\frac{K_0}{T_0\sqrt{R}}\,\frac{1}{T_0}\left(\sqrt{4 + \frac{2K_0T_0}{\sqrt{R}}} - 2\right)\right].$$

(b) The optimal performance index is

$$J_0 = \frac{1}{2}r_1{}^2\left[\frac{\sqrt{R}}{K_0} + 1 + \frac{\sqrt{R}}{K_0}(w - 2R)\right].$$

(c) The performance-index sensitivities are

$$J_K = \frac{1}{2} r_1^2 \left[\frac{T_0^2}{K_0^2} \sqrt{R}\left(\frac{w}{R} - 1\right) + \frac{RT_0^2}{K_0(w - R)}\right],$$

$$J_T = \frac{1}{2} r_1^2 \left[-\frac{2}{\sqrt{R K_0 T_0}} - \frac{RT_0}{w - R} + \frac{\sqrt{R}}{K_0} (w - 2R)\right],$$

with the abbreviation $w = \sqrt{4R^2 + 2R^{3/2}K_0T_0}$. Note that r_1 disappears in the relative performance-index sensitivities $\bar{J}_K$ and $\bar{J}_T$.

8.5 (a) The feedback gain factors for nominal parameters are

$$K_{10} = w + \frac{1}{K_0T_0} - \sqrt{\frac{1}{K_0^2T_0^2} + \frac{q_{22}^2}{K_0^2R} + \frac{2w_0}{K_0T_0}} = 0.2168,$$

$$K_{20} = -\frac{1}{T_0} + \sqrt{\frac{1}{T_0^2} + \frac{q_{22}^2}{K_0} + 2\frac{K_0}{T_0} w_0} = 1.1974,$$

with

$$w_0 = R^{-1/2}\sqrt{q_{11} + 2\frac{q_{12}}{K_0} + \frac{q_{22}}{K_0^2}} = 1.414.$$

(b) The optimal performance index is given by

$$J_0 = \tfrac{1}{2}\boldsymbol{x}^{\mathrm{T}}(0)\boldsymbol{M}_0\boldsymbol{x}(0),$$

where

$$\boldsymbol{M}_0 = \begin{bmatrix} \frac{T_0}{2}(q_{11} - RK_{10}^2) & RK_{10} \\ RK_{10} & RK_{20} \end{bmatrix} = \begin{bmatrix} 0.48 & 0.217 \\ 0.217 & 1.19 \end{bmatrix}.$$

(c) The absolute performance-index sensitivities are

$$J_T = \tfrac{1}{2}\, \boldsymbol{x}^{\mathrm{T}}(0)\boldsymbol{P}_T\boldsymbol{x}(0),$$

with

$$\boldsymbol{P}_T = \begin{bmatrix} \frac{q_{11}}{2} - RK_{10}^2 \frac{3 + T_0K_{20}}{2(1 + T_0K_{20})} & \frac{RK_{10}}{T_0(1 + T_0K_{20})} \\ \frac{RK_{10}}{T_0(1 + T_0K_{20})} & -\frac{RK_{10}}{T_0(1 + T_0K_{20})} \end{bmatrix} = \begin{bmatrix} 0.455 & 0.1 \\ 0.1 & -0.1 \end{bmatrix},$$

$$J_K = \tfrac{1}{2}\, \boldsymbol{x}^{\mathrm{T}}(0)\, \boldsymbol{P}_K\boldsymbol{x}(0)$$

with

$$\boldsymbol{P}_K = \begin{bmatrix} -T_0K_{10}\,p & p \\ p & \frac{1}{T_0K_{20}}(K_0p + RK_{10}) \end{bmatrix} = \begin{bmatrix} -0.03 & 0.136 \\ 0.136 & 0.3 \end{bmatrix}$$

and

$$p = \frac{T_0 K_0 q_{11} - RK_{10}^2(T_0 K_{20} + 2)}{2(T_0 K_{20} + 1)(K_0 K_{10} + K_{20})} = 0.136.$$

(d) The dependences of $\bar{J}_T$ and $\bar{J}_K$ upon n are

$$\bar{J}_T = \frac{0.227 + 0.1n - 0.05n^2}{0.24 + 0.217n + 0.595n^2}, \qquad \bar{J}_K = \frac{-0.015 + 0.136n + 0.15n^2}{0.24 + 0.217n + 0.595n^2}.$$

(e) $\bar{J}_T = 0$ for $n = 3.3537$.

(f) The matrix $\boldsymbol{M}^*$ is found from $\boldsymbol{M}_0$ by replacing the nominal by the actual parameters. Thus

$$\boldsymbol{M}^* = \begin{bmatrix} \frac{1}{2}T(q_{11} - RK_1^2) & RK_1 \\ RK_1 & RK_2 \end{bmatrix} = \begin{bmatrix} 0.5216 & 0.2262 \\ 0.2262 & 1.188 \end{bmatrix},$$

with

$$K_1 = w + \frac{1}{KT} - \sqrt{\frac{1}{K^2T^2} + \frac{q_{22}^2}{K^2R} + \frac{2w}{KT}} = 0.226,$$

$$K_2 = -\frac{1}{T} + \sqrt{\frac{1}{T^2} + \frac{q_{22}^2}{R} + \frac{2Kw}{T}} = 1.188,$$

$$w = R^{-1/2}\sqrt{q_{11} + \frac{2q_{12}}{K} + \frac{q_{22}}{K^2}} = 1.4142.$$

The matrix $\boldsymbol{M}$ of the actual performance index $J(\boldsymbol{\alpha}, u)$ is found from $\boldsymbol{MF} + \boldsymbol{FM} + \tilde{\boldsymbol{Q}} = \boldsymbol{0}$. The result is

$$\boldsymbol{M} = \begin{bmatrix} m_{11} & m_{12} \\ m_{21} & m_{22} \end{bmatrix} = \begin{bmatrix} 0.5219 & 0.2262 \\ 0.2262 & 1.188 \end{bmatrix},$$

with

$$m_{12} = \frac{T}{2}\,\frac{RK_{10}K_{20}(KK_{10} + K_{20}) + KK_{20}q_{11} - K_{10}q_{22} + 2K_{20}q_{12}}{(1 + TK_{20})(K_{20} + KK_{10})}$$

$$m_{11} = -K_{10}Tm_{12} + \frac{T}{2}RK_{10}^2 + \frac{T}{2}q_{11},$$

$$m_{22} = \frac{K}{TK_{20}}m_{12} + \frac{1}{2}RK_{20} + \frac{q_{22}}{2K_{20}}.$$

Thus the optimality loss $\bar{S}(\boldsymbol{\alpha}, u)$, according to Eq.(8.5-22), becomes

$$\underbrace{1.03 \times 10^{-4}}_{(n=0)} \geq \bar{S}(\boldsymbol{\alpha}, u) \geq \underbrace{1.77 \times 10^{-5}}_{(n=\infty)}.$$

(g) The matrix $\boldsymbol{Z}$ becomes

$$\boldsymbol{Z} = \begin{bmatrix} 5.35 & -2.22 \\ -2.22 & 2.21 \end{bmatrix} \times 10^{-3}$$

whence

$$\underbrace{1.03 \times 10^{-4}}_{(n=0)} \le \bar{S}(\boldsymbol{\alpha}, u) \le \underbrace{1.86 \times 10^{-4}}_{(n=\infty)}$$

and

$$\underbrace{1.11 \times 10^{-4}}_{(n=0)} \le \bar{S}_0(\boldsymbol{\alpha}, u) \le \underbrace{1.85 \times 10^{-5}}_{(n=\infty)}.$$

8.6 (a) The optimal feedback gains are

$$K_1 = a_0^2 + z - a_0 \sqrt{a_0^2 + 1 + 2z + \frac{q_{22}}{R}},$$

$$K_2 = -a_0 - 1 + \sqrt{a_0^2 + 1 + 2z + \frac{q_{22}}{R}},$$

where

$$z = \sqrt{a_0^2 + \frac{q_{11}}{R} + 2a_0 \frac{q_{12}}{R} + a_0^2 \frac{q_{22}}{R}}.$$

$\boldsymbol{M}_0$ is found to be

$$\boldsymbol{M}_0 = \begin{bmatrix} \frac{1}{2a_0}(q_{11} - RK_1^2) & RK_1 \\ RK_1 & RK_2 \end{bmatrix}.$$

(b) The performance-index sensitivity is

$$J_a = \tfrac{1}{2}\, \boldsymbol{x}^{\mathrm{T}}(0) \boldsymbol{P}_a \boldsymbol{x}(0),$$

with

$$\boldsymbol{P}_a = \begin{bmatrix} -\frac{K_1}{a_0} w + \frac{RK_1^2}{2a_0^2} - \frac{q_{11}}{2a_0} & w \\ w & \frac{w}{1 + K_2} \end{bmatrix}$$

and

$$w = -\frac{1}{2a_0} \frac{2a_0^2 RK_1 - RK_1^2 + q_{11}}{K_1 + a_0(1 + K_2) + a_0^2 + a_0 K_1/(1 + K_2)}$$

(c) (1) $K_1 = 0.1896$, $K_2 = 0.5425$,
$n = 0$: $\bar{J}_a = -0.91$, $n = \infty$: $\bar{J}_a = -0.28$.

(2) $K_1 = 0.1059$; $K_2 = 0.3083$
$n = 0$: $\bar{J}_a = -1$; $n = \infty$: $\bar{J}_a = -0.7$

(3) $K_1 = -0.1097$; $K_2 = 0.3344$
$n = 0$: $\bar{J}_a = -1$; $n = \infty$: $\bar{J}_a = -0.22$

(4) $K_1 = 0.2168$; $K_2 = 0.1974$
$n = 0$: $\bar{J}_\alpha = -0.88$; $n = \infty$: $\bar{J}_a = -1.13$

8.7 The optimal feedback gains are $K_1 = 0.6447$, $K_2 = 0.4950$, $K_3 = 1.2586$. The optimal performance index J_0 and the performance-index sensitivities $\bar{J}_{T_1}$, $\bar{J}_{T_2}$ are (a) $J_0 = 0.7305$, $\bar{J}_{T_1} = 0.5960$, $\bar{J}_{T_2} = 0.1887$, (b) $J_0 = 0.1713$, $\bar{J}_{T_1} = -1.006$, $\bar{J}_{T_2} = 1.338$.

8.8 The gains are $K_1 = 0.8139$, $K_2 = 0.3472$, $K_3 = 0.0516$. The required values are listed in the tabulation.

$x_1(0)$	$x_2(0)$	$x_3(0)$	J_0	$\bar{J}_{T1}$	$\bar{J}_{T2}$	$\bar{J}_{T4}$	$\bar{J}_{T1}$	$\bar{J}_{V2}$	$\bar{J}_{V3}$
1	0	0	0.42	0.58	0.35	0.07	—0.37	—0.37	—0.37
0	1	0	0.04	—0.65	1.5	0.15	0.78	—1.22	—1.22
0	0	1	0.01	—0.65	—0.43	1.99	0.68	0.68	—1.32

8.9 (a) The feedback gains are $K_1 = 0.9244$, $K_2 = 0.1711$, $K_3 = 0.0161$, $K_4 = 0.0492$, $K_5 = 0.2644$. The optimal performance index is $J_0 = 0.182$. The performance-index sensitivities are $\bar{J}_{V_1} = -0.28$, $\bar{J}_{V_2} = -0.28$, $\bar{J}_{V_4} = -0.28$, $\bar{J}_{V_5} = -0.28$, $\bar{J}_{T_1} = 0.36$, $\bar{J}_{T_2} = 0.28$, $\bar{J}_{T_3} = 0.12$, $\bar{J}_{T_4} = 0.08$, $\bar{J}_{T_5} = 0.16$.

(b) The optimality loss is

$$\Delta T_1/T_{10} = 20\%: \quad \bar{S} = 1.38\%, \quad \bar{S}_0 = 1.06\%;$$
$$\Delta T_1/T_{10} = 50\%: \quad \bar{S} = 4.84\%, \quad \bar{S}_0 = 8.63\%;$$
$$\Delta T_4/T_{40} = 50\%: \quad \bar{S} = 0.42\%, \quad \bar{S}_0 = 0.41\%.$$

8.10 The desired performance-index sensitivities are $\bar{J}_{T_1} = 0.307$, $\bar{J}_{T_2} = 0.305$, $\bar{J}_{T_3} = 0.133$, $\bar{J}_{T_4} = 0.086$, $\bar{J}_5 = 0.168$, $\bar{J}_{V_1} = \cdots = \bar{J}_{V_5} = -0.3$.

Index

L

M

N

O

P

Q

R

S

T

U

V

Z

A B C 8 D 9 E 0 F 1 G 2 H 3 I 4 J 5